Managing Human and Social Systems

Environmental Management Handbook, Second Edition

Edited by
Brian D. Fath and Sven E. Jørgensen

Volume 1
Managing Global Resources and Universal Processes

Volume 2
Managing Biological and Ecological Systems

Volume 3
Managing Soils and Terrestrial Systems

Volume 4
Managing Water Resources and Hydrological Systems

Volume 5
Managing Air Quality and Energy Systems

Volume 6
Managing Human and Social Systems

Managing Human and Social Systems

Second Edition

Edited by
Brian D. Fath and Sven E. Jørgensen

Assistant to Editor
Megan Cole

CRC Press
Taylor & Francis Group
Boca Raton London New York

CRC Press is an imprint of the
Taylor & Francis Group, an **informa** business

Cover photo: Znojmo, Czech Republic, B. Fath

Second edition published 2021
by CRC Press
6000 Broken Sound Parkway NW, Suite 300, Boca Raton, FL 33487-2742

and by CRC Press
2 Park Square, Milton Park, Abingdon, Oxon, OX14 4RN

© 2021 Taylor & Francis Group, LLC

First edition published by CRC Press 2013

CRC Press is an imprint of Taylor & Francis Group, LLC

ISBN: 978-1-138-34268-2 (hbk)
ISBN: 978-0-367-51361-0 (pbk)
ISBN: 978-1-003-05351-4 (ebk)

Typeset in Minion
by codeMantra

Contents

SECTION I   APC: Anthropogenic Chemicals and Activities

SECTION II   COV: Comparative Overviews of Important Topics for Environmental Management

SECTION III   CSS: Case Studies of Environmental Management

SECTION IV   DIA: Diagnostic Tools: Monitoring, Ecological Modeling, Ecological Indicators, and Ecological Services

SECTION V   ELE: Focuses on the Use of Legislation or Policy to Address Environmental Problems

SECTION VI ENT: Environmental Management Using Environmental Technologies

SECTION VII PRO: Basic Environmental Processes

Preface

Given the current state of the world as compiled in the massive Millennium Ecosystem Assessment Report, humans have changed ecosystems more rapidly and extensively during the past 50 years than in any other time in human history. These are unprecedented changes that need certain action. As a result, it is imperative that we have a good scientific understanding of how these systems function and good strategies on how to manage them.

In a very practical way, this multivolume *Environmental Management Handbook* provides a comprehensive reference to demonstrate the key processes and provisions for enhancing environmental management. The experience, evidence, methods, and models relevant for studying environmental management are presented here in six stand-alone thematic volumes, as follows:

VOLUME 1 – Managing Global Resources and Universal Processes
VOLUME 2 – Managing Biological and Ecological Systems
VOLUME 3 – Managing Soils and Terrestrial Systems
VOLUME 4 – Managing Water Resources and Hydrological Systems
VOLUME 5 – Managing Air Quality and Energy Systems
VOLUME 6 – Managing Human and Social Systems

In this manner, the handbook introduces in the first volume the general concepts and processes used in environmental management. The next four volumes deal with each of the four spheres of nature (biosphere, geosphere, hydrosphere, and atmosphere). The last volume ties the material together in its application to human and social systems. These are very important chapters for a wide spectrum of students and professionals to understand and implement environmental management. In particular, the features include the following:

- The first handbook that demonstrates the key processes and provisions for enhancing environmental management.
- Addresses new and cutting-edge topics on ecosystem services, resilience, sustainability, food–energy–water nexus, socio-ecological systems, etc.
- Provides an excellent basic knowledge on environmental systems, explains how these systems function, and gives strategies on how to manage them.
- Written by an outstanding group of environmental experts.

Since the handbook covers such a wide range of materials from basic processes, to tools, technologies, case studies, and legislative actions, each handbook entry is further classified into the following categories:

APC: Anthropogenic chemicals: The chapters cover human-manufactured chemicals and activities
COV: Indicates that the chapters give comparative overviews of important topics for environmental management

CSS: The chapters give a case study of a particular environmental management example

DIA: Means that the chapters are about diagnostic tools: monitoring, ecological modeling, ecological indicators, and ecological services

ELE: Focuses on the use of legislation or policy to address environmental problems

ENT: Addresses environmental management using environmental technologies

NEC: Natural elements and chemicals: The chapters cover basic elements and chemicals found in nature

PRO: The chapters cover basic environmental processes.

Volume 6, *Managing Human and Social Systems*, applies the cumulative knowledge of environmental science and systems specifically into managing human and social systems. There are over 50 entries covering a wide area from environmental legislation and policy to human health, economics, sustainable development, and green technologies. New entries are included to cover environmental accounting, limits to growth, and urban agriculture. Case studies investigate the impact of cell tower placement, health consequences of pesticides in developing countries, and the promise of community-based monitoring. This culminating volume gives guidance for effective environmental management and a glimpse into future challenges and opportunities.

Brian D. Fath
Brno, Czech Republic
December 2019

Editors

Brian D. Fath is Professor in the Department of Biological Sciences at Towson University (Maryland, USA) and Senior Research Scholar at the International Institute for Applied Systems Analysis (Laxenburg, Austria). He has published over 180 research papers, reports, and book chapters on environmental systems modeling, specifically in the areas of network analysis, urban metabolism, and sustainability. He has co-authored the books *A New Ecology: Systems Perspective* (2020), *Foundations for Sustainability: A Coherent Framework of Life–Environment Relations* (2019), and *Flourishing within Limits to Growth: Following Nature's Way* (2015). He is also Editor-in-Chief for the journal Ecological Modelling and Co-Editor-in-Chief for *Current Research in Environmental Sustainability*. He was the 2016 recipient of the Prigogine Medal for outstanding work in systems ecology and twice a Fulbright Distinguished Chair (Parthenope University, Naples, Italy in 2012 and Masaryk University, Czech Republic in 2019). In addition, he has served as Secretary General of the International Society for Ecological Modelling, Co-Chair of the Ecosystem Dynamics Focus Research Group in the Community Surface Modeling Dynamics System, and member and past Chair of Baltimore County Commission on Environmental Quality.

Sven E. Jørgensen (1934–2016) was Professor of environmental chemistry at Copenhagen University. He received a doctorate of engineering in environmental technology and a doctorate of science in ecological modeling. He was an honorable doctor of science at Coimbra University (Portugal) and Dar es Salaam (Tanzania). He was Editor-in-Chief of *Ecological Modelling* from the journal inception in 1975 until 2009. He was Editor-in-Chief for the *Encyclopedia of Environmental Management* (2013) and *Encyclopedia of Ecology* (2008). In 2004, he was awarded the Stockholm Water Prize and the Prigogine Medal. He was awarded the Einstein Professorship by the Chinese Academy of Sciences in 2005. In 2007, he received the Pascal Medal and was elected a member of the European Academy of Sciences. He has published over 350 papers and has edited or written over 70 books. He gave popular and well-received lectures and courses in ecological modeling, ecosystem theory, and ecological engineering worldwide.

Contributors

Diana Aga
Department of Chemistry
State University of New York at Buffalo
Buffalo, New York

Jirapat Ananpattarachai
Center of Excellence for Environmental
    Research and Innovation
Faculty of Engineering
Naresuan University
Phitsanulok, Thailand

Massimo Antoninetti
Institute for Electromagnetic Sensing of the
    Environment (IREA)
National Research Council of Italy (CNR)
Milan, Italy

Seungyun Baik
Department of Chemistry
State University of New York at Buffalo
Buffalo, New York

Simone Bastianoni
Ecodynamics Group
Department of Chemistry
University of Siena
Siena, Italy

Anders Baun
Department of Environmental Engineering
Technical University of Denmark
Kongens Lyngby, Denmark

Richard W. Bell
School of Environmental Science
Murdoch University
Perth, Western Australia, Australia

Sanford V. Berg
Director of Water Studies
Public Utility Research Center
University of Florida
Gainesville, Florida

W.E.H. Blum
Institute of Soil Research
University of Natural Resources and Life Sciences
Vienna, Austria

Ben Boer
School of Law
University of Sydney
Sydney, Australia

J.D. Booker
Environmental Compliance Department
Consolidated Nuclear Security LLC
Amarillo, Texas

Elvira Buonocore
Laboratory of Ecodynamics and Sustainable
    Development
Department of Science and Technology
Parthenope University of Naples
Napoli, Italy
and
CoNISMa
Piazzale Flaminio, Rome, Italy

Ni-Bin Chang
Department of Civil, Environmental, and
    Construction Engineering
University of Central Florida
Orlando, Florida

Guangnan Chen
Faculty of Engineering and Surveying
University of Southern Queensland
Toowoomba, Australia

Angelique Chettiparamb
School of Real Estate and Planning
University of Reading
Reading, United Kingdom

Richard Cowell
School of Geography and Planning
Cardiff University
Cardiff, United Kingdom

Gemma Cranston
Global Footprint Network
Geneva, Switzerland

Christina D. DiFonzo
Department of Entomology
Michigan State University
East Lansing, Michigan

Barbara Dinham
Eurolink Center
Pesticide Action Network UK
London, United Kingdom

George Ekström
Swedish National Chemicals
    Inspectorate (KEMI)
Solna, Sweden

Brian D. Fath
Department of Biological Sciences
Towson University
Towson, Maryland
and
Advanced Systems Analysis Program
International Institute for Applied System
    Analysis
Laxenburg, Austria

Natalia Fath
Department of Geography and
    Environmental Planning
Towson University
Towson, Maryland

Maria R. Finckh
Department of Ecological Plant Protection
University of Kassel
Witzenhausen, Germany

Pier Paolo Franzese
Laboratory of Ecodynamics and Sustainable
    Development
Department of Science and Technology
Parthenope University of Naples
Napoli, Italy
and
CoNISMa
Piazzale Flaminio, Rome, Italy

W. Friesl-Hanl
Environmental Analysis
Environment Agency Austria
and
Institute of Soil Research
University of Natural Resources and
    Life Sciences
Vienna, Austria

Alessandro Galli
Global Footprint Network
Geneva, Switzerland

M.H. Gerzabek
Institute of Soil Research
University of Natural Resources and
    Life Sciences
Vienna, Austria

Timothy S. Goebel
Agricultural Research Service (USDA-ARS)
U.S. Department of Agriculture
Lubbock, Texas

Khara D. Grieger
Department of Environmental Engineering
Technical University of Denmark
Kongens Lyngby, Denmark

Lisa Guan
School of Chemistry, Physics and Mechanical
    Engineering
Science and Engineering Faculty
Queensland University of Technology
Brisbane, Australia

Denis Hamilton
Queensland Department of Primary Industries
Brisbane, Australia

Ian Hannam
Center for Natural Resources
Department of Infrastructure
Planning and Natural Resources
Sydney, Australia

Steffen Foss Hansen
Department of Environmental Engineering
Technical University of Denmark
Kongens Lyngby, Denmark

Kelsey Hart
College of Veterinary Medicine
University of Georgia
Athens, Georgia

James G. Hewlett
Energy Information Administration
U.S. Department of Energy
Washington, District of Columbia

Rusty T. Hodapp
Energy and Transportation Management
Dallas/Fort Worth International Airport Board
Dallas/Fort Worth Airport, Texas

Tim Jackson
University of Surrey
United Kingdom

Ike Jeon
Department of Animal Science and Industry
Kansas State University
Manhattan, Kansas

Sven E. Jørgensen
Institute A, Section of Environmental Chemistry
Copenhagen University
Copenhagen, Denmark

Puangrat Kajitvichyanukul
Center of Excellence for Environmental Research
    and Innovation
Faculty of Engineering
Naresuan University
Phitsanulok, Thailand

Robert J. Lascano
Agricultural Research Service (USDA-ARS)
U.S. Department of Agriculture
Lubbock, Texas

Leslie London
Occupational and Environmental Health
    Research Unit
University of Cape Town
Observatory, South Africa

Graça Martinho
Department of Environmental Sciences and
    Engineering
Faculty of Sciences and Technology
New University of Lisbon
Caparica, Portugal

Thomas J. Mbise
Tanzania Association of Public Occupational and
    Environmental Health Experts
Dar-es-Salaam, Tanzania

Alexandra Navrotsky
Department of Chemical Engineering and
    Materials Science
University of California, Davis
Davis, California, U.S.A.

Elena Neri
Ecodynamics Group
Department of Chemistry
University of Siena
Siena, Italy

Aiwerasia V.F. Ngowi
Tanzania Association of Public Occupational and
   Environmental Health Experts
and
Department of Environmental and Occupational
   Health
Muhimbili University of Health and Allied
   Sciences (MUHAS)
Dar-es-Salaam, Tanzania

Valentina Niccolucci
Ecodynamics Group
Department of Chemistry
University of Siena
Siena, Italy

Egide Nizeyimana
Department of Agronomy
and
Environmental Resources Research Institute
Pennsylvania State University
University Park, Pennsylvania

Jacob Opadeyi
Department of Surveying and Land Information
Faculty of Engineering
University of the West Indies
St. Augustine, Trinidad and Tobago

Margareta Palmborg
Swedish Poisons Information Center
Stockholm, Sweden

Mark A. Peterson
Sustainable Success LLC
Clementon, New Jersey

David Pimentel
Department of Entomology
Cornell University
Ithaca, New York

Ana Pires
Department of Environmental Sciences and
   Engineering
Faculty of Sciences and Technology
New University of Lisbon
Caparica, Portugal

Wendell A. Porter
Department of Agricultural and
   Biological Engineering
University of Florida
Gainesville, Florida

Larama M.B. Rongo
Muhimbili University of Health and
   Allied Sciences
Dar-es-Salaam, Tanzania

Stephen A. Roosa
Energy Systems Group, Inc.
Louisville, Kentucky

Giovanni F. Russo
Laboratory of Ecodynamics and
   Sustainable Development
Department of Science and
   Technology
Parthenope University of Naples
Napoli, Italy
and
CoNISMa
Piazzale Flaminio, Rome, Italy

Alka Sapat
School of Public Administration
Florida Atlantic University
Boca Raton, Florida

Jan Henrik Schmidt
Department of Ecological Plant
   Protection
University of Kassel
Witzenhausen, Germany

Joshua Steinfeld
Old Dominion University
Norfolk, Virginia

Evamarie Straube
University Professor Emerita
Rostock, Germany

Sebastian Straube
Professor and Division Director
Division of Preventive Medicine
Department of Medicine
University of Alberta
Alberta, Canada

Praful Suchak
Sneha Plastics Pvt. Ltd.
Suchak's Consultancy Services
Mumbai, India

Abhishek Tiwary
School of Engineering and Sustainable
    Development
De Montfort University
United Kingdom

Peter A. Victor
Professor Emeritus
York University
Toronto, Canada

Nemi Vora
Biological Systems and Engineering Division
Lawrence Berkeley National Lab
Berkeley, California

Mathis Wackernagel
Global Footprint Network
Oakland, California

Apichon Watcharenwong
School of Environmental Engineering
Suranaree University of Technology
Nakhon Ratchasima, Thailand

W.W. Wenzel
Institute of Soil Research
University of Natural Resources and Life Sciences
Vienna, Austria

Catharina Wesseling
Central American Institute for Studies on Toxic
    Substances (IRET)
National University
Heredia, Costa Rica

Wanpen Wirojanagud
Department of Environmental Engineering
Faculty of Engineering
KhonKaen University
Khon Kaen, Thailand
and
Center of Excellence on Hazardous Substance
    Management
National Centers of Excellence (PERDO)
Bangkok, Thailand

I

APC: Anthropogenic Chemicals and Activities

1

1

Food: Pesticide Contamination

Denis Hamilton

Introduction

How much pesticide residue did I eat today?

No more than necessary, and less than would be detrimental to your health.

Government authorities must be able to support the answer to the consumer's question with scientific data and valid scientific studies.[1] The "no more residues than necessary" concept originates from the principle of good agricultural practice, which implies that the desired effect (pest control) will be achieved without leaving more residues than necessary in the food.

Before Registration

Pesticide residue evaluation and risk assessment prior to registration are summarized in Figure 1.[2,3]

Risks to the environment and to the user are also evaluated but are not considered further under the present topic—*food contamination with pesticide residues*.

Metabolism studies on a pesticide in crops and farm animals identify the nature of the residue. The residue may consist of a parent compound or metabolites or a mixture. In some cases, different pesticides produce the same metabolites; in other cases, the metabolite of one pesticide is another pesticide. Some crops genetically modified for herbicide resistance achieve their resistance by metabolizing the herbicide to a derivative with no herbicidal activity.

The acceptable daily intake (ADI) of a chemical is the daily intake, expressed on a body-weight basis, which, during an entire lifetime, appears to be without appreciable risk to the health of the consumer on the basis of all the known facts at the time. The ADI is based on animal feeding studies that find the daily dose over a lifetime resulting in no observable adverse effect on the most sensitive animal species tested. Then, a margin of safety (safety factor, commonly 100) is applied to allow for extrapolation from animals to humans and the variability in responses between average and highly sensitive humans.

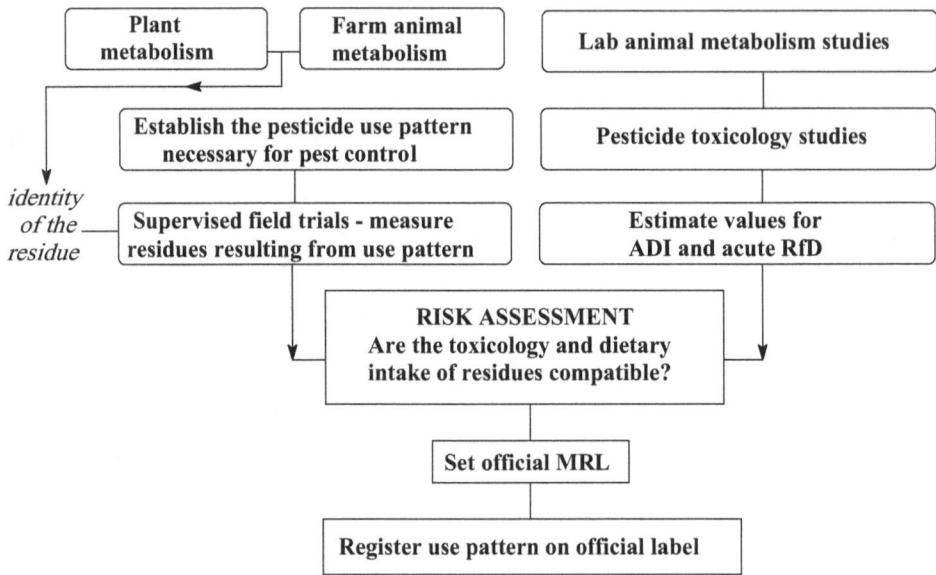

FIGURE 1   Risk assessment process before registration for pesticide residues in food. ADI: acceptable daily intake; Acute RfD: acute reference dose; MRL: maximum residue limit or tolerance.
Source: Hamilton DJ, Food contamination with pesticide residues, in *Encyclopedia of Pest Management,* 2002, p 287.

The acute reference dose (acute RfD or ARfD) of a chemical is an estimate of the amount normally expressed on a body-weight basis, which can be ingested in a period of 24 hours or less without appreciable health risk to the consumer on the basis of all known facts at the time of the evaluation. The acute RfD is also based on the results of animal dosing studies with a suitable safety factor.

The maximum residue limit (MRL), synonymous with "tolerance," is the maximum concentration of pesticide residue legally permitted in or on food commodities. The MRL usually applies to the commodity of trade, which may or may not be the same as the edible portion. For a fruit such as apples, it is the same, while for bananas, the MRL applies to the whole banana, but only the pulp is eaten. An MRL provides a division between food that is legally acceptable or unacceptable. Foods derived from commodities complying with the relevant MRLs are intended to be toxicologically acceptable, but the MRL is not a dividing line between safe and unsafe.

Supervised residue trials on animal feed commodities and livestock feeding studies with pesticide residues generate the information required to support MRLs for meat, milk, and eggs.

Risk assessment tells us whether or not the amounts of residue are likely to be safe for consumers.[4,5] We estimate dietary intake (also referred to as "dietary exposure") of pesticide residues by multiplying the level of residue in the food ready for consumption by the amount of the food consumed. For chronic risk assessment, we compare the sum for all foods of expected long-term average intake with the ADI for the pesticide. For acute risk assessment, we compare possible intake from high consumption of a food, in a period of 24 hours or less, with the acute RfD.

After Registration

The design of monitoring studies for residues in food commodities depends on the purpose: random survey of food consignments (surveillance), targeted enforcement sampling where a residue problem is suspected, export monitoring to meet trade requirements, and total diet studies.

Government authorities regularly survey agricultural and animal products for levels of pesticide residues. If the label directions were based on reliable and representative field trials and if users are faithfully following label directions, then residues will be within the legal MRLs. Most surveys have demonstrated a high level of compliance.

Total diet studies identify which pesticides and measure in what quantities people are actually consuming. Food purchased in the marketplace is prepared by peeling and cooking as in the normal household and is then subjected to residue analysis. Amounts of foods consumed are known from specially designed food surveys for subpopulations such as adult males and females, children, toddlers, and infants, as well as for ethnic groups and regions or localities. Dietary intakes for populations and subpopulations are calculated from the diets and the residue levels found by analysis. Commonly, total diet studies demonstrate intakes much less than the ADI.

Food Processing

Food processing usually reduces pesticide residue levels because of the washing or cleaning, peeling, milling, juicing, cooking, or baking. Residue levels may increase in some processed commodities because the residue tends more to one fraction than another.[6] For example, residues on the surface of a wheat grain will find their way into the bran fraction with little in the flour. Residues of oil-soluble pesticides will find their way mainly into the vegetable oil fraction from an oilseed such as soybean.

In particular cases, a food process can change the nature of the residue. For example, ethylenebisdithiocarbamate fungicides are converted, on cooking, to ethylenethiourea, which is more toxic than the parent pesticide. Fortunately, ethylenebisdithiocarbamates are essentially surface residues, and their levels can substantially be reduced by thorough washing before a cooking or blanching step.

Trade Issues

MRL values derived from good agricultural practice are, by their nature, local. A pesticide is used in the best way within local cultural practices to control a specific pest, and the rate of pesticide disappearance depends on local environmental conditions. Comparisons among countries of national MRLs and tolerances will frequently reveal substantial differences. Table 1 shows the range of MRLs for ethephon in 17 countries for each of four commodities.

TABLE 1 National Ethephon MRLs and Tolerances (mg/kg) in 1999[7]

	Peppers	Tomatoes	Pineapples	Grapes
Argentina	2	2		
Australia		2	2	10
Brazil		1.5	0.5	
Canada		2		
France				0.05
India		2	2	
Ireland	3	3		
Italy		3		0.05 Wine grapes 3 Table grapes
Korea		3	1	2
Netherlands	3	3		
New Zealand		1		
Poland		3		
Portugal	3	3		
South Africa			1	5
Taiwan		2	2	2
United Kingdom	3	3		
United States	30	2	2	2

The differences pose problems for international trade in food commodities. The importing country may reject shipments of food that do not comply with its national MRLs. It is attractive for some lobby groups and some governments to use national differences in MRLs as a barrier to trade.

Where no MRL or tolerance has been set for a pesticide on a food, some national governments apply a "zero tolerance," that is, the MRL is assumed to be zero unless otherwise stated. The reason no MRL is set could simply be that the pest problem does not occur or that the crop is not produced locally; for example, cold temperate countries do not produce pineapples, so there will be no local uses or local MRLs.

The Codex Alimentarius Commission was established in 1961 to implement the FAO/WHO Food Standards Program. A purpose of the program is to protect the health of consumers and to ensure fair practices in the food trade. The Codex Committee on Pesticide Residues (CCPR) has the responsibility to establish Codex MRLs for food commodities in international trade.

CCPR relies on the data supplied by member governments and has established many MRLs. The methods of data evaluation in Codex are very similar to the methods in countries with regulatory control of pesticide use; Codex draws on the expertise of scientists from such countries around the world. Member government acceptance of Codex MRLs for food commodities in international trade is reducing the incidence of trade barriers based on national MRLs.

Developing countries have sometimes suffered pesticide residue trade difficulties because a lack of resources has made it difficult for them to monitor their exports effectively to ensure compliance with the importing country MRL requirements.

Analytical Methods for Pesticide Residues

Analytical methods for pesticide residues in food typically rely on gas-liquid chromatography (GLC) or high-performance liquid chromatography (HPLC) in the final measurement step following extraction from the sample and a sequence of clean-up steps.[2,8] Multiresidue methods include many residues in one procedure for the sake of economy. Monitoring usually requires the detection and quantitative measurement of residue levels down to concentrations of around 0.01–0.05 mg/kg. Laboratories must validate their procedures down to the required level, that is, prove that the procedures can identify and measure with a specified precision residues down to a required "limit of quantification" (LOQ).

The LOQ is important in the interpretation of monitoring data. An analytical result reported as "less than LOQ" or sometimes as "no detectable residue" could possibly mean no residue or a residue at a level too low for the method.

Not all pesticide residues are amenable to inclusion in multiresidue methods; they may need separate analysis, which becomes expensive. Reports of monitoring data should state explicitly which residues would have been detected if present above stated LOQs.

Reliable high-quality data are essential for correct interpretation during registration, investigation, and control of residues in food. Highly skilled analysts using good laboratory practices, standard procedures, and other measures are generating valid data to support those requirements.

Future

The science of risk assessment will be further developed. Food safety and food security will continue to be important for consumer, government, and industry.[9] Trade issues will continue to be problematic with specific incidents of residues in foods arising from time to time. National governments will develop strategic approaches to deal with trade issues related to pesticide residues. Knowledgeable people in government and industry and experienced workers in functioning laboratories will be needed to support those strategic approaches. Exporters will need to monitor residues in a high percentage of their exports to meet the requirements of their customers. We might expect more developments with biopesticides. Relevant impurities in biopesticides are more likely to be biological than chemical, posing new challenges for analytical and test methods.

References

1. Frehse, H. ed. 1991. Pesticide Chemistry, Advances in International Research, Development, and Legislation, *Proceedings of the Seventh International Congress of Pesticide Chemistry (IUPAC), Hamburg, 1990.* VCH Verlagsgesellschaft mbH: Weinheim, Germany; 361–601.
2. FAO. 2016. Submission and Evaluation of Pesticide Residues Data for the Estimation of Maximum Residue Levels in Food and Feed; third edition. *FAO Plant Production and Protection Paper,* 225:1–286.
3. Dishburger, H.J., Ballantine, L.G., McCarthy, J., Murphy, J. and Tweedy, B. G., eds. 1991. *Pesticide Residues and Food Safety: A Harvest of Viewpoints. ACS Symposium Series 446,* American Chemical Society: Washington, DC, 1–348.
4. Hamilton, D.J., Holland, P.T., Ohlin, B. et al. 1997. Optimum use of available residue data in the estimation of dietary intake of pesticides. *Pure Appl. Chem.* 69:1373–1410.
5. WHO. 2009. Principles and Methods for the Risk Assessment of Chemicals in Food, IPCS. *Chapter 6, Dietary Exposure Assessment of Chemicals in Food.* Environmental Health Criteria 240.
6. Holland, P.T., Hamilton, D., Ohlin, B. and Skidmore, M.W. 1994. Effects of storage and processing on pesticide residues in plant products. *Pure Appl. Chem.* 66:335–356.
7. FAO. 2000. Pesticide Residues in Food. Ethephon. Evaluations 1999. FAO Plant Production and Protection Paper, 157:210.
8. Ambrus, Á. 1999. Quality of Residue Data. In *Pesticide Chemistry and Bioscience: The Food-Environment Challenge,* Brooks, G.T. and Roberts, T.R., eds., 339–360. The Royal Society of Chemistry: Cambridge.
9. Ambrus, Á. and Hamilton, D. 2017. Chapter 12, Future Directions. In *Food Safety Assessment of Pesticide Residues,* Ambrus, Á. and Hamilton, D., eds., 507–510. World Scientific Press: Singapore.

2

Human Health: Consumer Concerns to Pesticides

George Ekström and
Margareta Palmborg

Introduction

A recent European survey of risk perception and food safety showed, in line with previous research findings, that consumers tend to worry most about risks caused by external factors over which they have little or no control. Consequently, consumers appear to be less worried about risks possibly associated with their own behavior or practices. Physicians and scientists are the most trusted information sources with regard to serious food risks, followed by public authorities and mass media. Economic operators (food manufacturers, farmers, and retailers) are cited as being among the least trusted.[1]

Interviews conducted with over 1,000 consumers in a survey done in 2001 by the British Co-op Group showed that consumers were concerned about the effects of pesticides. Consumers who took part in this survey, when prompted with a series of questions, were concerned that pesticides are harmful to wildlife, leave residues in food, pollute water courses, are harmful to growing children, are harmful to the respondents themselves, and damage the health of farm workers.[2,3] According to a personal communication with David Pimentel, in the United States, the Food and Drug Administration has reported that 97% of people prefer foods without pesticides.

Particular causes of consumer concern are the potential for 'cocktail' effects from multiple residues (see Table 1),[4–6] and the fact that children may exceed health- related acute reference doses even at legally acceptable residue levels (see Table 2).[6] In Australia, the Food Standards code contains provisions for an additional, overall limit for pesticides belonging to the same chemical group (see Table 3).[7]

TABLE 1 Multiple Residues Found in a Single Sample of Pears 2004

Pesticides Found	Residue Level (mg/kg)	Maximum Residue Limit (mg/kg)	Residue Level in % of Maximum Residue Limit
Dithiocarbamates	0.305	3	10
Chlorpropham	0.155	0.05	310
Azinphosmethyl	0.084	0.5	17
Procymidone	0.071	1	7
Dichlofluanid	0.060	5	1
Chlorpyriphos	0.059	0.5	12
Bromopropylate	0.055	0.05	110
Cyprodinil	0.022	—	—
		Combined total residues	466

Source: Swedish food residue monitoring report (see Andersson and Jansson[6]).

TABLE 2 Food Residues Potentially Leading to Short Time Intake in Excess of the Acute Reference Dose (ARfD) for Toddlers 2004

Pesticide	Food Commodity	Highest Residue Found (mg/kg)	Maximum Residue Limit (mg/kg)	ARfD (mg/kg Body weight)	Intake, % of ARfD for Toddlers
Dicrotophos	Chinese broccoli	4.14	—	0.0017	1,763
Lambda-cyhalothrin	Lettuce	0.92	1	0.0075	106
Oxamyl	Cucumber	0.42	—	0.009	135
Endosulfan	Melon	0.21	0.3	0.02	110
Monocrotophos	Zuccini	0.14	—	0.002	381
Aldicarb	Potatoes	0.035	0.5	0.003	122

Source: Swedish food residue monitoring report (see Andersson and Jansson[6]).

TABLE 3 Approaches to the Limitation of Organophosphorus Pesticide Residues in Food in Australia and by the British Co-operative Group, Respectively

Group Tolerance in Australia	Group Tolerance in Australia and Co-op Zero Tolerance	Co-op Zero Tolerance
Azamethiphos, azinphos-ethyl, azinphos-methyl, coumaphos, demeton, diazinon, dichlorvos, dimethoate, disulfoton, dithianon, ethion, famphur, fenchlorphos, fenitrothion, fenthion, formothion, maldison,[a] methamidophos, methidathion, mevinphos, naphtalophos,[b] parathion-methyl, phosmet, pirimiphos- ethyl, pirimiphos-methyl, pyrazophos, sulprophos, temephos, tetrachlorvinphos, thiometon, S.S.S-tributylphosphorotrithioate, trichlorfon, vamidothion	Ethoprophos, fenamiphos, omethoate, phorate, prothiofos	Cadusafos, chlorfenvinphos, demeton-S-methyl, phosphamidon, tebupirimfos, terbufos

Source: Maximum residue limits—Chemical groups[7] and The Co-operative Group.[11]
[a] ISO common name is malathion.
[b] WHO INN, no ISO common name available.

Pests and Pesticide Safety in Homes and Gardens

Consumers use a range of pesticides in their homes and gardens:

- *Herbicides* against weeds in vegetables, moss in turf, brush, etc.
- *Fungicides* against mold, mildew, etc.

- *Insecticides* against aphids, greenflies, ants, wasps, pests on potted plants, moths, pantry pests, cockroaches, flies, etc.
- *Rodenticides* against moles, rats, mice, voles, etc.
- *Repellents* against mosquitoes, black flies, ticks, game, and pests on dogs, cats, and horses
- *Wood preservatives* against rot on timber or furniture

For the general public, ingestion is the most common route of pesticide exposure. Accidental, single, high- level exposures can lead to acute pesticide poisoning, often in children, and may result from mistakenly swallowed pesticides stored in unlocked cabinets or in unmarked bottles or containers. With regard to long-term, low-level exposure of the general public, the main route of exposure, ingestion through food and drinking water, is followed by inhalation through air or dust. This exposure results in an unknown number of people with diverse chronic health effects.[8]

A comparison of pesticide poisoning cases in 1984, 1994, and 2004, performed by the Swedish Poisons Information Centre, showed that there was an increase in the overall number of human cases related to pesticide exposure—from 493 in 1984, and 774 in 1994, to 1,071 in 2004.

The proportion of pesticide-related inquiries, however, remained constant at 3%.

Most incidents were due to accidental exposure at home. Ingestion was the most frequent route of exposure, followed by inhalation. Data from Swedish hospitals reported to the Poisons Information Centre showed the same pattern, that is, an increase in the proportion of cases related to accidental exposure at home. Children were involved in about 60% of all cases in 1984 as well as in 1994 and 2004. With the exception of 'superwarfarins' found in some rodenticides, most pesticides involved in incidents at home were of low toxicity and present at low concentrations in the formulated products. The few severe cases are mainly intentional poisonings.[9]

Thirty percent of the total number of inquiries to the Centre in 2004 was due to children's ingestion of insecticides intended for control of ants, containing low concentrations of borax, organophosphorus compounds or pyrethroids. No symptoms were recorded. Therefore, the considerable number of inquiries may reflect anxiety about pesticide exposure among the Swedish population.

Food Residues

Pesticide Residues in Foods from Organic, Integrated and Conventional Production[6]

In the Swedish monitoring program for 2004, no residues were detectable in 57% of the samples. Residues at or below maximum residue limits (MRLs) were found in 39% of the samples. 3.5% of all samples contained residues above the MRLs. Of foods from organic production, 4%–5% (import and domestic, respectively) contained detectable residues. Foods from integrated production were free from detectable residues in 91% of domestic produce and 50% of imported product. Foods from conventional production contained no detectable residues in 83% of domestic foods and 46% of imported foods. Residues below the MRLs were found in foods from all three production categories. Residues above the MRLs were found only in imported products from conventional production. No residues were found in any of the 92 samples of foods intended for infants and young children.

Although produced without pesticides, organically produced foods sometimes contain residues. The reason for this may be unintentional mix-up of foods from different sources (organic and conventional), environmental contamination of soils and plants, or fraud. The organic foods that contained pesticides in the United States have been shown to come mostly from soils treated many years ago with DDT or arsenical compounds, according to a personal communication with David Pimentel.

In 11 food commodities (22 samples) from ten countries, residues of ten different pesticides were found at levels 10–37 times the MRL. Multiple residues were found in 492 samples of which 279 samples with two residues, 127 samples with three residues, 54 samples with four residues, 25 samples with five residues, five samples with six residues, and two samples with eight residues (see Table 1 and Figure 1).

FIGURE 1　European Union organic logo. (Available at http://europa.eu.int/comm/agriculture/qual/organic/logo/index_en.htm.)

Towards Residue-Reduced Food Crops

Government Action Plans

The British Food Standards Agency has recognized that while levels of pesticide residues typically found in food are not normally a food safety concern, consumer preference is for food that does not contain residues. Sixty-eight percent of consumers consider that reducing residue levels further than the current level is important. As a result, the Agency has developed an action plan for pesticide residue minimization with a goal of enabling consumers to make informed choices, and promoting best practice within the food industry. The overall action plan includes, among other things, development of crop specific action plans to achieve pesticide residue minimization for five priority crops: apples, pears, potatoes, tomatoes, and cereal grains.[10]

Retailer Initiatives[11,12]

Retailers may employ a range of strategies to reduce pesticide use and residues in the foods they produce and put on the market:

- Monitoring pesticide usage and residues
- Consulting with growers, including advice on integrated pest management and on alternative pest control systems
- Designing and providing decision tools, including crop-specific advisory sheets and frameworks for pesticide selection
- Prohibiting or restricting the use of certain pesticides
- Publishing monitoring results, for example, on corporate websites
- Promoting organically grown foods

NGO Initiatives—Ranking Residue Contents

In the United States, the Environmental Working Group has designed a report card to score pesticide residues in food products.[13] The report card, which is based on government agency monitoring and published monitoring data, shows scores for each analyzed commodity based on a number of residue characteristics, and a combined (or total) score in these categories:

- Percentage of samples with detectable residues
- Percentage of samples with two or more pesticides
- Average number of pesticides found on each sample
- Average total concentration of pesticides found
- Maximum number of pesticides found on a single sample
- Total number of pesticides found on a single commodity

In the Netherlands, Natuur and Milieu (a Dutch NGO), has designed and used a ranking system based on the following components, and a calculated total score[14]:

- For any residue not exceeding an MRL: 1 penalty point
- For each residue of a pesticide with neurotoxic effects: 2 penalty points
- For each residue exceeding an MRL or resulting from the use of a pesticide not authorized for use in the Netherlands (applicable also to imported foods): 4 penalty points
- For each residue exceeding an MRL and resulting from the use of a pesticide not authorized for use in the Netherlands (applicable also to imported foods): 8 penalty points

Conclusions

Polls have shown repeatedly that consumers are concerned about pesticide residues in food. Maximum residue limits are trading standards, which prescribe the maximum amount of particular pesticides legally permitted. These limits (MRLs) are generally based on the level of pesticides expected if good agricultural practice is followed.[10] Other MRLs may reflect only that a pesticide is no longer authorized for use, leading to zero tolerance, and to potential problems for food exporters overseas. Since many consumers feel that current good agricultural practice is not good enough for their own or their childrens' safety, governments, retailers and NGOs have initiated actions to reduce residues. Strategies focus on production methods (integrated, organic) as well as product quality (residue-reduced or residue-free foods).

Acknowledgments

Contributions from Stephanie Williamson, Pesticide Action Network U.K., are gratefully acknowledged.

References

1. European Food Safety Authority. Risk perception and food safety: Where do European consumers stand today? February 2006. http://www.efsa.eu.int/press_room/press_release/1340_en.html (accessed May 2006).
2. Food Standards Agency. Consumer concern over the use of pesticides to grow food, March 2004. http://www.food.gov.uk/multimedia/pdfs/pestresconsumeresearch.pdf (accessed January 2005).
3. Croft, D. Removing pesticides from the food chain, December 2002. http://www.pan-uk.org/pestnews/pn58/pn58p9.htm (accessed January 2006).
4. Committee on Toxicity of Chemicals in Food, Consumer Products and the Environment. Risk assessment of mixtures of pesticides and similar substances, September 2002. http://www.food.gov.uk/multimedia/pdfs/reportindexed.pdf (accessed January 2006).
5. Beaumont, P.; Buffin, D. A cocktail of problems, March 2002. http://www.pan-uk.org/pestnews/pn55/pn55p10.htm (accessed March 2006).
6. Andersson, A.; Jansson, A. The Swedish monitoring of pesticide residues in food of plant origin 2004. National (Swedish) Food Administration. http://www.slv.se/upload/dokument/Rapporter/Bekampningsmedel/2005Z2005_17eng_Livsmedelsverket.pdf (accessed January 2006).

7. Maximum residue limits—Chemical groups. Australian Food Standards No 1.4.2. http://www.foodstandards.gov.au/foodstandardscode/ (accessed January 2006).
8. World Health Organization. Public health impact of pesticides used in agriculture. 1990, ISBN 92-4-1561-39-4.
9. Ekstrom, G.; Hemming, H.; Palmborg, M. Swedish pesticide risk reduction 1981–1995: food residues, health hazard, and reported poisonings. *Rev. Environ. Contam. Toxicol.* **1996**, *147*, 119–147.
10. Food Safety Authority. Progress on an agency action plan to minimize pesticide residues in food, May 2004. http://www.food.gov.uk/multimedia/pdfs/fsa040502.pdf (accessed January 2006).
11. The Co-operative Group. Co-op and the responsible use of pesticides. Consumer issues website. http://www.co-op.co.uk (accessed February 2006).
12. Barker, K. Co-operative retail's pesticide reduction program, July 2005. http://www.pan-europe.info/conferences/PURE%20workshop%202005/Kevin%20Barker%20Co-op.pdf (accessed March 2006).
13. Environmental Working Group. Report card: Pesticides in produce 2005. http://www.foodnews.org/reportcard.php (accessed December 2005).
14. Muilerman, H. Dutch supermarket residue campaign. *Pestic. News* **2006**, *71*, 6–7.

3

Human Health: Endocrine Disruption

Evamarie Straube
and Sebastian
Straube

Introduction

In general, endocrine disruption in humans can result from genetic disorders,[1] diseases,[2] medical treatments, mental and physical[3] stress, and chemical exposures. Endocrine disruption through chemical agents forms the focus of the present chapter.

Substances relevant for such endocrine-disrupting chemical exposures include pesticides (organochlorines such as DDT,[4] other organohalogens such as dibromochloropropane,[5] some organophosphates, carbamates, dithiocarbamates, phthalates), polychlorinated biphenyls,[6] some solvents,[7] metals such as cadmium, lead, and manganese,[8] phytoestrogens, and isoflavonoids.[9] Furthermore, endocrine disruption can be caused by smoking and alcohol use,[3] and by certain drugs, for example, glucocorticoids, hypnotics, antihypertensives, neuroleptics, and H_2-antihistaminies.[2]

Endocrine disruption can affect various endocrine systems. For example, thyroid hormone inhibition has been reported in humans after occupational exposure to amitrol and mancozeb.[10] Insulin levels can be affected by streptozotocin, which is toxic to pancreatic beta cells.[11]

Arguably the most extensive knowledge exists for chemical exposure affecting the reproductive system, which is discussed in further detail below.

Of note, for some endocrine-disrupting chemicals, non-monotonic dose–response relationships have been described with hormonal disruption occurring at relatively low levels of exposure.[12]

Mechanisms of Endocrine Disruption

Diverse mechanisms of endocrine disruption by chemical agents have been described. Endocrine disruption can be caused by xenohormones. Xenoestrogens, such as endosulfan, toxaphene, dieldrin, DDT, bisphenol A, nonylphenols, and dibutylphthalates,[7] mimic the physiological effects of estrogens. Xenoantiestrogens have effects opposite to those of xenoestrogens. For example, dioxin exerts its inhibitory effect by enhancing the expression of enzymes that degrade the estrogen receptors.[13] Antiandrogenic effects may result from competitive antagonism at androgen receptors. This was demonstrated for vinclozolin and DDE, the stable metabolite of the DDT.[7] Sometimes a xenobiotic and its metabolite (such as DDT and DDE) can exert their effects at different targets in the organism.

Pesticide-induced enzymes such as UDP-glucuronyl transferase and monooxygenases can degrade hormones (e.g., testosterone). Furthermore, the pesticides endosulfan, mirex, and DDT can increase the elimination of androgens by stimulating cytochrome P450.[14,15] Pesticide exposure can also disrupt hormonal status by inhibiting enzymes. For example, inhibition of the aromatase system can lead to an increase in testosterone levels and a decrease in the formation of estradiol from testosterone.[16] Inhibitors of the aromatase system include prochloraz, imazalil, propiconazole, fenarimol, triadimenol, triadimefon, and dicofol.[17]

Endocrine disruption can furthermore involve hormone transport proteins: for example, polychlorinated biphenyl can induce thyroid disruption, and this may involve its sulfated metabolites which bind to the thyroid hormone transport protein transthyretin with high affinity.[18]

Moreover, recent research demonstrates that endocrine disruption can involve epigenetic alterations, which may be preserved into the third generation, as has been shown for DNA methylation changes that are associated with lead exposure.[19]

Endocrine Disruption with Occupational Exposure in Women

Pesticide exposure has been linked with reproductive difficulties and menstrual abnormalities. For example, prolonged time to pregnancy, reduced fecundability, reduced fertility as well as infertility have been described for pesticide-exposed women and for women working in agriculture and greenhouses, presumably related to pesticide exposure.[20–24] Furthermore, pesticide exposure has been associated with long cycles, missed periods, and intermenstrual bleeding.[25]

Endocrine Disruption with Occupational Exposure in Men

In men, pesticides may adversely affect sperm count and quality as well as impacting the levels of testosterone, follicle-stimulating hormone, and luteinizing hormone.[26]

We have found changes in sex hormone concentrations after low-dose occupational exposure to pesticides (mainly pyrethroids, carbamates, and organophosphates). With chronic occupational pesticide exposure, we found a higher level of testosterone in comparison to control persons. There also was a reduction in estradiol levels during and after the application season in pesticide applicators.[16] Another study[27], however, found an increase in estradiol concentration in pesticide-exposed men; it may be that variations in the nature and timing of the exposure can account for these differences.

Endocrine Disruption with *In Utero* Exposure

Historically, diethylstilbestrol (DES) was the first recognized example of a xenobiotic eliciting a hormonal effect. This now very well-described paradigm serves to illustrate the potential consequences of *in utero* exposure to endocrine-disrupting chemicals. The treatment of pregnant women with DES leads to an increase in the incidence of adenocarcinoma of the vagina in their daughters[28] and malformations of the external genitals in their sons[29] and grandsons.[30] The risk of breast cancer after age 40 was also increased in women with prenatal DES exposure, as were the risks for spontaneous abortions, premature births, and ectopic pregnancies in later pregnancies of women exposed *in utero*.[31–33] Treatment with DES has, furthermore, been reported to have an effect on sexual orientation[34] and handedness.[35]

Acknowledgment

We acknowledge the contribution of our previous co-author, Wolfgang Straube, to whose memory this chapter is dedicated.

References

1. Achermann, J.C.; Ozisik, G.; Meeks, J.J.; Jameson, J.L. Genetic causes of human reproductive disease. *J. Clin. Endocrinol. Metab.* **2002**, *87* (6), 2447–54.

2. Turner, H.E.; Wass, J.A.H. Gonadal function in men with chronic illness. *Clin. Endocrinol.* **1997**, *47* (4), 379–403.

3. Vermeulen, A.; Kaufman, J.M. The age associated decline in testicular function: Partial androgen deficiency of the ageing male (PADAM). *Menopause Rev.* 1999, *IV* (2), 23–35.

4. Welch, R.M.; Levin, W.; Conney, A.H. Estrogenic action of DDT and its analogs. *Toxicol. Appl. Pharmacol.* **1969**, *14* (2), 358–67.

5. Potashnik, G.; Porath, A. Dibromochloropropane (DCBP): A 17-year reassessment of testicular function and reproductive performance. *J. Occup. Environ. Med.* **1995**, *37* (11), 1287–92.

6. Bitman, J.; Cecil, H.J. Estrogenic activity of DDT analogs and polychlorinated biphenyls. *J. Agric. Food Chem.* **1970**, *18* (6), 1108–12.

7. Massaad, C.; Entezami, F.; Massade, L.; Benahmed, M.; Olivennes, F.; Barouki, R.; Hamamah, S. How can chemical compounds alter human fertility? *Eur. J. Obstet. Gynecol. Reprod. Biol.* **2002**, *100* (2), 127–37.

8. Gennart, J.P.; Buchet, J.P.; Roeis, H.; Ghyselen, P.; Ceulemans, E.; Lauwerys, R. Fertility of male workers exposed to cadmium, lead or manganese. *Am. J. Epidemiol.* **1992**, *135* (11), 1208–19.

9. Jacobs, M.N.; Lewis, D.F. Steroid hormone receptors and dietary ligands: A selected review. *Proc. Nutr. Soc.* **2002**, *61* (1), 105–22.

10. Cocco, P. On the rumors about the silent spring. Review of the scientific evidence linking occupational and environmental pesticide exposure to endocrine disrupting health effects. *Cad. Saude Publica* **2002**, *18* (2), 379–402.

11. Bolzan, A.D.; Bianchi, M.S. Genotoxicity of streptozotocin. *Mutat. Res.* **2002**, *512* (2–3), 121–34.

12. Schmidt, C.W. The lowdown on low-dose endocrine disruptors. *Environ. Health Perspect.* **2001**, *109* (9), 420–21.

13. Yonemoto, J. The effects of dioxin on reproduction and development. *Ind. Health* **2000**, *38* (3), 259–68.

14. Savas, U.; Griffin, K.J.; Johnson, E.F. Molecular mechanisms of cytochrome P-450 induction by xenobiotics: An expended role for nuclear hormone receptors. *Mol. Pharmacol.* **1999**, *56* (5), 851–57.

15. Waxman, D.J. P-450 gene induction by structurally diverse xenochemicals: Central role of nuclear receptors CAR, PXR, and PPAR. *Arch. Biochem. Biophys.* **1999**, *369* (1), 11–23.

16. Straube, E.; Straube, W.; Krüger, E.; Bradatsch, M.; Jacob-Meisel, M.; Rose, H.-J. Disruption of male sex hormones with regard to pesticides: Pathophysiological and regulatory aspects. *Toxicol. Lett.* **1999**, *107* (1–3), 225–31.

17. Vinggaard, A.M.; Hnida, C.; Breinholt, V.; Larsen, J.C. Screening of selected pesticides for inhibition of CYP19 aromatase activity *in vitro. Toxicol. In Vitro* **2000**, *14* (3), 227–34.

18. Grimm, F.A.; Lehmler, H.J.; He, X.; Robertson, L.W.; Duffel, M.W. Sulfated metabolites of polychlorinated biphenyls are high-affinity ligands for the thyroid hormone transport protein transthyretin. *Environ. Health Persp.* **2013**, *121*, 657–62.

19. Sen, A.; Heredia, N.; Senut, M-C.; Land, S.; Hollocher, K,; Lu, X.; Dereski, M.O.; Ruden, D.M. Multigenerational epigenetic inheritance in humans: DNA methylation changes associated with maternal exposure to lead can be transmitted to the grandchildren. *Sci. Rep.* **2015**, *5*, 14466–82.

20. Fuortes, L.; Clark, M.K.; Kirchner, H.L.; Smith, E.M. Association between female infertility and agricultural work history. *Am. J. Ind. Med.* **1997**, *31* (4), 445–51.

21. Smith, E.M.; Hammonds-Ehlers, M.; Clark, M.K.; Kirchner, H.L.; Fuortes, L. Occupational exposures and risk of female fertility. *J. Occup. Environ. Med.* **1997**, *39* (2), 138–47.

22. Abell, A.; Juul, S.; Bonde, J.P. Time to pregnancy among female greenhouse workers. *Scand. J. Work. Environ. Health* **2000**, *26* (2), 131–6.

23. Idrovo, A.J.; Sanin, L.H.; Cole, D.; Chavarro, J.; Caceres, H.; Narvaez, J.; Restrepo, M. Time to first pregnancy among women working in agricultural production. *Int. Arch. Occup. Environ. Health* **2005**, *78* (6), 493–500.

24. Bretveld, R.W.; Thomas, Ch.M.G.; Scheeepers, P.T.J.; Zielhuis, G.A.; Roeleveld, N. Pesticide exposure: The hormonal function of the female reproductive system disrupted? Reprod. *Biol. Endocrinol.* **2006**, *4*, 30–54.

25. Farr, S.L.; Cooper, G.S.; Cai, J.; Savitz, D.A.; Sandler, D.P. Pesticide use and menstrual cycle characteristics among premenopausal women in the agricultural health study. *Am. J. Epidemiol.* **2004**, *160* (12), 1194–204.

26. Mehrpour, O.; Karrari, P.; Zamani, N.; Tsatsakis, A.M.; Abdollahi, M. Occupational exposure to pesticides and consequences on male semen and fertility: A review. *Toxicol. Lett.* **2014**, *230* (2), 146–56.

27. Oliva, A.; Spira, A.; Multigner, L. Contribution of environmental factors to the risk of male infertility. *Hum. Reprod.* **2001**, *16* (8), 1768–76.

28. Herbst, A.L.; Ulfelder, H.; Poskanzer, D.C. Adenocarcinoma of the vagina. Association of maternal diethylstilbestrol therapy with tumor appearance in young women. *N. Engl. J. Med.* **1971**, *284* (15), 878–81.

29. Bibbo, M.; Gill, W.B.; Azizi, F.; Blough, R.; Fang, V.S.; Rosenfield, R.L.; Schuhmacher, G.F.P.; Sleeper, K.; Sonek, M.G.; Wied, F.; Wied, G.L. Follow-up study of male and female offspring of DES-exposed mothers. *Obstet. Gynecol.* **1977**, *49* (1), 1–8.

30. Klip, H.; Verloop, J.; van Gool, J.D.; Koster, M.E.; Burger, C.W.; van Leeuwen, F.E. Hypospadias in sons of women exposed to diethylstilbestrol in utero: A cohort study. *Lancet* **2002**, *359* (9312), 1102–7.

31. Kaufman, R.H.; Adam, E.; Hatch, E.E.; Noller, K.; Herbst, A.L.; Palmer, J.R.; Hoover, R.N. Continued follow-up of pregnancy outcomes in diethylstilbestrol-exposed offspring. *Obstet. Gynecol.* **2000**, *96* (4), 483–9.

32. Palmer, J.R.; Wise, L.A.; Hatch, E.E.; Troisi, R.; Titus-Emshoff, L.; Strohsnitter, W.; Kaufmann, R.; Herbst, A.L.; Noller, K.L.; Hyer, M.; Hoover, R.N. Prenatal diethylstilbestrol exposure and risk of breast cancer. *Cancer Epidemiol. Biomarkers Prev.* **2006**, *15* (8), 1509–14.

33. Gaskins, A.J.; Missmer, S.A.; Rich-Edwards, J.W.; Williams, P.L.; Souter, I.; Chavarro, J.E. Demographic, lifestyle, and reproductive risk factors for ectopic pregnancy. *Fertil. Steril.* **2018**, *110* (7), 1328–37.

34. Ehrhardt, A.; Meyer-Bahlburg, H.; Rosen, L.; Feldman, J.; Veridiano, N.; Zimmermann, I.; McEwen, B.S. Sexual orientation after prenatal exposure to exogenous estrogen. *Arch. Sex. Behav.* **1985**, *14* (1), 57–77.

35. Scheirs, J.G.M.; Vingerhoets, A.J.J.M. Handedness and other laterality indices in women prenatally exposed to DES. *J. Clin. Exp. Neuropsychol.* **1995**, *17* (5), 725–30.

4

Human Health: Pesticides

Kelsey Hart and
David Pimentel

Introduction

Since the first use of DDT for crop protection in 1945, the total amount of pesticides used in agriculture worldwide has been staggering. In 1945, about 50 million kg of pesticides were applied worldwide. Today, global usage is currently at about 2.5 billion kg/yr, an approximate 50-fold increase since 1945.[1] Unfortunately, the toxicity of most modern pesticides is more than 10-fold greater than those used in the early 1950s,[1] so the potential hazards have increased as well. In fact, studies have linked exposure to pesticides with a variety of human health problems, from asthma to cancer.[2] High levels of exposure to toxic pesticides can even result in fatal poisonings. In addition, we are discovering that we can be unknowingly exposed to pesticides and pesticide residues through the food we eat, water we drink, and air we breathe,[3] and that both short- and long-term exposure to pesticides can lead to chronic health effects. Based on the available data, estimates are that human pesticide poisonings and related illnesses in the United States cost about $933 million each year.[3]

Exposure to Pesticides

While farmers and pesticide applicators typically are exposed to higher levels and more kinds of pesticides, the general public is also exposed—often unwittingly—to pesticides and pesticides residues in their daily lives. For example, about 35% of the foods purchased by American consumers have detectable levels of pesticide residues.[4] This estimate is, in fact, conservative because we currently test for only about one-third of the pesticides in use.

In addition to food contamination, the public can also be exposed to pesticides in other ways. The principle exposure of the general public in the United States occurs in the home; about 90% of all U.S. households use pesticides on their lawn, garden, and/or the inside of the home.[5] Drinking water can also contain significant chemical and pesticide residues; at present, more than 10% of U.S. rivers and 5% of U.S. lakes are measurably polluted with pesticides.[5] Groundwater supplies can also be polluted when pesticides seep into aquifers or wells.[6] Finally, public exposure to pesticides can occur through accidents or spills, or even through the air during application, when pesticides drift from the target area into more populated towns and cities. In fact, only 25%–50% of pesticides applied by aircraft under ideal weather conditions actually reach the target area.[1]

Acute Effects: Pesticide Poisonings

In 1945, when synthetic pesticides were first used, few pesticide poisonings were reported. But by the late 1960s, both pesticide usage and toxicity had increased so dramatically that the number of human pesticide poisonings was substantial.[1] Unfortunately, this trend continued into the year 2000 (Figure 1). Just in the last decade, the total number of pesticide poisonings in the United States has increased from 67,000 in 1989, to the current level of 110,000 per year.[2] Worldwide, the increased use of pesticides results in approximately 26.5 million cases of occupational pesticide poisonings each year, and an unknown number of non-occupational pesticide poisonings.[7] Of all these estimated poisoning episodes, about 3 million cases are hospitalized, resulting in approximately 220,000 fatalities and about 750,000 cases of chronic illness every year.[8]

Poisonings can occur when pesticides contact the skin or eyes, are inhaled, or are ingested. Typically, pesticides can have acute local effects on the area they directly contact—skin, eye, or respiratory tract irritation—in addition to acute systemic effects. Different pesticides act on different systems in the body in a variety of ways, but most common insecticides—organophosphates and organo-chlorines (i.e., Parathion)—have acute neurotoxic effects. This means that they impair normal functioning of the brain and/or the spinal cord, which can result in tremors, paralysis, seizures, and other systemic effects. Large degrees of exposure or exposure to highly toxic pesticides can be fatal.

Chronic Effects: Cancer and Other Health Concerns

Exposure to pesticides, though, does not always occur at a level sufficient to produce these acute symptoms. Many people are exposed to low levels of pesticides over a long period of time through their occupations or in the food they eat or water they drink. If no overt symptoms occur close to the time of exposure, we often assume that no damage is being done. However, pesticides have been associated with numerous chronic illnesses and health problems, especially in some highly sensitive individuals.

Chronic effects of pesticides are diverse and can affect most systems of the human body. The major types of chronic health effects that pesticides can have are neurological effects, respiratory and

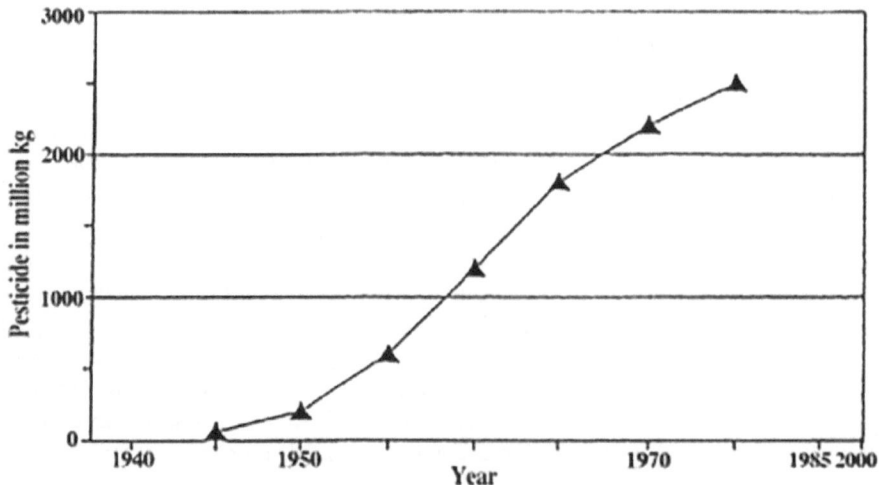

FIGURE 1   Trend in annual world pesticide use.[2]

reproductive effects, and carcinogenic effects. The chronic neurotoxic effects of pesticide exposure are not well understood, but there is some evidence linking pesticide exposure with symptoms like fatigue, muscle weakness, and sensory disturbances as well as cognitive effects such as memory loss, language problems, and learning impairment. The malady organophosphate-induced delayed poly-neuropathy (OPIDP) is well documented and includes irreversible neurological effects.

In addition to their neurotoxic effects, pesticides can have adverse effects on the respiratory and reproductive systems. For example, 15% of a group of professional pesticide applicators suffered asthma, chronic sinusitis, or chronic bronchitis as compared with 2% of people who used pesticides infrequently.[9] Studies have also linked pesticides with reproductive effects such as infertility and fetal deformities, but the data are still inconclusive.[10] Some pesticides have been shown to cause testicular dysfunction or sterility in animals, and similar effects are suspected in humans, but are less well understood at present.[10] Sperm counts in males in Europe and the United States, for example, declined by about 50% between 1938 and 1990[11]—a time period during which the use of synthetic pesticides increased about 30-fold and toxicity per pound increased about 10-fold.[1] At present, there is evidence that human sperm counts continue to decrease by about 2% per year.[3]

U.S. data indicate that 18% of all insecticides, and about 90% of all fungicides, are carcinogenic,[12] and many studies have shown that risks for certain types of cancers are higher in people—such as farmworkers and pesticide applicators—who are more frequently exposed to certain pesticides. Certain pesticides have been shown to induce tumors in lab animals; there is some evidence to suggest that they may have similar effects in humans.[10]

Many pesticides are also estrogenic—they mimic or interact with the hormone estrogen—linking them to the increased breast cancer rate among some groups of women in the United States. The breast cancer rate rose from 1 in 20 in 1960 to 1 in 8 in 1995.[13] There was a concurrent increase in pesticide use during that time period, and, although it has not been concretely linked to the increase in breast cancer rates, some studies suggest that exposure to pesticides is related to breast cancer incidence. Pesticides that interfere with the body's endocrine—hormonal—system can also have reproductive, immulogic, or developmental effects; these effects are well documented in animals and just starting to be understood in humans.[10] While endocrine disrupting chemicals may appear less dangerous at first glance—hormonal effects rarely result in acute poisonings or sudden death—their effects on reproductive and developmental processes may prove to have far-reaching and even more sinister consequences.[10]

Effects on Children

The negative health effects that pesticides can have—both acute and chronic—can be more significant in children than in adults, for several important reasons. First, children have much higher metabolic rates than adults, and their ability to activate, detoxify, and excrete toxic compounds is different from that of adults. Also, because of their smaller physical size, children are exposed to higher levels of pesticides per unit of body weight. In addition, certain types of pesticides are inherently more dangerous for children than for adults.[2,10] For example, the organophosphate and carbamate classes of pesticides adversely affect the nervous system by inhibiting cholinesterase, a critical enzyme, and can produce difficulty breathing, partial or total paralysis, convulsions, coma, or death. This problem is particularly significant for children since their brains are more than five times larger in proportion to their body weight than an adult's brain, making cholinesterase even more vital. In a California study, 40% of the children working in agricultural fields had blood cholinesterase levels below normal, a strong indication of organophosphate and carbamate pesticide poisoning.[14] In addition, a study in England and Wales has shown that 50% of all pesticide poisoning incidents in those countries involved children less than 10 years of age.[15] Use of pesticides in the home is also linked to childhood cancer.[16]

Because these studies and others have demonstrated that children's increased sensitivities to toxicants is a significant concern, some pesticide regulations have recently been reevaluated to provide special protections for children and infants. The Food Quality Protection Act of 1996 requires the EPA to specifically address risks that pesticides pose to infants and children and provides for an additional safety factors to take into account the potentially greater exposure and/ or sensitivity to pesticide effects on infants and children.

Economic Costs and Conclusions

Although no one can place a precise monetary value on a human life, the economic "costs" of human pesticide poisonings have been estimated. For our assessment, we use the conservative estimate of $2.2 million per human life—the average value that the surviving spouse of a slain New York City police-man receives.[17] Available estimates suggest that human pesticide poisonings and related illnesses in the United States cost about $933 million per year.[3] Pesticide use, though, provides a substantial net agricultural return of $12 billion per year.[2] Are the public health risks associated with pesticide use a great enough concern to warrant a reduction in pesticide use?

Given the known—and suspected—adverse effects of pesticide use on public health discussed in previous sections, it seems fairly clear that our present levels and methods of pesticide use present significant public health dangers and concerns. However, the abrupt and complete cessation of syn-thetic pesticide use is not without substantial costs of its own. Termination of pesticide use would result in huge crop losses—the food supply would be severely reduced and a huge number of people would starve in a short time,[18] an especially dire consequence given the 3 billion people—half the world's population—currently malnourished worldwide.[19]

Clearly, it is essential that all the costs and benefits—economic, environmental, and social/health—of pesticide use be considered when current and future pest control programs are being developed and evaluated. A recent study estimated that the environmental and social costs related to U.S. pesticide use—crop losses, public health effects, pesticide resistance, water pollution, and other environmental effects—total $8.3 billion each year.[3] Furthermore, another study has shown that U.S. pesticide use can be reduced up to one-half without any reduction in crop yields and cosmetic standards, and only a mini-mal 0.6% increase in food costs.[20] Therefore, it is clear that our current methods of chemical pesticide use need to be reconsidered and evaluated with the above-outlined public health effects and the related economic costs in mind, toward the goal of the development of sound, sustainable pest management practices that maximize the benefits of pesticide use while at the same time minimizing the adverse effects that pesticides can have on human health.

References

1. Pimentel, D. Protecting crops. In *The Literature of Crop Science*; Olsen, W.C., Ed.; Cornell University Press: Ithaca, NY, 1995; 49–66.
2. Pimentel, D.; Hart, K.A. Ethical, environmental, and public health implications of pesticide use. In *Perspectives in Bioethics*; Galston, A., Ed.; Johns Hopkins University Press: Baltimore, MD, 2000, in press.
3. Pimentel, D.; Greiner, A. Environmental and socio economic costs of pesticide use. In *Techniques for Reducing Pesticide Use*; Pimentel, D., Ed.; John Wiley & Sons: Chichester, 1997; 51–78.
4. FDA. Food and drug administration pesticide program residues in foods—1989. *J. Assoc. Anal. Chem.* **1990**, *73*, 127A–146A.
5. Pimentel, D.; Wilson, C.; McCullum, C.; Huang, R.; Dwen, P.; Flack, J.; Tran, Q.; Saltman, T.; Cliff, B. Economic and environmental benefits of biodiversity. *BioSci.* **1997**, *47* (11), 747–757.
6. Pimentel, D.; Acquay, H.; Biltonen, M.; Rice, P.; Silva, M.; Nelson, J.; Lipner, V.; Giordano, S.; Horowitz, A.; D'Amore, M. Assessment of environmental and economic impacts of pesticide use. In *The Pesticide Question: Environment, Economics and Ethics*; Pimentel, D., Lehman, H., Eds.; Chapman & Hall: New York, 1993; 47–84.
7. UNEP. *Global Environmental Outlook*; United Nations Environment Program: Nairobi, 1997.

8. WHO. *Our Planet, Our Health: Report of the WHO Commission on Health and Environment*; World Health Organization: Geneva, 1992.

9. Weiner, B.P.; Worth, R.M. Insecticides: Household use and respiratory impairment. In *Adverse Effects of Common Environmental Pollutants*; MSS Information Corporation: New York, 1972; 149–151.

10. Colborn, T.; Dumanoski, D.; Myers, J.P. *Our Stolen Future: Are We Threatening Our Fertility, Intelligence, and Survival? A Scientific Detective Story*; Dutton: New York, 1996.

11. Carlsen, E.; Giwercman, A.; Keiding, N.; Skakkebaek, N.E. Evidence for decreasing quality of semen during past 50 years. *Br. Med. J.* **1992**, *305*, 609–613.

12. NAS. *Regulating Pesticides in Food*; National Academy of Sciences: Washington, DC, 1987.

13. McCarthy, S. Congress takes a look at estrogenic pesticides and breast cancer. *J. Pestic. Reform* **1993**, *13* (4), 25.

14. Repetto, R.; Baliga, S.S. *Pesticides and the Immune System: The Public Health Risks*; World Resources Institute: Washington, DC, 1996.

15. Casey, P.; Vale, J.A. Deaths from pesticide poisoning in England and Wales: 1945–1989. *Human Exp. Toxicol.* **1994**, *13*, 95–101.

16. Leiss, J.K.; Savitz, D.A. Home pesticide use and childhood cancer: A case-control study. *Am. J. Public Health* **1995**, *85*, 249–252.

17. Nash, E.P. What's a Life Worth; New York Times: New York, 1994.

18. Lehman, H. Values, ethics, and the use of synthetic pesticides in agriculture. In *The Pesticide Question: Environment, Economics and Ethics*; Pimentel, D., Lehman, H., Eds.; Chapman & Hall: New York, 1993; 347–379.

19. WHO. *Micronutrient Malnutrition—Half the World's Population Affected*; World Health Organization: Geneva, 1996; 78, 1–4.

20. Pimentel, D.; McLaughlin, L.; Zepp, A.; Lakitan, B.; Kraus, T.; Kleinman, P.; Vancini, T.; Roach, W.J.; Graap, E.; Keeton, W.S.; Selig, G. Environmental and economic impacts of reducing U.S. agricultural pesticide use. In *Handbook on Pest Management in Agriculture*; Pimentel, D., Ed.; CRC Press: Boca Raton, FL, 1991; 679–718.

5

Nanoparticles

Alexandra
Navrotsky

Introduction

Nanoparticles have no exact definition, but they are aggregates of atoms bridging the continuum between small molecular clusters of a few atoms and dimensions of 0.2–1 nm and chunks of solid containing millions of atoms and having the properties of macroscopic bulk material. In water, nanoparticles include colloids; in air, they include aerosols. Nanoparticles are ubiquitous. We pay to have them. We pay more to not have them. They occur as dust in the air, as suspended particles that make river water slightly murky, in soil, in volcanic ash, in our bodies, and in technological applications ranging from ultratough ceramics to microelectronics. They both pollute our environment and help keep it clean. Microbes feast on, manufacture, and excrete nanoparticles.

Understanding nanoparticle formation and properties requires sophisticated physics, chemistry, and materials science. Tailoring nanomaterials to specific applications requires both science and Edisonian inventiveness. Applying them to technology is state-of-the-art engineering. Tracing their transport and fate in the environment invokes geology, hydrology, and atmospheric science. Applying them to improving soil fertility and water retention links soil science and agriculture to surface chemistry. Understanding their biological interactions brings in fields ranging from microbiology to medicine. Probing the impact of nanoparticles on humans and of human behavior on the production and control of nanoparticles requires the behavioral and social sciences, e.g., in dealing with issues of automotive pollution. The purpose of this review is to describe some of the unique features of nanoparticles and to discuss their occurrence and importance in the natural environment.

Although we often think of the natural environment as that part of the planet which we can see, a somewhat broader definition includes the "critical zone": the atmosphere, hydrosphere, and shallow portion of the solid earth that exchange matter on a geologically short time scale, on the order of tens to thousands of years. This critical zone affects us directly, and our activities influence it. Because of the active chemical reactions continuously taking place in the critical zone, and because its temperatures and pressures are relatively low and it is dominated by water, solids are constantly being formed and decomposed. Many of these solids start out as nanoparticles; many remain so. In a yet broader sense, our entire planet from crust to core, the solar system, and the galaxy are part of our environment.

Physical Chemistry of Nanoparticles

A major feature of nanoparticles is their high surface-to-volume ratio. Figure 1 shows the volume fraction within 0.5 nm of the surface for a spherical particle of radius r. One can think of this fraction either as the fraction of atoms likely to be influenced by processes at the surface, or as the fraction of the volume of a material that could be taken up by a 0.5-nm coating of another material. In the first case, because the surface dominates chemical reactivity, the increased surface to volume ratio means that nanoparticles dominate chemical reactions. In the second case, the ability to carry a substantial coating offers a mechanism for the transport of nutrients or pollutants.

Many oxides are *polymorphic,* exhibiting several crystal structures as a function of pressure and temperature. Often, nanosized oxide particles crystallize in structures different from that of large crystals of the same composition.[1] Examples are γ-Al_2O_3, a defect spinel rather than α-Al_2O_3, corundum, γ-Fe_2O_3, the defect spinel maghemite rather than α-Fe_2O_3, hematite, and the anatase and brookite forms of TiO_2 rather than rutile. From arguments based on transformation sequences and the occurrence of phases, it was long argued that there may be a crossover in phase stability at the nanoscale if the structure which is metastable for large particles has a significantly lower surface energy.[2] This has been proven for alumina and titania in recent calorimetric studies (Figure 2).[3,4] The resulting transformation enthalpies and surface energies, and those of other related systems are shown in Table 1. Another interesting feature is that the hydrous phases AlOOH boehmite and FeOOH goethite have significantly lower surface energies than their anhydrous counterparts, Al_2O_3 and Fe_2O_3.[5,6] Whether this is a general feature of hydrous minerals with hydroxylated surfaces is not yet known.

As particles become less than about 10 nm in size, their x-ray diffraction patterns are broadened sufficiently that they begin to appear "x-ray amorphous" (Figure 3). This term lacks exact definition. High-resolution electron microscopy may still detect periodicity, and short-range order is certainly present.[7] The identification of structure in 1–10 nm particles is very difficult, and phases are empirically described as, for example, "two line ferrihydrite," based on x-ray diffraction patterns.[8]

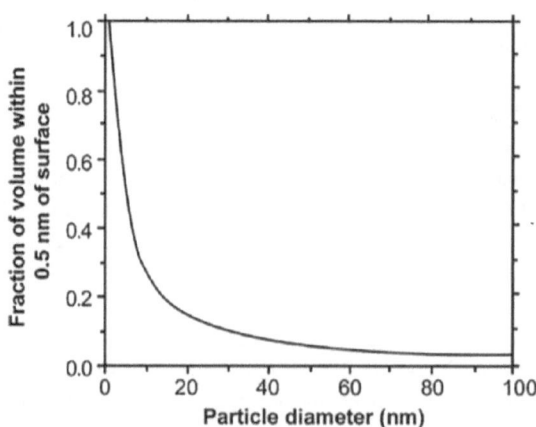

FIGURE 1 Volume fraction of a nanoparticle within 0.5 nm of the surface as a function of particle radius.
Source: Navrotsky,[41] Kluwer Academic Publishers.

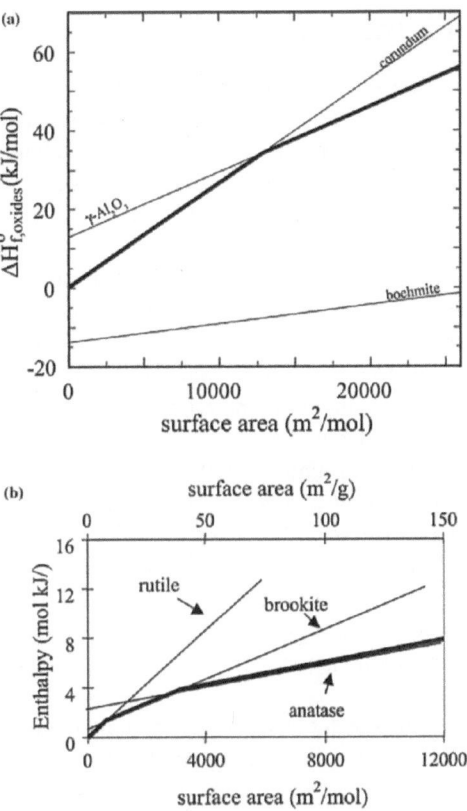

FIGURE 2 (a) Enthalpies of alumina polymorphs as a function of surface area. (**Source:** From McHale et al.[3]) (b) Enthalpies of titania polymorphs as a function of surface area. The heavy lines show the stable polymorphs in each size range. (**Source:** Ranade,[4] PNAS.)

TABLE 1 Energetic Parameters for Oxide and Oxyhydroxide Polymorphs

Formula	Polymorph	Metastability (kJ/mol)	Surface Energy (J/m²)
Al_2O_3[a]	Corundum (α)	0	2.6
	Spinel (γ)	13.4	1.7
Fe_2O_3[b]	Hematite	0	0.8
	Maghemite	20	0.8
TiO_2[c]	Rutile	0	2.2
	Brookite	0.7	1.0
	Anatase	2.6	0.4
$AlOOH$[d]	Diaspore	0	?
	Boehmite	4.9	0.5
$FeOOH$[b]	Goethite	0	0.3
	Lepidocrocite		0.3

[a] From McHale et al.[3]
[b] From Majzlan.[6]
[c] From Ranade.[4]
[d] From Majzlan.[5]

Nanoparticles in Soil and Water

Soil is a complex aggregate of inorganic, organic, and biological material.[9] Its constituents of largest size are rocks and gravel, small animals, plant roots, and other debris. Smaller mineral grains, clumps of organic matter, and microorganisms make up an intermediate size fraction. The smallest particles, ranging into the nanoscale, are clays, iron oxides, and other minerals. These are often heterogeneous and coated by other minerals and organic matter. The entire composite is porous and hydrated. The percolation of water in soil transports both nanoparticles and dissolved organic and inorganic species. The texture and porosity, as well as the chemical composition and pH, are crucial to biological productivity. The surfaces of nanoparticles provide much of the chemical reactivity for both biological and abiotic processes.

Major aluminosilicate minerals in soils include clays, zeolites, and poorly crystalline phases (Table 2). These can change their water content in response to ambient conditions, often swelling in wet seasons, and shrinking in dry seasons. These nanophase materials are major controllers of soil moisture and permeability. Iron and manganese oxides are another class of major sol minerals. Their extensive polymorphism at the nanoscale makes them highly variable. They sequester and/or transport and make available the essential plant nutrient iron, as well as other essential transition metals (cobalt, copper, zinc, etc.). They frequently carry coatings of other metal oxides and oxyhydroxides, including toxic metals such as lead and chromium. They also frequently have organic coatings. Sulfates, including the jarosite–alunite family of hydrated [(K, Na), (Al, Fe)] sulfates, are another important constituent. In alkaline and arid environments, other sulfates and halides form, and their formation, dissolution, and transport is a major issue in heavily irrigated regions. How much these processes are controlled by nanoscale phenomena is not known.

Groundwater is constantly in touch with soil and rock, and minerals are dissolving and precipitating as it flows. The load of fine sediments in streams and groundwater can be substantial, especially during

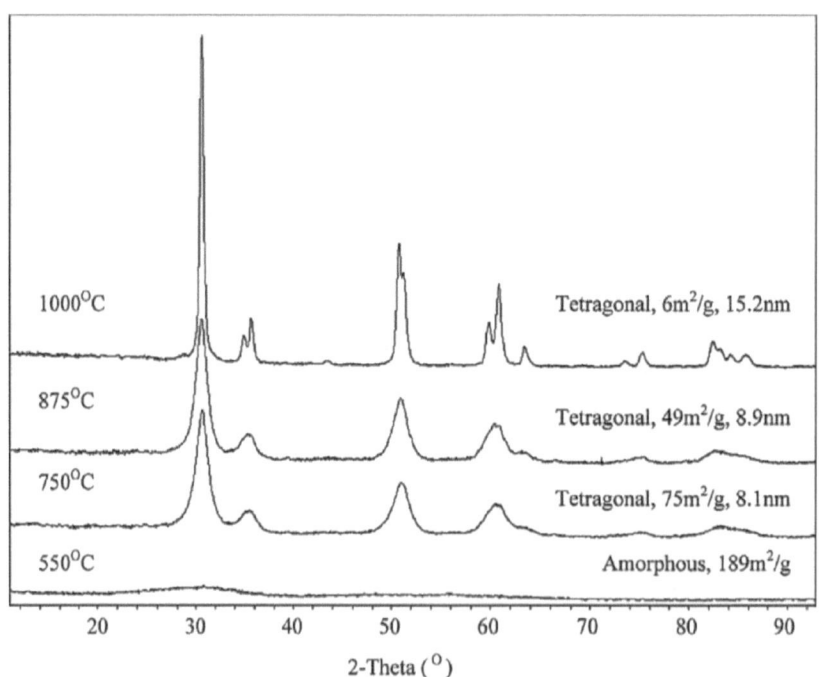

FIGURE 3   Powder x-ray diffraction patterns of sol–gel zirconia heated at various temperatures. The structure and average particle diameters are indicated.
Source: Pitcher and Navrotsky.[42]

TABLE 2 Major Soil Minerals and Constituents

Type	Composition	Structures
Clay	Hydrated aluminosilicate	Layered
Zeolite	Hydrated aluminosilicate	Three-dimensional porous
Salts	$NaCl$, Na_2SO_4, $CaSO_4$	Ionic crystals
Carbonates	$CaCO_3-MgCO_3-FeCO_3$	Calcite, dolomite, others
Allophane	Hydrous aluminosilicate gel	Amorphous
Iron oxides	$Fe2O_3$, $FeOOH$	Various polymorphs
Aluminum oxides	$AlOOH$, $Al(OH)_3$	Various polymorphs
Quartz	SiO_2	Quartz
Manganese oxides	Mn_2O_3, $MnOOH$, MnO_2	Various polymorphs
H_2O	H_2O	Water, ice, vapor
Organics	$C-H-N-O$	Large surface area amorphous colloids
Jarosite-alunite	Alkali (Fe, Al) sulfates	Ionic double salts

spring floods. The Missouri River is called "the Big Muddy" because of its load of particulate matter, a large fraction of which is of nanoscale dimensions. The yearly flooding of the Nile, depositing fertile soil with its large nanoparticle content, made ancient Egyptian civilization flourish. Today, one of the major concerns of our system of dams, especially in the arid western United States, is interference with the normal cycle of sediment transport and "silting up" of the lakes behind the dams. Silt is partly nanoparticles.

Contaminants and pollutants in water can be transported as aqueous ions (dimensions < 0.5 nm), as molecular clusters (0.5–2 nm), as nanoparticles (2–100 nm), as larger colloids (100–1000 nm), and as macroscopic particles (>1 µm). These size range distinctions are rather arbitrary and serve to illustrate the continuity between the dissolved and the solid state. Several examples illustrate this complexity. Aluminum oxyhydroxide particles can transport transition metals such as nickel, cobalt, and zinc, seemingly as adsorbed coatings. Initially thought to be loosely bound metal complexes at the surface of the aluminum oxyhydroxide mineral grain, these are now realized to be precipitates, only a few atomic layers thick, of mixed double hydroxides of the hydrotalcite family, in which anions such as carbonate play an essential role.[10] The transport of plutonium through groundwater is a concern in old plutonium processing facilities such as the Hanford, WA atomic energy reservation, in the Nevada nuclear test site, and in the planned nuclear waste repository at Yucca Mountain, Nevada. There remain questions of permeability and the adhesion of particles to the rock and engineered barrier walls, of colloid transport, of biological transport, and of mineral precipitation which can change the rate of progress of a contamination plume. Linking laboratory scale, field scale, and simulation studies of nanoparticle transport is an essential area of research for understanding radioactive and chemical contamination and geologic processes involving uranium and other actinides.[11]

When particles are below 5 nm in size, several other effects must be considered. Whereas for larger particles, most of the atoms are in specific planes or faces, for smaller ones, an increasing number of surface atoms must sit at the intersection of facets, in presumably even higher energy sites. An alternate, more macroscopic way of describing this is to consider the surface as curved, rather than as a series of planes. Then the surface energy per unit area is no longer a constant, but potentially increases quite rapidly with decreasing particle size. This unfavorable energy may be relaxed by the adsorption of various molecules on the surface, and there is evidence that the adsorption coefficient of organics rises steeply at very small particle size.[12]

The flocculation of colloids depends on the surface charge; the pH of which the surface is neutral is the "point of zero charge."[13] Does this depend on particle size? This is an area of active research.

How do nanocrystals form from solution? The classical picture of nucleation and growth by addition of single atoms or ions is probably inadequate.[14] There is increasing evidence for clusters of atoms or

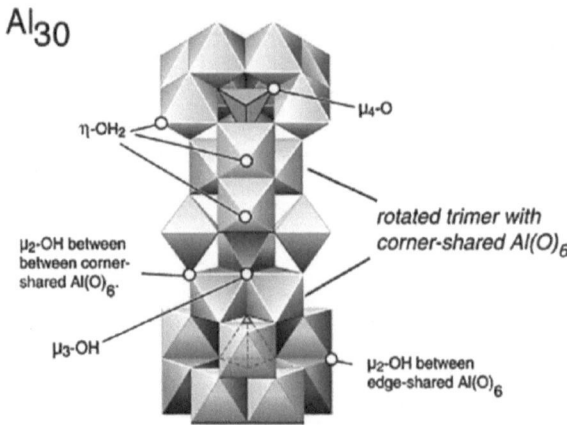

FIGURE 4 $Al_2O_8Al_{28}(OH)_{56}(H_2O)_{24}^{18+}$(aq) (often called Al_{30}) cluster of 2 nm dimensions, which is intermediate in structure and properties between isolated ions and solid aluminum.
Source: W.H. Casey, personal communications. From Rowsell and Nazar.[15]

ions in solution which contain 5–50 atoms and clearly show some of the structural features of the solid. An example is the Keggin-molecular cluster containing 13 aluminum atoms shown in Figure 4.[16] It appears stable over a wide range of neutral to basic pH, and is probably a major precursor to and a dissolution product of aluminum oxyhydroxides.[17] The growth of TiO_2 anatase may occur by the oriented attachment of ~3-nm particles.[18] The growth of zeolites templated by organics may involve 3-nm cuboctahedral clusters.[19] Nanoclusters have been invoked in the growth of sulfides in ore-forming solutions.[20] Characterization of such nanoscale precursors in aqueous solutions remains a major challenge.

The coarsening and phase transformation of nanoscale precipitates upon heating is equally important to the sol–gel synthesis of ceramics and the geologic compaction and diagenesis of buried sediments. Loss of water, loss of surface area, and phase transformations to the stable bulk polymorph are closely interlinked.[14] A nanoparticle with dimensions below 5 nm probably cannot maintain defects or dislocations; they can migrate to the surface and be annihilated.[21] An aggregate of such single domain nanocrystals, with disorder and impurities at their surfaces, may be a common morphology in nature. Such aggregates give smaller average particle size from x-ray peak broadening than from gas adsorption [Brunauer–Emmett–Teller (BET)] measurements.

Magnetic nanoparticles cannot hold a direction of magnetization for a long time because of thermal fluctuations.[22] The magnetic iron oxides found in magnetotactic bacteria, which are single domain particles, neither too large nor too small, provide orientation in the Earth's magnetic field (see below). On a geologic time scale (millions of years), magnetization of largely nanophase iron oxides provides a record of the variation of the Earth's magnetic field through time, including periodic reversals of north and south poles. Thus the ability or inability of nanoparticle oxides to retain magnetization is of critical importance.

Nanoparticles in the Atmosphere

Atmospheric particles include dust (rock and soil), sea salt, acids including sulfuric, organics (especially carbon), inorganics, and, of course, water and ice (Table 3). The atmosphere can carry particles of spherical equivalent diameters from 1 to 10^5 nm. Often, a trimodal distribution of particle sizes, with peaks in number density near 5, 50, and 300 nm, is seen.[23] The smaller particles account for most of the reactive surface area but little of the total mass.

Atmospheric particles affect the environment in many ways.[23,24] They reduce visibility (smog, haze) primarily through light scattering. They act as nuclei for water vapor condensation and cloud formation.

TABLE 3   Atmospheric Nanoparticles

Liquid droplets

Water

Sulfuric acid

Nitric acid

Sea water and other salt solutions

Organics

Solid particles

Ice (H_2O)

NaCl

Na_2SO_4

$CaSO_4 \cdot 2H_2O$

$NaNO_3$

$H_2SO_4 \cdot 4H_2O$

$HNO_3 \cdot 3H_2O$

C (graphite, amorphous, fullerenes, nanotubes)

SiO2

Iron oxides

Clays

Organics

Many particles have core–shell structures and coatings

They are involved in radiative forcing, changing the ratio of absorbed sunlight to reflected sunlight. Thus they are implicated in global climate change. Their effect on radiative forcing can be positive (more energy adsorbed) or negative (more energy reflected), leading to warming or cooling, respectively.[23,24] Their net effect is a subject of vigorous research and controversy.

Anthropogenic particles contribute disproportionately to the fine particle fractions.[23] These may have very significant effects on climate and (see below) health. Soot (carbon) from burning coal and oil and from automobile (especially diesel) emissions contributes greatly to the atmospheric load of nanoparticles.

Particles in the atmosphere travel a long way. Dust from Africa is seen in Florida; industrial emissions from China are detected in North America. Particles are removed from the atmosphere by diffusion and gravitational settling (aided by small particles coalescing into larger ones) and by rain. The residence time of nanoparticles in the atmosphere ranges from minutes to days.[23]

Atmospheric nanoparticles are more involved in gas phase reactions than particles in soil and water.[24] Their formation may involve combustion synthesis, as in industrial or automobile emission. Mineral nanoparticle surfaces may catalyze the oxidation of SO_2 and NO_2, leading to sulfuric and nitric acid. These acids can exist as gaseous species, liquids, or solid hydrates at low temperature. Nanoparticles are invoked in the depletion of atmospheric ozone by catalytic production of reactive chlorine compounds. Changes of phase (liquid to solid) are critical to the chemistry of sodium chloride and sodium nitrate particles, with their water content being controlled by available humidity.

Mineral dust particles may provide critical nutrients (e.g., iron) to the surface of the ocean far from land. The ocean's biological productivity is often limited by the availability of these nutrients; thus such inorganic nanoparticles may significantly influence the global cycling of carbon through ocean biomass.[25]

Nanoparticles in Sediments, Rocks, and the Deep Earth

The debris of rock weathering is brought down river to the ocean in sediments consisting of nanoscale particles of clay, small quartz grains, and other minerals. Indeed, the terms "clay" and "silt" have a classic connotation of size fraction, although the former also implies a structural group of minerals, the layered aluminosilicates. In the ocean, carbonates precipitate, dissolve, and reprecipitate as a complex function of depth.[26] Both silica and various polymorphs of calcium carbonate (calcite, aragonite, vaterite) are produced by organisms such as diatoms, foraminifera, and corals. Their debris rains down on the ocean bottom, forming sediments which often show annual cycles in composition and texture and which bear records of climate change, shifts in ecosystems, and catastrophic events such as meteor impacts.[27] These sediments start off largely nanoscale. They coarsen and dehydrate with time and depth of burial. The evolution of their organic matter leads to petroleum. The evolution of their minerals, involving coarsening and compaction, called diagenesis, leads to rocks such as limestones and shales. The nanoscale processes that take place (dehydration and organic loss, phase transformation, coarsening and densification) are natural analogs of ceramic processing which starts with nanoscale precipitates or gels.

Natural processes involving changes in temperature, pressure, acidity, and oxygen fugacity cause the concentration of trace metals into ore deposits. These often occur in hydrothermal systems, spatially contained circulations of hot, pressurized, metal-rich aqueous solutions. Our ability to mine low-grade deposits by chemical leaching techniques brings us into the world of nanoparticles and reactions at mineral surfaces. There is increasing evidence that microorganisms play an active role in ore deposition.[28,29] Hot springs at the surface produce deposits of nanoscale amorphous silica and other minerals, which may also be closely linked to microbial activity.[30]

At temperatures above a few hundred degrees Centigrade and pressures above a few kilobars, coarse-grained metamorphic and igneous rocks predominate. The interior of the Earth is layered, with seismic discontinuities delineating the crust, upper mantle, transition zone, lower mantle, and core. These discontinuities represent regions of rapidly changing density, mineralogy, and chemistry.[31] Ongoing phase transitions and chemical reactions can decrease the grain size of a material and render it easier to deform.[32] Thus nanoscale phenomena, occurring at specific locations, may play a disproportionate role in processes such as subduction, plate tectonics, earthquake generation, and volcanism. Shock processes, (e.g., meteor impact, nuclear detonation) also produce nanoparticles.

When a volcano erupts explosively, a plume of dust particle is sent into the atmosphere, sometimes reaching the stratosphere. These particles make beautiful sunsets but they also exert a significant cooling effect on climate for several years and pose a significant aviation hazard. Combining sedimentation, coarsening, subduction, volcanism, and weathering, there is an ongoing global geochemical cycle of nanoparticles, analogous in some ways to global geochemical cycles of elements such as carbon. However, the mass balances, or imbalances, in global nanoparticle production and consumption through time have not been characterized.

Nanoparticles beyond the Earth

In the early stages of planet formation, dilute and more or less uniform gas condensed to form a series of mineral particles, with the order of condensation described by thermodynamic calculations based on the volatility and stability of these phases.[37] The more refractory oxides condensed earlier than those with higher volatility. These particles accreted, under the influence of gravity, to form our solar system. What was the nature of these initial particles? What was their size distribution? Were they crystalline or amorphous? Were metastable polymorphs formed? While the initial high temperatures might argue against such metastability, the low pressures and condensation from a vapor argue for it. In technological processes, chemical vapor deposition produces nanoscale amorphous silica "snow," and combustion produces soot and inorganic nanoparticles. The role of nanoparticles in planetary accretion has not yet been explored. The change in stability at the nanoscale, which will be different for various compositions and polymorphs, may alter the sequence of condensation of phases. Are the particles now present in space as interplanetary dust partly or mostly nanoparticles?

The surfaces of the Moon and Mars, subject to "space weathering" by bombardment with meteorites of all sizes, contain an extensive fine grained dust or soil layer.[32] Samples of lunar soil, brought back by the Apollo missions, contain a distribution of particle sizes of spherules and irregular shards. Their particle size distribution appears not to have been a subject of active interest, but clearly a significant number are in the nanoregime. The red surface of Mars appears to be dominated by various fine-grained or nanophase iron oxides. Until Martian sample return missions, planned to occur in the next decade or two, bring some of this material to Earth, we must rely on remote sensing technology (spectroscopic techniques) and instrumentation on Martian landers (possibly Mossbauer spectroscopy, x-ray fluorescence, and x-ray diffraction) to obtain information on the composition and structure of Martian soil. Considering the difficulty of characterizing iron oxide nanoparticles in the best laboratories on Earth, definitive conclusions about the nature of Martian soil are unlikely until we have some samples in hand. Meteorites believed to be from Mars contain micron-sized spherules, which were proposed to be biological in origin. This sparked much recent controversy and it is by no means settled whether these structures are fossil microorganisms or the product of inorganic nanoscale crystal growth processes.[33,34]

Nanoparticles and Life

Microbial communities are rich in the production and utilization of nanoparticles.[35] Table 4 lists some examples. In addition to aerobic respiration (the enzymatic oxidation of carbohydrates and other organics with molecular oxygen to produce water, carbon dioxide, and energy stored as high-energy phosphate linkages) organisms use many other strategies to extract energy from the environment. The following biological reactions produce or consume nanoparticles. Dissolved Mn(II) or Fe(II) can be oxidized by oxygen, producing Mn(III), Mn(IV), or Fe(III) oxide nanoparticles, while organics can be oxidized by manganese or iron oxides, producing soluble Mn(II) and Fe(II) species. Some bacteria can also utilize the U(IV)–U(VI) couple as an energy source. Because hexavalent uranium is much more soluble than tetravalent, biological processes that accelerate its production are of concern in modeling nuclear waste leaching. The sum of these two groups of redox processes is the oxidation of organics by oxygen, akin to respiration. The important difference is that the organic food source and the oxygen source can be spatially separated in the sharp gradients in oxygen and organic contents that frequently occur in sediments, and different communities of organisms participate in the two processes. In marine sediments, sulfate is the dominant biological electron acceptor and is more important than oxygen. Bacterial sulfate reduction produces sulfide which often precipitates as nanophase metal sulfide minerals. Sulfide and sulfur oxidizing bacteria typically live in specialized environments where there is enough oxygen to oxidize sulfur but not so much that chemical oxidation swamps biological oxidation. This oxidation consumes solid sulfur and sulfides, and produces soluble sulfate.

TABLE 4   Example of Interaction of Microorganisms and Nanoparticles

Class of Organisms	Example	Nanoparticle Interaction
Iron and manganese oxidizing bacteria	*Thiobacillus*	Oxidize soluble Mn^{2+} and Fe^{2+} to insoluble higher oxides
Iron and manganese reducing bacteria	*Shewenella*	Reduce insoluble Mn and Fe oxides to soluble forms oxidize sulfide to sulfate to sulfide precipitate
Sulfur reducing bacteria	*Thiobacillus*	Reduce sulfate to sulfur in sulfide
Magnetotactic bacteria	*Aquaspirillum magnetotacticum*	Nanoparticles of Fe_2O_3, Fe_3O_4, and/or iron sulfides
Uranium reducing bacteria	*Geobacter, shewenella*	Soluble U^{6+} → insoluble U^{4+}
Fungi	Specific strains unknown	Oxidize Mn^{2+}, precipitate MnO_2
Diatoms	Various	Precipitate silica
Foraminifera	Various	Precipitate $CaCO_3$ calcite and aragonite

Organisms utilize nanoparticles in processes other than respiration. Bacterial precipitation of sulfide minerals, e.g., ZnS and UO_2, may also be a mechanism of detoxification.[36] Similar detox processes may occur in plants. Magnetotactic bacteria synthesize and align single domain magnetic iron oxide and iron sulfide particles in structures called magnetosomes.[37] Such bacteria align themselves both north–south and vertically in the Earth's magnetic field. The navigational (homing) capabilities of bees, pigeons, and probably other higher organisms utilize magnetic field orientation sensed by magnetic iron oxide particles in their brains. Similar particles, although at lower abundance, occur in many mammals, including *Homo sapiens*.[38] There has been a debate in the public sector whether the magnetic fields produced by high-voltage power lines are potentially dangerous to human health. In contrast, the use of magnets in alternative medicine, and the market for magnetic pillows, back supports, etc., suggests, or at least hopes for, a beneficial effect of the interaction of magnetic fields with animals. Key to either harmful or helpful biological effects is a mechanism for the magnetic field to interact with living cells. Interaction with biological magnetic nanoparticles may provide such a mechanism, but very little is known at present.

Nanoparticles have other documented health effects.[39] When inhaled into the lungs, particles cause an inflammatory response, which contributes to allergies, asthma, and cancer.[48] The detailed mechanism of this response, and how it depends on surface area, particle size, or specific particle chemistry, is not clear. The harmful effects of inhaled particles may be enhanced by other pollutants, particularly ozone, typically present in smog. Nanoparticles penetrate deep into the lungs. Many are returned with exhaled air, some stick to the surfaces of the alveoli, and some may even penetrate into general blood circulation and be transported to other organs. Studies linking detailed nanoparticle characterization, biochemical and physiological processes, and health effects are just beginning to be carried out. It is likely that not all particles have comparable effects, and understanding which are the most dangerous could lead to rational, rather than arbitrary, emission standards for automotive and industrial particulates.

In the early Earth, prebiotic processes culminated in the origin of life.[40] Because the synthesis of complex organic molecules competes with their destruction by hydrolysis and other degradation, it is possible that the most successful synthesis could have occurred in sheltered and catalytic environments, such as those provided by mineral surfaces, nanoparticle surfaces, and pores within mineral grains. Present-day organisms utilize a wide variety of elements (e.g., Fe, Co, Ni, Cr, Zn, Se) in specific enzymes. Although large amounts of such elements are toxic, trace amounts are essential. The active centers in enzymes utilizing these trace elements often consist of clusters of metal atoms, sometimes associated with sulfide. Are these fine-tuned by evolution from earlier simpler metal clusters and nanoparticles existing in the environment? Thus nanoparticles may play a role not just in the sustenance of life but in its origin.

Conclusions

Nanoparticles play diverse roles in the environment and are involved in both abiotic and biologically mediated chemical and physical processes. Their high surface area, chemical reactivity, polymorphism, and unique properties involve nanoparticles in a disproportionately large fraction of the chemical reactions occurring on and in the Earth and other planets. Understanding this involvement is itself evolving into a new field of study in the environmental and Earth sciences, which is beginning to be called "nanogeoscience." Nanogeoscience will take its place alongside other new areas such as astrobiology and biogeochemitry, fields that link physical, chemical, and biological processes viewed in the context of the long time and distance scales natural to the Geosciences.

References

1. Navrotsky, A. Thermochemistry of Nanomaterials. In *Nanoparticles and the Environment: Reviews in Mineralogy and Geochemistry*; Banfield, J.F., Navrotsky, A., Eds.; Mineralogical Society of America: Washington, DC, 2001; Vol. 44, 73–103.

2. Garvie, R.C. The occurrence of metastable tetragonal zirconia as a crystallite size effect. *J. Phys. Chem.* **1965**, *69*, 1238–1243.

3. McHale, J.M.; Auroux, A.; Perrotta, A.J.; Navrotsky, A. Surface energies and thermodynamic phase stability in nanocrystalline alumina. *Science* **1997**, *277*, 788–791.

4. Ranade, M.R.; Navrotsky, A.; Zhang, H.Z.; Banfield, J.F.; Elder, S.H.; Zaban, A.; Borse, P.H.; Kulkarni, S.K.; Doran, G.S.; Whitfield, H.J. Energetics of nanocrystalline TiO_2. *Proc. Natl. Acad. Sci.* **2002**, *99* (Suppl. 2), 6476–6481.

5. Majzlan, J.; Navrotsky, A.; Casey, W.H. Surface enthalpy of boehmite. *Clays Clay Miner.* **2000**, *48*, 699–707.

6. Majzlan, J. Ph.D. Thesis; University of California at Davis, 2002.

7. Janney, D.E.; Cowley, J.M.; Buseck, P.R. Structure of synthetic 2-line ferrihydrite by electron nano-diffraction. *Am. Mineral.* **2002**, *85* (9), 1180–1187.

8. Schwertmann, U.; Cornell, R.M. *Iron Oxides in the Laboratory*, 2nd Ed.; Wiley-VCH: Weinheim, 2000; 188.

9. Singer, M.J.; Munns, D.N. *Soils: An Introduction*, 3rd Ed.; Simon and Schuster Company: Upper Saddle River, NJ, 1991.

10. Thompson, H.A.; Parks, G.A.; Brown, G.E., Jr. Ambienttemperature synthesis, evolution, and characterization of cobalt–aluminum hydrotalcite-like solids. *Clays Clay Mater.* **1999**, *47*, 425–438.

11. Ragnarsdottir, K.V.; Charlet, L. Uranium behavior in natural environments. *Environ. Mineral.* **2000**, *9*, 245–289.

12. Zhang, H.; Penn, R.L.; Hamers, R.J.; Banfield, J.F. Enhanced adsorption of molecules on surfaces of nanocrystalline particles. *J. Phys. Chem. B.* **1999**, *103*, 4656–4662.

13. Hunter, R.J. *Foundations of Colloid Science*; Oxford University Press: New York, 1993; Vol. 1.

14. Banfield, J.F.; Zhang, H. Nanoparticles in the Environment. In *Nanoparticles and the Environment: Reviews in Mineralogy and Geochemistry*; Banfield, J.F., Navrotsky, A., Eds.; Mineralogical Society of America: Washington, DC, 2001; Vol. 44, 2–58.

15. Rowsell, J.; Nazar, L.F. Speciation and thermal transformation in alumina sols: Structures of the polyhydroxyoxoaluminum cluster $[Al_{30}O_8(OH)_{56}(H_2O)_{26}]^{18+}$ and its Keggin moeté. *J. Am. Chem. Soc.* 2000, *122*, 3777–3778.

16. Casey, W.H.; Phillips, B.L.; Furrer, F. Aqueous Aluminum Polynuclear Complexes and Nanoclusters: A Review. In *Nanoparticles and the Environment: Reviews in Mineralogy and Geochemistry*; Banfield, J.F., Navrotsky, A., Eds.; Mineralogical Society of America: Washington, DC, 2001; Vol. 44, 167–190.

17. Furrer, G.F.; Phillips, B.L.; Ulrich, K.U.; Poethig, R.; Casey, W.H. The origin of aluminum flocs in polluted streams. *Science* 2002, *297* (5590), 2245–2247.

18. Penn, R.L.; Banfield, J.F. Oriented attachment and growth, twinning, polytypism, and formation of metastable phases: Insights from nanocrystalline TiO2. *Am. Mineral.* **1998**, *83*, 1077–1082.

19. de Moor, P.P.E.A.; Beelen, T.P.M.; Komanschek, B.U.; Beck, L.W.; Wagner, P.; Davis, M.E.; Van Santen, R.A. Imaging the assembly process of the organic-mediated synthesis of a zeolite. *Chem. Eur. J.* **1995**, *5*, 2083–2088.

20. Luther, G.W.; Theberge, S.M.; Richard, D.T. Evidence for aqueous clusters as intermediates during zinc sulfide formation. *Geochim. Cosmochim. Acta* **1999**, *64*, 579.

21. Jacobs, K.; Alivisatos, A.P. Nanocrystals as model systems for pressure-induced structural phase transitions. In *Nanoparticles and the Environment: Reviews in Mineralogy and Geochemistry*; Banfield, J.F., Navrotsky, A., Eds.; Mineralogical Society of America: Washington, DC, 2001; Vol. 44, 59–104.

22. Rancourt, D.G. Magnetism of earth, planetary, and environmental nanomaterials. In *Nanoparticles and the Environment: Reviews in Mineralogy and Geochemistry*; Banfield, J.F., Navrotsky, A., Eds.; Mineralogical Society of America: Washington, DC, 2001; Vol. 44, 217–292.

23. Anastasio, C.; Martin, S.T. Atmospheric Nanoparticles. In Nanoparticles and the Environment: *Reviews in Mineralogy and Geochemistry*; Banfield, J.F., Navrotsky, A., Eds.; Mineralogical Society of America: Washington, DC, 2001; Vol. 44, 293–349.

24. Ramanathan, V.; Crutzen, P.J.; Kiehl, J.T.; Rosenfeld, D. Aerosols, climate, and the hydrological cycle. *Science* **2001**, *294*, 2119–2124.

25. Martin, J.H.; Coale, K.H.; Johnson, K.S.; Fitzwater, S.E.; Gordon, R.M.; Tanner, S.J.; Hunter, C.N.; Elrod, V.A.; Nowicki, J.L.; Coley, T.L.; Barber, R.T.; Lindley, S.; Watson, A.J.; Vanscoy, K.; Law, C.S.; Liddicoat, M.I.; Ling, R.; Stanton, T.; Stockel, J.; Collins, C.; Anderson, A.; Bidigare, R.; Ondrusek, M.; Latasa, M.; Millero, F.J.; Lee, K. Testing the iron hypothesis in ecosystems of the equatorial Pacific Ocean. *Nature* **1994**, *371*, 123–129.

26. Millero, F.J. *Chemical Oceanography*, 2nd Ed.; CRC: Boca Raton, FL, 1996.

27. Broecker, W.S. *The Great Ocean Conveyor*; AIP Conference Proceedings; Columbia University Palisades: New York, 1992; Vol. 347, 129–161.

28. Labrenz, M.; Druschel, G.K.; Thomsen-Ebert, T.; Gilbert, B.; Welch, S.A.; Kemner, K.M.; Logan, G.A.; Summons, R.E.; de Stasio, G.; Bond, P.L.; Lai, B.; Kelly, S.D.; Banfield, J.F. Formation of sphalerite (ZnS) deposits in natural biofilms of sulfate-reducing bacteria. *Science* **2000**, *290*, 1744–1745.

29. Ehrlich, H.L. Microbes as geologic agents: Their role in mineral formation. *Geomicrobiol. J.* **1999**, *16*, 135–153.

30. Konhauser, K.O.; Phoenix, V.R.; Bottrell, S.H.; Adams, D.G.; Head, I.M. Microbial–silica interactions in Icelandic hot spring sinter: Possible analogues for some Precambrian siliceous stromatolites. *Sedimentology* **2001**, *48*, 415–433.

31. Anderson, D.L. *Theory of the Earth*; Blackwell Scientific Publications: Brookline Village, MA, 1989.

32. Karato, S.; Li, P. Diffusion creep in perovskite: Implications for the rheology of the lower mantle. *Science* **1992**, *255*, 1238–1240.

33. Sasaki, S.; Nakamura, K.; Hamabe, Y.; Kurahashi, E.; Hiroi, T. Production of iron nanoparticles by laser irradiation in a simulation of lunar-like space weathering. *Nature* **2001**, *410*, 555–557.

34. Buseck, P.R.; Dunin-Borkowski, R.E.; Devouard, B.; Frankel, R.B.; McCartney, M.R.; Midgley, P.A.; Posfai, M.; Weyland, M. Magnetite morphology and life on Mars. *Proc. Natl. Acad. Sci. U. S. A.* **2001**, *98* (24), 13490–13495.

35. Gibson, E.K.; McKay, D.S.; Thomas-Keprta, K.L.; Wentworth, S.J.; Westall, F.; Steele, A.; Romanek, C.S.; Bell, M.S.; Toporski, J. Life on mars: Evaluation of the evidence within Martian meteorites ALH84001, Nakhla, and Shergotty. *Precambrian Res.* **2001**, *106* (1–2), 15–34.

36. Nealson, K.H.; Stahl, D.A. Microorgansims and Biogeochemical Cycles: What Can We Learn from Layered Microbial Communities? In *Geomicrobiology: Interaction Between Microbes and Minerals; Review in Mineralogy*; Banfield, J.F., Navrotsky, A., Eds.; Mineralogical Society of America: Washington, DC, 1997; Vol. 35, 5–34.

37. Suzuki, Y.; Banfield, J.F. Geomicrobiology of Uranium. In. *Uranium: Mineralogy, Geochemistry and the Environment; Review in Mineralogy*; Burns, P.C., Finch, R., Eds.; Mineralogical Society of America: Washington, DC, 1999; Vol. 38, 388–432.

38. Stolz, J.F. Magnetotactic Bacteria: Biomineralization, Ecology, Sediment Magnetism, Environmental Indicator. In *Biomineralization Process of Iron and Manganese: Modern and Ancient Environments*; Catena Supplement 21; Skinner, H.C.W., Fitzpatrick, R.W., Eds.; Catena: Destedt. Germany, 1992; 133–146.

39. Kirschvink, J.L.; Walker, M.M.; Diebel, C.E. Magnetitebased magnetoreception. *Curr. Opin. Neurobiol.* **2001**, *11* (4), 462–467.

40. Guthrie, G.D.; Mossman, B.T. *Health Effects of Mineral Dusts: Reviews in Mineralogy*; Mineralogical Society of America: Chelsca, MI, 1993; Vol. 28.

41. Navrotsky, A. Nanomaterials in the environment, agriculture, and technology (NEAT). *J. Nanopart. Res.* **2000**, *2*, 321–323.
42. Pitcher, M.; Navrotsky, A. unpublished data.

Bibliography

Nakashima, S.; Ikoma, M.; Shiota, D.; Nakazawa, K.; Maruyama, S. Geochemistry and the origin and evolution of life: A tentative summary and future perspectives. *Precursors Chall. Investig. Ser.* **2001**, *2*, 329–344.

6

Pharmaceuticals: Treatment

Diana Aga and
Seungyun Baik

Introduction

The presence of pharmaceutical chemicals and their byproducts in soil, wastewater effluents, surface water, and drinking water sources has become a growing concern over the past two decades. Improvements in analytical methods coupled with large-scale surveys have revealed the broad range of persistent pharmaceuticals that are cycling through our wastewater-to-drinking water cycle. Non-metabolized active ingredients and transformation products of veterinary and human pharmaceuticals are introduced into the environment through the effluents of municipal wastewater treatment plants (WWTPs), pharmaceutical formulation facilities, and through the land application of animal waste and sewage sludge. Approximately 10 million tons of sewage sludge and manure containing residues of pharmaceuticals are used to fertilize croplands each year. Public awareness of the potential problems related to pharmaceutical pollution, widely known as "emerging contaminants," has brought this issue to the forefront in the water and wastewater treatment industries. Hence, advanced water treatment systems are being evaluated to potentially eliminate these emerging contaminants from effluents of WWTPs and from drinking water sources. This entry aims to provide an overview on the occurrence of pharmaceuticals in the environment and recent studies that investigate promising treatment technologies to eliminate emerging contaminants from wastewater and drinking water systems.

Although the human risk associated with chronic exposure to pharmaceuticals in the environment remains unclear, evidence of detrimental ecological impacts is growing. The environmental contamination by pharmaceutical residues, especially antibiotics, may have profound environmental effects at several levels. While the promotion of antibiotic resistance in pathogenic microorganisms has been the major concern associated with the presence of antibiotics in the environment, other issues such as endocrine disruption in fish and wildlife, plant uptake, and phytotoxicity are also significant and warrant discussion. Therefore, the second goal of this entry is to summarize current knowledge on the ecological impacts of pharmaceutical pollution.

Occurrence and Impacts of Pharmaceuticals in the Environment

The presence of pharmaceutical residues in terrestrial and aquatic systems, resulting largely from discharges of municipal WWTPs and the land application of animal wastes, is now well documented in the literature.[1–3] As depicted in Figure 1, residues of human pharmaceuticals and their metabolites may eventually enter surface water, groundwater, and drinking water systems after passing through WWTPs. While most active ingredients of drugs are metabolized in the body, or removed during wastewater treatment, others remain intact and persist in the environment. Low levels of persistent pharmaceuticals can eventually end up in finished drinking water and distribution system (tap) water, when using source waters that have been affected by effluents from WWTPs.[4]

Veterinary pharmaceuticals, particularly antibiotics used for therapeutic purposes and for growth promotion, are also finding their ways into the environment. While some antibiotics used in animal production are decomposed quickly after being excreted, others remain stable during manure storage and end up in agricultural fields upon manure application. Additionally, antibiotics are widely used in fish farms and may enter the aquatic environment via direct discharge. For example, the antibiotic oxytetracycline has been detected in groundwater that has been affected by fish farming at sub-parts-per-billion concentrations.[5] Highly polar pharmaceuticals are susceptible to leaching and may therefore reach the groundwater aquifer. For example, the high frequency of detection of sulfonamide antibiotics in groundwater from various sites in the United States can be attributed to the relatively high water solubility and poor biodegradability of these drugs.[6,7] The detection of sulfonamides in groundwater from wells deeper than 50 ft, located downgradient from an animal feeding operation in the United States, indicates the persistence and mobility of this class of antibiotics.[8] Similar results were observed in Germany where sulfonamides were detected in groundwater located downgradient from an agricultural field where sewage sludge was used for irrigation.[9] The concentrations of pharmaceuticals in groundwater are typically lower than in surface water because infiltration through the soil profile removes some

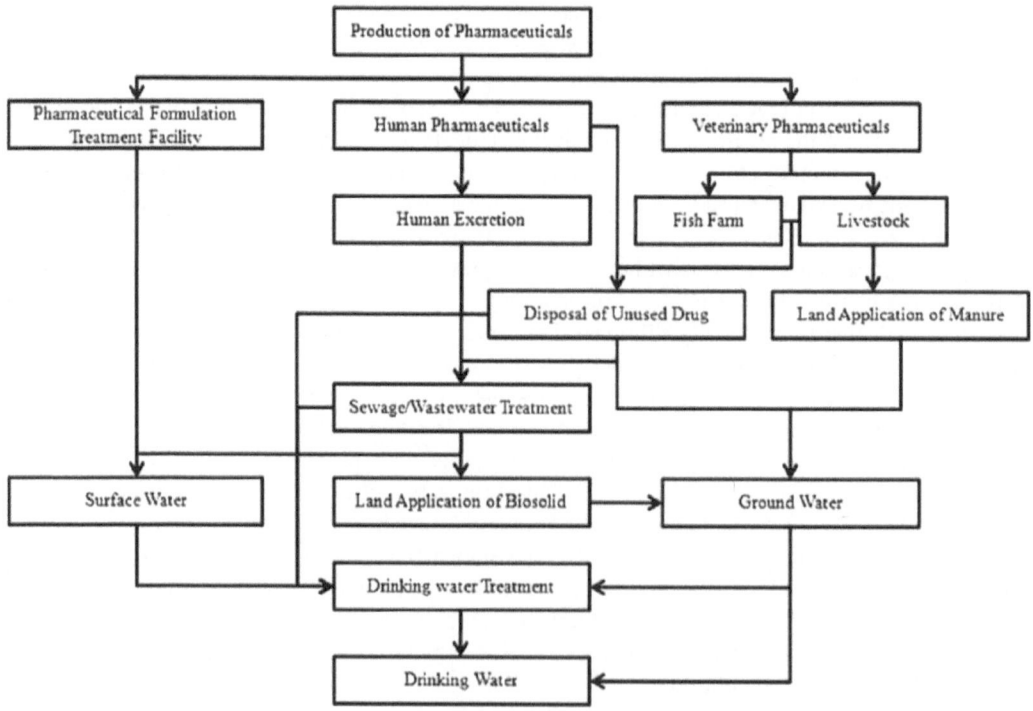

FIGURE 1 Exposure routes of pharmaceuticals in the environment.

fraction of the organic pollutants before entering groundwater systems.[10] Nevertheless, contamination of groundwater by pharmaceuticals is a significant issue because these compounds are not readily biodegradable under anoxic conditions.[9] The persistence of pharmaceuticals in the groundwater aquifer is a concern because groundwater is a major source of drinking water in many areas in the United States and around the world.[8] If the contaminated groundwater is used as drinking water source, pharmaceuticals may eventually reach finished drinking water systems[4] and may not be degraded during conventional drinking water treatment processes.

In the first comprehensive study on the occurrence of pharmaceuticals in the United States surface waters, Kol-pin et al.[1] reported detection of 95 organic contaminants, which included 30 antibiotics, 12 prescribed drugs, 4 nonprescribed drugs, and 6 drug metabolites. A later report summarized the occurrence of 80 pharmaceuticals and drug metabolites in the aquatic systems for eight different countries.[11] The concentrations of pharmaceuticals in surface water and groundwater depend on several factors, such as removal processes employed in the WWTPs, source variability, dilution, retardation, and weather events.[12]

Table 1 shows examples of the types of pharmaceuticals frequently detected in the environment and their typical removal rates (high, medium, low) in conventional activated sludge (CAS) systems during wastewater treatment.[13,14] Consequently, these pharmaceuticals are often detected in receiving surface waters at a wide range of concentrations as depicted in Figure 2. The variability in pharmaceutical concentrations in

TABLE 1 Classification of Pharmaceuticals Based on Their Typical Removal Rates in CAS Systems

Group	Pharmaceutical	Usage
High removal	Acetaminophen (ACE)	Analgesic (non-NSAID)
(>65%)	Ibuprofen (IBP)	Analgesic (NSAID)
	Naproxen (NAP)	Analgesic (NSAID)
	Paroxetine (PRX)	Antidepressant
	Iopamidol (IOM)	Iodinated contrast agent
	Caffeine (CAF)	Psychoactive stimulant
Medium removal	Sulfamethoxazole (SMX)	Antibiotic
(30%–65%)	Gemfibrozil (GFB)	Lipid regulator
	Atenolol (ATN)	β-Blocker
	Propranolol (PRN)	β-Blocker
	Ranitidine (RTD)	Antihistamine (Zantac)
	Fluoxetene (FXT)	Antidepressant
	Iopromide (IOP)	Iodinated contrast agent
Low removal	Diclofenac (DCF)	Analgesic (NSAID)
	Mefenamic acid (MFN)	Analgesic (NSAID)
(<30%)	Ciprofloxacin (CIP)	Antibiotic
	Erythromycin (ERY)	Antibiotic—macrolide
	Roxythromycin (ROX)	Antibiotic—macrolide
	Trimethoprim (TMP)	Antibiotic
	Clofibric acid (CLO)	Lipid regulator
	Carbamazepine (CBZ)	Anticonvulsant
	Dilantin (DLT)	Anticonvulsant
	Meprobamate (MPB)	Anti-anxiety drug

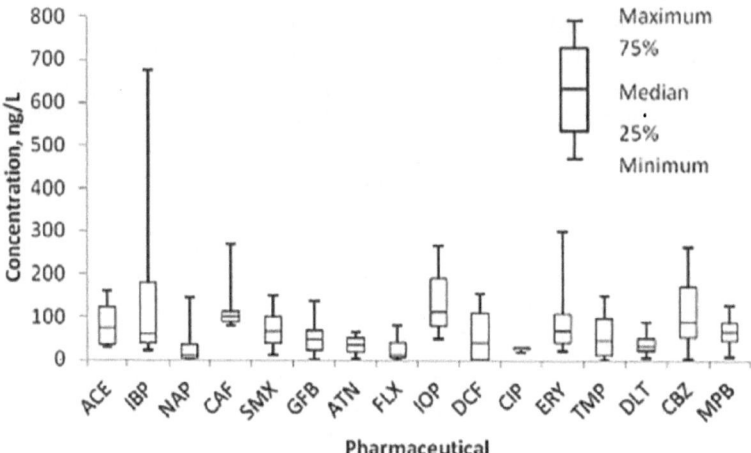

FIGURE 2 Typical concentrations of selected pharmaceuticals in surface waters in the U.S. The middle line in the box plot indicates median value of the data; The upper edge (hinge) of the box indicates 75th percentile of the data set; the lower hinge indicates 25th percentile. The whiskers indicate minimum and maximum data values.
Source: Kolpin, D.W. et al.[1]; Benotti, M.J. et al.[4]; Snyder, S.A. et al.[15]; Westerhoff, P. et al.[16]; Focazio, M.J. et al.[17]; Nodler, K. et al.[18]; Boyd, G.R. et al.[19]; Ferrer, I. et al.[20].

the environment can be attributed to the types of WWTPs, the frequency and time of sampling, and the biodegradability of the pharmaceuticals in a given environmental conditions, among others.

Despite the relatively low concentrations of pharmaceuticals typically found in the environment, the ecological effects of pharmaceutical pollution cannot be ignored. Several standard toxicity tests reveal that while acute toxicity is not a concern because of the high effective concentrations needed to elicit observable acute effects, chronic toxicity may be important at environmentally relevant concentrations.

Acute and chronic toxicity of several pharmaceuticals, including non-steroidal anti-inflammatory drugs (NSAIDs) and β-blockers, against phytoplankton, zooplankton, and other aquatic organisms[21–23] have been reported. Similarly, chronic toxicity studies toward aquatic organisms of lipid-lowering agents,[24,25] neuroactive pharmaceuticals (antidepressants),[22,26] and anti-epileptic drugs[22] have been conducted at concentrations typically found in the environment. Some of the documented ecological effects of pharmaceuticalsin the environment are summarized in Table 2. Both the U.S. Environmental Protection Agency (USEPA) and European Union (EU) have recognized residues of pharmaceuticals and personal care products as emerging contaminants of concern that may require future environmental regulation if their persistence in the environment proves to be ecologically significant. In fact, diclofenac, ibuprofen, triclosan, and clofibric acid have been identified as future emerging priority candidates by the EU water framework directive, which is a priority substance list that is updated every 4 years.[27]

TABLE 2 Reported Ecotoxicity for Selected Pharmaceuticals Using Various Test Organisms

Pharmaceutical	Effect Observed	Effective Concentration	Environmentally Relevant[a]	References
Atenolol	Fathead minnow	1.0 mg/L (21 days)—NOEC	No	[28]
		3.2 mg/L (21 days)—LOEC	No	
Carbamazepine	*Daphnia magna*	0.492 μg/L—multigeneration (up to 6 generations) effect	Yes	[29]

(Continued)

TABLE 2 (*Continued*)   Reported Ecotoxicity for Selected Pharmaceuticals Using Various Test Organisms

Pharmaceutical	Effect Observed	Effective Concentration	Environmentally Relevant[a]	References
	Japanese medaka	6.15 mg/L—swim speed effect (9 days)	No	[30]
	Oryzias latipes	45.87 mg/L—LC50 (96 hours)	No	[31]
	Thamnocephalus platyurus	>100 mg/L—LC50 (24 hours)	No	[31]
Clarithromycin	*Oryzias latipes*	>100 mg/L—LC50 (96 hours)	No	[31]
	Thamnocephalus platyurus	94.23 mg/L—LC50 (24 hours)	No	[31]
Diclofenac	*Daphnia magna*	0.355 µg/L—multigeneration (up to 6 generations) effect	Yes	[29]
	Hyalella azteca	0.467 mg/kg—LC50 (72 hours) with sediment	No	[32]
	Rainbow trout	25 µg/L—accumulation test on gall (21 days)	Yes	[33]
Fluoxetene	Sheepshead minnow	>2.0 mg/L—LC50 2.0 mg/L—LOEC 1.87 mg/L—NOEC (All for 96 hours)	No	[34]
	Western mosquitofish	0.546 mg/L—LC50 (7 days)	No	[35]
	Western mosquitofish	0.5 µg/L—increasing lethargy in 59–159 days	Yes	[35]
Ibuprofen	*Oryzias latipes*	>100 mg/L—LC50 (96 hours)	No	[31]
	Planorbis carinatus	17.1 mg/L—LC50 (48 and 72 hours)	No	[36]
	Planorbis carinatus	>5.36 mg/L—LOEC 5.36 mg/L—NOEC (All for 21 days)	No	[36]
	Thamnocephalus platyurus	19.59 mg/L—LC50 (24 hours)	No	[31]
Metoprolol	*Daphnia magna*	1.170 µg/L—multigeneration (up to 6 generations) effect	Yes	[29]
Paracetamol (Acetaminophen)	Wheat	22.4 mg/L—damage in 21 days exposure	No	[37]
Propranolol	Fathead minnow	1.0 mg/L—NOEC (female) 3.4 mg/L—LOEC (female) 0.1 mg/L—NOEC (male) 1.0 mg/L—LOEC (male) (All for 21 days)	No (Yes for NOEC male)	[38]
	Rainbow trout	1.0 mg/L—NOEC 10 mg/L—LOEC (All for 10 days)	No	[39]
	Oryzias latipes	11.40 mg/L—LC50 (96 hours)	No	[31]
	Thamnocephalus platyurus	10.31 mg/L—LC50 (24 hours)	No	[31]
Verapamil	Juvenile rainbow trout	2.72 mg/L—LC50 (96 hours)	No	[40]

Note: NOEC, no observed effect concentration; LOEC, lowest observed effect concentration; LC50, 50% lethal concentration.

[a] Environmentally relevant—"Yes" for concentrations of compounds are in the ranges in surface water and WWTP effluent reported through 2011.

Treatment of Pharmaceuticals in the Aquatic Systems

Alternative Systems for Water Treatment

Activated sludge treatment, which relies on microbes to biodegrade contaminants in wastewater, is the most widely used waste water treatment system in the United States and around the world. While many pharmaceuticals are removed partially or completely during CAS treatment, there are a significant number of pharmaceuticals that have very little to no removal. Therefore, alternative treatment systems are being explored to improve removal efficiencies for these trace organic contaminants. Examples of treatment systems that are suspected to be more efficient are membrane bioreactors (MBRs)[41–44]; membrane treatment with nano-, micro-, and ultrafiltration (NF/MF/UF); and reverse osmosis (RO).[45] Granular activated carbon (GAC) is also used as another alternative system in the Unites States and Canada for removal of trace organic matters through filtration and adsorption.[45] Recent studies have also shown that advanced oxidation processes, such as ozonation and UV–H_2O_2 disinfection, which are employed during tertiary treatment before discharging the treated wastewater, could have additional benefits in removing pharmaceuticals from water.[15,46,47]

For drinking water treatment systems, the removal of many pharmaceuticals during the first three steps (coagulation, flocculation, and sand filtration) before disinfection is typically incomplete.[16,48] On the other hand, similar to wastewater treatment, pharmaceuticals may be effectively removed by activated carbon adsorption, ozone oxidation, and membrane filtration.[49] Chlorination, the most common disinfection step, may also remove pharmaceuticals but produce unwanted by-products.[16]

Alternative water treatment systems are summarized and compared in Table 3.[50,51] These systems are advantageous because they offer increased removal of pharmaceuticals and other organic contaminants that cause undesirable odor and taste in water. To enhance the efficiency of treatment, combining systems such as MBRs followed by activated carbon adsorption system maybe applied.

TABLE 3 Summary and Comparison of Alternative Systems: Membrane Process, Activated Carbon Adsorption Process, and Advanced Oxidation Process for Contaminant Removal

Process		Operation and Application	Important Notes
Membrane process	Reverse osmosis (RO)	• Pressure driven (150–400 psi)	• Best removal performance
			• Possible for desalination
		• Removes organics with molecular weight >100	• Needs pretreatment such as UF or MF to prevent plugging and fouling; hence, high cost
		• Demineralization	• Low product recovery (30%–85%)
	Nanofiltration (NF)	• Pressure driven (80–200 psi)	• Good removal performance
		• Removes organics with molecular weight between 100 and 500	• Needs pretreatment such as UF or MF to prevent plugging and fouling; hence, high cost
		• Removes NOM	• Medium product recovery (70%–90%)
	Ultrafiltration (UF)	• Pressure driven (15–60 psi)	• Replacement for CT
			• High product recovery (80%–95%)
			• Higher cost than CT
		• Pretreatment for NF or RO	• Sensitive with water temperature and viscosity

(Continued)

TABLE 3 (*Continued*) Summary and Comparison of Alternative Systems: Membrane Process, Activated Carbon Adsorption Process, and Advanced Oxidation Process for Contaminant Removal

Process		Operation and Application	Important Notes
	Microfiltration (MF)	• Macromolecule removal includes viruses	• Needs other process to increase treatment performance
		• Pressure driven (5–40 psi)	• Replacement for CT
			• High product recovery (95%–98%)
			• Higher cost than CT
		• Pretreatment for NF or RO	• Sensitive with water temperature and viscosity
		• Bacteria removal	• Needs other process to increase treatment performance
Activated carbon adsorption	Powder activated carbon (PAC)	• Used in early steps of treatment process	• Good performance in combination with MF or UF membrane systems
			• Usually used for drinking water treatment
			• Good for emergency situations, with high organic pollutants
			• Handling problem—dry PAC causes dust problem
		• Controls taste or odor	• May pass through filters and enter the final treated water
		• Applied for small or mid-sized plants with moderate or severe taste, odor, or organic contaminants problem	• Cannot be mixed with chlorine
	Granular activated carbon (GAC)	• Used in adsorption beds or tanks	• Replacement of conventional gravity filter
			• Possible for WWTPs
			• Fouling by chemicals—need to consider breakthrough for design
			• Backwash requirement
		• Gravity or pressure driven	• Need to monitor carbon bed depth
		• Applied for plants with moderate or severe taste, odor, or organic contaminants problem	• Needs carbon regeneration
Advanced oxidation process	Ultraviolet (UV)	• Chemical treatment	• Disinfection benefit
		• Low pressure ($\lambda = 253.7\,nm$) or medium pressure ($\lambda = 180$–$1370\,nm$)	• Controls taste and odor
		• UV radiation with H_2O_2 addition (hydroxyl radical oxidation)	• Relatively high cost and complexity
		• Combined with ozonation	• Treatment effectiveness is weakened by turbidity or color in water
		• NOM in water may inhibit or promote oxidation	• Unwanted by-products

Note: NOM, natural organic matter; CT, conventional treatment.

Removal Efficiencies in Municipal WWTP Systems

Several reports have shown that MBRs generally have higher removal efficiencies for NSAIDs than CAS systems. For example, it was demonstrated that up to 98%–100% ketiprofen and 86%–89% naproxen were removed in MBRs.[41,44] In addition, MBRs also have better removal efficiencies for the antibiotics roxythromycin, sulfamethoxazole, trimethoprim, and diclofenac relative to the CAS systems.[43,44] However, certain compounds, such as carbamazepine, remain mostly undegraded in both conventional WWTP and MBRs.[43] A study compared different filter systems such as MBR, MF/UF, and RO in removing pharmaceuticals using WWTP primary effluent as the feed water.[45] In contrast to other studies, most pharmaceuticals were removed well (except for phenytoin and meprobamate) in the MBR, including a 90% removal of carbamazepine. These differences in results suggest that design and operation of MBRs can be optimized to increase removal efficiencies of pharmaceuticals in wastewater. Additionally, it was demonstrated that more than 90% of all pharmaceuticals tested were removed in the RO system alone; however, the use of combined systems such as UF/RO, MF/RO, and MBR/RO resulted in more than 99% removal for all pharmaceuticals. However, membrane systems are expensive and may not be affordable to many municipalities. In addition, membrane systems require that the brine, in which the rejected pharmaceuticals are concentrated after removing pharmaceuticals from mother water, must be appropriately treated.

Granular activated carbon (GAC) adsorption systems were also investigated in two facilities, one with regular regeneration of GAC and the other without regular replacement/regeneration of GAC.[45] As summarized in Table 4, the latter facility had poor removal efficiencies for most of the pharmaceuticals tested. Hence, it is important to determine the breakthrough for individual pharmaceuticals so that regeneration or replacement GAC can be scheduled on a timely fashion to be most effective in removing pharmaceutical residues in the water.

Ozone oxidation has been investigated for sulfonamides, macrolides, and iodinated contrast media, as well as other acidic pharmaceuticals spiked in MBR-treated wastewater effluent.[46] As may be expected, increased ozone concentrations resulted in increased degradation of pharmaceuticals. However, while the sulfonamides and macrolides were very sensitive toward ozone degradation and were completely removed with high ozone concentration, iodinated contrast media were only 40% degraded. Ozone oxidation experiments have been performed to investigate the removal of 17 pharmaceuticals, as well as other personal care products and endocrine-disrupting chemicals, in bench-top pilot plants and in one full-scale WWTP system. The removal efficiencies of pharmaceuticals by ozonation can be classified into four groups: >80% removal, 80%–50% removal, 50%–10% removal, and <20% removal.[15] Most of the pharmaceuticals exhibited >80% removal; diazepam, phenytoin, and ibuprofen had 80%–50% removal; and iopromide and meprobamate were in the group of 50%–10% removal. In a separate

TABLE 4 Comparing Removal Efficiencies of Pharmaceuticals for GAC Facilities with and without Regular Regeneration or Replacement of Activated Carbon

Pharmaceuticals	Removal (%) in Facility 1 (with Regular GAC Regeneration)	Removal (%) in Facility 2 (without Regular GAC Regeneration)
Caffeine	>41.2	16.3
Carbamazepine	>54.5	15.6
Phenytoin (Dilantin)	>44.4	22.7
Erythromycin–H_2O	>44.4	7.9
Gemfibrozil	>16.7	8.2
Ibuprofen	>9.1	16.4
Iopromide	>69.7	72.0
Meprobamate	>16.7	13.3
Sulfamethoxazole	>83.3	83.8

study using two different wastewater matrices, one with effluent water from a conventional WWTP and the other with effluent from an MBR system, it was shown that the removal efficiencies by ozonation, ozone–UV, and H_2O_2–UV may be slightly lower in the presence of higher dissolved organic carbon.[47]

The removal of 13 selected pharmaceuticals in full-scale WWTP with CAS system, followed by advanced treatment by ozonation, was evaluated.[52] Interestingly, most of the pharmaceuticals that are typically poorly removed (<20% removal) in the secondary clarifier, such as crotamiton, sulfapyridine, and roxithromycin, were almost completely-degraded upon ozonation. An exception to the pharmaceuticals effectively removed by ozonation was carbamazepine (which was only <5% degraded by ozonation).

In a separate study, the efficiency of UV oxidation was examined with and without H_2O_2 for atenolol, carbamazepine, phenytoin, meprobamate, primidone, and trimethoprim, using three different wastewater matrices.[53] Results of this study indicated that the nature of the aqueous matrix, most likely defined by the amount and composition of the natural organic matter, is an important factor in optimizing the removal efficiencies of pharmaceuticals by advanced oxidation. For example, removal efficiencies between 16% and 95% were observed for the pharmaceuticals tested in one WWTP using a high power of UV (fluence of 700 mJ/cm²) with H_2O_2. However, removal efficiencies of the same pharmaceuticals in water from other sampling locations with high organic matter content were relatively lower (10%–85%), although the same UV treatment conditions were used.

Removal Efficiencies in Drinking Water Treatment Systems

Polar and persistent pharmaceuticals may eventually enter the drinking water systems.[49] Since conventional processes used in drinking water treatment may not completely remove pharmaceuticals, other advanced oxidation processes have been considered for treatment of pharmaceuticals in drinking water. Ozonation appears to be the most effective way to disinfect the water and at the same time oxidize organic chemicals via direct reaction with ozone, or through reactions with hydroxyl radicals (OH) formed during ozonation.[15] For disinfection purposes, monochromatic UV at 254 nm is used for drinking water treatment. In UV advanced oxidation process, the oxidation of pharmaceuticals is enhanced by addition of H_2O_2 to facilitate formation of hydroxyl radicals and promote indirect photolysis.[54] The oxidation of diclofenac, a frequently detected NSAID in the environment, was evaluated using ozone, UV, and UV–H_2O_2. Ozonation showed 100% removal, while UV–H_2O_2 oxidation showed only 52% removal of diclofenac at conditions that corresponded to a 35% decrease in total organic carbon, after 90 minutes of oxidation time.[55] The antidepressant pharmaceutical fluoxetene (trade name of Prozac), which is typically detected at parts per billion levels in U.S. streams,[1] was used as test compound to evaluate the efficiency of direct and indirect photolysis in removing pharmaceuticals in water. The degradation rate for fluoxetene reached up to 9.60×10^9/M.s for indirect photolysis.[56] It has been shown that the use of TiO2 catalyst can significantly increase the photolysis rate constants for NSAIDs, including diclofenac, naproxen, and ibuprofen in water.[57] Experimental oxidation rates of the hydroxyl radical for β-blockers, atenolol, metoprolol, and propranolol, were found to be 7.05×10^9, 8.39×10^9, and 1.07×10^9/M/s, respectively.[58] Ozonation of tetracycline showed the complete removal within 4–6 minutes of ozonation despite the maximum degradation of 40% total organic carbon after 2 hours of ozonation.[59] Finally, photo-Fenton (Fe(II)) reaction is another oxidation method that can be applicable to drinking water treatment systems owing to its cost effectiveness and ease of operation.[60] It was demonstrated that diclofenac can be 100% degraded within 60 minutes of photo-Fenton oxidation.[60]

Organic pollutants in drinking water sources, including pharmaceuticals, can be removed by activated carbon adsorption.[16] Laboratory-scale tests mimicking full-scale drinking water treatment system using activated carbon-showed good removal efficiencies of trace organic compounds, except for the more hydrophilic pharmaceuticals (e.g., clofibric acid and ibuprofen were only 40% removed) after 3 hours of contact time.[61]

Water filters can provide an additional step, at a point-of-use consumption treatment, to remove trace levels of pharmaceuticals and other drinking water contaminants from tap water. The removal efficiencies of chlorination by-products, including trihalomethanes, bromodichloromethane, dibromochloromethane, and bromoform, by "Envirofilter" (made of nutshell carbon and commercial water filters) have been reported.[62] Water filters are made with activated carbon to remove dissolved organic matter,[62,63] together with polar and ion-exchange resins to remove charged species such as metal ions.[64,65] Since activated carbon has been shown to have high removal efficiencies for many pharmaceuticals in wastewater,[16,66–71] it is reasonable to expect that water filters will be effective in removing trace levels of pharmaceuticals from drinking water.

Conclusion

The decreasing amount of clean water resources for drinking water and for food production has become one of the most challenging problems in the world.[72] To alleviate the shortage in water supply, reuse of treated wastewater for irrigation[72–74] and as drinking water source[75] is becoming more and more common. Therefore, pharmaceuticals can enter the groundwater and drinking water systems, as shown in Figure 1 depicting the routes of entry of pharmaceuticals into the environment. It is encouraging to see a recent survey reporting that only very few pharmaceuticals from biosolid application and wastewater irrigation can be transported to the groundwater aquifer.[72] Because only lipophilic pharmaceuticals tend to sorb onto the biosolids, and these pharmaceuticals in turn are not mobile in soil, contamination of groundwater aquifer from biosolids application may not pose a significant source of pharmaceutical pollution. With the exception of four pharmaceuticals (two iodinated contrast media of diatrizoate and iopamidol, carbamazepine and sulfamethoxazole) most of the 52 pharmaceuticals being targeted for analysis were not detected in the said study. Nevertheless, optimization of the design and operating conditions of wastewater treatment systems is key to eliminating pharmaceuticals and their deleterious ecological effects in the environment. Advanced oxidation processes are very promising treatment technologies that are waiting to be tested and implemented under full-scale treatment plants.

References

1. Kolpin, D.W.; Furlong, E.T.; Meyer, M.T.; Thurman, E.M.; Zaugg, S.D.; Barber, L.B.; Buxton, H.T. Pharmaceuticals, hormones, and other organic wastewater contaminants in US streams, 1999–2000: A national reconnaissance. *Environmental Science & Technology* **2002**, *36* (6), 1202–1211.
2. Aga, D.S.; O'Connor, S.; Ensley, S.; Payero, J.O.; Snow, D.; Tarkalson, D. Determination of the persistence of tetracycline antibiotics and their degrades in manure-amended soil using enzyme-linked immunosorbent assay and liquid chromatography-mass spectrometry. *Journal of Agricultural and Food Chemistry* **2005**, *53* (18), 7165–7171.
3. Batt, A.L.; Bruce, I.B.; Aga, D.S. Evaluating the vulnerability of surface waters to antibiotic contamination from varying wastewater treatment plant discharges. *Environmental Pollution* **2006**, *142* (2), 295–302.
4. Benotti, M.J.; Trenholm, R.A.; Vanderford, B.J.; Holady, J.C.; Stanford, B.D.; Snyder, S.A. Pharmaceuticals and endocrine disrupting compounds in US drinking water. *Environmental Science & Technology* **2009**, *43* (3), 597–603.
5. Avisar, D.; Levin, G.; Gozlan, I. The processes affecting oxytetracycline contamination of groundwater in a phreatic aquifer underlying industrial fish ponds in Israel. *Environmental Earth Sciences* **2009**, *59* (4), 939–945.
6. Lindsey, M.E.; Meyer, M.; Thurman, E.M. Analysis of trace levels of sulfonamide and tetracycline antimicrobials, in groundwater and surface water using solid-phase extraction and liquid chromatography/mass spectrometry. *Analytical Chemistry* **2001**, *73* (19), 4640–4646.
7. Karthikeyan, K.G.; Meyer, M.T. Occurrence of antibiotics in wastewater treatment facilities in Wisconsin, USA. *Science of the Total Environment* 2006, *361* (1–3), 196–207.

8. Batt, A.L.; Snow, D.D.; Aga, D.S. Occurrence of sulfonamide antimicrobials in private water wells in Washington County, Idaho, USA. *Chemosphere* **2006**, *64* (11), 1963–1971.

9. Richter, D.; Massmann, G.; Taute, T.; Duennbier, U. Investigation of the fate of sulfonamides downgradient of a decommissioned sewage farm near Berlin, Germany. *Journal of Contaminant Hydrology* **2009**, *106* (3–4), 183–194.

10. Katz, B.G.; Griffin, D.W.; Davis, J.H. Groundwater quality impacts from the land application of treated municipal wastewater in a large karstic spring basin: Chemical and microbiological indicators. *Science of the Total Environment* **2009**, *407* (8), 2872–2886.

11. Heberer, T. Occurrence, fate, and removal of pharmaceutical residues in the aquatic environment: A review of recent research data. *Toxicology Letters* **2002**, *131* (1–2), 5–17.

12. Musolff, A.; Leschik, S.; Moder, M.; Strauch, G.; Reinstorf, F.; Schirmer, M. Temporal and spatial patterns of micropollutants in urban receiving waters. *Environmental Pollution* **2009**, *157* (11), 3069–3077.

13. Jones, O.A.H.; Voulvoulis, N.; Lester, J.N. Human pharmaceuticals in wastewater treatment processes. *Critical Reviews in Environmental Science and Technology* **2005**, *35* (4), 401–427.

14. Sipma, J.; Osuna, B.; Collado, N.; Monclús, H.; Ferrero, G.; Comas, J.; Rodriguez-Roda, I. Comparison of removal of pharmaceuticals in MBR and activated sludge systems. *Desalination* **2010**, *250* (2), 653–659.

15. Snyder, S.A.; Wert, E.C.; Rexing, D.J.; Zegers, R.E.; Drury, D.D. Ozone oxidation of endocrine disruptors and pharmaceuticals in surface water and wastewater. *Ozone-Science & Engineering* **2006**, *28* (6), 445–460.

16. Westerhoff, P.; Yoon, Y.; Snyder, S.; Wert, E. Fate of en-docrine-disruptor, pharmaceutical, and personal care product chemicals during simulated drinking water treatment processes. *Environmental Science & Technology* **2005**, *39* (17), 6649–6663.

17. Focazio, M.J.; Kolpin, D.W.; Barnes, K.K.; Furlong, E.T.; Meyer, M.T.; Zaugg, S.D.; Barber, L.B.; Thurman, M.E. A national reconnaissance for pharmaceuticals and other organic wastewater contaminants in the United States-II) Untreated drinking water sources. *Science of the Total Environment* **2008**, *402* (2–3), 201–216.

18. Nodier, K.; Licha, T.; Bester, K.; Sauter, M. Development of a multi-residue analytical method, based on liquid chromatography-tandem mass spectrometry, for the simultaneous determination of 46 micro-contaminants in aqueous samples. *Journal of Chromatography A* **2010**, *1217* (42), 6511–6521.

19. Boyd, G.R.; Palmeri, J.M.; Zhang, S.Y.; Grimm, D.A. Pharmaceuticals and personal care products (PPCPs) and endocrine disrupting chemicals (EDCs) in stormwater canals and Bayou St. John in New Orleans, Louisiana, USA. *Science of the Total Environment* **2004**, *333* (1–3), 137–148.

20. Ferrer, I.; Zweigenbaum, J.A.; Thurman, E.M. Analysis of 70 Environmental Protection Agency priority pharmaceuticals in water by EPA Method 1694. *Journal of Chromatography A* **2010**, *1217* (36), 5674–5686.

21. Huggett, D.B.; Brooks, B.W.; Peterson, B.; Foran, C.M.; Schlenk, D. Toxicity of select beta adrenergic receptor-blocking pharmaceuticals (β-blockers) on aquatic organisms. *Archives of Environmental Contamination and Toxicology* **2002**, *43* (2), 229–235.

22. Ferrari, B.; Paxeus, N.; Lo Giudice, R.; Pollio, A.; Garric, J. Ecotoxicological impact of pharmaceuticals found in treated wastewaters: Study of carbamazepine, clofibric acid, and diclofenac. *Ecotoxicology and Environmental Safety* **2003**, *55* (3), 359–370.

23. Ferrari, B.; Mons, R.; Vollat, B.; Fraysse, B.; Paxeus, N.; Lo Giudice, R.; Pollio, A.; Garric, J. Environmental risk assessment of six human pharmaceuticals: Are the current environmental risk assessment procedures sufficient for the protection of the aquatic environment? *Environmental Toxicology and Chemistry* **2004**, *23* (5), 1344–1354.

24. Donohue, M.; Baldwin, L.A.; Leonard, D.A.; Kostecki, P.T.; Calabrese, E.J. Effect of hypolipidemic drugs gemfibrozil, ciprofibrozil, ciprofibrate, and clofibric acid of peroxisomal beta-oxidation in primary cultures of rainbow-trout hepatocytes. *Ecotoxicology and Environmental Safety* **1993**, *26* (2), 127–132.

25. Nunes, B.; Carvalho, F.; Guilhermino, L. Acute and chronic effects of clofibrate and clofibric acid on the enzymes acetylcholinesterase, lactate dehydrogenase and catalase of the mosquitofish, *Gambusia holbrooki*. *Chemosphere* **2004**, *57* (11), 1581–1589.

26. Brooks, B.W.; Foran, C.M.; Richards, S.M.; Weston, J.; Turner, P.K.; Stanley, J.K.; Solomon, K.R.; Slattery, M.; La Point, T.W. Aquatic ecotoxicology of fluoxetine. *Toxicology Letters* **2003**, *142* (3), 169–183.

27. Ellis, J.B. Pharmaceutical and personal care products (PPCPs) in urban receiving waters. *Environmental Pollution* **2006**, *144* (1), 184–189.

28. Winter, M.J.; Lillicrap, A.D.; Caunter, J.E.; Schaffner, C.; Alder, A.C.; Ramil, M.; Ternes, T.A.; Giltrow, E.; Sumpter, J.P.; Hutchinson, T.H. Defining the chronic impacts of atenolol on embryo-larval development and reproduction in the fathead minnow (*Pimephales promelas*). *Aquatic Toxicology* **2008**, *86* (3), 361–369.

29. Dietrich, S.; Ploessl, F.; Bracher, F.; Laforsch, C. Single and combined toxicity of pharmaceuticals at environmentally relevant concentrations in Daphnia magna-A multigenerational study. *Chemosphere* **2010**, *79* (1), 60–66.

30. Nassef, M.; Matsumoto, S.; Seki, M.; Khalil, F.; Kang, I.J.; Shimasaki, Y.; Oshima, Y.; Honjo, T. Acute effects of triclosan, diclofenac and carbamazepine on feeding performance of Japanese medaka fish (*Oryzias latipes*). *Chemosphere* **2010**, *80* (9), 1095–1100.

31. Kim, J.W.; Ishibashi, H.; Yamauchi, R.; Ichikawa, N.; Takao, Y.; Hirano, M.; Koga, M.; Arizono, K. Acute toxicity of pharmaceutical and personal care products on freshwater crustacean (*Thamnocephalus platyurus*) and fish (*Oryzias latipes*). *Journal of Toxicological Sciences* **2009**, *34* (2), 227–232.

32. Oviedo-Gomez, D.G.C.; Galar-Martinez, M.; Garcia-Medina, S.; Razo-Estrada, C.; L.Gomez-Olivan, M. Diclofenac-enriched artificial sediment induces oxidative stress in Hyalella azteca. *Environmental Toxicology and Pharmacology* **2010**, *29* (1), 39–43.

33. Mehinto, A.C.; Hill, E.M.; Tyler, C.R. Uptake and biological effects of environmentally relevant concentrations of the nonsteroidal anti-inflammatory pharmaceutical Diclofenac in Rainbow Trout (*Oncorhynchus mykiss*). *Environmental Science & Technology* **2010**, *44* (6), 2176–2182.

34. Winder, V.L.; Sapozhnikova, Y.; Pennington, P.L.; Wirth, E.F. Effects of fluoxetine exposure on serotonin-related activity in the sheepshead minnow (*Cyprinodon variegatus*) using LC/MS/MS detection and quantitation. *Comparative Biochemistry and Physiology C-Toxicology & Pharmacology* **2009**, *149* (4), 559–565.

35. Henry, T.B.; Black, M.C. Acute and chronic toxicity of fluoxetine (selective serotonin reuptake inhibitor) in western mosquitofish. *Archives of Environmental Contamination and Toxicology* **2008**, *54* (2), 325–330.

36. Pounds, N.; Maclean, S.; Webley, M.; Pascoe, D.; Hutchinson, T. Acute and chronic effects of ibuprofen in the mollusc *Planorbis carinatus* (*Gastropoda: Planorbidae*). *Ecotoxicology and Environmental Safety* **2008**, *70* (1), 47–52.

37. An, J.; Zhou, Q.X.; Sun, F.H.; Zhang, L. Ecotoxicological effects of paracetamol on seed germination and seedling development of wheat (*Triticum aestivum* L.). *Journal of Hazardous Materials* **2009**, *169* (1–3), 751–757.

38. Giltrow, E.; Eccles, P.D.; Winter, M.J.; McCormack, P.J.; Rand-Weaver, M.; Hutchinson, T.H.; Sumpter, J.P. Chronic effects assessment and plasma concentrations of the betablocker propranolol in fathead minnows (*Pimephales promelas*). *Aquatic Toxicology* **2009**, *95* (3), 195–202.

39. Owen, S.F.; Huggett, D.B.; Hutchinson, T.H.; Hetheridge, M.J.; Kinter, L.B.; Ericson, J.F.; Sumpter, J.P. Uptake of propranolol, a cardiovascular pharmaceutical, from water into fish plasma and its effects on growth and organ biometry. *Aquatic Toxicology* **2009**, *93* (4), 217–224.

40. Li, Z.H.; Li, P.; Randak, T. Ecotoxocological effects of short-term exposure to a human pharmaceutical Verapamil in juvenile rainbow trout (*Oncorhynchus mykiss*). *Comparative Biochemistry and Physiology C-Toxicology & Pharmacology* **2010**, *152* (3), 385–391.

41. Kimura, K.; Hara, H.; Watanabe, Y. Removal of pharmaceutical compounds by submerged membrane bioreactors (MBRs). *Desalination* **2005**, *178* (1–3), 135–140.

42. Quintana, J.B.; Weiss, S.; Reemtsma, T. Pathway's and metabolites of microbial degradation of selected acidic pharmaceutical and their occurrence in municipal wastewater treated by a membrane bioreactor. *Water Research* **2005**, *39* (12), 2654–2664.

43. Celiz, M.D.; Perez, S.; Barcelo, D.; Aga, D.S. Trace Analysis of Polar Pharmaceuticals in Wastewater by LC-MS-MS: Comparison of Membrane Bioreactor and Activated Sludge Systems. *Journal of Chromatographic Science* **2009**, *47* (1), 19–25.

44. Tambosi, J.L.; de Sena, R.F.; Favier, M.; Gebhardt, W.; Jose, H.J.; Schroder, H.F.; Moreira, R. Removal of pharmaceutical compounds in membrane bioreactors (MBR) applying submerged membranes. *Desalination* **2010**, *261* (1–2), 148–156.

45. Snyder, S.A.; Adham, S.; Redding, A.M.; Cannon, F.S.; DeCarolis, J.; Oppenheimer, J.; Wert, E.C.; Yoon, Y. Role of membranes and activated carbon in the removal of endocrine disruptors and pharmaceuticals. *Desalination* **2007**, *202* (1–3), 156–181.

46. Huber, M.M.; Gobel, A.; Joss, A.; Hermann, N.; Loffler, D.; McArdell, C.S.; Ried, A.; Siegrist, H.; Ternes, T.A.; von Gunten, U. Oxidation of pharmaceuticals during ozonation of municipal wastewater effluents: A pilot study. *Environmental Science & Technology* **2005**, *39* (11), 4290–4299.

47. Gebhardt, W.; Schroder, H.F. Liquid chromatography-(tan-dem) mass spectrometry for the follow-up of the elimination of persistent pharmaceuticals during wastewater treatment applying biological wastewater treatment and advanced oxidation. *Journal of Chromatography A* 2007, 1160 (1–2), 34–43.

48. Vieno, N.; Tuhkanen, T.; Kronberg, L. Removal of pharmaceuticals in drinking water treatment: Effect of chemical coagulation. *Environmental Technology* **2006**, *27* (2), 183–192.

49. Vieno, N.M.; Harkki, H.; Tuhkanen, T.; Kronberg, L. Occurrence of pharmaceuticals in river water and their elimination a pilot-scale drinking water treatment plant. *Environmental Science & Technology* **2007**, *41* (14), 5077–5084.

50. HDR Engineering, ed. *Handbook of Public Water Systems*, 2nd Ed.; John Wiley and Sons: New York, 2001.

51. Baruth, E.E., ed. *Water Treatment Plant Design*, 4th Ed.; McGraw-Hill: New York, 2005.

52. Nakada, N.; Shinohara, H.; Murata, A.; Kiri, K.; Managaki, S.; Sato, N.; Takada, H. Removal of selected pharmaceuticals and personal care products (PPCPs) and endocrine-disrupting chemicals (EDCs) during sand filtration and ozonation at a municipal sewage treatment plant. *Water Research* **2007**, *41* (19), 4373–4382.

53. Rosario-Ortiz, F.L.; Wert, E.C.; Snyder, S.A. Evaluation of UV/H_2O_2 treatment for the oxidation of pharmaceuticals in wastewater. *Water Research* **2010**, *44* (5), 1440–1448.

54. Pereira, V.J.; Weinberg, H.S.; Linden, K.G.; Singer, P.C. UV degradation kinetics and modeling of pharmaceutical compounds in laboratory grade and surface water via direct and indirect photolysis at 254 nm. *Environmental Science & Technology* **2007**, *41* (5), 1682–1688.

55. Vogna, D.; Marotta, R.; Napolitano, A.; Andreozzi, R.; d'Ischia, M. Advanced oxidation of the pharmaceutical drug diclofenac with UV/H_2O_2 and ozone. *Water Research* **2004**, *38* (2), 414–422.

56. Lam, M.W.; Young, C.J.; Mabury, S.A. Aqueous photochemical reaction kinetics and transformations of fluoxetine. *Environmental Science & Technology* **2005**, *39* (2), 513–522.

57. Mendez-Arriaga, F.; Esplugas, S.; Gimenez, J. Photocatalytic degradation of non-steroidal anti-inflammatory drugs with TiO2 and simulated solar irradiation. *Water Research* **2008**, *42* (3), 585–594.

58. Song, W.H.; Cooper, W.J.; Mezyk, S.P.; Greaves, J.; Peake, B.M. Free radical destruction of beta-blockers in aqueous solution. *Environmental Science & Technology* **2008**, *42* (4), 1256–1261.

59. Khan, M.H.; Bae, H.; Jung, J.Y. Tetracycline degradation by ozonation in the aqueous phase: Proposed degradation intermediates and pathway. *Journal of Hazardous Materials* **2010**, *181* (1–3), 659–665.

60. Perez-Estrada, L.A.; Malato, S.; Gernjak, W.; Aguera, A.; Thurman, E.M.; Ferrer, I.; Fernandez-Alba, A.R. Photo-fenton degradation of diclofenac: Identification of main intermediates and degradation pathway. *Environmental Science & Technology* **2005**, *39* (21), 8300–8306.

61. Simazaki, D.; Fujiwara, J.; Manabe, S.; Matsuda, M.; Asami, M.; Kunikane, S. Removal of selected pharmaceuticals by chlorination, coagulation-sedimentation and powdered activated carbon treatment. *Water Science and Technology* **2008**, *58* (5), 1129–1135.

62. Ahmedna, M.; Marshall, W.E.; Husseiny, A.A.; Goktepe, L.; Rao, R.M. The use of nutshell carbons in drinking water filters for removal of chlorination by-products. *Journal of Chemical Technology and Biotechnology* **2004**, *79* (10), 1092–1097.

63. Humbert, H.; Gallard, H.; Suty, H.; Croue, J.P. Natural organic matter (NOM) and pesticides removal using a combination of ion exchange resin and powdered activated carbon (PAC). *Water Research* **2008**, *42* (6–7), 1635–1643.

64. An, H.K.; Park, B.Y.; Kim, D.S. Crab shell for the removal of heavy metals from aqueous solution. *Water Research* **2001**, *35* (15), 3551–3556.

65. Rengaraj, S.; Moon, S.H. Kinetics of adsorption of Co(II) removal from water and wastewater by ion exchange resins. *Water Research* **2002**, *36* (7), 1783–1793.

66. Ternes, T.A.; Meisenheimer, M.; McDowell, D.; Sacher, F.; Brauch, H.J.; Gulde, B.H.; Preuss, G.; Wilme, U.; Seibert, N. Z. Removal of pharmaceuticals during drinking water treatment. *Environmental Science & Technology* **2002**, *36* (17), 3855–3863.

67. Snyder, S.A.; Westerhoff, P.; Yoon, Y.; Sedlak, D.L. Pharmaceuticals, personal care products, and endocrine disruptors in water: Implications for the water industry. *Environmental Engineering Science* **2003**, *20* (5), 449–469.

68. Kim, S.D.; Cho, J.; Kim, I.S.; Vanderford, B.J.; Snyder, S.A. Occurrence and removal of pharmaceuticals and endocrine disruptors in South Korean surface, drinking, and waste waters. *Water Research* **2007**, *41* (5), 1013–1021.

69. Yu, Z.R.; Peldszus, S.; Huck, P.M. Adsorption characteristics of selected pharmaceuticals and an endocrine disrupting compound-Naproxen, carbamazepine and nonylphenol-on activated carbon. *Water Research* **2008**, *42* (12), 2873–2882.

70. Yu, Z.; Peldszus, S.; Huck, P.M. Adsorption of selected pharmaceuticals and an endocrine disrupting compound by granular activated carbon. 1. Adsorption Capacity and Kinetics. *Environmental Science & Technology* **2009**, *43* (5), 1467–1473.

71. Yu, Z.; Peldszus, S.; Huck, P.M. Adsorption of Selected Pharmaceuticals and an Endocrine Disrupting Compound by Granular Activated Carbon. 2. Model Prediction. *Environmental Science & Technology* **2009**, *43* (5), 1474–1479.

72. Ternes, T.A.; Bonerz, M.; Herrmann, N.; Teiser, B.; Andersen, H.R. Irrigation of treated wastewater in Braunschweig, Germany: An option to remove pharmaceuticals and musk fragrances. *Chemosphere* **2007**, *66* (5), 894–904.

73. Chefetz, B.; Mualem, T.; Ben-Ari, J. Sorption and mobility of pharmaceutical compounds in soil irrigated with reclaimed wastewater. *Chemosphere* **2008**, *73* (8), 1335–1343.

74. Xu, J.; Chen, W.P.; Wu, L.S.; Green, R.; Chang, A.C. Leach-ability of some emerging contaminants in reclaimed municipal wastewater-irrigated turf grass fields. *Environmental Toxicology and Chemistry* **2009**, *28* (9), 1842–1850.

75. Reungoat, J.; Macova, M.; Escher, B.I.; Carswell, S.; Mueller, J.F.; Keller, J. Removal of micropollutants and reduction of biological activity in a full scale reclamation plant using ozonation and activated carbon filtration. *Water Research* **2010**, *44* (2), 625–637.

II

COV:
Comparative
Overviews
of Important
Topics for
Environmental
Management

7

Buildings: Climate Change

Lisa Guan and
Guangnan Chen

Introduction

The greenhouse effect is a natural warming process of the earth. It is caused by the greenhouse gases, such as water vapor, carbon dioxide (CO_2), and methane (CH_4), which trap long-wave radiation and then radiate the energy in all directions, warming the earth's surface and atmosphere (Figure 1).[1] Without the heat-trapping greenhouse gases, scientists estimate that the average earth surface temperature would likely to be some 30°C colder than it is today, or −18°C instead of the present mild average 15°C.[1]

The enhanced greenhouse effect is additional to the natural process of greenhouse effect and is mainly due to human activities (human induced), including burning fossil fuels, land clearing, and agriculture, which change the makeup of the atmosphere and lead to an increased concentration of greenhouse gases.[1] This has the potential to cause significant changes in the global climate system (referred to as climate change), including increased temperature, changed patterns of rainfall, tropical cyclone activities, and other extreme climatic events.

Climate change has now become one of the most important global environmental issues facing the world today. It is now widely recognized as having significant potential to seriously affect the integrity of our ecosystems and human welfare.[2] The effects, or impacts, of climate change may be in physical, ecological, social, and/or economic areas.

In this entry, the likely future climate change is first presented. The cycling interaction between climate change and buildings is then discussed, which includes both aspects of the implication of climate change on building performance and the contribution of buildings to the process of human-induced climate change. The potential strategies for building design and operation are then highlighted, in order to reduce the greenhouse gas emissions from buildings and to prepare the buildings to withstand a range of possible climate change scenarios.

Likely Future Climate Change

Due to uncertainties in future emissions and concentrations of greenhouse gases, as well as the climate system's response to the changing conditions and the natural influences (e.g., changes in the sun and volcanic activity), it may be difficult to accurately predict the extent of climate changes.[2] However, the advancements in climate model simulations, combined with more and more observed data on climate changes, led the Intergovernmental Panel on Climate Change (IPCC) to predict the following likely scenarios of climate changes[2]:

- Average global surface temperature will likely rise a further 1.1–6.4°C (2.0–11.5°F) during the 21st century. It is expected that the average rate of warming is very likely to be at least twice as large as that experienced during the 20th century.
- Warming will not be evenly distributed around the globe and will vary with different seasons. For example, land areas will warm more than oceans, high latitudes will warm more than low latitudes, and winters will be warming more than summers in most areas.
- There will be significant changes to the amount and pattern of precipitation, including an increase in droughts, tropical cyclones, and extreme high tides. The changes in precipitation, either an increase or a decrease, will vary from region to region.
- The global average sea level is estimated to rise by 18–59 cm (7.2 to 23.6 in.) by 2100 relative to 1980–1999. Current model projections also indicate substantial variability in future sea level rise between different locations.
- Increases in the intensity of extreme weather events, such as storms, heat waves, and drought, are also predicted.
- In regard to the projection of future climate change, it is noted that there are different levels of confidence (e.g., in terms of accuracies and reliability) for different climate parameters. For

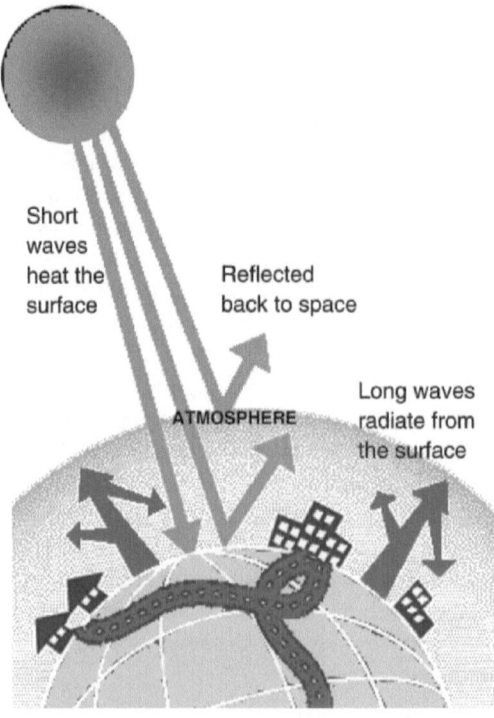

FIGURE 1   Natural/enhanced greenhouse effect.

TABLE 1   List of Climate and Associated Scenario Variables, Ranked Subjectively in Decreasing Order of Confidence

Climate Variable	Confidence
Atmospheric CO_2 concentration	High
Global—mean sea level	
Global—mean temperature	
Regional seasonal temperature	
Regional temperature extremes	
Regional seasonal precipitation and cloud cover	
Regional potential evapotranspiration	
Changes in climatic variability (e.g., El Niño, daily precipitation regimes)	
Climate surprises (e.g., disintegration of the West Antarctic Ice Sheet)	Low
	Very low or unknown

Source: Hulmer and Sheard.[3]

example, it is believed that there is higher confidence in the projection of increases in carbon dioxide concentrations and rises in sea level than in storminess or intense precipitation events. A list of climate and associated scenario variables, ranked subjectively in decreasing order of confidence, had been recommended by Hulme and Sheard[3] and is shown in Table 1.

Implication of Climate Change on Buildings

Climate change is likely to affect both the performance of existing building stock and the design of new buildings. For example, under climate change, buildings will have more overheating hours in summer and less underheating hours in winter, thus use more cooling energy and less heating energy. Where heating and cooling are provided by different fuels, this could have a significant influence on the design and operations of energy delivery systems.

Because climate change entails new climatic conditions for the building industries, it is expected that climate change will have a significant impact on the design, construction, and performance of buildings, as well as the health and productivity of people living and working inside them.[4] These impacts may include the following:

- *Higher building energy consumption*: Climate change may require higher capacity and more uses of air-conditioning equipment to provide comfort indoor environment. For example, it has been predicted that for air-conditioned office buildings in Australia, the cooling load may increase by 2%–47%, depending on the assumed future climate scenarios, as well as different locations.[5] The increases of total building energy use would range from 0.4% to 15.1%. However, due to the potential decrease of heating energy in winter, skin-dominated buildings located in cold regions could receive some benefits from the climate change. Moreover, the expected increased stringencies in building energy codes around the world would also offset some increases of building energy use due to climate change.[6,7]
- *Deteriorating internal thermal environment*, such as more overheating in summer. It has been found that for air-conditioned office buildings in Australia, when the annual average temperature increase exceeds 2°C "threshold," the risk of current office buildings subjected to overheating will be significantly increased.[5] When the increase of external air temperature is more than 5°C, all the Australian office buildings would suffer from the overheating problem regardless where they are located. This could have significant implications on people's health and capacity to work and productivity. It has been estimated that global warming is currently contributing to the death of about 160,000 people every year.[8]

- *Structural integrity*, such as more severe wind and snow loading, and foundation movement. For example, shrinkage or expansion in clay soils can lead to foundation movement and cracking of walls, which may cause damage in building structure. In Northwest Norway, severe damage to buildings was caused by the hurricane on January 1, 1992, and several buildings collapsed due to heavy snow loads on roofs during the 1999/2000 winter.[9]
- *External fabric*: Durability of external fabric becomes shorter due to increased storm, rain, flood, and other weather conditions. For example, the strength and durability properties of concrete may be influenced by the changes in its environment (e.g., temperature, humidity).
- *Construction process* may be disrupted due to adverse weather condition. This may have implication in the project planning and the associated challenges to complete the project on time and within the budget.
- *Service infrastructure* may become inadequate. For example, changing weather patterns and more frequent and intense storm may lead to the drainage problem. The existing drainage system in many parts of the world may not be able to cope with increased storm loads.

Contribution of Buildings to Human-Induced Climate Change

The climate system is a dynamic system in transient balance.[10] A change of external and/or internal climate forcing imposed on the planetary energy balance would cause a corresponding change in global temperature.[10] Overall, scientists have now been able to reach a broad consensus and provide overwhelming evidence to suggest that human activities are having a discernible influence on the global climate.[2] In particular, it has been found that the recent rapid increase in global temperature is closely aligning with the strong growth in use of fossil fuel over the past 50 years.

The construction and operation of modern buildings consume a considerable amount of energy and materials and therefore contribute significantly to the process of human-induced climate change. Figure 2 shows the world greenhouse gas emissions by sector, end use, activity, and gas types.[11] It can be seen that buildings, including both commercial buildings and residential buildings, account for 15.3% of world greenhouse gas emission, which is greater than the sectors of transportation, agriculture, and waste. Indeed, buildings are one of the most significant infrastructures in modern societies.

Worldwide, the Worldwatch Institute estimates that the construction and operation of buildings is responsible for 40% of the world's total energy use, 30% of raw materials consumption, 55% of timber harvests, 16% of freshwater withdrawal, 35% of global carbon dioxide (CO_2) emissions, and 40% of municipal solid waste sent to landfill.[12] In 2004, it was estimated that worldwide, the total emissions from the building sector, including the electricity consumed, were 8.6 Gt CO_2, 0.1 Gt CO_2–eq N_2O, 0.4 Gt CO_2–eq CH_4, and 1.5 Gt CO_2–eq halocarbons (including CFCs and HCFCs).[13]

Basically, the impact of buildings on the human-induced climate change is through three routes, including building operational energy, building embodied energy, and building-related refrigerants. Operational energy includes all energy used for mechanical services [e.g., heating, ventilation, and air conditioning (HVAC) systems], electrical services (e.g., lighting and other office appliances), and hydraulic services (e.g., pumping system). Embodied energy includes all energy used for the production of building materials, their transportation and handling, and building construction processes. Building-related refrigerants include the refrigerants used in air-conditioning systems (e.g., Freon used in compressors and chillers) and other building appliances (e.g., refrigerators and freezers), which may have ozone depletion potential (e.g., depleting the ozone in the upper atmosphere) and/or global warming potential that persists in the upper atmosphere, trapping the radiation emitted by the earth.

Previous studies of Life Cycle Assessment (LCA) have shown that the CO_2 emission from the sources of building embodied energy and building-related refrigerants is often fairly small for commercial office buildings. For instance, based on the assumption of a building having a 40 years life span, it has been

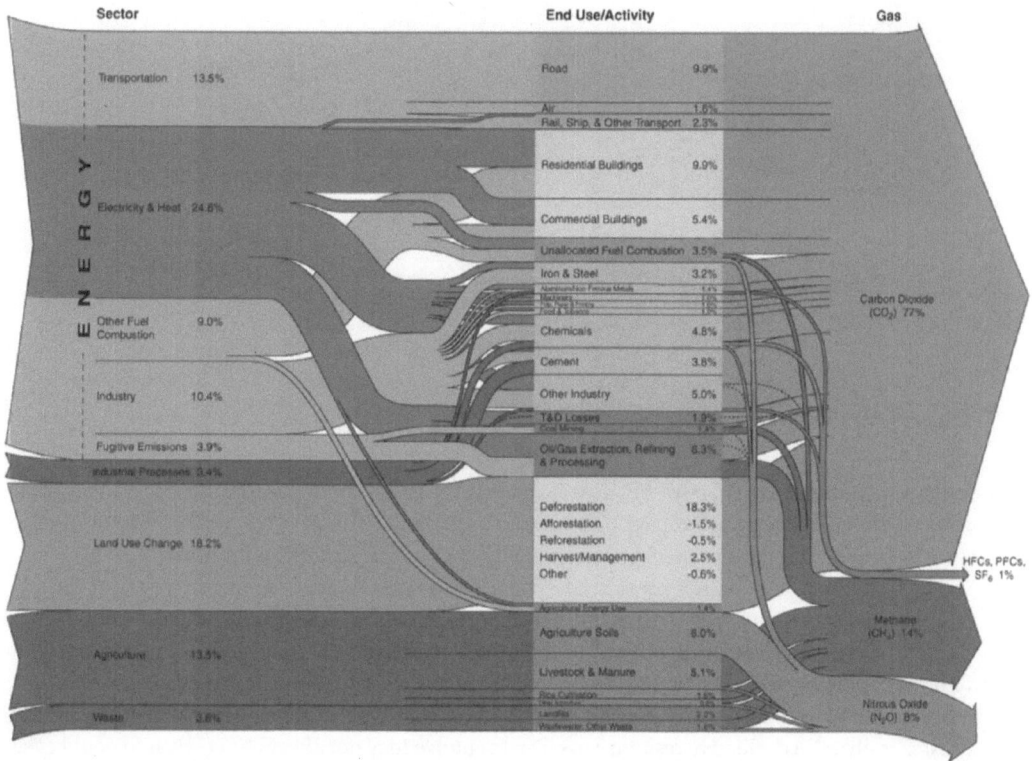

FIGURE 2 World greenhouse gas emissions by sector, end use, activity, and gas types.
Notes: All data are for 2000. All calculations are based on CO_2 equivalents, using 100 years global warming potentials from the IPCC (1996), based on a total global estimate of 41,755 MtCO$_2$ equivalent. Land use change includes both emissions and absorptions. Dotted lines represent flows of less than 0.1% of total greenhouse gas emissions.
Source: World Resources Institute.[11]

shown that the embodied energy emissions contributed only approximately 8%–10% of building total emission.[14] More than 90% of building energy consumption occurs during the use/operational phase, so the energy performance of the buildings is particularly important, especially for buildings with 24 hours occupancy.

Relationship between Climate Change and Buildings

As discussed above and also illustrated in Figure 3, the cycling interaction between climate change and buildings is of dynamic nature. Climate change, for instance, would generally lead to more use of air conditioning, which leads to more greenhouse gas emission and then climate change. Therefore, both climate change and buildings are essentially the cause and the effect of each other.[15] On one hand, climate change is expected to impact on many aspects of building design, construction, and operation. On the other hand, buildings have also contributed significantly to the human-induced climate change. They have produced more greenhouse gas emission than the sectors of transportation, agriculture, and waste.

Because greenhouse gas concentrations are still continuing to increase, the process of climate change will continue and may accelerate. This requires the building sector to develop suitable strategies, including the enhancement in building energy codes, to mitigate the greenhouse gas emissions, which should

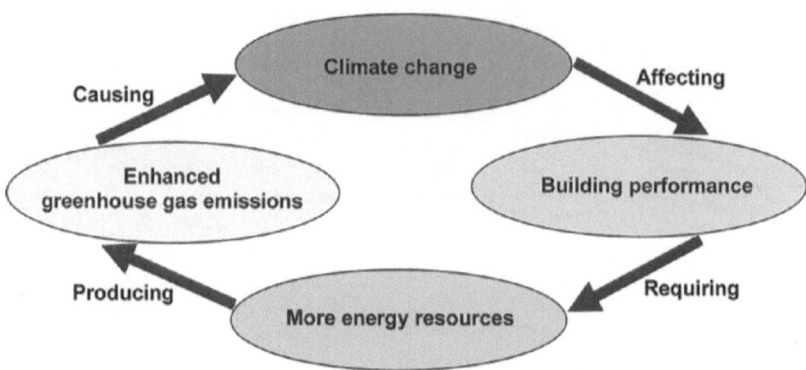

FIGURE 3 The cycling interaction between global warming and buildings.

be not only environmentally effective but also economically efficient. In parallel, appropriate adaptation strategies are also needed to prepare buildings to withstand the future inevitable climate change. Adaptation to climate change has now become one of the key requirements for buildings.

It is noted that there may be both potential synergy and conflict between adaptation and mitigation measures and strategies. This will require the development of integrated rather than separate responses.[16–18] Moreover, it seems that there is a tight intertwining between the issues of adaptation and mitigation.[19,20] The energy-efficient and renewable energy technologies, for instance, often can play dual roles in mitigating greenhouse gas emissions while increasing adaptive capacity by making buildings more disaster resilient. An effective response to climate change may not always be costly and could even be both profitable and socially beneficial.

Potential Strategies for Building Design and Operation

Buildings are one of the most significant infrastructures in modern societies, having significant impact on health and productivity of people living and working inside them. In particular, as buildings have a very long life span, typically 50–100 years, it is very important that all the current and future building stocks be designed and maintained to perform satisfactorily in future climates.

For both new and existing buildings, potential mitigation may include the utilization of renewable energy and low emission energy, energy efficiency, and energy conservation. Appropriate adaptation may also be achieved by focusing on major building design parameters for new buildings or retrofit of building envelope and/or internal heat sources such as lighting systems for existing buildings. In particular, those improved building energy codes currently adopted around the world, including the requirement of thermal conductance values for building envelope, lighting power densities for internal heat sources, and HVAC system efficiency for building mechanical services,[6,7] could play key roles in both the potential mitigation and adaptation of buildings. Overall, the aim is to ensure that all the existing and new buildings not only perform and operate satisfactorily in the new environment but also satisfy the environmental performance criteria of sustainability.

Building Envelope

The building envelope is a significant determinant of energy required to heat and cool a building. Building designers can use the building envelope as a "filter" to accept or reject solar radiation and outside air, depending on the need for heating, cooling, ventilation, and lighting.[13] The heat storage capacity of the building envelope can also be used to reduce peak thermal loads. As a result, a well-designed building shell can substantially reduce the building cooling and heating demands and lighting energy through the provision of daylighting, particularly at the perimeter zones. This can result in not only substantial capital cost savings by enabling smaller sizing of the air-conditioning plant to be used but also significant operating cost saving by reducing the quantity of air circulation and fan power, and partial load situation for the HVAC systems.

It is particularly important that building designers employ energy-efficient principles, as well as integrated building design approaches at design and construction of the building envelope. Overall, for the building envelope, the potential mitigation and adaptation strategies in the face of climate change may include the following:

- Installation of insulation in foundation, walls, and/or roofs to minimize conductive heat loss/gain.
- Optimum design of windows, including careful determination of window-to-wall ratio (WWR) and selection of types of glass and shading to optimize use of solar heat and light.
- Sealing gaps to prevent draughts and heat infiltration to minimize uncontrolled movement of air into the building.
- Utilizing thermal mass, reflective roofs, and trees for shade, etc., if it is appropriate.

Internal Heat Sources

The internal heat sources include the occupants, lights, electric appliances, and machines. Their influence may be not only directly on the total energy consumption of the building but also on the building thermal loads, which indirectly influence the energy consumption of HVAC systems. The selection of lights, for example, can have significant impact on the building energy and thermal performance. Advances in higher-efficiency light sources, lower-loss ballasts and control gear, and more efficient luminaries will therefore enable lighting energy usage to be significantly reduced. Developments in the technology of daylighting will further increase the potential lighting energy savings. The potential mitigation and adaptation strategies to manage internal heat sources may include the following:

- Optimum integration of lamp (lighting sources), ballasts (for electric-discharge lamps), and fixtures (e.g., reflectors, diffusers, and/or polarizing panels). It was found that to maintain the same lighting level of 320 lux, lighting power densities can vary from 2 to 5 W/m^2 (averaged over time) for having state-of-the-art lighting technology to 25–35 W/m^2 for having only mediocre level of efficiency.[21]
- Purchasing energy-efficient office equipment and running it wisely. Energy-efficient office equipment can use less than half the energy as the standard models—at no or little additional capital cost.[21] Recommendations may include the use of "energy star" office equipment and "energy labeling" appliance. It was found that every dollar invested in energy efficiency, at a 20%–30% saving rate, is equivalent to increasing net operating income by 3%–4% and net asset value by \$2.50–\$3.75.[22]

Using the building simulation technique, it has been shown that for Australian office buildings, if the building total internal load density, which includes all internal heat sources of lighting, plug load, and occupants, is reduced from 43 to 21.5 W/m^2, the building total energy use under the future 2070 high scenario (the worst case) can be reduced by up to 89–120 kWh/m^2 per annum and the overheating problem could be completely avoided and the office building will perform as good as at the current climate scenario.[23]

Building Mechanical Services

Building mechanical service or HVAC systems are used to control factors affecting thermal comfort, with the ultimate purpose being to provide a clean, noise-free, and efficient working environment.[24] Since air-conditioning systems account for around one-third of the total energy use in typical office buildings, the proper design and selection of HVAC systems are critical for the performance of these buildings.

Generally, according to the purpose of the building, as well as local site information and climate condition, a building may be designed as naturally ventilated, mechanically ventilated, air conditioned, or a combination of these three options (hybrid mixed-mode system). Potential mitigation and adaptation strategies for design and selection of building service systems include the following:

- Using passive design for heating, cooling, and ventilation. A building relying entirely on air conditioning should be avoided as much as possible.
- Utilization of renewable energy and/or low emission energy for the operation of the building. It was found that on clear days throughout the year, reductions of conventional power use of at least 60% can be achieved with optimum photovoltaic cladding densities targeted to lighting and small power load demands.[25]
- Selecting energy-efficient technologies. Such technologies may include evaporative cooling, desiccant cooling, chilled ceilings, displacement ventilation, cogeneration and district heating and cooling, active solar and heat pump systems, and underground earth pipes.[26]
- Good control of HVAC systems. It was found that the performance of the HVAC system is subject to the significant influence of its control strategy and maintenance. Building energy management systems (BEMSs) can also be a useful tool for energy saving in the range of 5%–40%.[26]
- Good commissioning, operation, and maintenance of HVAC systems. It was found that of the deficiencies found in the retro-commissioning of buildings, 85% were related to HVAC systems.[22] In the same study, it was also reported that energy savings between 7% and 29% may be achieved with paybacks ranging from 0.2 to 2.1 years.

Recommendations on Implementation of Strategies

Overall, it is recommended that the potential mitigation and adaptation strategies should be taken at the early planning and design stage. This is because the thermal loads gained/lost from the building envelope and internal heat sources are the "base load" for the selection of HVAC equipment. Appropriate selection and design of these systems are therefore vital for the energy efficiency of the building. Without a suitable choice, all later work will be built on an unsatisfactory foundation. It is also noted that the mitigation and adaptation strategies taken at the early design stage will future-proof the building and be more effective in the longer term. Although the initial capital cost of taking the measures may be higher, the overall cost in many cases would actually be much lower than having to do retrofit strengthening at a later stage.[16]

Building regulation and design codes can also play a crucial role in the anticipation and avoidance of risk.[27] Because climate change would have a major impact on the frequency of extreme weather events, building codes need to be reviewed regularly in order to maintain a proper level of reliability and be adhered to in practice.[9] For instance, changes in the frequency of storms will have building code implications.[28] However, it is argued that simply raising performance standards as a response to climate change, without dealing with the issue of noncompliance with existing standards, may not be desirable. It may run the risk of undermining the legitimacy of regulation generally.[27]

In the face of inevitable climate change, the shifting from a reliance on historical data to a reliance on predicted future data may be also necessary, as it is increasingly important to precisely predict the conditions under which buildings and other infrastructure will need to withstand in the future. It is understood that using different sets of design conditions will have significant implication on the cost and performance of buildings. For example, using the building simulation technique, it has been shown that for typical Australian office buildings, the required cooling capacity may increase from 28% to 59% if the new buildings were designed using the future project climate (e.g., 2070 high scenario) rather than the current climate condition.[5] In order to maintain comfort indoor condition, a further increase of 4%–10% may be required in addition to the possible increase of 27%–47% cooling load if the buildings were designed at current climate conditions.

Conclusion

The climate change induced by the emissions of greenhouse gases is one of the most important global environmental issues facing the world today. Buildings are one of the most significant infrastructures in modern societies, and they need to be prepared to withstand climate change. On one hand, climate change is going to impact on many aspects of buildings, including both building design and building operation. On the other hand, buildings have also contributed significantly to the process of human-induced climate change. In this entry, both aspects of knowledge have been discussed.

It has been suggested that the potential mitigation and adaptation strategies should focus on major energy-related factors, such as design and construction of building envelope, design and careful selection of the air-conditioning system, and selection of management of internal heat sources including both lighting systems and electrical equipment. In many cases, it has been demonstrated that the energy-efficient and renewable energy technologies can play dual roles in mitigating greenhouse gas emissions while increasing adaptive capacity by making buildings more disaster resilient.

It has also been recommended that energy efficiency principles, such as adopting passive design principles for heating, cooling, ventilation, and lighting, and integrated design approaches should be adopted at the early design stage and be maintained in the whole life of buildings. These strategies have the potential of not only being environmentally effective but also economically efficient and socially beneficial.

References

1. Healey, J., Ed. *Climate Change—Issues in Society*; The Spinney Press: Sydney, 2003; Vol. 184, ISBN 1-876811-93-5.
2. IPCC. Climate change 2007: The physical science basis. *Contribution of Working Group I to the Fourth Assessment Report of the Intergovernmental Panel on Climate Change*; Solomon, S., Qin, D., Manning, M., Chen, Z., Marquis, M., Averyt, K.B., Tignor, M., Miller, H.L., Eds.; Cambridge University Press: Cambridge and New York, 2007, 996.
3. Hulme, M.; Sheard, N. *Climate Change Scenarios for Australia*. Climatic Research Unit; Norwich, 1999, 6 pp.
4. Guan, L. Global warming: Impact on building design and performance. *Encycl. Energy Eng. Technol.* **2009**, *1* (1), 1–6.
5. Guan, L. Implication of global warming on air-conditioned office buildings in Australia. *Build. Res. Inf.* **2009**, *37* (1), 1–12.
6. ANSI/ASHRAE/IESNA Standard 90.1-2007. Energy standard for buildings except low-rise residential buildings.
7. Building Code of Australia (BCA)—Australian Building Codes Board (ABCB), 2010.
8. Bhattacharya, S. Global warming kills 160,000 a year. *New-Scientist.com News Service*, 17, 01 October 2003, available at http://www.newscientist.com/article.ns?id=dn4223.
9. Lisø, K.R.; Aandahl, G.; Eriksen, S.; Alfsen, K. Preparing for climate change impacts in Norway's built environment. Build. Res. Inf. **2003**, *31* (3), 200–209.
10. McGuffie K.; Henderson-Sellers, A. A Climate Modeling Primer, 3rd Ed.; John Wiley and Sons, Ltd.: Chichester, 2005.
11. *World GHG Emissions Flow Chart*; World Resources Institute, Retrieved on 09 February 2011, available at http://cait.wri.org/figures.php?page=/World-FlowChart.
12. Fenner, R.A.; Ryce, T. *A comparative analysis of two building rating systems. Part 1: Evaluation. Proc. Inst. Civil Eng., Eng. Sustain,* **2008**, *161* (ES1), 55–63. doi: 10.1680/ensu.2008.161.1.55.
13. Levermore, G.J. A review of the IPCC assessment report four, Part 1: The IPCC process and greenhouse gas emission trends from buildings worldwide. *Build. Serv. Eng. Res. Technol.* **2008**, *29* (4), 349–361.

14. Australian Greenhouse Office. *Australian Commercial Building Sector Greenhouse Gas Emissions 1990–2010*; Canberra, 1999; ISBN 18-76536-195.
15. Degelman, L.O. Which came first—Building cooling loads or global warming? A cause and effect examination. *Build. Serv. Eng. Res. Technol.* **2002**, *23* (4), 259–267.
16. Lowe, R. Really rethinking construction. *Build. Res. Inf.* **2001**, *29* (5), 409–412.
17. Steemers, K. Towards a research agenda for adapting to climate change. *Build Res. Inf.* **2003**, *31*, 291–301.
18. Lowe, R. Lessons from climate change: A response to the commentaries. *Build. Res. Inf.* **2004**, *32* (1), 75–78.
19. Camilleri, M.; Jaques, R.; Isaacs, N. Impacts of climate change performance on building in New Zealand. *Build. Res. Inf.* **2001**, *29* (6), 440–450.
20. Mills, E. Climate change, insurance and the buildings sector: Technological synergisms between adaptation and mitigation. *Build Res. Inf.* **2003**, *31* (3–4), 257–277.
21. Sustainable Energy Development Authority (SEDA). Tenant energy management handbook—Your guide to saving energy and money in the workplace, 2000, ABN 80-526-465-581.
22. Cull, S.L. Commissioning: Retrocommissioning. *Encycl. Energy Eng. Technol.* **2007**, *1* (1), 200–206.
23. Guan, L. Adaptation to global warming by changing internal loads of buildings. *Proceedings of International Conference on Building Energy and Environment (COBEE)*, Dalian, China, July 13–16, 2008.
24. ASHRAE. *ASHRAE Handbook: Fundamentals*; American Society of Heating, Ventilating and Air-Conditioning Engineers (ASHRAE): Atlanta, 2001.
25. Jones, A.D.; Underwood, C.P. Cladding strategies for building-integrated photovoltaics. *Build. Serv. Eng. Res. Technol.* **2002**, *23* (4), 243–250.
26. Levermore, G.J. A review of the IPCC assessment report four, Part 2: Mitigation options for residential and commercial buildings. *Build. Serv. Eng. Res. Technol.* **2008**, *29* (4), 363–374.
27. Lowe, L. Preparing the built environment for climate change. *Build. Res. Inf.* **2003**, *31* (3), 195–199.
28. Larsson, N. Adapting to climate change in Canada. *Build. Res. Inf.* **2003**, *31* (3), 231–239.

8

Economic Growth: Slower by Design, Not Disaster

Peter A. Victor
and Tim Jackson

Introduction

Economic growth is a very recent phenomenon, dating back two or three centuries and limited in its extent to only some parts of the world. While economic growth has dramatically improved the lives of billions of people, billions more remain in abject poverty. The mainstream view is that all countries should strive for economic growth, and through a process of convergence where poorer countries grow the fastest, all the people of the world will eventually enjoy a high and secure material standard of living.

There are many problems with this vision of growth for all. Foremost is the mounting evidence that the world's economy is already bumping up against and even surpassing the biophysical limits of the planet. This is showing up in terms of global and local environmental degradation and resource scarcity. Consequently, a different vision is emerging, one in which rich countries no longer pursue economic growth as a primary objective, leaving room for the economies of poorer countries and regions to expand, at least temporally. An additional motivation for low and no growth is to reduce competition with other species with which humans share the planet.

This chapter examines the possibility of rich countries managing without growth, by which is meant achieving improvements in well-being without economic growth. That such an outcome is possible is illustrated with the use of LowGrow SFC, an interactive macroeconomic model of the Canadian economy. The chapter concludes with a discussion of policy directions implied by such a scenario with special emphasis on employment policies and funding public services in a low-/no-growth economy.

A Brief History of Economic Growth

In terms of human history, economic growth is a very recent phenomenon. World gross domestic product (GDP) was only 14% higher at the end of the first millennium than when it began. GDP is the "total unduplicated value of the goods and services produced in the economic territory of a country or region during a given period" (Statistics Canada). GDP per capita was no higher at all, slightly less even, owing to population growth. From the year 1000 to 1400, world GDP increased by 140%, but GDP per capita grew only 60% and at a barely perceptible rate, as population expanded. Over the next 400 years, world GDP increased a further 140%, and although GDP per capita increased by 30% by 1700, it fell back again in 1800 to the same level as in 1400. Average world GDP per capita declined further in the early 1800s, after which economic growth accelerated. By 1900, world GDP was over five times greater than in the previous 100 years, growing at an average annual rate of 1.7%, and GDP per capita was more than three times greater. The rate of economic growth increased in the 20th century such that annual world GDP grew almost ninefold at an average annual rate of 2.2% and GDP per capita rose 2 1/2 times.

These figures,[1] approximate as they are especially early on, show that economic growth has been exceptional in human history. The vast majority of people have lived in circumstances where they had no reason to believe that their children's lives would be materially any different from their own. To think otherwise is a very modern idea dating back perhaps a dozen or so generations, and even then, only in some parts of the world.

Of course, global averages conceal much of the huge variation in experiences from country to country and region to region. In the 18th and 19th centuries, most of the world's economic growth was confined to Europe and ex-colonies of European powers. Some regions were impoverished in the process, 19th-century India being a prime example through the deliberate destruction of cotton manufacturing at the behest of British industrialists. The rest of the world continued to experience lives in which economic growth had little or no impact and was not part of the lived experience of most people. Even in places where growth rates were highest, there were substantial areas of poverty. In the 20th century and into the 21st century, economic growth spread to more parts of the globe but unevenly with 735 million people in 2015 still living in extreme poverty (less than US$1.90 per day) and 3.4 billion living below US$5.50 per day, about the price of a single cup of coffee in the United States. While the number living in extreme poverty declined by 40% since 1990, the number living below $5.50 only declined by 5%, and the rate of reduction in extreme poverty is slowing down.[2] Contrary to earlier expectations that global economic growth would close the gap between rich and poor, the spread has increased, just as it has done within even the richest countries.[3]

All this economic growth has required a massive increase in the use of natural resources. In the 20th century, the global use of construction materials, ores and industrial materials, fossil fuels, and biomass increased eightfold, almost as much as GDP.[4] Since 2000, the global use of these resources has risen even faster than GDP. The much-vaunted "green growth" which depends fundamentally on increasing GDP with declining use resources is nowhere to be seen.[5] The increasing extraction, use, and disposal of materials have resulted in large-scale, adverse environmental impacts. One prominent list includes the following: climate change, ocean acidification, stratospheric ozone depletion, overloading of the nitrogen and phosphorus cycles, global freshwater uses, land system change, biodiversity loss, atmospheric aerosol loading, and chemical pollution.[6] Then, there is the looming threat of peak oil from conventional sources[7] and the more general concern that the age of cheap energy is coming to an end,[8] as well as an emerging scarcity of "critical" minerals for new technologies.[9] All this plus serious doubts about future prospects for increases in productivity[10,11] and concerns over secular stagnation,[12] make for a bleak outlook for real, long-term, comprehensive, global economic growth. With the world human population approaching 8 billion, projected to rise to nearly 10 billion by mid-century, the question arises as to whether economic growth for all is a viable option in the 21st century and if not, what are the alternatives?

One alternative is to shift the over-riding economic priority of the richest economies away from the single-minded pursuit of growth and to reduce "growth dependency." After all, economic growth has only held this position as the single most important economic policy objective since about 1960. "There is in fact hardly a trace of interest in economic growth as a policy objective in the official or professional literature of western countries before 1950."[13] And even when it was introduced, it was as a means to fulfill other policy objectives such as full employment, rather than as an end in itself. Now the pursuit of economic growth is deemed so important that it is customary for policy proposals across many domains including environment, education, and the arts to be judged in terms of their implications for growth, or one of its surrogates such as competitiveness or productivity.

If the poor countries of the world are to benefit from economic growth, and the world economy is to function within the "safe operating space" of the planet,[6] then rich countries, those that have benefited the most from economic growth in the past, must be prepared to make room for them. Otherwise, disaster threatens, brought about by the excessive pressure on Earth systems.

One approach to this predicament based on principles of distributive justice would be to determine fair shares of access to global resources, accounting also for the interests of other species. This is the kind of dialogue that is underway in the slow progression of international climate negotiations, illustrating most profoundly the difficulties in reaching and then implementing agreement in a global world divided into national and regional power blocs. Nonetheless, a country or group of countries could adopt a view of its fair share independently of an international agreement and set these as boundaries within which their economy must function. Depending on the specification of the boundaries and the capability of economies to adjust, economic growth, measured in the conventional way as an increase in real (inflation) adjusted GDP, could conceivably continue for a time, but incidentally rather than as a primary objective.

What would such development entail? Total environmental (including resource) impact is the product of GDP and impact per unit of GDP (e.g., GDP×greenhouse gas (GHG) emissions per unit of GDP or GDP×energy used per unit of GDP). Therefore, if environmental impact is to decline as GDP grows, environmental impact per unit GDP must decline faster than the rate of economic growth. This is one meaning of "green" growth, and it underlies the downward-sloping portion of the Environmental Kuznets Curve.[14] (The Environmental Kuznets Curve is an inverted "U" with a measure of environmental impact plotted on the y-axis and GDP plotted on the x-axis. The hypothesis is that in the early stages of economic growth, environmental impact rises to a maximum after which it declines as growth continues.) There are examples of obvious, local problems such as urban air quality whose history can be described by an Environmental Kuznets Curve, but it is a poor description of global materials and energy use or global environmental impacts over the past century.

Whether green growth defined in this way (i.e., impact/GDP declining faster than GDP increases) is possible at the global level remains an open question. It is complicated by the fact that many environmental and resource depletion problems relate to stocks such as accumulating GHGs in the atmosphere, diminishing rain forests, and declining fish stocks. Reducing the flows that determine these stocks may be insufficient to bring the stocks to the required levels fast enough or at all. If it is not possible to sufficiently decouple economic growth from its environmental and resource impacts, growth will have to cease and even turn negative for a time as proponents of degrowth argue.[15] Otherwise, it will not be possible to bring the global economy back within the planetary boundaries that are already being exceeded and others which will be exceeded if present trends are not reversed. Given what we know about the state of the environment, and concern over supplies of low-cost energy and critical materials, there is a very strong case on ethical and practical grounds for rich countries to take the lead in managing without growth.

Economists Question Growth

At the same time as economic growth was reaching the pinnacle of policy objectives, dissenting voices were beginning to be heard. One of the most widely prominent was John Kenneth Galbraith. In *The Affluent Society*[16] published in 1958 and revised through multiple editions, Galbraith compared private affluence in the United States with public squalor. He also questioned the efficacy of dealing with poverty through a general rise in incomes. Many academic economists regarded Galbraith as more of a political commentator than a serious economist because of his disdain for theoretical economics, and on these tenuous grounds, they resisted his arguments. The same could not be said of British economist Ezra Mishan who published *The Costs of Economic Growth* in 1967.[17] Mishan was a highly regarded and well-published expert in "welfare economics," the field within mainstream economics that is concerned with the relationship between economic activity and well-being. Thus, although Mishan's analysis of the costs of economic growth was aimed at a broad audience, no one could dismiss the author as not really understanding modern economic theory.

Perhaps this is one reason why Mishan's critique of economic growth, unlike Galbraith's, ignited a heated debate that went on for several years between him and Wilfred Beckerman, another well-established British economist. Beckerman wrote "Why We Need Economic Growth"[18] and *In Defence of Economic Growth*.[19] Later, Beckerman wrote *Small Is Stupid*[20] in response to Schumacher's widely read *Small Is Beautiful*,[21] Schumacher's critique of modern industrialized economies. Many of Schumacher's arguments about the optimal scale of an economy were anticipated, echoed, and augmented by other economists such as Kenneth Boulding in his seminal essay "The Economics of the Coming Spaceship Earth,"[22] N. Georgescu-Roegen,[23] who explored the implications of the second law of thermodynamics for economics and economic growth, and Herman Daly who has promoted a steady-state economy for more than three decades.[24] The publication of *The Limits to Growth*[25] in 1972 addressed similar themes using systems dynamics but was roundly and largely unfairly criticized especially by economists (e.g. Maddison,[1] pp. 90–94). It remains influential to this day. A useful summary of the state of the growth and no-growth debate, largely where it was left in the 1970s, can be found in Olson and Landsberg's collection of essays.[26] (For a contemporary assessment of the Limits to Growth debate see the review by Jackson and Webster.[27])

After this flurry of publications in the 1960s and 1970s, the growth debate subsided. In the late 1990s, the criticisms of growth resurfaced stronger than ever with economists such as Douthwaite, *The Growth Illusion*,[28] Daly, *Beyond Growth: The Economics of Sustainable Development*,[29] and Booth, *The Environmental Consequences of Growth: Steady-State Economics*[30] leading the charge. By this time, the transdisciplinary, ecological economics was almost 20 years old, with a dedicated peer-reviewed journal, *Ecological Economics*, publishing 12 times a year including many papers dealing with problematic aspects of economic growth. In the first part of the 21st century, it is impossible to keep up with the many papers, reports, blogs, conferences, media entries, and YouTube videos questioning economic growth. And there are many books such as Hamilton, *Growth Fetish*,[31] Booth, *Hooked on Growth*,[32] Victor, *Managing without Growth*,[33] Speth, *The Bridge at the End of the World*,[34] Brown, *Right Relationship*,[35] Jackson, *Prosperity without Growth*,[36] Schor, *Plenitude*,[37] Dietz and O'Neill,[38] *Enough Is Enough*, Czech, *Supply Shock*,[39] and von Weizsäcker and Wijkman, *Come On!*[40] just to name a few in the English language alone.

The remainder of this chapter describes an investigation into what might be possible in an economy in which economic growth ceases. In particular, the following question is addressed: is it possible to have low unemployment, reduced GHG emissions and other environmental pressures, reduced income inequality, and reduced hours of paid work while maintaining reasonable debt levels in the public and household sectors, all in the absence of economic growth? And, if so, what policy frameworks or initiatives would be required? LowGrow SFC, an interactive macroeconomic systems dynamics model, was developed for the Canadian economy specifically to help answer these questions. In the next section, LowGrow SFC is described in fairly general terms and three very different scenarios for Canada to the year 2067 are presented. Brief comments on policy directions suggested by the simulation of low/no growth are followed by a more detailed consideration of employment in a no-/low-growth economy and the implications for government finance and the provision of public services.

Exploring Low and No Growth in Canada with LowGrow SFC

LowGrow SFC is a quantitative model of the Canadian economy designed to explore future scenarios. LowGrow SFC has been calibrated using Canadian data with simulation results reported from 2017 to 2067. The scenarios are not predictions. Rather, they are offered as consistent and plausible future possibilities intended to feed discussions about current choices and alternative futures. The model integrates three primary spheres of interest within a system dynamics framework: (1) the environmental and resource constraints on economic activity; (2) a full account of production, consumption, employment, and public finances in the "real economy" at the level of the nation state; and (3) a comprehensive account of the financial economy, including the main interactions between financial agents. LowGrow SFC conforms quite closely to standard economic frameworks. Data are drawn directly from the Canadian national accounts and some of the behavioral relationships in the model are estimated econometrically on the basis of time-series data from the Canadian economy. At the same time, LowGrow SFC departs from more conventional economic modeling approaches by incorporating time-lags, feedbacks and expectations in the model, and also by allowing for some potentially radical variations on "typical" macroeconomic policy.

The theoretical basis for LowGrow SFC draws heavily on the post-Keynesian macroeconomic approach of Godley and Lavoie (2012), which places a particular emphasis on a full and consistent account of the relationships between monetary stocks and flows within and between different financial sectors. In the aftermath of the 2008 financial crisis, so-called stock-flow consistent (SFC) modeling has gained a particular traction because of its ability to provide a comprehensive account of financial transactions in the economy and to map the impact of these on financial balance sheets—something that was conspicuously missing in the run-up to the crisis. LowGrow SFC is articulated in terms of six interrelated sectors: households, firms, banks, government, a central bank and the "rest of the world" (or "foreign" sector). It models a range of financial assets and liabilities including deposits, loans, mortgages, government bonds, and firms' equities.

In LowGrow SFC, as in the economy that it represents, economic growth is driven primarily by net investment which increases the capital stock and hence labor productivity, combined with growth in the labor force. The extent to which the full productive capacity of the economy is employed depends on aggregate demand comprising expenditures by the private and public sectors on consumption and investment, and international trade. LowGrow SFC incorporates numerous features related to the overall environmental performance of the economy. For instance, a key focus of the model is on climate change and GHG emissions. LowGrow SFC includes a sub-model of the electricity sector. This is useful for assessing the economic and environmental effects of a transition to renewable sources of electricity and the widespread electrification of the economy.

One of the most important elements in the model relates to investments undertaken to reduce environmental impact. Key to a future in which economies reduce the burden they place on the biosphere is a shift from "brown" investment in activities that increases environmental impacts to "green" investment in activities that reduces them. Some green investments such as cost-competitive, energy-efficient equipment can add to the productive capacity of the economy. Other green investments such as seawalls built to protect coastlines from rising sea levels, protect productive capital but do not add to it, reducing the rate of economic growth. A further consideration that is important for determining macroeconomic outcomes is whether green investment is "additional" or "non-additional." Additional green investment increases total investment expenditures whereas "non-additional" green investment simply displaces brown investment without adding to total investment. Only additional green investment increases GDP.

LowGrow SFC generates the values of many variables relevant to an assessment of the performance of the economy. Seven of these are incorporated in a Sustainable Prosperity Index (SPI) for evaluating the scenarios:

- GDP per capita (more relevant to well-being than GDP).
- The rate of unemployment.
- The ratio of government debt to GDP.
- The ratio of household debt to household net worth.
- The Gini coefficient on household incomes (a measure of inequality).

- The average hours of paid work.
- The Environmental Burden Index (EBI).

The EBI is designed to capture the environmental impacts of economic activity notably absent from GDP. While comprehensive in scope, the EBI lacks specificity other than with respect to GHG emissions. Both the SPI and EBI are preliminary and could be improved with better data. However, they share with GDP the redeeming feature that they emerge from a model of the system in whose performance we are interested and so can be used to measure the effect of measures designed to make the system work better.

Scenarios for the Canadian Economy

Former British Prime Minister, Margaret Thatcher, insisted that "there is no alternative" to the market economy and economic growth. She was mistaken. There are in fact many alternatives. We describe three scenarios for the Canadian economy. None of them is a prediction of the future. Rather they are intended to illustrate some of the possibilities facing Canada, to inform discussion and debate, and to suggest the kinds of choices available, not just in Canada but in other advanced economies. The three scenarios are summarized in Table 1.

The Base Case scenario is a benchmark against which other scenarios can be compared. It is a description of what would happen, broadly speaking, at the national level, if current trends continue through and beyond mid-century. The GHG Reduction scenario includes several measures specifically to reduce GHG emissions: a price on GHG emissions from the electric power sector rising from $0 in 2017 to $300 per ton over 15 years, gradual electrification of road and rail, and substantial expenditures to reduce GHG emissions from other sources. In addition to these GHG reducing measures, the Sustainable Prosperity scenario includes more green investment aimed at reducing a wider set of environmental impacts, transfer payments rising to $20 billion per year to reduce inequality and poverty, a shorter work year and a slower rate of population growth.

Figure 1 shows GDP per capita for the three scenarios. Under the Base Case, per capita GDP increases from $50,000 in 2017 to $97,000 in 2067, with an average growth rate of 1.3%. (Unless otherwise stated, values are in 2007$.) This is essentially a conventional, growth-based view of the future, in which the economy as a whole (taking into account population growth of around 44%) increases 2.8 times by the year 2067. The GHG Reduction scenario has a somewhat lower average growth rate in GDP per capita of 1.1%, with incomes in 2067 achieving a level of $87,000 per year. In this scenario, some of the green investment is "unproductive" in the sense that it does not add to the capital stock with which labor produces output. Hence, labor productivity is not as high as in the Base Case and so neither is GDP. The

TABLE 1   Three Scenarios for Canada 2017–2067

Scenario	Main Features
1. Base case	Scenario 1 • Continuation of current trends and relationships
2. GHG reduction	Scenario 1 plus: • Carbon price on GHG emissions from electricity generation • GHG abatement by non-electric industrial sources • Electrification of road and rail transport
3. Sustainable prosperity	Scenario 2 plus: • Switch from brown to green investment • Increased transfer payments to reduce income inequality and reduce poverty • Lower rate of population growth • Reduced average hours worked

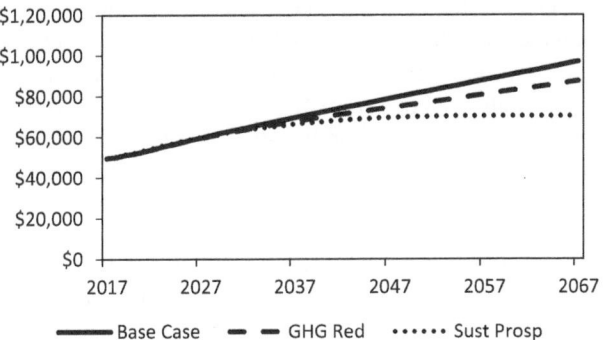

FIGURE 1 GDP per capita in the LowGrow SFC scenarios.

projected reduction in GDP growth is at the high end of the range of published estimates.[41] It also runs counter to the view that a green economy grows faster than a brown one.[42]

The most marked difference, however, is between the Base Case and the Sustainable Prosperity scenario. The latter illustrates a stabilization of per capita income at a level above current income levels. Specifically, the GDP per capita in 2067 is $70,000, an average annual increase of 0.7% over the period. More significantly, GDP and GDP per capita are essentially stable over the final 20 years of the scenario because of the stabilization of population. This scenario thus illustrates a transition from a growth-based economy to an economy managing without growth. The lower rate of economic growth and ultimately its cessation altogether result from the reduced investment in brown capital and the consequential lower increase in labor productivity combined with a reduction in average hours of paid work. Conventional wisdom suggests that such a transition toward what is effectively a steady-state economy is impossible without causing irreparable damage to prosperity and well-being in society. But Figure 2 suggests that this undesirable outcome can be avoided. In fact, the composite SPI described above rises significantly in the Sustainable Prosperity scenario despite falling in both the other two scenarios. Starting from a base of 100 in 2017, the SPI falls precipitously by more than 50% in the Base Case. Even in the GHG Reduction scenario, the SPI declines 10%. In the Sustainable Prosperity scenario, by contrast, the SPI increases 35% from 2017 to 2067.

To understand these differences, all of the components of the SPI must be considered. One of these is GDP per capita, which tends to push the SPI upwards, the higher the level of GDP. Principal among the factors that favor the Sustainable Prosperity scenario over the Base Case is the EBI, which includes the negative impact of GHG emissions and other environmental pressures which are significantly reduced in the Sustainable Prosperity scenario. The EBI for the Base Case more than triples over

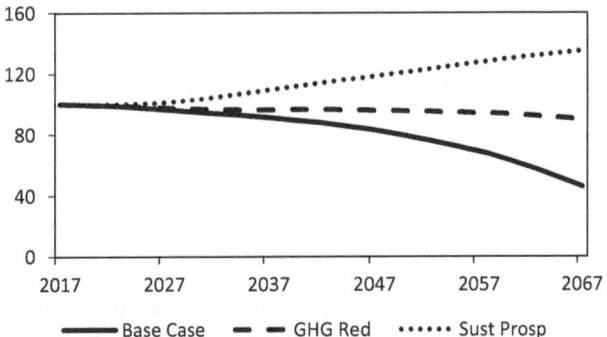

FIGURE 2 SPI in the LowGrow SFC scenarios.

the period of the scenario, as GHGs continue to rise and little is done to offset other environmental impacts from the economy. The EBI for the GHG Reduction scenario performs significantly better. The main reason for this is a significant decline in GHG emissions, which suppresses the rise in the EBI and in doing so has a notably positive effect on the SPI as compared with the Base Case. However, because of the continued expansion of economic output in the GHG Reduction scenario, the reduction in GHG emissions while substantial falls far short of the Canadian government's target of 80% reduction by 2050 from 2005 levels. By comparison, the Sustainable Prosperity scenario more than meets this target. Some of the decline in the EBI, in particular for the Sustainable Prosperity scenario, is also due to the deliberate policy of shifting the electricity sector toward renewable energy sources induced by the carbon price on GHG emissions. The rest of the decline comes from the lower level of economic activity in this scenario.

Two specific social measures adopted in the Sustainable Prosperity scenario also contribute to the improved performance of this latter case over the other two scenarios. The first of these is the redistributive fiscal policy in which transfer payments are progressively increased from 2020 and distributed preferentially to the lower-income categories. Also in the Sustainable Prosperity scenario, annual average hours of paid work decline from more than 1,700 h/yr in 2017 to less than 1,400 h/yr in 2067. Fewer hours working offers more opportunities for people to enjoy time with their families and friends, perhaps volunteering in the community or taking advantage of increased leisure, all of which contribute to people's well-being and quality of life and help increase the SPI.

LowGrow SFC keeps track of the financial flows among sectors as well as sector balance sheets. From this information, two measures of financial prudence included in the SPI are generated by the model: government debt and household loans to value (i.e., net worth). In the Base Case and GHG Reduction scenarios, government debt to GDP increases slightly but more so in the Sustainable Prosperity scenario where it rises from 55% to 70% by the end of the run. This has a negative impact on the SPI. Nonetheless, even at the end of the run, the debt-to-GDP ratio remains at a level that has been far surpassed by many countries without the collapse of their economies reaching 250% in Japan in 2016.

With government running a deficit, it follows from the SFC of the model and the financial behaviors of the other sectors that the overall net lending position of the household sector is positive in all three scenarios, leading to a healthy position in terms of household net worth. There are minor increases in the ratio of household loans to net worth in all three scenarios, with the smallest being in the SP scenario.

In summary, the results discussed in this section suggest that the Sustainable Prosperity scenario remains a realistic alternative to the conventional wisdom of continual exponential growth, outperforming the Base Case in several important ways over the next half a century. Even the financial indicators of a low-growth economy can, under the right conditions, remain relatively sustainable. Investment portfolios change, productivity growth declines, consumption expenditures stabilize, but the economy is nonetheless still financially resilient, its social outcomes improve, and its environmental burden on the planet is dramatically reduced.

Recognizing that the results described here are exploratory, it is fair to conclude that they support the following three conclusions:

1. The pursuit of economic growth at the expense of a deepening environmental crisis has a very high probability of catastrophe.
2. Substantial reductions in GHG emissions can clearly be achieved without massive changes to the structure of society. But the impact on the rate of growth, while modest, is larger than others have suggested and at odds with those who see a "green" economy as growing faster than a "brown" one. Furthermore, this "green growth" scenario falls well short of the Government of Canada's GHG target reduction of 80% by 2050.
3. Sustainable prosperity is attainable, but it will require a major reorientation of society's priorities toward improvements in social equity, economic security, and environmental quality. These changes may well lead to low- or no-growth economy, but they will also deliver a better quality

of life. The result may not be entirely incompatible with capitalism, but it will look very different from the over-financialized consumer capitalism of the early 21st century and may well be worthy of a different name altogether.

Policy Directions for a Low-/No-Growth Economy

The achievement of sustainable prosperity will require some very significant changes in the economy and in society at large. Policy and policy-related measures that drive the Sustainable Prosperity scenario include the following:

- *Investment*: a shift from brown to green investment and from private to public goods through changes in taxation and expenditures.
- *Labor force*: stabilization through changing age structure of the population and population stabilization.
- *Population*: stabilization through a declining fertility rate and changes to immigration policy.
- *Poverty and inequality*: trickle down replaced with focused antipoverty programs that address the social determinants of illness and provide more direct income support.
- *Technological change*: slower, more discriminating, preventative rather than end-of pipe, through technology assessment and changes in the education of scientists and engineers.
- *Government expenditures*: a declining rate of increase until stabilization is reached.
- *Trade*: a neutral net trade balance and diversification of markets.
- *Workweek*: shorter, more leisure through changes in compensation, work organization and standard working hours, and active market labor policies.
- *Environment and energy*: a comprehensive environmental program of pricing, regulation, green investment, land use planning, and education to achieve a rapid transition from fossil fuels to renewable sources of energy, significant reductions in material and energy throughput, and a reversal of habitat destruction conversion.
- *Consumption and lifestyles*: more public goods and fewer positional (status) goods through changes in taxation and marketing.
- *Localization*: fiscal and trade policies to strengthen local economies.

The next two sections look more closely at two specific policy areas in relation to the low/no-growth scenario: strategies for full employment and funding government programs.

Economic Growth and Employment

In 1960, the UN World Economic Survey stated that "the reinterpretation of the objective of full employment under the United Nations Charter to embrace the goal of economic growth marks a second fundamental change in public policy thinking."[43] This statement from the UN is based on the insight derived from the early work on economic growth by Harrod, Domar, and others that if aggregate expenditure required for full employment in the short run expands the productive capacity of the economy, further increases in aggregate expenditure will be required in the future if full employment is to be maintained. This relationship between growth and employment is accentuated if the size of the labor force is increasing as well.

Equation 1 expresses the relationship between GDP, productivity, the labor force, and unemployment:

$$\text{GDP} = P(1-u)L \qquad (1)$$

where GDP is the real gross domestic product, P is the productivity (real GDP per employed person), L is the labor force (employed plus unemployed persons), and u is the unemployment rate (unemployed/labor force).

Between 1971 and 2016, Canadian real GDP grew 234%, productivity increased by 50%, the labor force rose by 125%, and the unemployment rate increased from 6.2% to 7.1%. While the increase in GDP had a positive impact on employment (and vice versa), it was more than offset by the increase in productivity (P). The net effect was an increase in the rate of unemployment (u).

This is a classic dilemma. A growing economy stimulates employment, an increase in productivity reduces it. How can the advantages of increased productivity be realized in an economy that is not growing without causing high unemployment? One way is to reap the benefits of increased productivity as more leisure rather than more goods and services. This can be accomplished by reducing the average number of hours worked by an employed person so that unemployment for a few becomes more leisure for the many. If more people worked fewer hours, it should be possible to have high levels of employment without relying on economic growth.

Some counterfactuals from the past show how this could work. From 1971 to 2016, the average hours worked per year by a Canadian employee decreased by 10.9%. If the decrease in average hours worked had been 13.8%, the rate of unemployment would have been 4% not 7.1% in 2016 given the same increases in GDP and the labor force. At an average of 1647 hours of work per employed Canadian, employees in Canada would still have been working about the same number of hours per year as the average employee in Sweden and the United Kingdom and more than the average employee in seven other OECD countries, in some cases substantially more. Had there been no decrease in the average hours worked between 1971 and 2016, the rate of unemployment would have been 17.3% not 7.1% for the same increases in GDP, productivity, and the labor force.

These calculations show that the average length of the work year, which includes vacation days, can have a marked impact on the rate of unemployment. By spreading the same amount of work among a larger number of employees, the unemployment rate can be lowered and the relationship, as shown by the above examples, is strong. For this reason, researchers have examined the potential for reductions in the average number of hours worked per employee to contribute to full employment. From the standpoint of managing without growth, the benefits of increased productivity would be experienced as increases in leisure and reduced impacts on the environment rather than as increases in output, consumption, and environmental impacts.

A shorter average work year is one of the factors included in the Sustainable Prosperity scenario illustrated in Figures 1 and 2. Over the 50 years of the scenario, the work year declines by 28% to 1400 in 2067. This compares with levels already approached or surpassed in 2015 in Belgium (1427), the Netherlands (1420), Norway (1408), and Germany (1370). In general, European countries have been more proactive than Canada and the United States in reducing the working time as an instrument of employment policy. During the 2008/2009 recession, some countries, Germany for example, mitigated the impact on employment by relying more heavily on reductions in work time.

The arithmetic of reducing the rate of unemployment by reducing the average hours each employed person works is compelling. Achieving such gains in employment in the real world is another matter, but in a review of studies of the employment effects of working time reductions, Bosch finds that most show a gain of "25–70 percent of the arithmetically possible effect."[44] Bosch has examined the European experience, and the six conditions he identified as particularly important for the success or failure of this policy are summarized in Table 1. He points out that the general political conditions must be suitable for a policy of reducing work time to reduce unemployment. There must be acceptance from employees, trade unions, and employers and support of the State.

Looking at working time policy in the future, Bosch concludes that "shorter working hours are an indicator of prosperity."[44] They have been in the past, though more recently we have seen the emergence of a sector of the labor force that is "overemployed," working long hours and "failing to achieve a desired balance in their lives between paid work, family life, personal, and civic time."[45] These are usually men with higher levels of education in management positions. Simultaneously, there are people who are underemployed and poorly paid, more often than not women. These circumstances contribute to and accentuate rising income inequality.

TABLE 2   Policies for Reducing the Workweek

1.	Wage compensation—"If working time reductions and pay increases are negotiated as a total package, then the compensatory increase for the working time reduction can be offset by lower pay rises." This could become more difficult with no or low growth.
2.	Changes in work organization—"Larger reductions in working time generally have to be accompanied by changes in work organization"; otherwise, firms will rely on overtime and the employment effects will not materialize.
3.	Shortages of skilled labor—"An active training policy is an indispensable supplement to working-time policy" to ensure that there are people with the necessary skills to pick up the slack when skilled workers reduce their hours.
4.	Fixed cost per employee—Such as benefits paid on a per-employee basis rather than an hourly basis are an obstacle to reducing working hours because it is costly to employers. Canada shares with most Western European countries the practice of financing statutory social programs through contributions that are usually a proportion of earnings or through taxation, minimizing this fixed cost problem.
5.	The evolution of earnings—"The decreasing rate of real wage rises in most industrialized countries has reduced the scope for implementing cuts in working time and wage increases simultaneously." This would be a serious obstacle unless there is widespread support for seeking prosperity without growth though it can be mitigated by a more equal distribution of income. "One fundamental precondition for the working time policy pursued in Germany and Denmark, for example, was a stable and relatively equal earning distribution".
6.	The standardization of working hours—Any reduction in standard working hours must strongly influence actual hours worked. If it merely generates more overtime for those already with jobs, it will fail to increase employment. Work reorganization will be required to allow more flexibility in hours worked.

Source: Summarized from Bosch.[44]

Layard in his work on economics and happiness concludes "that people over-estimate the extra happiness they will get from extra possessions" because of habituation. "The required correction is towards lower work effort and thus lower consumption."[46] This means that a shorter work year would not only contribute to reducing unemployment but may also increase the general level of happiness for employees who find themselves better off working fewer hours, for less income and consuming at lower levels.

Funding Public Services in a Low-/No-Growth Economy

Economic growth provides government with increasing resources without increasing tax rates. In times of rapid growth, receipts from corporation profits taxes, personal income taxes, and value-added taxes tend to increase faster than the economy as a whole, allowing governments to provide more services, invest more in infrastructure, redeem outstanding debt, reduce tax rates, or some combination of all these. Governments welcome these circumstances. They have as much to gain from economic growth as anyone. How might this be different in an economy that eschewed economic growth as a policy objective?

Insight into this matter can be gained by considering in some detail the Sustainable Prosperity scenario in which economic growth slows and eventually ceases. In particular, we look at the projected ratio of government debt to GDP, the government's net lending position, and government expenditure in total and per capita. We have already seen that in this scenario, the ratio of government debt to GDP rises from 55% to 75% over the 50-year simulation, which though not necessarily desirable, is well within the bounds of viability based on the experience of other countries and in Canada as well. It is interesting to note here that modern money theorists advise against the use of the debt-to-GDP measure as an indicator of long-run resilience, on the grounds that in countries with sovereign monetary systems such as Canada, the United Kingdom, and the United States, the state does not have a budget constraint comparable to that of a household.[45] The argument is that government can always pay debts denominated in their own currency. In the Sustainable Prosperity scenario, government borrowing continues while GDP stabilizes leading to a rise in the debt-to-GDP ratio. Another reason for thinking that this is unlikely to be a significant problem is that government borrowing is projected to decline, approaching zero by the end of the simulation. This evolution of the government's net lending position coincides, as it must under SFC, with a decline in lending by the other sectors of the economy.

Government expenditures on consumption and investment are assumed in the model to be constant proportions of GDP, modified by countercyclical expenditures to moderate increases in unemployment above a target rate. These expenditures represent the utilization by government of the output of the economy. In the Sustainable Prosperity scenario, the sum of these government expenditures increases gradually from 25% in 2017 to 29% by 2040, remaining at that level as GDP stabilizes. As well as these expenditures, government supplements those with low incomes and supports selected business activities. These "transfer" payments affect the incomes of non-government sectors to pay for expenditures, which are included in GDP in household consumption and private sector investment. It would be double-counting to also include them in government expenditure in the calculation of GDP. However, they are included in the calculation of government net lending.

In the Sustainable Prosperity scenario, government expenditure per capita stabilizes about 70% higher than in 2017 in real, inflation-adjusted terms. This substantial increase should allay any concern of insufficient government funding for services in the absence of economic growth.

Conclusion

There are many reasons for considering how rich economies might manage without growth: growth rates are slowing down, biophysical constraints to continued growth are becoming more apparent, mounting evidence indicates that higher incomes do not make people happier beyond a level of per capita incomes far surpassed in rich countries, economic inequality continues to rise, and despite decades of substantial economic growth many social and environmental problems remain. It is past time to set aside contributions to economic growth as a criterion for assessing measures to improve well-being for humans and other species over the long term and to replace it with the pursuit of sustainable prosperity in all its various dimensions.

Acknowledgments

This chapter is based on Victor, P.A. *Managing without Growth: Slower by Design, Not Disaster*, 2nd edition 2019 and in particular, Chapter 11 which is co-authored with Tim Jackson.

References

1. Maddison, A. *The World Economy: A Millennial Perspective*; OECD: Paris, 2001.
2. World Bank. *Poverty and Shared Prosperity, Piecing Together the Poverty Puzzle*; World Bank Group: Washington, DC, 2018.
3. Alvaredo, F. et al., *World Inequality Report 2018*; World Inequality Lab: Paris, 2018.
4. Krausmann, F.; Gingrich, S.; Eisenmenger, N.; Erb, K.-H.; Haberl, H.; Fisher-Kowalski, M. Growth in global materials use, GDP and population during the 20th century. *Ecol. Econ.* **2009**, *68* (10), 2696–2705.
5. Ekins, P.; Hughes, P. et al. *Resource Efficiency: Potential and Economic Implications. A Report of the International Resource Panel*; UNEP: Paris, 2016.
6. Rockstrom, J. et al. A safe operating space for humanity. *Nature* **2009**, *24* (461), 472–475.
7. Sorrell, S. *Global Oil Depletion*; Technology and Policy Research Centre: London, 2009.
8. Ayres, R.U.; Warr, B. *The Engine of Economic Growth*; Edward Elgar: Cheltenham, 2009.
9. Coulomb, R.; Dietz, S.; Godunova, M.;Nielsen, T.B. 'Critical Minerals Today and in 2030: An Analysis for OECD Countries', Environment Working Paper No. 91; ESRC Centre for Climate Change Economics and Policy, Grantham Research Institute on Climate Change and the Environment: London, 2015.
10. Gordon, R.J. *The Rise and Fall of American Growth: The US Standard of Living since the Civil War*; Princeton University Press: Princeton, NJ, 2016.

11. Jackson, T.; Victor, P. Productivity and work in the 'green economy. Some theoretical reflections and empirical tests'. *Environ. Innov. Soc. Trans.* **2011**, *1*, 101–108

12. Jackson, T. The post-growth challenge: Secular stagnation, inequality and the limits to growth. *Ecol. Econ.* **2019**, *156*, 236–246.

13. Arndt, H.W. *The Rise and Fall of Economic Growth: A Study in Contemporary Thought*; Longman Cheshire: Melbourne, 1978. 13.

14. Dinda, S. Environmental Kuznets curve hypothesis: A survey. *Ecol. Econ.* **2004**, *49* (4), 431–455.

15. Schnieder, F.; Kallis, G.; Martinez-Alier, M. Crisis or opportunity? Economic degrowth for social equity and ecological sustainability. Introduction to this special issue. *J. Cleaner Prod.* **2010**, *18* (6), 511–518.

16. Galbraith, J.K. *The Affluent Society*; Houghton Mifflin: Boston, MA, 1958.

17. Mishan, E.J. *The Costs of Economic Growth*; F.A. Praeger: New York, 1968.

18. Beckerman, W. *Why we need economic growth. Lloyds Bank Rev.* **1971**, *102*, 1–15.

19. Beckerman, W. *In Defence of Economic Growth*; J. Cape: London, 1974.

20. Beckerman, W. *Small Is Stupid*; Duckworth: London, 1995.

21. Schumacher, E.F. *Small Is Beautiful: Economics As If People Mattered*; Harper and Row: London, 1973.

22. Boulding, K.E. The economics of the coming spaceship earth. In *Environmental Quality in a Growing Economy*; Jarrett, H., Ed.; Johns Hopkins Press: Baltimore, MD, 1966, 3–14.

23. Georgescu-Roegen, N. *The Entropy Law and the Economic Process*; Harvard University Press: Cambridge, MA, 1971.

24. Daly, H. *Steady-State Economics: The Economics of Biophysical Equilibrium*; W.H. Freeman: San Francisco, CA, 1977.

25. Meadows, D.H.; Meadows, D.L.; Randers, J.; Behrens III, W.W. *The Limits to Growth*; Earth Island Limited: London, 1972.

26. Olson, M.; Landsberg, H.H., Eds. *The No-Growth Society*; W.W. Norton and Company: New York, 1973.

27. Jackson, T.; Webster, R., *Limits revisited. A review of the limits to growth debate.* A report to the APPG on limits to growth published under Creative Commons, CC BY-NC-ND 4.0., 2016.

28. Douthwaite, R. *The Growth Illusion*; New Society Publishers: Gabriola Island, 1994 (also, a second edition in 1999).

29. Daly, H. *Beyond Growth: The Economics of Sustainable Development*; Beacon Press: Boston, MA, 1996.

30. Booth, D.E. *The Environmental Consequences of Growth: Steady-State Economics*; Routledge: London, 1998.

31. Hamilton, C. *Growth Fetish*; Allen and Unwin: Sydney, 2003.

32. Booth, D.E. *Hooked on Growth: Economic Addictions and the Environment*; Rowman and Littlefield: Lanham, MD, 2004.

33. Victor, P.A. *Managing without Growth: Slower by Design, Not Disaster*; Edward Elgar Publishing: Camberley, 2008 (2nd edition, 2019).

34. Speth, J.G. *The Bridge at the End of the World: Capitalism, the Environment, and Crossing from Crisis to Sustainability*; Yale University Press: New Haven, CT, 2008.

35. Brown, P.; Carver, G. *Right Relationship*; Berrett-Koehler Publishers Inc.: San Francisco, CA, 2009.

36. Jackson, T. *Prosperity without Growth*; Earthscan: London, 2009 (2nd edition, 2017).

37. Schor, J. *Plenitude: The New Economics of True Wealth*; The Penguin Press: New York, 2010.

38. Dietz, R.; O'Neill, D. *Enough is Enough*; Berrett-Koehler Publishers: San Francisco, CA, 2013.

39. Czech, B. *Supply Shock: Economic Growth at the Crossroads and the Steady State Solution*; New Society Publishers: Gabriola Island, 2013.

40. Von Weizsäcker, E.; Wijkman, A. *Come On! Capitalism, Short-Termism, Population and the Destruction of the Planet—A Report to the Club of Rome*; Springer: New York, 2018.

41. Ekins, P. Ecological modernisation and green growth: Prospects and potential. In *Handbook on Growth and Sustainability*; Victor, P.A.; Dolter, B. Eds.; Edward Elgar Publishing: Cheltenham and Northampton, MA, 2017, 107–137.

42. Bassi, A., *UNEP GER Modeling Work. Technical Background Material, V.4*; Millennium Institute: Arlington, VA, 2011.

43. United Nations Development Programme. *Human Development Report 2006*; UNDP: New York, 2006.

44. Bosch, G. Working time reductions, employment consequences and lessons from Europe. In *Working Time: International Trends, Theory and Policy Perspectives*; Golden, L., Figart, D.M., Eds.; Routledge: London, 2000, 177–211, p. 192.

45. Figart, D.M.; Golden, L. Introduction and overview, understanding working time around the world. In *Working Time: International Trends, Theory and Policy Perspectives*; Golden, L., Figart, D.M., Eds.; Routledge: London, 2000, 1–17.

46. Layard, R. Happiness and public policy: A challenge to the profession. *Econ. J.* **2006**, *116*, C24–C33.

9

Food–Energy–Water Nexus

Nemi Vora

Introduction

The United Nations established 17 distinct goals (referred to as the sustainable development goals or SDGs) for the 2030 agenda for global sustainable development ranging from climate action to economic growth. It has been noted that the agenda can only be achieved successfully if their interconnections are recognized and incorporated into planning (Nilsson, Griggs, and Visbeck 2016). For instance, SDG 2 outlines ending hunger, providing nutrition, achieving food security, and promoting sustainable agriculture. It is closely related to SDG 14 of life below water and SDG 15 of life on land, which in turn requires planning for quality and quantity supply of water (SDG 6), renewable, affordable energy (SDG 7), and sustainable consumption and production (SDG 12). The concept of interconnections between Food–Energy–Water (referred to as the FEW nexus) is one such piece to a much larger puzzle of managing for the SDGs. Recently, international and national funding calls on the FEW nexus (National Science Foundation, National Science Foundation (U.S.–China), European Commission 2015) have mobilized research around the topic. These organizations recognize the challenges and benefits of managing the FEW nexus, as individual management has often resulted in unintended consequences.

The connections between food, energy, and water resources are apparent in daily lives, although numerous examples exist of policies focusing on managing one resource but resulting in straining the others deeply connected with them. One such an example is the well-intended, but damaging irrigation policies in Gujarat two decades ago. Gujarat, a relatively arid state in western India, saw a rise in agriculture productivity through investments in irrigation infrastructure, land reforms, and advances in seed technology since the 1980s (Mathur and Kashyap 2000). Due to erratic precipitation patterns, groundwater irrigation became central to the success of Gujarat's farmers. The government supported groundwater use through a series of initiatives including access to cheap electricity for pumping water out (Narula et al. 2011). However, decades of excessive pumping coupled with low recharge levels resulted in severe groundwater depletion, posing a threat for the future of agriculture in the region. As the water levels declined, farmers pumped deeper, putting undue pressure on the power grid and the government for continuing to subsidize expensive electricity (Narula et al. 2011). Thus, managing electricity for promotion of agriculture without considering impacts on water use resulted in severe and long-lasting consequences. In 2012, a drought in eastern India resulted in farmers turning to groundwater for irrigation. Sudden excessive pumping and lack of water for power plants caused a massive grid failure. Comparable situations were reported during California's recent 7-year drought (Webber 2015). From an energy policy perspective, the promotion of biofuels from food and oilseed crops, and their effects on water, land, and biodiversity have been a center of debate and contention for a while (Searchinger et al. 2008). The context of these issues may be different, but they demonstrate the potential tradeoffs that may occur if resources are managed without regarding their connections.

The concept of the FEW nexus is not new; similar calls for integrated water resource management (IWRM; an approach to promote simultaneous development of water and associated resources for social and economic benefits) date back to 1962 (Lloyd 1963). Similarly, the integrated natural resource management concept was put forward to couple agrarian objectives with ecosystem services (Twomlow, Love, and Walker 2008). In 2011, Hoff et al. highlighted the concept of "nexus approach" to simultaneously target food, energy, and water security for the background paper at the Bonn Conference (Hoff 2011). With the FEW nexus, a clearly defined scope of focusing on three essential resources was put forth. In the scientific literature, many variations of the term FEW nexus exist with studies coining the terms Energy–Water–Food (EWF) nexus, Water–Energy–Food (WEF) nexus, Climate–Land–Energy–Water (CLEW nexus), etc.; however, the underlying message of the interlinkages remains the same.

Defining the Food–Energy–Water Nexus

There is no single agreed-upon definition of FEW nexus in the literature, apart from recognizing the FEW nexus as linkages. It often makes the scope of the work more complex as numerous connections exist between these three resources. For example, growing food requires water for agriculture and energy to employ machinery on farm and transport food. Global agriculture is the largest consumer of water amounting to 69% of total withdrawals, while food supply chain accounts for 30% of global energy consumption (Dubois 2011). Water is integral in generating energy from renewable and fossil resources alike: water is required in thermoelectric power plants, and for hydraulic fracturing to obtain natural gas. Renewable resources, such as hydropower, and biofuels production also heavily depend on water availability. Energy is

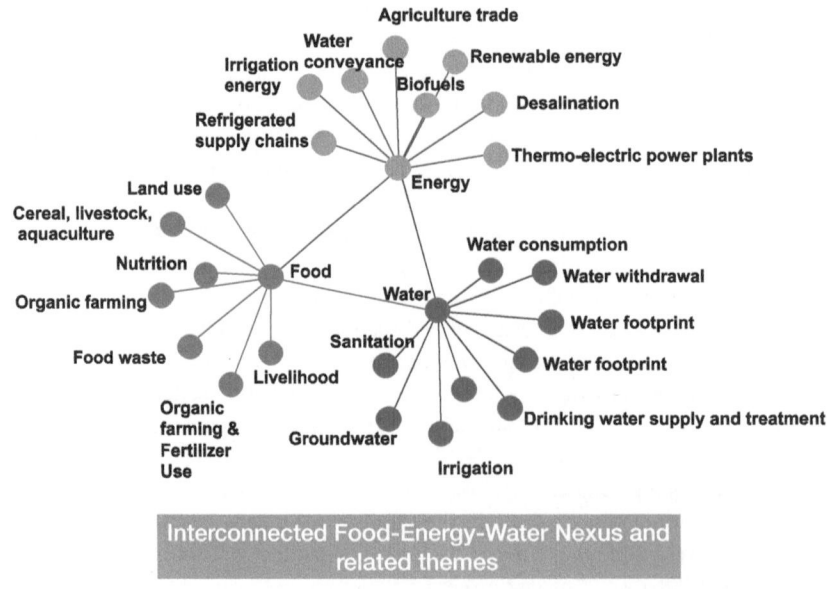

FIGURE 1   Common research and policy themes under the umbrella of Food–Energy–Water nexus.

required to withdraw, treat, and convey water for municipal, agriculture, and industrial use. Desalination, an important alternative for fresh water in water-stressed countries, is also highly energy intensive.

While still relevant to the nexus, some studies are often not included in the FEW nexus if they focus on less than three resources (i.e. food–water, energy–water), or if they go beyond the three to include climate, livelihoods, and biodiversity in their analysis. Additionally, the focus of FEW nexus studies is often on physical interactions and environmental concerns alone, while the crucial social component is overlooked. It has also been argued that calling a study "a nexus approach" does not increase its scientific merit, and therefore FEW nexus does not add anything new to the literature (Wichelns 2017). Here, I adopt the perspective that establishing universal criteria for inclusion in the FEW nexus is not as important as recognizing the existence of interconnections, both hidden and visible, between resources, and resources and the rest of the environment. Thus, managing one resource requires effort to identify and evaluate the influence on/of other resources.

Admittedly, merely recognizing the interconnections may not simplify their management, it does initiate a discourse. In the aforementioned case of Gujarat, the issue of expensive subsidies and groundwater depletion was tackled by separating and tightly rationing farm electricity supply from the rest of the rural areas, while providing superior quality supply on a pre-announced schedule. The "Jyotigram (electrified village)" scheme was instrumental in overhauling Gujarat's rural electrification while reducing groundwater overdraft and boosting non-farm rural economy (Shah et al. 2008). Using the irrigation–energy nexus to tackle not only energy-water tradeoffs but also to benefit rural economy has been touted a win–win scheme and is now being implemented across other Indian states.

The scientific community has largely agreed upon managing resources by recognizing their interconnections, and it has been widely acknowledged to adopt a systems perspective in tackling the complex issue (Bazilian et al. 2011, Liu et al. 2018). Systems analysis recognizes that real-world issues do not strictly arise from a linear chain of cause and effect but are rather a result of complex feedback loops between individual components. As such, it recognizes that a system is often greater than the sum of its parts. Thus, rather than focusing on individual components, systems analysis examines how each component interacts with other components and the external environment. The result of these multiple interactions, often unobservable by studying just one component, can be understood through a holistic perspective. In a more practical sense, this is done by employing a collection of models and assumptions from different disciplines to represent a simplified version of reality. This approach helps identify potential side effects and address the issue of considering biophysical and socio-economic interactions together. The following section briefly talks about already existing models and methods used in the FEW nexus context.

Quantitative Analysis of the FEW Nexus

Much has been written regarding the FEW nexus and models to quantify linkages (Bazilian et al. 2011, Newell, Goldstein, and Foster 2019). As the research under the wide umbrella of FEW nexus studies is highly interdisciplinary, a series of methods and their combinations have been used to model one or more FEW interactions. Studies have used methods such as life cycle assessment, network analysis, footprint approach, optimization, input–output analysis, computable general equilibrium models, and hydrological models to model FEW interactions. These methods are adopted from disparate disciplines including ecology, social science, probability and statistics, thermodynamics, and hydrology. In their review articles, Albrecht et al. and Newell et al. provide a detailed list enumerating the studies of various methods that have been used to tackle FEW challenges (Albrecht, Crootof, and Scott 2018, Newell, Goldstein, and Foster 2019). Therefore, instead of repeating such a list, a few examples are provided here that show the use of interdisciplinary methods to help answer specific FEW issues at different geographical scales.

Vora et al. used social network analysis combined with environmental life cycle assessment to assess energy and greenhouse gas emissions associated with irrigated food trade in the United States (Vora et al. 2017). Davis et al. developed an optimization framework using spatially explicit data and crop water use models to assess redistribution of crops at an international scale with the goal to reduce global water use, without compromising nutritional benefits or loss of crop diversity (Davis et al. 2017). Ramaswami et al. developed a framework around cities as emerging demand centers and quantified FEW nexus impacts associated with meeting current demand from cities (Ramaswami et al. 2017). Using regression models, Herrera-Estrada et al. assessed the effect of drought on air emissions from electricity sector with a focus on regions heavily relying on hydropower and thermoelectric power plants (Herrera-Estrada et al. 2018). Keskinen et al. utilized a combination of mathematical models pertaining to hydrology and climate to understand tradeoffs between new hydropower development and its effect on soil fertility and local agriculture in Mekong Basin (Keskinen et al. 2015). Not all of the studies cited here framed their work explicitly within the FEW nexus contexts, nevertheless their focus on inter-linkages makes them nexus studies. The key takeaway is that there is no singular method and discipline that can provide a comprehensive quantification of FEW linkages, but requires interdisciplinary approach spanning both natural and social systems.

While these studies piece together different methods to develop a FEW nexus frameworks, there are also already existing models that help quantify some of these linkages. Pollitt et al. provide a detailed assessment of large-scale models along with their drawbacks and advantages (Pollitt et al. 2010), and Bazilian et al. provide a brief list of sectoral and integrated models and methods specifically pertaining to the FEW nexus (Bazilian et al. 2011). Some of the notable and widely used models include Climate, Land-Use, Energy and Water Strategies (CLEW) by International Atomic Energy Agency (Howells et al. 2013), Global Agro Ecological Zones (GAEZ) by Food and Agriculture Organization (Fischer et al. 2012), and the Integrated Solutions for Water, Energy, and Land (ISWEL) project that links multiple open-source models such as MESSAGE for energy modeling and GLOBIOM for land-use modeling (Willaarts et al. 2018).

Method and Model Selection

As there is no dearth of different methods and models that can be used for FEW nexus modeling, selecting an appropriate one depends on several factors: (1) scope and objective of the work, (2) availability and quality of data, (3) appropriate spatial scale, and (4) usefulness of analyzing long-, medium- and short-term temporal trends. Table 1 lists each factor and relevant questions for choosing appropriate models and methods.

TABLE 1 Decision Factors Influencing Model and Method Selection for the FEW Nexus

Scope	• Goal of the study (e.g. new policy formulation, infrastructure decisions, disaster preparedness)
	• Already known information about resource linkages (e.g. groundwater levels affecting irrigation, fossil heavy grid)
	• Defining system boundary of the work (e.g. production-based impacts vs. consumption impacts, inclusion/exclusion of other resource connections)
Data	• Availability of open data and time/money budget if none available (i.e. budget for purchasing data or gathering yourself)
	• Sensitivity of the chosen method to the size of data (e.g. certain methods may only work for larger datasets)
	• Quality of the datasets and quantification of uncertainty (precision of data collection methods, survey methodologies used, representativeness of data)
	• Appropriate assumptions and relying on substitutable data to fill data gaps (e.g. using national averages when local not available)
	• Age of the data (are 10-year old datasets appropriate to represent the current situation)?
Spatial and temporal scales	• Spatial scale of the study (local, metropolitan, state/province, basin level)
	• Integrating multiple datasets and models available for different spatial scale (e.g. hydrology models developed at basin scale vs. energy models for specific electricity markets and grids)
	• Available data reconciliation methods and their complexity (e.g. downscaling aggregate global data to local through cross-entropy, spatial estimation)
	• Relevance of long-term impacts on study results (e.g. climate-induced change on water availability, increase in crop yields)
	• Usefulness of incorporating historical trends

Scope

Although FEW nexus is circular, there is generally a particular resource of interest that is established (e.g. reservoir water management for agriculture, municipal, and energy supply) and subsequent linkages are then assessed. If there are previous assessments and known information about historical connections between resources, it should be incorporated in the study. For example, decision on upgrading irrigation technologies should consider impact on groundwater levels. Additionally, depending on the work, it should be determined whether the study should be limited to FEW nexus linkages alone or FEW nexus-everything studies that consider economy-wide impacts as well (Lant et al. 2018). This relates to setting a system boundary to contain the complexity of the system. A system boundary delineates impacts and resources to be considered, and those that will be out of the scope for the work. However, system boundary should be selected carefully as to not leave out relevant and important impacts. This can be done based on expert judgment, literature review of similar case studies, and use of screening tools such as economic input–output life cycle assessment (Hendrickson et al. 1998).

In the FEW nexus discourses, choosing a systems boundary is particularly important due to the connectedness of the products and services through global supply chains. Global trade and goods exchange are driven by competitive advantage and endowment of resources between regions, domestic and international policies and agreement, and politics. Depending on the production practices and environmental regulations in the production region, trade can alleviate or increase environmental impacts associated with consumption (Dalin and Rodríguez-Iturbe 2016). Studies have shown that domestic policies promoting human and environmental health have driven production out of the country causing environmental damages elsewhere (Nesme et al. 2016, Plevin et al. 2010). Therefore, careful consideration needs to be made to assess whether distant impacts should be incorporated as to not under/overestimate the resource use and impacts. FEW nexus studies have used the concept of embodied (virtual) resource/impacts trade to incorporate the role of global supply chains and connect distant locations (Kastner, Erb, and Haberl 2014, Konar et al. 2016). Here, the embodied trade refers to not the physical trade but indirect use of resources and resulting impacts from the production of specific goods and services in distant locations. Finally, the scope of the analysis may need to be adjusted depending on whether there are existing models that can be used to establish and quantify linkages or new models need to be built from scratch and required investment of time, effort, and interdisciplinary expertise.

Data

Freely available quality data are a necessity for carrying out insightful quantitative analysis and modeling. Numerous international agencies (UN Statistics division, International Energy Agency, Food and Agriculture Organization, World Resource Institute, etc.) and country-specific agencies (US Department of Agriculture, US Geological Survey, Government of India Planning Commission, Statistics Canada, etc.) collect and host data (see Data.Gov, data.gov, data.europa.eu, data.un.org for a single point of access to disparate databases) pertaining to food supply, agriculture, energy use, trade, water withdrawal statistics. Data collection is a time-consuming and money-intensive process, and therefore necessary data may not be available for a specific year, a time period, or a specific location. To this end, appropriate assumptions and substitutions may need to be made to fill in data gaps. These assumptions can be made with the help of scientific principles such as mass–energy balance, engineering equations, or based on peer-reviewed literature. A common method for data substitutions is based on similarity, be it geographical or technological: for example, substituting lack of data for one developed country with another developed country with similar gross domestic product (GDP).

However, many of the datasets are usually limited at country level and/or by time scale. Therefore, success of any integrated management approach hinges on data being available at an appropriate spatial scale and for a longer time frame. Therefore, new data collection and compilation techniques are needed to analyze systems at specific scales with recent/current data and for impoverished countries that may not be able to spend money on census surveys. Currently, the research work is driven toward bypassing traditional government surveys and using novel techniques such as satellite imagery, remote sensing, and cellphone data to gain important insights on living conditions (Azzari, Jain, and Lobell 2017, Blumenstock, Cadamuro, and On 2015). Although, more work is needed before such methods can replace the need for traditional surveys. Finally, methods such as sensitivity analysis and uncertainty quantification should be used to help convey quality of the data. Sensitivity analysis is useful in understanding which parameters affect the results the most and indicates where data collection improvements can be made. Uncertainty quantification helps in conveying variability and reliability of the results. Due to limited scope, these concepts are mentioned here in brief. However, the open-source life cycle assessment textbook from Matthews et al. is a good source for introductory material on conducting both types of analysis (Matthews, Hendrickson, and Matthews 2015).

Spatial and Temporal Scales

As shown with aforementioned examples of quantitative studies, the spatial scale for FEW analyses has varied from local to international. However, the appropriate scale in which these systems should be analyzed for effective policymaking and decisions is still ambiguous. While local challenges of drought and crop failure remain, the global supply chains have extended the system boundary and made it more fragile as distant events can cause cascading failures. Marston et al. noted in their analysis that depletion of major ground water aquifers in the United States would not only impact domestic food consumption but also affect distant countries such as Japan relying on food exports from these regions (Marston et al. 2015). Boundaries for water management can also be unclear as interconnectivity between water basins can go beyond political boundaries. There are numerous examples of water conflicts within a country such as water wars in the Western United States between different states (Tory 2018) and international issues such as the Brahmaputra river basin spanning China, India, Bangladesh, and Bhutan (Yang et al. 2016). The interconnectedness of economic and political tensions and extreme weather events to energy supply has been evident in numerous examples of energy crisis across the world (Hamilton 2011). Thus, the connections between places are not only based on political or physical boundaries but also through common pool of resources, infrastructure, and social systems. To integrate this complexity in analysis will require redefining how distant FEW systems are connected as regulations may stop at political boundaries, but the actual environmental impact may be larger beyond the traditional perceived boundaries. Recently, Lant et al. (2019) discussed the need of "mesoscale" analysis to capture differences that cannot be captured by analysis at two extremes of local scale vs. international scale (Lant et al. 2019). Based on the chosen temporal aspect (long term vs. short term), climate narratives provided by the shared socio-economic pathway projections and national climate assessments can be incorporated.

There is also ambiguity regarding the best approach to analyze such complex interactions: specifically pertaining to a bottom-up approach that builds up from a smaller system or a top-down approach that looks at interactions between different sectors and economy. We need integration of both approaches (top-down, bottom-up) and multiple spatial and temporal scales as decisions are made at various levels from household and local community level actions to national and international policies and agreement. As one size does not fit all, analyzing such complex systems will require both multi-sector and multilevel actions.

The Criticism and Cautions for the Food–Energy–Water Nexus

A few have criticized the overuse of the word FEW nexus indicating that while it is undisputable that these resources are interconnected, the phrase is ambiguous in terms of prescribing a framework in tackling these issues (Cairns and Krzywoszynska 2016, Wichelns 2017). The critics point out that by setting such arbitrary bounds on only looking at three resources, the FEW nexus misses the opportunity to consider other factors such as human health, livelihood, and farm chemicals. However, as mentioned earlier, the study of FEW nexus should be looked at in the much larger context of assessing connections between SDGs. While FEW nexus has indeed become a buzzword (Cairns and Krzywoszynska 2016), the interest garnered can be successful in raising awareness at both policy and stakeholder levels. Additionally, having a general umbrella under which such case studies, policies, and research can be compiled provides a huge opportunity to mobilize large-scale funding, exchange knowledge, and support interdisciplinary and interagency collaboration.

An important consideration that nexus proponents should keep in mind is that when managing for tradeoffs, one interest lobby may be stronger than the others and argue for their interest over an optimal solution (Jensen 2013). Furthermore, the economic cost, time, and expertise required for an integrated approach will be much higher than traditional silos approach and would involve substantial work toward capacity building, making the decision process lengthy (Liu et al. 2018). Additionally, an integrated approach may not be the panacea as considering too many factors at once may make a system unsolvable or at best provide a solution that may not be agreeable to all.

Conclusion

It is important to note that beyond the three resources, the usefulness of the concept lies in the term "nexus thinking." A nexus approach involves addressing the interlinkages and feedback between different systems, so synergies are promoted and tradeoffs minimized. However, in some instances, even a FEW nexus approach may not provide optimal outcomes and overlook unintended consequences on interdependent systems such as human livelihood or biodiversity. Therefore, a context-specific analysis that prioritizes local stakeholders' benefits should be adopted in policymaking with an overarching goal of achieving the sustainable development agenda. Going forward, the FEW nexus literature should explicitly recognize that there is no one suitable method, model, or spatial/temporal scales to conduct such complex analysis. But interdisciplinary multi-scale effort is needed to understand the issues in entirety. Good quality open data is at heart of such analysis and future work and studies should focus on novel data collection and data treatment approaches that take care of bridging the spatial and temporal resolution gaps, conducting hybrid analysis that incorporate both top-down and bottom-up approaches, and adopt/develop methods that provide a robust statistical framework for empirical data analysis.

Despite its drawbacks, FEW nexus has been successful in mobilizing the scientific and policy community to go beyond the "silos thinking" approach and recognize the need for interdisciplinary work. By incorporating three large policy and scientific communities of agriculture, hydrology, and energy analysis, FEW nexus has also merged expertise within these communities, with studies formulating novel frameworks and techniques to overcome issues of data availability, spatial and temporal scope, and model integration. FEW nexus has also drawn the scientific communities' attention back to SDG interlinkages and driven the policy community to engage in quantitative prospective analysis before making large-scale decisions. The penetration of FEW nexus in humanities has also incorporated more dialogue through stakeholder engagement. Finally, the primary takeaway lesson from the FEW nexus for environmental management is to adopt holistic approach and recognize that there may be hidden or visible links connected to the system being managed, and therefore, these links need to be identified and incorporated in the management plan.

References

Albrecht, Tamee R., Arica Crootof, and Christopher A. Scott. 2018. "The water-energy-food nexus: A systematic review of methods for nexus assessment." *Environmental Research Letters* 13 (4):043002.

Azzari, George, Meha Jain, and David B. Lobell. 2017. "Towards fine resolution global maps of crop yields: Testing multiple methods and satellites in three countries." *Remote Sensing of Environment* 202:129–141.

Bazilian, Morgan, Holger Rogner, Mark Howells, Sebastian Hermann, Douglas Arent, Dolf Gielen, Pasquale Steduto, Alexander Mueller, Paul Komor, and Richard S.J. Tol. 2011. "Considering the energy, water and food nexus: Towards an integrated modelling approach." *Energy Policy* 39 (12):7896–7906.

Blumenstock, Joshua, Gabriel Cadamuro, and Robert On. 2015. "Predicting poverty and wealth from mobile phone metadata." *Science* 350 (6264):1073–1076.

Cairns, Rose, and Anna Krzywoszynska. 2016. "Anatomy of a buzzword: The emergence of 'the water-energy-food nexus' in UK natural resource debates." *Environmental Science & Policy* 64:164–170.

Dalin, Carole, and Ignacio Rodríguez-Iturbe. 2016. "Environmental impacts of food trade via resource use and greenhouse gas emissions." *Environmental Research Letters* 11 (3):035012.

Davis, Kyle Frankel, Maria Cristina Rulli, Antonio Seveso, and Paolo D'Odorico. 2017. "Increased food production and reduced water use through optimized crop distribution." *Nature Geoscience* 10 (12):919–924. doi: 10.1038/s41561-017-0004-5.

Dubois, Olivier. 2011. *The State of the World's Land and Water Resources for Food and Agriculture: Managing Systems at Risk*. Earthscan: London.

European Commission. 2015. "Integrated approaches to food security, low-carbon energy, sustainable water management and climate change mitigation Horizon 2020 funding call." https://cordis.europa.eu/programme/rcn/664566/en.

Fischer, Günther, Freddy O. Nachtergaele, Sylvia Prieler, Edmar Teixeira, Géza Tóth, Harrij Van Velthuizen, Luc Verelst, and David Wiberg. 2012. *Global Agro-Ecological Zones (GAEZ v3.0) – Model Documentation*. IIASA; FAO: Laxenburg; Rome.

Hamilton, James D. 2011. *Historical Oil Shocks*. National Bureau of Economic Research: Washington, DC.

Hendrickson, Chris, Arpad Horvath, Satish Joshi, Octavio Juarez, Lester Lave, H. Scott Matthews, Francis C. McMichael, and Elisa Cobas-Flores. 1998. "Economic input-output-based life-cycle assessment (EIO-LCA)." *Mental*. https://www.researchgate.net/profile/H_Matthews/publication/242142910_Economic_Input-Output-Based_Life-Cycle_Assessment_EIO-LCA/links/0c96053b2c2bd76615000000.pdf

Herrera-Estrada, Julio E, Noah S. Diffenbaugh, Fabian Wagner, Amy Craft, and Justin Sheffield. 2018. "Response of electricity sector air pollution emissions to drought conditions in the western United States." *Environmental Research Letters* 13 (12):124032.

Hoff, Holger. 2011. *Understanding the Nexus. Background Paper for the Bonn 2011 Conference: The Water, Energy and Food Security Nexus*. Stockholm Environment Institute: Stockholm.

Howells, Mark, Sebastian Hermann, Manuel Welsch, Morgan Bazilian, Rebecka Segerström, Thomas Alfstad, Dolf Gielen, Holger Rogner, Guenther Fischer, and Harrij Van Velthuizen. 2013. "Integrated analysis of climate change, land-use, energy and water strategies." *Nature Climate Change* 3 (7):621.

Jensen, Kurt Mørck. 2013. "Viewpoint–swimming against the current: Questioning development policy and practice." *Water Alternatives* 6 (2):276–283.

Kastner, Thomas, Karl-Heinz Erb, and Helmut Haberl. 2014. "Rapid growth in agricultural trade: Effects on global area efficiency and the role of management." *Environmental Research Letters* 9 (3):034015.

Keskinen, Marko, Paradis Someth, Aura Salmivaara, and Matti Kummu. 2015. "Water-energy-food nexus in a transboundary River Basin: The case of Tonle Sap Lake, Mekong River Basin." *Water* 7 (10):5416–5436.

Konar, Megan, Jeffrey J. Reimer, Zekarias Hussein, and Naota Hanasaki. 2016. "The water footprint of staple crop trade under climate and policy scenarios." *Environmental Research Letters* 11 (3):035006.

Lant, Christopher, Jacopo Baggio, Megan Konar, Alfonso Mejia, Benjamin Ruddell, Richard Rushforth, John L. Sabo, and Tara J. Troy. 2018. "The US food–energy–water system: A blueprint to fill the mesoscale gap for science and decision-making." *Ambio*:1–13.

Lant, Christopher, Jacopo Baggio, Megan Konar, Alfonso Mejia, Benjamin Ruddell, Richard Rushforth, John L. Sabo, and Tara J. Troy. 2019. "The US food–energy–water system: A blueprint to fill the mesoscale gap for science and decision-making." *Ambio* 48 (3):251–263.

Liu, Jianguo, Vanessa Hull, Hugh Charles, Jonathan Godfray, David Tilman, Peter Gleick, Holger Hoff, Claudia Pahl-Wostl, Zhenci Xu, Min Gon Chung, and Jing Sun. 2018. "Nexus approaches to global sustainable development." *Nature Sustainability* 1 (9):466.

Lloyd, Emlyn Howard. 1963. "Design of water-resource systems; new techniques for relating economic objectives, engineering analysis, and government planning." *Journal of the Royal Statistical Society: Series A (General)* 126 (3):480–481.

Marston, Landon, Megan Konar, Ximing Cai, and Tara J. Troy. 2015. "Virtual groundwater transfers from overexploited aquifers in the United States." *Proceedings of the National Academy of Sciences* 112 (28):8561–8566.

Mathur, Niti, and Surendra P. Kashyap. 2000. "Agriculture in Gujarat: Problems and prospects." *Economic and Political Weekly* 35:3137–3146.

Matthews, H. Scott, Chris T. Hendrickson, and Deanna H. Matthews. 2015. "Life cycle assessment: Quantitative approaches for decisions that matter." Retrieved June 1, 2016. https://www.lcatextbook.com/

Narula, Kapil Kumar, Ram Fishman, Vijay Modi, and Lakis Polycarpou. 2011. "Addressing the water crisis in Gujarat, India." Columbia Water Center White Paper, March 2011.

National Science Foundation. 2018. "Innovations at the nexus of food, energy and water systems (INFEWS)." https://www.nsf.gov/funding/pgm_summ.jsp?pims_id=505241.

National Science Foundation (U.S.-China). "Innovations at the nexus of food, energy, and water systems (INFEWS: U.S.-China)." https://www.nsf.gov/pubs/2018/nsf18096/nsf18096.pdf.

Nesme, Thomas, Solène Roques, Geneviève S Metson, and Elena M. Bennett. 2016. "The surprisingly small but increasing role of international agricultural trade on the European Union's dependence on mineral phosphorus fertiliser." *Environmental Research Letters* 11 (2):025003.

Newell, Joshua Peter, Benjamin Paul Goldstein, and Alec Foster. 2019. "A 40-year review of food-energy-water nexus literature with a focus on the urban." *Environmental Research Letters* 14: 073003.

Nilsson, Måns, Dave Griggs, and Martin Visbeck. 2016. "Policy: Map the interactions between sustainable development goals." *Nature News* 534 (7607):320.

Plevin, Richard J., Andrew D. Jones, Margaret S Torn, and Holly K. Gibbs. 2010. "Greenhouse gas emissions from biofuels' indirect land use change are uncertain but may be much greater than previously estimated." *Environmental Science & Technology* 44 (21):8015–8021.

Pollitt, Hector, Anthony Barker, Jennifer Barton, Elke Pirgmaier, Christine Polzin, Stephan Lutter, Friedrich Hinterberger, and Andrea Stocker. 2010. *A Scoping Study on the Macroeconomic View of Sustainability. Final Report for the European Commission, DG Environment, SERI (Sustainable Europe Research Institute).* Cambridge Econometrics. http://seri.at/wp-content/uploads/2009/12/SERICE-2010_Macroeconomic-view-of-sustainability.pdf.

Ramaswami, Anu, Dana Boyer, Ajay Singh Nagpure, Andrew Fang, Shelly Bogra, Bhavik Bakshi, Elliot Cohen, and Ashish Rao-Ghorpade. 2017. "An urban systems framework to assess the transboundary food-energy-water nexus: Implementation in Delhi, India." *Environmental Research Letters* 12 (2):025008.

Rothberg, Daniel. 2018. "States accuse Arizona water agency of gaming Lake Mead, undermining Colorado River drought plans." *The Nevada Independent*, April 17, 2018.

Ruddell, Benjamin L., Elizabeth A. Adams, Richard. Rushforth, and Vincent C. Tidwell. 2014. "Embedded resource accounting for coupled natural-human systems: An application to water resource impacts of the western US electrical energy trade." *Water Resources Research* 50 (10):7957–7972.

Searchinger, Timothy, Ralph Heimlich, Richard A. Houghton, Fengxia Dong, Amani Elobeid, Jacinto Fabiosa, Simla Tokgoz, Dermot Hayes, and Tun-Hsiang Yu. 2008. "Use of US croplands for biofuels increases greenhouse gases through emissions from land-use change." *Science* 319 (5867):1238–1240.

Shah, Tushaar, Sonal Bhatt, Ramesh K. Shah, and Jayesh Talati. 2008. "Groundwater governance through electricity supply management: Assessing an innovative intervention in Gujarat, western India." *Agricultural Water Management* 95 (11):1233–1242.

Tory, Sarah. 2018. "A Southwest water dispute reaches the Supreme Court." *High Country News*, January 23, 2018.

Twomlow, Stephen, David Love, and Sue Walker. 2008. "The nexus between integrated natural resources management and integrated water resources management in southern Africa." *Physics and Chemistry of the Earth, Parts A/B/C* 33 (8–13):889–898.

Vora, Nemi, Apurva Shah, Melissa M. Bilec, and Vikas Khanna. 2017. "Food-energy-water nexus: Quantifying embodied energy and GHG emissions from irrigation through virtual water transfers in food trade." *ACS Sustainable Chemistry & Engineering* 5: 2119–2128.

Webber, Michael E. 2015. "A puzzle for the planet." *Scientific American* 312 (2):62–67.

Wichelns, Dennis. 2017. "The water-energy-food nexus: Is the increasing attention warranted, from either a research or policy perspective?" *Environmental Science & Policy* 69:113–123.

Willaarts, Barbara, Simon Langan, Juraj Balkovic, Peter Burek, Edward Byers, Arnulf Deppermann, Stefan Frank, Matthew Gidden, Peter Greve, and Petr Havlik. 2018. *Integrated Solutions for Water, Energy and Land Progress Report 3*. United Nations Industrial Development Organization (UNIDO) and International Institute for Applied Systems Analysis (IIASA): Laxenburg.

Yang, Y. C. Ethan, Sungwook Wi, Patrick A. Ray, Casey M. Brown, and Abedalrazq F. Khalil. 2016. "The future nexus of the Brahmaputra River Basin: Climate, water, energy and food trajectories." *Global Environmental Change* 37:16–30.

10

Geographic Information System (GIS): Land Use Planning

Egide Nizeyimana
and Jacob Opadeyi

Introduction

Agricultural scientists are required to provide information needed to address land degradation and land use conflicts confronting the world today. As the population increases and land becomes a commodity in many parts of the world, careful planning of the use of land must be undertaken to accommodate conflicting people's needs and preserve and/or protect the environment. The decision about the use of land must be made based on analyses of each potential use in terms of its economic and biophysical suitability to the specific tract of land and possible impact to environmental degradation.

Geographic information system (GIS) and related technologies (e.g., remote sensing, global position system) have proven to be a valuable tool in land use planning activities. The GIS approach is important in this area because it provides functions to capture, store, organize, and analyze spatially referenced data. Moreover, GIS has been coupled to a variety of models (e.g., crop productivity, hydrology, and water quality simulations) and may be an important component of spatial decision support systems (SDSS) and land resource information systems (LRIS). GIS enhances model flexibility and efficiency in these systems where it is often regarded as a centralized data analysis, management, and planning system. As a result, decision makers and land use planners in private companies, universities, and at various levels of the government are using GIS to develop spatial environmental databases, perform land evaluations, and analyze and manage resources. There is no doubt that the use of GIS and GIS-based systems in land use planning activities will continue to increase in the future, as more detailed digital environmental datasets become available and the capability of computers to handle large volumes of data increases.

GIS Use in Land Use Planning Activities

The goal of land use planning is to make decisions about the use of land and resources.[1] Its implementation is often driven by future people's needs in terms of productivity and environmental sustainability. Land use planning is important in highly populated communities primarily due to conflicts between competing uses and interests of users.[2] In this case, planning activities are tailored to make the optimal use of the limited land resources. In general, the land use planning involves sequentially an organization of thoughts and establishment of long-term goals, a land evaluation that includes appraising alternatives, and finally designing and implementing the plan. Land evaluation is the most important aspect of this process.

The information within a GIS consists of a spatial component represented by points (e.g., well locations), lines (e.g., streams, road networks), or polygons (e.g., soil delineations) and attribute data or information that describes characteristics of the spatial features. The spatial entity is referenced to a geographic coordinate system and is stored either in a vector or raster format.[3] GIS is primarily used in the development of spatial databases and land evaluation, a procedure in the land use planning process that is aimed at determining the suitability of land units to current and alternative uses and the potential impact of each on the environment.

Development of Spatial Databases

GIS stores, retrieves, and allows efficient manipulation of database information. It provides powerful analysis and relational database facilities to modify and/or integrate spatial data from different sources and resolutions as well as advanced visualization functions to display output data in the form of interpretative maps. In this case, land attributes and qualities are derived from geographic databases and used to determine land suitability, limitations, or ratings for various land use types. The analysis results may be presented in tabular or graphical form and are intended to provide key information necessary for land users or planners to make meaningful decisions about land management and conservation and/ or land use planning.

Site Suitability Analyses

A site suitability analysis typically involves the assessment of the level of affinity a specific land has for a particular use. Soil information available in soil databases is rarely enough for site evaluations. In addition to soils, the analysis often integrates local information on landforms, current land uses, the relative location of the land, and associated social and political restrictions. The proposed use may have additional limitations that should be taken into account. For example, an analysis for suitable sites for land application of sewage sludge should consider the physical, chemical, and biological properties of the waste in soil and water.[4,5]

GIS is efficient in identifying and ranking sites for various land use planning activities. First, site-specific analyses often require many and detailed data sources. Second, GIS overlay features, logical operations, and display functions are tailored to speed up data processing and therefore allow efficiently suitability class assignment and graphical display of results. A good example is the use of GIS for locating appropriate sites for forestland application of sewage waste.[6] The authors derived physical site suitability ratings for an area in Vermont based on EPA guidelines[7] and merged them with social and political restrictions of the state and counties to derive a land applicability classification. Similar GIS-based approaches have been used to locate sites for solid waste disposals.[8]

Linking GIS and Models

The spatial databases and associated attribute data described above may be part of a GIS/model graphical user interface (GUI). These models may be those that determine land suitabilities or those that simulate water quality for environmental impact assessments. As an entity of GIS/model interface, GIS allows easy access to database attributes by various algorithms, statistical software, and environmental models for a variety of land use analyses. In addition to model parameterization, a full integration of GIS and models also allows the user to interact with various modules, select data input and module utilities, and display graphical and/or tabular representations of modeling results.[9] The use of this type of interface reduces significantly the processing time and resources required to develop input data and run the model. GIS/model linkage has been accomplished for field- and watershed-scale hydrology/ water quality models such as the leaching and chemistry estimation (LEACHM)[10] and agricultural non-point source pollution (AGNPS)[11] models. Results of these analyses are used to support management and land use planning activities in the farm or watershed.

The GIS/model interface may also be part of an SDSS. In addition to modeling parameterization, SDSS offer the users with functions to evaluate different land use scenarios necessary for making management recommendations and/or land use planning decisions. Results may be used by farmers to adjust management practices of distinct fields in a given farm or may be used by field officers to help farmers set priorities while providing technical assistance for nutrient management planning. Figure 1 shows a flow diagram indicating data integration in a GIS with models for water quality assessment in watersheds.

GIS has been incorporated in many land evaluation systems so that planners and public officials can take advantage of its spatial modeling and visualization capabilities. An example of such a system is the USDA–NRCS land evaluation and site assessment (LESA).[12] The land evaluation portion of LESA determines soil productivity levels, farm size and agricultural sales volume; the site assessment portion deals with factors such as location, amount of nonagricultural land, zoning restrictions, etc.[13] GIS, in GIS-based LESA systems, is used in both modules and at all levels of the analysis.

Development of Land Resource Information Systems

A number of land information delivery systems have been developed in recent years to make spatial data available to users for application in various aspects of land use planning. These systems, also called LRIS, are multipurpose systems that integrate geographic databases and GIS tools to analyze, record, report, and display spatial data relationships. Some of these systems have been embedded into the World Wide Web (WWW) for quick and easy access and analysis via Internet browsers.

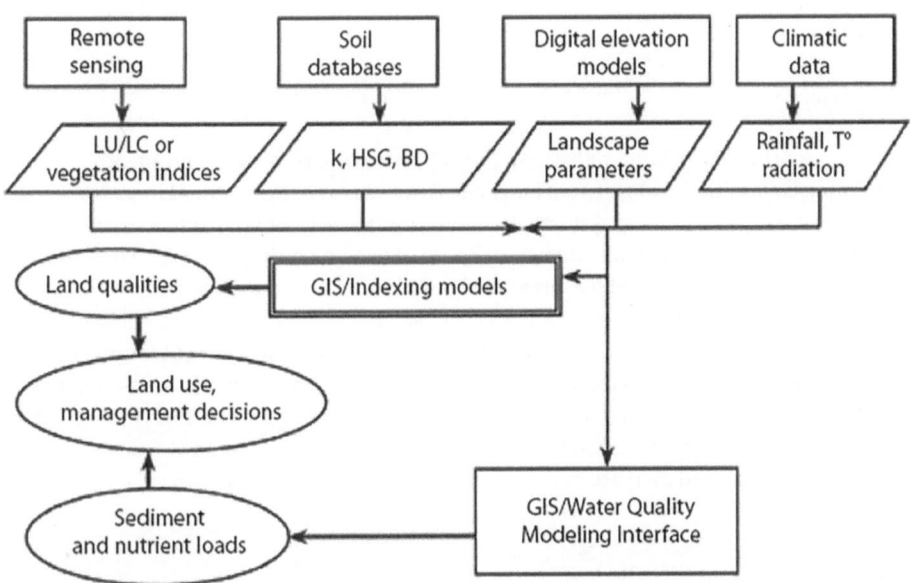

FIGURE 1   Integration of GIS, spatial database, and models.

Reliability of GIS-Based Analysis Results

GIS and GIS-based systems used in spatial data analyses for land use planning have grown in recent years. GIS is particularly attractive in these areas because it allows overlay of spatial data sets and the merging and analysis of attribute data from different sources. The resulting data are obtained using data from digital and paper maps of different scales or acquired at different resolutions such as in the case of digital elevation models (DEM) and remote sensing imagery. The combination of such data layers may produce unrealistic data and consequently lead to erroneous predictions. The question is how reliable are these results when used for developing land use and management plans. A discussion on the propagation of error and uncertainty in GIS-based systems is provided elsewhere.[14]

The accuracy of GIS-based land evaluations is a function of the quality of attribute data and mapping delineations of databases, and the type of model or assessment scheme used in the analyses. Various algorithms for assessing the quality of GIS analysis results as affected by error and uncertainty in GIS layers, and their propagation are part of popular GIS and image analysis software (e.g., Arc/Info, ERDAS Imagine). The results, which are often in the form of reliability diagrams are not used by average GIS practitioners. This is probably due to the fact that they are not easy to understand or interpret. Similarly, models vary depending on how each represents various processes of the system. Lumped-parameter models and indexing/ ranking schemes are mostly used in land evaluations because they are easy to parameterize. However, these models ignore spatial variations of parameters throughout the field, watershed, or region of study. Furthermore, models originally designed for fields and watersheds are often applied to regional analyses, thus adding some level of uncertainty in modeling predictions. For example, most land use planning programs use conventional methods of land evaluation. Each land parameter is given a range of values with corresponding ratings showing its suitability to crop production. These indexes are added or multiplied to create a single index that is to rank land units. The method is simple, but carries a high uncertainty because breaks between two ranges or ranks are subjective.

The effect of map scale and resolution on environmental assessment and modeling output data has been subject to many studies. Raster-based GIS systems require that a grid cell size be defined prior to the analysis. However, as pixel size increases above the resolution of the original data, the spatial variability decreases. This causes a decrease of the predictive power of generated input parameters particularly for small land areas.

Conclusions

GIS has been used primarily in land evaluation, a procedure in the land use planning process that deals with determining land suitability to current and alternative uses and the potential impact of each use on the environment. GIS and GIS-based systems are often considered as integrated spatial information systems (ISIS) for data analysis needed in land use management and planning activities. Despite the advantages of GIS outlined above, the land use planner should be aware of the error and uncertainty in GIS and GIS-based analyses resulting from digitizing and scaling inaccuracies, data conversion between vector and raster formats, and others. Finally, GIS implementation in land use planning activities can be expensive. In addition to costs associated with hardware and software purchase and maintenance, a high level of technical expertise is required to perform complex modeling tasks in land evaluations for alternative uses and to sustain databases. Nonetheless, the demand for GIS and GIS-based analysis systems in land use planning is to increase in the future as more detailed spatial datasets become available.

References

1. FAO. *Guidelines for Land Use Planning*; FAO Development Series 1; Soil Resources, Management and Conservation Service; FAO: Rome, 1993; 96 pp.

2. Brinkman, R. Recent developments in land use planning, with special reference to FAO. In *The Future of the Land: Mobilizing and Integrating Knowledge for Land Use Options*; Fresco, L.O., Stroosnijder, L., Bouma, J., van Keulen, H., Eds.; Wiley: Chichester, 1994; 13–21.

3. Burrough, P.A. Matching databases and quantitative models in land resource assessment. *Soil Use Manage.* **1989**, *5*, 3–8.

4. Nizeyimana, E.; Petersen, G.W.; Looijen, J.C. Land use planning and environmental impact assessment using GIS. In *Environmental Modeling with GIS and Remote Sensing*; Skidmore, A., Prins, H., Eds.; Taylor and Francis Ltd.: London, 2002; in press.

5. Petersen, G.W.; Nizeyimana, E.; Miller, D.A.; Evans, B.M. The use of soil databases in resource assessments and land use planning. In *Handbook of Soil Science*; Summes, M.E., Ed.; CRC Press: Boca Raton, FL, 2000.

6. Hendrix, W.G.; Buckley, J.A. Use of a geographic information system for selection of a sites for land application of sewage waste. *J. Soil Water Conserv.* **1992**, *92*, 271–275.

7. U.S. Environmental Protection Agency. *Process Design Manual for Land Treatment of Municipal Waste*; Environmental Protection Agency: Cincinnati, OH, 1981, PB 299655/1.

8. Weber, R.S.; Jenkins, J.; Leszkiewicz, J.J. Application of geographic information system technology to landfill site selection. *Proceedings of the Application of Geographic Information Systems, and Knowledge-Based Systems for Land Use Management*, November 12–14; VPI and State University: Blacksburg, VA, 1990.

9. Petersen, G.W.; Nizeyimana, E.; Evans, B.M. Application of GIS to land degradation assessments. In Methods for Assessment of Soil Degradation; Lal, R., Blum, W.H., Valentine, C., Stewart, B.A., Eds.; CRC Press: Boca Raton, FL, 1997; 377–391.

10. Inskeep, P.P.; Wraith, J.M.; Wilson, J.P.; Snyder, R.D.; Ma-cur, R.E.; Gaber, H.M. Input parameter and model resolution effects on predictions of solute transport. *J. Environ. Qual.* **1996**, *25*, 453–462.

11. Tim, U.S.; Jolly, R. Evaluating agricultural nonpoint source pollution using integrated geographic information systems and hydrology/water quality models. *J. Environ. Qual.* **1994**, *23*, 25–35.

12. Wright, L.E.; Zitzmann, W.; Young, K.; Googins, R. LESA—Agricultural land evaluation and site assessment. *J. Soil Water Conserv.* **1983**, *38*, 82–86.

13. Daniels, T. Using LESA in a purchase of development rights program. *J. Soil Water Conserv.* **1990**, *45*, 617–621.

14. Heuvelink, G.B.M. *Error Propagation in Environmental Modeling with GIS*; Taylor & Francis: London, 1998; 127 pp.

11

Industrial Networks

Sven E. Jørgensen

The Importance of System Networks

An important property known from ecosystems, operation in networks, can be implemented in industries by building a network of several industries. This would facilitate the possibilities to find a better matter and energy (or rather, work energy or exergy) efficiency, understood as the entire efficiency, equal to the sum of all outputs (products) divided by the sum of all inputs. Inputs and outputs may be either matter or work energy (exergy; see the entries about exergy and eco-exergy). The matter can eventually be multiplied by the price or costs of products and of inputs to express matter efficiency in economic terms.

Industrial symbiosis in the form of a network is similar to the ecological network of an ecosystem, i.e., the network consists of flows of energy, matter, and information.

An Example of an Industrial Network

The use of industrial networks has been tested in the Danish town Kalundborg. The network (for further details, see Jørgensen[1]) is shown in Figure 1.

The heart of the system is the Asnæs Power Station (APS), the largest power plant in Denmark. Half of the plant is fueled by coal and half by a new fuel called orimulsion, a bituminous product made from Venezuelan tar sand. The use of orimulsion was introduced in 1998 and gave an 18% reduction in carbon dioxide emission compared with the use of 100% coal but, due to the sulfur content of Venezuelan tar, gave a higher production of gypsum by the sulfur dioxide scrubber. By exporting half of the former waste energy, APS has reduced the fraction of available energy directly discharged by about 80%. Since 1981, the municipality of Kalundborg has eliminated the use of 3,500 oil-fired residential furnaces by distributing heat from the power plant through a network of well-insulated underground pipes. The homeowners have paid for the pipeline but receive cheap reliable heat in return.

The power plant also supplies cooling water that is warmed 7–8°C to an on-site fish farm producing about 200 tons of trout per year. The APS also delivers process steam to its neighbors, Novo Nordisk (a pharmaceutical and biotechnological industry) and Statoil (an oil refinery). The APS produces 70,000 tons of ash that is sold for road building and cement production.

In 1993, APS installed a sulfur dioxide scrubber that produces calcium sulfate or industrial gypsum. Gypsum is the primary ingredient of wallboard, and ASP is the primary supplier of gypsum to the

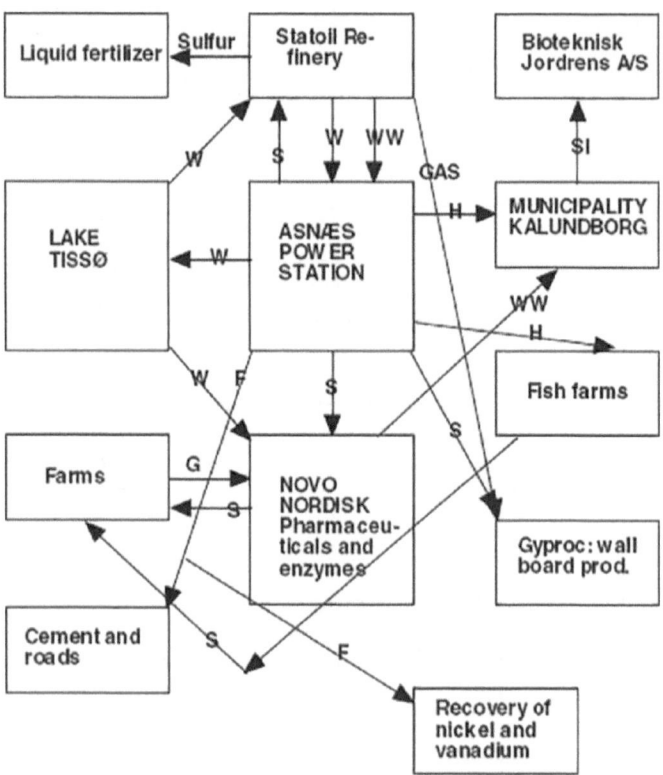

FIGURE 1 The industrial symbiosis at Kalundborg, Denmark. The flows of energy and matter make up an "ecological" network that implies that the overall efficiency of the utilization of the exergy input to the network is high. W, water; F, fly ash; WW, wastewater; S, steam; Sl, sludge; H, heat; G, grains.

wallboard factory Gyproc. The refinery has refinery gas as a waste product that is sold after desulfurization as natural gas to be used in the town or elsewhere. The desulfurization process produces liquid sulfur that is used to produce a liquid fertilizer, ammonium thiosulfate.

Novo Nordisk and its sister firm Novozymes are world leaders in the production of insulin and enzymes. The plant employs 1,000 people, and the products are produced by fermentation processes. Based on agricultural crops, valuable products are produced by microorganisms. The waste is a nutrient-rich sludge that can be used by the nearby farmers as animal feed (the yeast sludge) or fertilizers (the microorganisms sludge). More than 3,000 tons of the two types of sludge are produced daily. The yeast sludge is sold, and the microorganism sludge is given away, due to the firm's concern for disposal security. Distribution of the sludge was the least cost-effective way to comply with environmental regulations.

A new partner, A/S Bioteknisk Jordrens (meaning Biotechnical Soil Purification), joined the industrial symbiosis in 1999. The company uses municipal sludge as a nutrient in a bioremediation process to decompose pollution by toxic substances in contaminated soil.

The Benefits of Industrial Networks

It is clear from a review of the industrial ecological network in Kalundborg that there is an enormous potential to save energy or, rather, work energy or exergy in the use of industrial symbiosis. In this context, the Nordiske Affaldsbørs (Nordic Waste Exchange) could also be mentioned, an Internet page that gives information about wastes produced in the Nordic countries of Finland, Iceland, Norway, Sweden, and Denmark. The idea is that what may be waste for one industry could maybe be used as a raw material for another. For instance, used solvent from the pharmaceutical industry may be applicable for the production of paint or dyestuff. It is very important to see such opportunities if we want to reduce the loss of exergy in the form of fuel or raw-material waste to the environment. The idea behind industrial symbiosis—an ecological-industrial network—has been used in agriculture, but it could be applied much more widely, which would make agriculture much more sustainable.[1]

The industries have already adopted several of the characteristic properties of ecosystems, because it has been profitable for the companies. The discipline behind this application of ecological principles in industries is called industrial ecology.[2] The most significant losses of work energy in the industries are waste and the ineffective use of energy. These losses can be reduced by recycling and reuse, which the industries are using increasingly, but only to the extent that it is profitable. If several companies, industries, and the local municipalities are forming an ecological symbiotic network, it is, however, possible to reduce the overall losses of work energy or exergy and at the same time increase the profit. The possibilities of finding an optimum solution are generally bigger for a large system than for a small system. The experience gained so far from an industrial symbiotic network is so positive that it is surprising that it has not found a wider application.

Changes to the rules, taxes, and legislation for discharge of wastes may also increase the motivation for companies to recycle and reuse.[1] If the industries have to pay for the discharge of wastes or pay more for raw materials, it is a question of economic optimization how much the company should recycle and reuse the waste. The conclusion is clear: A higher direct utilization efficiency of work energy or exergy coupled with increased use of recycling and reuse are the key methods to decrease the eco-exergy losses by industries. Moreover, the use of ecological symbiotic networks increases the possibilities for recycling and reuse.

References

1. Jørgensen, S.E. *Eco-Exergy as Sustainability*. WIT: Southampton, 2006; 220 pp.
2. Ayres, R.U.; Ayres, L.W. *Handbook of Industrial Ecology*. Edward Elgar: Cheltenham, 2002; 680 pp.

12

Land Restoration

Richard W. Bell

Introduction

Land degradation is widespread globally[1–3] and is devaluing the natural capital of land as well as compromising the provision of soil ecosystem services. Arresting the degradation of land should be a high-priority goal. However, restoration of land that has previously been degraded represents an untapped opportunity to increase the natural capital of land and the provision of key soil and plant ecosystem services, particularly those related to plant productivity and water supply. Many forms of disturbance such as overgrazing, excessive tillage, overirrigation lead to land degradation including water erosion, wind erosion, decline of soil structure, acidification, salinization, waterlogging, and decline in soil fertility. Land disturbance after mining is comparatively extreme, but research and adaptive management on these sites have fostered the development of sound principles and practices of land restoration. In the present entry, most examples of land restoration are drawn from mine land restoration to illustrate the principles of the emerging discipline of land restoration. While land restoration can be quite site specific in its practice, there are general principles that can be applied and these have a degree of relevance to most forms of land degradation.

Land as a Finite Resource

For sustainable environmental management, land should be regarded as a finite global resource. The critical surface layer of the finite land resource is the soil profile that varies from a few millimeters' thickness to several meters. While soils are continually forming, the rate is slow, equivalent to 1 mm every 100 years[4] or even less.[5] Land may be created by reclamation of submerged areas along shorelines or in wetlands, and river deltas create new land surfaces as they deposit sediments when they reach the sea or lakes. Locally, these processes that create new land surfaces may contribute significantly to available land resources, but at a global scale, the amounts created are small relative to the total area and to the areas being degraded. Moreover, there is already evidence that in low-lying coastal zones, land is being inundated by rising sea levels or eroded by storm surges associated with rising sea levels. These processes are offsetting gains from natural or anthropogenic land creation. There seems little alternative but to regard land as a finite resource, to slow and prevent degrading processes, and to implement restoration programs on land that has suffered degradation.

Natural Capital and Soil Services

While most emphasis in the past was on food and fiber production from land, the concept of its value is broadening. Land is a form of natural capital that underpins several critical ecosystem services.[6] The soil ecosystem services outlined in Table 1 comprise supporting, regulating, provisioning, and cultural services.[6] The emergence of the concept of natural capital is beginning to redefine the notion of land and its value. Production of food is still a critical function of land, indeed one whose importance is predicted to rise over the next several decades as the challenge of providing sufficient quantity and nutritional quality of food for a rising population becomes more evident.[7,8] In addition, the role of land and soil in regulating major elemental cycles, the hydrological cycle, and the disposal of wastes and dead organic matter is gaining increasing importance.

Land Degradation

Land as a finite resource is susceptible to a range of degrading processes that limit its productivity, its land use potential, and its value in providing ecosystem services. Conacher and Conacher[9] define land degradation broadly as "alteration of all aspects of the natural (biophysical) environment by human actions, to the detriment of vegetation, soils, landforms and water (surface and subsurface, terrestrial and marine) and ecosystems." More broadly, land degradation can be defined as those processes that lower the natural capital of land and compromise the provision of soil ecosystem services.

The processes of land degradation, their measurement, impact, and management have been dealt with else- where.[10–12] Most commonly, the impact of land degradation is described in terms of areas of land degraded.[13] However, this approach has several limitations. Apart from the difficulty of acquiring good quality data that are consistent across large areas to make such estimates, the different impacts of degrading processes makes it impossible to compare processes based only on area affected. For example, the degradation of soil structure by tillage may have an incremental effect on crop yields but cause a quantum increase in water runoff. The degradation of land by salinity, which raises soil salinity levels above a critical threshold for growth of most plants, prevents continuation of current land use options, while opening up others (such as salt-tolerant vegetation). By contrast, soil acidification may have no effect on plants until a critical threshold is crossed, leading to the release of toxic Al^{3+} concentrations that impair plant growth. Hence, the key to predicting the effects of land disturbance is to understand the threshold values of soil properties for plant growth and for

TABLE 1   Societal Soil Ecosystem Services

Supporting	Physical stability and support for plants
	Renewal, retention, and delivery of nutrients for plants
	Habitat and gene pool
Regulating	Regulation of major elemental cycles
	Buffering, filtering, and moderation of the hydrological cycle
	Disposal of wastes and dead organic matter
Provisioning	Building material
Cultural	Heritage sites, archaeological preserver of artifacts
	Spiritual value, religious sites, and burial grounds

Source: Adapted from Robinson et al.[6]

ecosystem services. The differences among degrading processes are also critically important in planning the restoration of degraded land.

Alternatives to assessing and comparing only the land areas suffering degradation are to estimate the value of lost production from the degradation or to estimate the cost of restoration. The latter is generally considered to be critical since restoration costs will be a major impediment to reversing the effects of degradation. The cost per hectare for land restoration on mine sites is much larger than that on production lands not only because of the more extreme degradation but also because of the greater capacity to pay for restoration. Costing the restoration works for different forms of degradation and different areas of degraded land provides a basis for prioritization of restoration works. However, costs of restoration commonly only consider land productivity, rather than biodiversity and the full range of ecosystem services. If the natural capital of land in total was considered, there may be less variation in costs of land restoration among different forms of disturbance.

Land Restoration Definition

The term *land restoration* has a generic meaning, first proposed by Hobbs and Norton,[14] to indicate "that restoration occurs along a continuum and that different activities are simply different forms of restoration." The term *land rehabilitation,* as defined by Aronson et al.,[15] is also appropriate when applied to land degraded by mining. In the United States, the term *reclamation* is commonly used to describe activities that replace ecological functions by planting different vegetation to what previously grew.

Land restoration will usually focus on re-establishing ecological functions such as nutrient cycling, hydrological balance, and ecosystem resilience.[16] In some situations, restoring the original flora is in addition a realistic and appropriate goal. A case study based on the application of this goal is outlined below for restoration of eucalyptus forest after bauxite mining. Further expansion of the concept of land restoration should consider its role in the recovery of natural capital and soil ecosystem services.

Land Restoration Principles

The process of land restoration comprises the following: determination of the end land use, definition of the main limiting factors for restoration and means of alleviating them, and, finally, planning and implementation of a restoration program including monitoring and evaluation against success or completion criteria. This entry focuses on the first two components, while the latter component is described in more detail by Hobbs.[16]

End Land Use

Development of a land restoration strategy is generally considered to need a well-defined end land use. The key constraints that need to be alleviated depend on the land use envisaged. Hence, definition of the end land use is often considered to be a prerequisite to the restoration of degraded land. From the defined end land use, it is possible to identify the stakeholders whose interests need to be considered, the scope of the restoration challenge, and the prime constraints that will have to be alleviated.[16,17] It is also a prerequisite for the establishment of the measurable goals and targets for restoration that are used in setting success or completion criteria. End land use will largely define the complexity and difficulty of the restoration task and the costs associated with achieving a successful outcome. At one end of the spectrum, the complete restoration of the pre-existing ecosystem is a challenging goal. It may require decades of systematic research and continuous improvement of the restoration procedures before that goal is achieved and before the success criteria can be defined and validated. On the other hand, simply achieving a stable land surface or creating a pasture suited to low-intensity grazing would be more easily achieved.

Existing land use on surrounding areas, or prior land use at the site, often determine the end land use. However, degradation such as salinization may alter the substrate for plant growth so radically as to make the pre-existing land use impossible or undesirable. Furthermore, on mine sites, the alteration in landforms may require a change in land use in order to achieve land stability. In the case where land use change is necessary, stakeholders will need to be engaged to arrive at an acceptable alternative end land use.

A great diversity of end land uses has been applied during land restoration. Apart from natural ecosystems, agriculture, forestry, nature conservation, grazing, housing, wetlands, amenity and recreational facilities, and waste disposal are all possible options. Sociopolitical and economic factors will generally determine the selection of end land use particularly if the land is in a protected area, on production land, or in a densely settled zone.

While a well-defined end land use is necessary to set realistic goals and targets for restoration, there are risks associated with setting a highly prescriptive end land use if the restoration technology has not been well developed and based on solid research. The development of best practice for restoration of a particular type of degraded land may require decades of research and adaptive management. Hence, the premature setting of the end land use and targets for achievement may lock in an inferior set of outcomes. Using an adaptive management approach will allow continuous improvements in restoration practice to be made and tested. As improved practices and outcomes become possible, new benchmarks for completion can be set and new possibilities for end land use emerge. The bauxite-mining case study below illustrates this process.

Diagnosis of Limiting Factors

Degradation of land takes many forms and is triggered by many agents.[10] The types of constraints and their severity and the consequences for the restoration plan will clearly vary from site to site. Correct diagnosis of the key constraints and identification of likely feasible solutions are a prerequisite for successful restoration of degraded land. Identification of the key agents causing degradation is also essential. At a mine site, the active cause of degradation is mining and its associated disturbance. In other cases, the degrading agent may be tillage, overgrazing, wind erosion, saline groundwater discharge, etc. Apart from the biophysical constraints related to substrate properties, landform, climate, and hydrology, limiting factors may be associated with socioeconomic factors that prevent the degrading processes from being arrested. In the present entry, the focus is on the biophysical constraints to restoration (Table 2).

Biophysical Factors Limiting Land Restoration

Climate

The growing conditions for plants when restoring degraded land are determined primarily by climate. Climate is the main limitation on potential land uses for each site. Species that are indigenous to the site will usually be well adapted to the rainfall, temperature, and extreme conditions that occasionally occur in a given climatic regime such as drought, heat stress, and frost. However, if the substrate is unsuitable for the indigenous species, or the end land use requires a different selection of species such as agricultural species, then the chosen species need to be well adapted to the site climate.

When considering climate-related constraints, it is not just the average conditions that should be examined, but also the frequency and severity of extreme events, such as drought, heavy rain storms (e.g., cyclones), frost, snow, hail, etc. The coincidence of extreme events with commencement of a land restoration project may prevent seed germination or cause the loss of seed and topsoil due to erosion.

TABLE 2   Biophysical Factors Limiting Land Restoration, Their Consequences for Site Stability and Plant Growth, and Common Treatment Methods

Factor	Constraint	Consequence	Treatment
Climate	Drought	Failure of germination, poor emergence or establishment, plant death	Irrigation, drought-tolerant or drought-avoiding species, adjust time of sowing
	High temperature	Poor germination and emergence, plant death	Mulching, retention of crop/plant residue
	Low temperature/ frost	Delayed emergence, plant death, poor seed set	Frost- or cold-tolerant species, adjust time of sowing
	Extreme rainfall/ wind events	Water or wind erosion episodes, loss of seeds	Contour banks, soil cover by foliage, mulch or stubble, windbreaks
Landform	Slope	Land slippage, soil creep, unsafe conditions for machinery operation	Deep-rooted plants, reshape to lower slope angle, contour banks, engineering design, prevent water run-on from upslope
	Runoff	Water loss, sediment loss, downslope deposition	Contour banks, reshape to lower slope angle, improve infiltration and soil water storage
	Exposure	Drought, high winds, extreme temperatures	Tolerant plant species especially as windbreaks
	Aspect	High winds, extreme temperatures	Tolerant plant species especially as shade plants
Hydrology	Runoff	Reduced soil water storage, waterlogging, flooding downslope	Contour banks, increased drainage intensity
	Limited profile water storage	Drought, runoff, increased groundwater recharge	Tolerant plant species, deep ripping, treatment of subsoil chemical constraints
	Groundwater discharge	Water-filled voids, acid mine drainage, waterlogging, salinity	Containment of water, wetland treatment ponds, drainage
Substrate properties	Acidity	Poor plant growth especially roots, nutrient deficiencies, plant death	Lime, acid-tolerant species, P fertilizer
	Alkalinity	Poor plant growth, nutrient deficiencies, plant death	Gypsum, leaching, alkaline tolerant species, acidifying materials
	Salinity	Poor plant growth, plant death	Leaching, salt-tolerant species, drainage
	Sodicity	Soil dispersion, crusting, poor seedling emergence, runoff, water erosion	Gypsum, leaching, organic matter addition
	Nutrient deficiency	Poor plant growth	Fertilizer (mineral and organic)
	Metal toxicity	Poor plant growth, plant death	Lime, tolerant species, burial or capping of substrate, removal of substrate, phytoremediation
	Low water availability	Poor plant growth, plant death	Irrigation, mulching, organic matter, deep ripping, adjust time of sowing, drought tolerant species
	Waterlogging	Poor plant growth, plant death	Drainage, tolerant plant species
	Poorly structured soils	Crusting, poor water holding capacity, poor root growth	Mulching, organic matter, gypsum
	Mycorrhiza	Poor plant growth, nutrient deficiency	Topsoil management, inoculation of nursery plants
	Rhizobium	Nitrogen deficiency	Topsoil management, inoculation, liming acid soils
	Soil microbes	Slow mineralization of soil organic matter	Topsoil management

Source: Modified from Bell.[18]

Landform

Mining creates voids below ground level and waste material that is stacked above ground level. Hence, the overall slope angle of land on a degraded mine site will be increased relative to the pre-existing landform. Hence, the creation of stable, non-eroding surfaces is generally the first major goal of restoration after mining. The key landform factors that limit the achievement of stable surfaces are slope angle, elevation, aspect, and surface drainage. Maximum slope angles are generally set by regulation and vary among jurisdictions. In West Australia, a maximum of 20° is the guideline for restored mine slopes such as the outer surface of waste rock dumps.[19] The change in landform is less of a consideration with other forms of disturbance apart from mining.

Hydrology

Any significant removal or disturbance of vegetation, change in surface soil properties, increase in average slope angle, or change in drainage density on a degraded site will alter the water balance. Reduced vegetation cover will increase runoff and there may also be increased deep drainage. The increase in slope angle on disturbed sites will generally increase the proportion of rainfall that becomes runoff. Erosion and downstream flooding and/or sedimentation are the likely consequences of altered hydrology unless precautions are taken to avoid these effects. Where sulfidic substrates are excavated by mining and stored in contact with moisture and oxygen, acid mine drainage may be discharged into groundwater or surface water. The discharge of acidic water alters downstream ecology and may damage the infrastructure it contacts, such as concrete structures in bridges.

Substrate Properties

Physical *properties:* Degrading processes such as erosion, overgrazing, or excessive tillage may alter the physical properties of the surface substrate in ways that decrease its suitability for plant establishment and growth, particularly by changing water storage and availability. Mining substrates and mineral processing residues are commonly poorly sorted, which alters physical properties such as available water capacity, porosity, soil strength, crusting, and susceptibility to wind or water erosion.

Water erosion strips away topsoil, decreases soil depth, and exposes subsoil material with different texture, lower organic matter levels, and degraded soil structure. The eroded soil has less favorable physical properties for water infiltration, water holding capacity, seed germination, and root growth. Wind erosion selectively removes clay and humus from the soil surface, increasing the prevalence of coarse materials. Fine sand deposits from wind erosion may bury topsoil and vegetation. Dust from bare surfaces can be a health hazard especially if it contains alkali salts or other toxic elements.

The passage of heavy machines on agricultural, forest, and mine sites causes compaction of the substrate, and this may be a major constraint to plant growth by restricting root depth. In drought-prone environments, the failure of roots to penetrate to depth may cause stunting or death of plants. Deep ripping to break the compacted layer is generally necessary in order to achieve deep root growth in mine pits.

Chemical *properties:* Plant growth may be hampered by low nutrient levels, acidity, alkalinity, salinity, sodicity, low organic matter levels, and excessive levels of toxic elements or compounds in the substrate. Therefore, effective land restoration depends on thorough chemical characterization of the substrate for these likely chemical constraints. Where feasible, any substrate that is likely to hinder plant growth should be isolated or buried under more benign materials so that root contact with it is avoided or minimized. This approach, which is practiced on mine sites, is clearly not possible on agricultural land. Soil and plant analysis are the most common methods for predicting likely nutrient deficiency or toxicity in the substrate.[20] Fertilizer applications can usually be effective in correcting nutrient deficiencies although determination of appropriate rates of application requires expert judgement or decision support systems and depends on the end land use. For acidity, alkalinity, salinity, or ion toxicities in the substrate, plant tolerance is the most cost-effective strategy for achieving successful plant growth. This may mean selecting a different suite of species or ecotypes to those that existed before mining unless the local species are already adapted to those constraints.

Biological *properties:* The disturbance of topsoil during mining, through stripping, transport, replacement, and/or storage, all have negative effects on soil biological activity.[21] Maximum soil microbial activity is retained when topsoil is used immediately after stripping without a period of storage. Storage for extended periods should be avoided, but in those mining operations where long-term storage is unavoidable, the stockpile should be uncompacted, should be <2 m deep, and should support a vegetative cover that maintains soil biological activity. Restoration of microbial biomass in replaced topsoil on revegetated mine sites may take 7–10 years.[21,22] In southwest Australia, the presence or absence of the pathogen *Phytophthora cinnamomi* in topsoil determines how it should be stored and reused.

Plant Establishment

Except in rare circumstances, vegetation cover is a requirement for land restoration. The alleviation of substrate constraints that limit plant establishment or growth is a prerequisite for successful vegetation establishment. Where a choice of substrate materials is available, the most benign of these should be placed at the surface to create a favorable root zone. This will minimize the need for expensive amelioration procedures for any adverse soil conditions. Topsoil, if available, is generally the most favorable substrate.

In land restoration, plant establishment is pursued through direct application of seed, the application of topsoil containing viable seed banks, transplanting of seedlings, or combinations of two or more approaches. Topsoil typically contains a substantial seed bank. Use of topsoil for revegetation can achieve large cost savings by avoiding seed collection and spreading. Topsoil is most commonly used for revegetation when the indigenous species of the area are to be replaced. However, where the topsoil contains seed of exotic species or has a large proportion of weeds in the seed bank, it may be preferable to avoid its use as the seedbed. In addition to the seed bank, topsoil contains soil microorganisms that are required for the restoration of soil functions such as organic matter decomposition and nutrient cycling. The nutrient stores in topsoil may be critically important for the production of adequate biomass in the restored ecosystem. The physical conditions in the topsoil are generally more favorable for seed germination and emergence than from other substrates. However, well-characterized subsoils or regolith materials may be suitable for revegetation when topsoil is absent or where topsoil is unsuitable due to chemical constraints or heavy weed infestation.

In some plant communities, seed is stored in the canopy of key species rather than in the topsoil. Placement of cut branches on the soil surface may be used in these cases to maximize recruitment of indigenous species.

Transplanting nursery-raised seedlings is a reliable approach for ensuring rapid plant establishment and canopy cover on a degraded site. Transplanted seedlings compete effectively against weeds and have greater survival under grazing by herbivores than plants recruited from direct seeding. For species that are difficult to propagate from seed, nursery-raised seedlings using tissue culture or other vegetative means of propagation may be the only reliable method for introducing those species during land restoration. However, raising nursery seedlings is relatively expensive, especially if only few plants of a large number of species need to be treated in this way.

Ecosystem Restoration

According to Hobbs and Norton,[14] the following are the ecosystem characteristics that should be considered when setting goals for land restoration: vegetation composition, structure, pattern, and heterogeneity; species interactions; and ecosystem function, dynamics, and resilience. A set of measures is needed to determine the success of restoration. They need to be not only low-cost and reliable indicators of present condition and function but also predictors of future trajectories for the restored ecosystem. Composition and structure of vegetation are the most commonly used indicators.[23] For pattern, heterogeneity, dynamics, and resilience, the indicators are less advanced in their development, in part because these indicators can only be identified and validated from long-term data. As a nutrient cycling indicator, microbial biomass has been proposed,[22] while Koch and Hobbs[23] concluded that soil organic matter was the best indicator of the restoration of nutrient cycling processes for bauxite mine restoration.

Increasingly, the land restoration activities in the mining industry are being assessed against completion criteria or success criteria.[17] These are legal instruments established following negotiation between regulators, mining companies, government advisers, and the community. Their aim is to provide certainty for all stakeholders about the restoration process to be followed and the expected outcome. They are designed to avoid future liability to government agencies or private landowners once mining has ceased and mine ownership has passed to new owners.

Case Study: Bauxite Mine Restoration

The restoration of land after bauxite mining in southwest Australia is recognized globally for its excellent practice.[24] Current practice is based on more than 40 years of research and development and adaptive management. There are important lessons to be learned about land restoration from this case study. Bauxite mining in the dry sclerophyll eucalyptus forest of southwest Australia is a shallow surface mining operation. The current restoration goal is to restore the forest values.[24] One hundred percent of the species are now routinely returned. Nutrient cycling appears to be on a trajectory towards restoring the nutrient stores and the fluxes of nutrients present in the premining forest.[22,25] Hydrological balance is disturbed for up to 12 years after land clearing for bauxite mining and subsequent revegetation.[26] Thereafter, water levels return to premining levels, but may drop below premining levels due to the increase in leaf area index relative to premining levels.[27] The restored forest is resilient to fire.[25] Completion criteria have been developed for bauxite mine restoration and several areas have been determined to meet the designated targets.[24]

The present restoration practice as described by Koch[28] is the result of four major revisions in the goals over 40 years and several other significant improvements in practice. The first rehabilitation simply planted *Pinus radiata* as a single species plantation. This was followed by a *Eucalyptus saligna* plantation and then by a goal to restore a diverse forest rather than a plantation. At this stage, it was thought that planting of local eucalyptus would not succeed because of the existence of *P. cinnamomi* in the soils and its threat to the survival of a wide range of native species. Further research demonstrated that the reconstructed profile would produce low risk of *P. cinnamomi* infection in susceptible species, and hence, it was decided to change the end land use goal to that of a forest compatible with the jarrah forest, using jarrah and marri as the overstory species.[24] Finally, it was decided to revise the end land use goal to achieve the restoration of the jarrah forest.[17,25] The selection of this goal was based on research breakthroughs that demonstrated that it was possible to stimulate the germination and emergence of recalcitrant species and hence reach close to 100% species return.

The main learning from this case study is that reaching the point when practices enable full ecosystem restoration takes several cycles of research and adaptive management. It will be based on a systematic program of research into biotic and abiotic constraints. A flexible approach from regulators enabled end land use goals to be revised over time as new research demonstrated the potential to achieve more challenging goals and targets. The end land use goals set in 1963 would have only resulted in exotic pine plantation on the mined bauxite pits. The present end land use goal is a fully functioning jarrah forest that can be integrated into existing forest management programs and achieve the multiple land use goals for the forest estate.[24]

A comprehensive report of the background research on bauxite mine restoration is found in *Restoration Ecology Special Issue* of 2007. It is a worthy model for study and emulation in land restoration on mine sites.

Case Study: Restoration of Land Affected By Dryland Salinity

Unlike the first case study of restoration on mined land, where disturbance is localized, dryland salinity has a more widespread impact and requires a landscape-scale response for land restoration. This case study is illustrated using examples of dryland salinity in southwest Australia and Northeast Thailand. By 2003, about 1 million ha in the wheatbelt of southwest Australia was affected by dryland salinity.[29] In Northeast Thailand, up to 30% of land could potentially become salt affected.[30]

Dryland salinity demonstrates how perturbation in water balance can have devastating consequences for the natural capital of landscapes. Whereas the native plants were predominantly deep-rooted perennial species, those agricultural species that replaced them were predominantly shallow rooted and annual. Annual plants use water only during their growing season, and their usage is limited by the fact that roots are generally confined to the surface 50–100 cm. Thus, the additional water under agricultural species is distributed to increased runoff, causing erosion and waterlogging, and to increased recharge to groundwater (Table 3). Williamson et al.[31] similarly concluded that dryland salinity in Northeast Thailand was triggered by deforestation of the uplands to produce crops like kenaf and cassava.

Two attributes of the landscape that gave rise to dryland salinity in southwest Australia were the deeply weathered regolith and the accumulation of salts from rainfall accretion.[33] Salt contents in extreme cases of up to 20,000 tons/ha have been reported,[34] most of it stored below 5 m depth.[32] Prior to clearing the native vegetation, plant roots in the upper 5 m of the regolith were largely separated from the salt bulge below, and the semipermeable aquifer at the base of the regolith was often dry.[32] However, with increased recharge, the aquifers have filled, causing water levels to rise at the rate of 0.2–1 m/yr. After a 20–30 years period of groundwater rise, saline groundwater discharge is observed commencing generally in valley floor land- forms. In Northeast Thailand, the origin of salt discharge is not rainfall accretion but salt mobilized from halite sequences in the Mesozoic sediments that underlie the Korat Plateau.[31]

Reversing Dryland Salinity

Ultimately, restoring the preclearing water balance is the only complete solution to the dryland salinity problem. This requires treatments in recharge zones of landscapes to decrease recharge rates. The species that can mimic recharge rates that existed before clearing will therefore probably need to be deep rooted and perennial. They will also have to be adapted to a variety of soil conditions and climatic regimes across the affected environment. Finally, it is imperative that the species chosen to fulfill the above functions are economically viable within the farm enterprise to accelerate their adoption by land managers.

In order to manage dryland salinity, it is necessary to understand the groundwater systems as well as water balance components. With intense winter rainfall, in landscapes extensively covered by shallow-rooted annual species, recharge can occur virtually anywhere in the landscape that is not

TABLE 3 Changes in Water Balances for Cleared Catchments before and after Clearing

Catchment	Year	Rainfall (mm)	Interception (mm)	Evapotranspiration (mm)	Change in Water Storage (mm)	Change in Groundwater Storage (mm)	Stream Flow (mm)
Wights forested	1975	1027	130	855	−28	−11	81
Wights cleared	1985	1147	0	565	–	−21	115
Lemon forested	1975	739	74	656	4	−1	5
Lemon cleared	1983	821	38	708	–	−19	56

Source: Williamson.[32]

actively discharging.[33] Computer modeling adds weight to the conclusion that the only fully effective revegetation solutions for salinity control in southwest Australia are with deep-rooted perennial vegetation over most of the whole catchment.[33] Even systems like agro-forestry that place a high density of woody shrubs and small trees in rows 30 m apart were insufficient in the modeling scenarios to restore water balance and achieve complete control of salinity. Continued reliance in the farming system on annual shallow-rooted crops such as cereals is problematic because these crops will allow continued recharge. Loss of these species is also problematic because they are the main source of income for farmers.

Until recharge control treatments start to decrease saline groundwater discharge, treatments are also needed in the discharge areas. These may include both engineering treatments to alleviate water-logging[35] as well as vegetation options that cope with saline waterlogged conditions.[36] In Northeast Thailand, there has been considerable investigation of salt-tolerant species that can be grown on various classes of salt-affected soil.[30]

The case study on dryland salinity may therefore serve as a useful model for landscape-scale restoration. As with the case of land restoration after bauxite mining, the present approach has involved background research to understand the underlying physical processes (landscape water balance, water fluxes, hydrogeology) and to develop effective solutions (the effect of land use and vegetation type on water balance). The present set of strategies to control dryland salinity has evolved out of several phases of research and adaptive management leading to current understanding and solutions.

Conclusions

Land is a form of natural capital that is essentially finite and non-renewable. Every effort must be exerted to avoid degradation of land because degradation diminishes its natural capital value and compromises the ecosystem services provided by soil. The restoration of degraded land has the potential to increase its natural capital value and enhance ecosystem services. Since large areas of land globally are degraded, there is substantial scope for increasing ecosystem services by restoring degraded land. Land restoration is a relatively new discipline, which, along with restoration ecology, consists of successful practices for land restoration at a site-specific scale. Examples of best practice can be found in restored mine sites. The challenge remaining is to scale up land restoration from site-specific cases on mine sites to regional or landscape scales for a variety of degrading land disturbances. Most success at both scales has, to date, been concerned with restoring key ecosystems functions such as organic matter accumulation, nutrient cycling, and water balance.

References

1. Lal, R.; Stewart, B.A., Eds. Soil Restoration. In *Advances in Soil Science;* CRC Press: Boca Raton, Florida, 1992; Vol. 17.
2. Kaiser, J. Wounding Earth's fragile skin. Science **2004**, *304*, 1616–1618.
3. Reich, P.; Eswaran, H. Soil and trouble. Science **2004**, *304;* 1614–1615.
4. McKenzie, N.; Jacquier, D; Isbell, R.F.; Brown, K. *Australian Soils and Landscapes. An Illustrated Compendium;* CSIRO Publishing: Collingwood, 2004; 416.
5. Pillans, B. Soil development at a snail's pace: Evidence from a 6 Ma soil chronosequence on basalt in north Queensland. Geoderma **1997**, *80,* 117–128.
6. Robinson, D.A.; Lebron, I.; Vereecken, H. On the definition of the natural capital of soils: A framework for description, evaluation, and monitoring. Soil Sci. Soc. Am. J. **2009**, *73*, 1904–1911.
7. Lal, R. Soil degradation as a reason for inadequate human nutrition. Food Secur. **2009**, *1*, 45–57.
8. *Schaffnit-Chatterjee, C. The Global Food Equation. Food Security in an Environment of Increasing Scarcity; Deutsche Bank Research: Frankfurt, Germany 2009.*

9. Conacher, A.J.; Conacher, J. *Rural Land Degradation in Australia;* Oxford Univ. Press: Melbourne, Australia, 1995.

10. Lal, R.; Blum, W.H.; Valentine, C.; Stewart, B.A., Eds. *Methods for Assessment of Soil Degradation;* CRC Press: Boca Raton, Florida, 1998.

11. Stocking, M. A. Land degradation. In *International Encyclopedia of the Social & Behavioral Sciences;* Smelser, N.J., Baltes, P.B., Eds.; 2001; 8242–8247.

12. Bossio, D.; Geheb, K.; Critchley, W. Managing water by managing land: Addressing land degradation to improve water productivity and rural livelihoods. Agric. Water Manage. **2010**, *97*, 536–542.

13. FAO; UNDP; UNEP. Land degradation in south Asia: Its severity, causes and effects upon the people; World Soil Resources Reports. Food and Agriculture Organization of the United Nations: Rome, 1994.

14. Hobbs, R.J.; Norton, D.A. Towards a conceptual framework for restoration ecology. Restor. Ecol. **1996**, *4*, 93–110.

15. Aronson, J.; LeFoc'h, E.; Floret, C.; Ovalle, C.; Pontanier, R. Restoration and rehabilitation of degraded ecosystems in arid and semiarid regions. II. Case studies in Chile, Tunisia and Cameroon. Restor. Ecol. **1993**, *1*, 168–187.

16. Hobbs, R.J. Restoration ecology. In *Encyclopedia of Soil Science;* Lal, R., Ed.; Marcel Dekker: New York, 2002; 1153–1155.

17. Ward, S. Success criteria. In *Encyclopedia of Soil Science;* Lal, R., Ed.; Marcel Dekker: New York, 2002; 1156–1160.

18. Bell, R.W. Principles of land restoration. In: *Encyclopedia of Soil Science;* Lal, R., Ed.; Marcel Dekker: New York, 2002; 766–769.

19. Department of Minerals and Petroleum. *Draft Guidelines for Preparing Mine Closure Plans;* Department of Minerals and Petroleum: Perth, 2010.

20. Bell, R.W. Diagnosis and prognosis of soil fertility constraints for land restoration (Ch.16). In *Remediation and Management of Degraded Lands;* Wong, M.H., Wong, J.W.C., Baker, A.J.M., Eds.; Lewis Publishers: Boca Raton, Florida, 1999, 163–173.

21. Jasper, D.A.; Sawada, Y.; Gaunt, E.; Ward, S.C. Indicators of reclamation success—Recovery patterns of soil biological activity compared to remote sending of vegetation. In *Land Reclamation: Achieving Sustainable Benefits;* Fox, H.R., Moore, H. M., McIntosh, A.D., Eds.; A.A. Balkema: Rotterdam, 1998; 21–24.

22. Jasper, D.A. Beneficial soil microorganisms of the Jarrah forest and their recovery in bauxite mine restoration in southwestern Australia. Restor. Ecol. **2007**, *15*, S74–S84.

23. Koch, J.M.; Hobbs, R.J. Synthesis: Is Alcoa successfully restoring a Jarrah forest ecosystem after bauxite mining in Western Australia? Restor. Ecol. **2007**, *15*, S137–S144.

24. Gardner, J.H.; Bell, D.T. Bauxite mining restoration by Alcoa World Alumina Australia in Western Australia: Social, political, historical, and environmental contexts. Restor. Ecol. **2007**, *15*, S3–S10.

25. Grant, C.D.; Ward, S.C.; Morley, S.C. Return of ecosystem function to restored bauxite mines in Western Australia. Restor. Ecol. **2007**, *15*, S94–S103.

26. Croton, J.T.; Reed, A.J. Hydrology and bauxite mining on the Darling Plateau. Restor. Ecol. **2007**, *15*, S40–S47.

27. Bari, M.A.; Ruprecht, J.K. *Water yield response to land use change in south-west Western Australia,* Salinity and Land Use Impacts Series Report No. SLUI 31; Department of Environment: Perth, Australia, 2003.

28. Koch, J.M. Alcoa's mining and restoration process in south western Australia. Restor. Ecol. **2007**, *15*, S11–S16.

29. McFarlane, D.; George, R.J.; Cacetta, P.A. The extent and potential area of salt-affected land in Western Australia estimated using remote sensing and digital terrain models. In *1st National Salinity Engineering Conference, 9–12 November 2004, Perth Australia,* Conference Proceedings; Dogramaci, S., Waterhouse, A., Eds.; Institution of Engineers: Australia, 2004; 55–60.

30. Yuvaniyama, A. Managing problem soils in northeast Thailand. In *Natural Resource Management Issues in the Korat Basin of Northeast Thailand: An Overview*; Kam, S.P., Ho- anh, C.T., Trebuil G., Hardy, B., Eds.; IRRI Limited Proceedings No. 7; 2001; 147–156.
31. Williamson, D.R.; Peck, A.J.; Turner, J.V.; Arunin, S. Groundwater hydrology and salinity in a valley in Northeast Thailand. Groundwater Contam. Int. Assoc. Hydrogeologists Publ. **1989**, *185*, 147–154.
32. Williamson, D.R. Land degradation processes and water quality effects: Waterlogging and salinisation (Ch. 17). In *Farming Action Catchment Reaction*; Williams, J., Hook, R.A., Gascoigne, H., Eds.; CSIRO: Melbourne, 1998; 162–190.
33. Clarke, C.J.; George, R.J.; Bell, R.W.; Hatton, T.J. Dryland salinity in southwestern Australia: Its origins, remedies, and future research directions. Aust. J. Soil Res. **2002**, *40*, 93–113.
34. Moore, G.W. Salinity. In *Soil Guide. A Handbook for Understanding and Managing Agricultural Soils*; Moore, G.W., Ed.; Bulletin 4343; Department of Agriculture: Western Australia, 2004; 146–158.
35. Bell, R.W.; Mann, S. Amelioration of salt and waterlogging-affected soils: Implications for deep drainage. In *1st National Salinity Engineering Conference, 9–12 November 2004, Perth Australia*, Conference Proceedings; Dogramaci, S., Waterhouse, A., Eds.; Institution of Engineers: Australia, 2004; 95–100.
36. Barrett-Lennard, E.G. *Saltland Pastures in Austalia. A Practical Guide*, 2nd Ed.; Land, Water and Wool Sustainable Grazing on Saline Lands Sub-program: Canberra, 2003.

13

Limits to Growth

Brian D. Fath

Introduction

In thermodynamic-based systems science terminology, the Earth is considered a closed system. This means that it is open to energy exchanges, but not matter exchanges. Regarding energy, the Earth receives on average 1,300 W/m² of solar radiation at the top of the atmosphere and radiates or reflects a near equal amount back to space (the current increase in greenhouse gases causes a small difference in the radiation balance enough to be responsible for the current observed warming). The amount of matter on the Earth is more or less fixed, with only negligible exchanges from receiving meteorite impacts or outgassing. The Earth's matter is distributed in the following way: The Earth has a polar radius of 6.37×10^3 km, giving it 501.1 million km² of surface area, of which 70.8% is covered with water. The overall volume of the Earth is 1.08×10^{12} km³, with a mass of 5.98×10^{24} kg. Those physical dimensions are hard evidence of the limits of the available space and material resources. The *Blue Marble* (Figure 1) shows this celestial body in all its complexities, opportunities, and beauty; and it also shows starkly the scale and boundaries of the one world that humans and all known species have. This photo, taken during the Apollo 17 Lunar Mission in 1972, coincided with the emergence of the modern environmental movement (more on that below). It was not the first environmental movement, but those ideas from the 1970s still cast a strong shadow on the discussions today that deal closely and explicitly about living in balance on the Earth as represented in among other things, the United Nations Sustainable Development Goals.

The ideas of bio-physical limits impacting human growth and resource consumption are not new. At the turn of the 19th century, Thomas Malthus was already writing about the convergence of population growth and food supply. He foresaw an inevitable turning point in which the finite resources of the planet could not support an exponentially growing population. His dour predictions later became the face of resource-limited, doom-and-gloom environmentalism, referred to as Malthusian. This discouraging perspective is a label that most modern environmentalists try to avoid, while still holding the reality of his concerns. At the time Malthus was writing, the global human population was around 1 billion, and today is approaching 8 billion, moving from abundant nature and scarce humanity to abundant humanity and scarce nature. Yet, there is not an absolute food shortage in terms of total calories produced, only hunger due to regional and distributional dilemmas. In fact, obesity is a more serious and growing problem in many countries than food deficiencies. Nonetheless, the effort that humans have taken to supply this food has meant the conversion of most arable land to agriculture leading to extensive direct impacts

111

FIGURE 1 The *Blue Marble*, view of the Earth from Apollo 17 (1972).

such as habitat loss, soil degradation, and biodiversity loss. It has also led to the doubling of nitrogen flows through industrial fixation (as a result of the Haber–Bosch process) and the tripling of phosphorus flows (benefitting from massive industrial, fossil fuel-based mining efforts). The proximate result of increasing the amount of these typically limiting nutrients expands food supply, but the production of both nitrogen and phosphorus is highly fossil fuel-intensive creating energy supply dependencies. Furthermore, excess nitrogen and phosphorus that runs off agricultural fields creates massive eutrophication problems that are evident in almost every heavily populated estuary and water body in the world. Agricultural production to meet the world's needs is also responsible for rapid groundwater withdrawals, pesticide applications, and abundant greenhouse gas emissions at numerous stages from the fossil fuel use, to soil alterations, to methane from rice and livestock production. So, while the verdict is not in regarding absolute constraint of food production on human population growth, it does appear clear that we are in fact not feeding the people of the planet in a sustainable manner, nor whether it is even possible to do so at this scale. Malthus' concerns should not be dismissed lightly.

Environmental Limits

Recognition of limits can also be found in the seminal writing of George Perkins Marsh. Marsh wrote *Man and Nature* in 1864, which was one of the first scientific treatises on the impacts humans have on nature and the consequential conditions that coincide. Marsh warned that we should take notice of the scale and extent to which humans can continue to modify nature for our own benefit. Presciently, he wrote:

> A certain measure of transformation of terrestrial surface, of suppression of natural, and stimulation of artificially modified productivity becomes necessary. This measure man has unfortunately exceeded.

Marsh (1964)

It is interesting that he recognized that humans will and must modify their environments similar to any species. There is a feedback in which the modifications, the other ecological interactions, and the environmental conditions reach a dynamic balance if the system is to sustain over time. Of course, this does not mean that the systems become rigid without change or variation because innovation, adaptation, evolution, information gain, and learning all allow for continual resilience and flexibility to meet the self-enhanced, recursive dynamic conditions. The path dependency between the interaction of the ecosystem constituents is one of the key features of living systems. Nonetheless, when humans erode too much land, deforest too many hectares, divert too much water, degrade too much habitat, etc., there are limits to which the ecosystem can recover and continue to provide the services we have come to expect. The diminishment first comes in the form of provisioning services when farms and fisheries fail, but then impacts cascade to supporting and regulating services, undermining the capacity of the land to regenerate and function sustainably. Continuation of an approach that overuses and abuses the natural resources will lead to eventual ruin, as Marsh noted in many earlier civilizations, which has been the topic of a recent plethora of research and books describing the collapse of complex societies (e.g., Tainter 1988, Diamond 2005, Kriwaczek 2010, Cline 2014).

The United States of America was settled on the notion of boundless space and opportunity, a land of cornucopia, which influenced not only the profligate physical resource use but also the psychological engagement that one had with resources. However, around the time that Marsh was writing, the transcontinental railroad was being completed. In 1869, the Union Pacific and Central Pacific railroads were joined in Promontory Point, Utah Territory, enabling rapid transportation access across the United States. Travel that previously took 4–6 months could be completed in 6 days. A few decades following both Marsh and transcontinental rail, another measure of limits was recognized in the closing of the Western frontier. The 1890 census showed the first time the disappearance of a contiguous frontier line of a migrating population. The westward wave of expansion was not endless as it had been (naively and myopically) perceived. Historian Frederick Jackson Turner used this moment to refer to the "closing of the American Frontier" (1893). At this bifurcation point, it should have been obvious that the solution to resource shortages, for example, in the form of eroded agricultural land, could not be found by simply relocating to the next plentiful area. While there was still plenty of "under-utilized land" (from a human economic perspective), the new reality meant filling in the middle states rather than experiencing an expansionary boundary. It should have been a time to rethink and reformulate our relationship with land and nature, thus confronting limits and working within constraints. However, the notion of boundless resources remains deeply held by many and institutionalized in many core economic practices (e.g., debt-based money supply, Ponzi-style retirement benefits, high future discount factors). Therefore, recognizing and accepting bio-physical limits was not the first impulse when faced with this new reality.

Ethical Limits

In the 1940s, limits were given another dimension by Aldo Leopold and his *Land Ethic* referring now not only to physical limits but also moral ones. He questioned, why would we pursue certain things (namely economic growth through resource consumption and technologies with unintended consequences) if the result is to destroy and degrade the life and ecosystems around us, loss of species and wild places. He eloquently wrote:

> Our grandfathers were less well-housed, well-fed, well-clothed than we are. The strivings by which they bettered their lot are also those which deprived us of [passenger] pigeons. Perhaps we now grieve because we are not sure, in our hearts, that we have gained by the exchange. The gadgets of industry bring us more comforts than the pigeons did, but do they add as much to the glory of the spring?

Leopold (1949)

He elaborated this new perspective in what he called a Land Ethic:

> A thing is right when it tends to preserve the integrity, stability, and beauty of the biotic community. It is wrong when it tends otherwise.

Leopold (1949)

His Land Ethic puts an environmental ethic clearly on the table as to what is right and wrong, expanding the boundaries of moral concern to the living environment and to the land itself, thus providing new limits of our actions. This ethic is informed by an ecological as well as a systems perspective. Leopold called on humans to constrain certain activities if those actions were detrimental to natural systems. He recognized, for example, the intricate balance cascading between the predator wolves and soils passing through the grass and deer. Too few wolves meant consequences for the soils. There are limits to the control which humans can exert over the ecosystem with a desire for favored species in certain times and places (e.g., the domestic chicken is the most abundant bird in the world, and *Zea mays* (corn) the most abundant plant). In the ensuing decades, we have better knowledge of complex systems, and with this we can have better management, but that does not obliviate the presence of limits imposed from outside and fostered from within. The Land Ethic allows one to see that beyond our own immediate sphere we interact with other bounded spheres, thus implying that the limit we sense is in fact an indication of another system beyond that we are pushing up against. In a zero-sum interaction, expanding our boundary, for example, into nature, diminishes nature. However, there are many non-zero-sum interactions and relations prevalent in ecosystems and complex adaptive systems. The challenge is not to completely disengage when we encounter another system, but to acknowledge it, respect it, work to understand it, and then try to find a way to engage it for the benefit of both parties (see, e.g., Fiscus and Fath 2019).

Socio-Economic Limits

These ideas of limits coalesced in a seminal work by Meadows et al. sponsored by the Club of Rome in the *Limits to Growth* report released in 1972. Their conclusion was based on the results of one of the first global systems dynamics models developed that included state variables representing human population, agricultural productivity, industrial production, resources, and pollution. The dire results were that under no scenario would unlimited growth be able to continue, and the only steady state was found by strict conditions of stable population, 100% use of renewable resources, and investment in business that equaled depreciation. The report was an important contribution to the debate about limits from a bio-physical perspective but was largely ignored by economists and politicians since the approach did not fit their existing growth-oriented mental model. The study was dismissed out-of-hand as having no relevance since their models—absent of any environmental resources or feedbacks—could grow forever. The Meadows et al. model was viewed as Malthusian and overly doom-and-gloom, a stigma that dogged the environmental community for decades. Nonetheless, the Club of Rome work continued as decadal update reports were published showing the projections were in line with reality.

Coincident with the release of the original *Limits to Growth*, the modern environmental movement was gaining traction (as stated above, spurred on by images such as the *Blue Marble* and the Spaceship Earth concept). In the United States, it was a period of aggressive federal policy protections for the environment (Table 1), most passing with very wide margins in both the House and Senate and signed by both Republican (Nixon and Ford) and Democratic (Carter) Presidents. Shy of 50 years later, the pendulum has swung far the other way. Concerted efforts within the United States Congress to dismantle or weaken these Acts, in particular the Endangered Species Act (ESA), is rampant; including news just today (August 13, 2019) of efforts to sign a Presidential Executive order to weaken ESA protections. It would be humorous if it were not sad that one prime argument against the ESA is the stated negative impact that it may have on the economy. However, measuring the success of this

TABLE 1   Major United States Legislation Passed in the 1970s during the Resurgent Environmental Movement

Coastal Management Zone (1972)
Clean Air Act (1972)
Clean Air Act Amendments (1977)
Comprehensive Environmental Response, Compensation and Liability Act (1980)
Endangered Species Act (1973)
Federal Land Policy and Management Act of 1976
Fishery Conservation and Management Act of 1976
Marine Mammal Protection Act (1972)
National Environmental Policy Act (1970)
National Forest Management Act (NFMA) of 1976
Forest and Rangeland Renewable Resources Planning Act of 1974
Noise Pollution and Abatement Act of 1972
Marine Protection, Research and Sanctuaries Act of 1972
Resource Conservation and Recovery Act (RCRA), enacted in 1976
Safe Drinking Water Act (SDWA) (1974)
Surface Mining Control and Reclamation Act of 1977 (SMCRA)
Toxic Substances Control Act of 1976

legislation by its economic contribution is fundamentally misunderstanding its purpose. Framing the ESA as a failure due to its hindrance of capital gain is historical revisionism. The language of the ESA was clear that protections were being put in place precisely because of a lack of concern for the environment and an alarming over-prioritization of economic growth at the expense of ecological limits. The first sentence of the ESA reads:

(a) Findings—The Congress finds and declares that—(1) various species of fish, wildlife, and plants in the United States **have been rendered extinct as a consequence of economic growth and development untempered by adequate concern and conservation** (emphasis added).

Endangered Species Act (1973)

The whole point of the legislation was to put a brake on growth and remember the limits of nature. It was also encouraging that the drafters of the legislation appreciated and recognized the interconnectedness of nature and that species are part of larger ecosystems.

(b) Purposes—The purposes of this Act are to provide a means whereby **the ecosystems upon which endangered species and threatened species depend may be conserved** … (emphasis added).

Endangered Species Act (1973)

The tension between pressures for economic growth and environmental constraints continued and elevated in the ensuing decades.

Flourishing within Limits

The idea of limits has been renewed but with a more positive outlook. In particular is the perspective that the presence of a limit does not need to invoke immediately negative connotations. For example, Jane Jacobs (2000) has focused on the opportunities that constraints and limits bring about. She insightfully remarked that

Natural principles of chemistry, mechanics and biology are not merely limits. They're invitations to work along with them.

In other words, our understanding of thermodynamics, the periodic table, and biological principles should help to design smarter and better performing systems than without this knowledge. Ecosystems and socio-economic systems are self-organizing and thrive even in the presence of constraints. Think back to the energy flows mentioned in the opening, around 1,300 W/m², yet that is enough energy to drive atmospheric circulation, global ocean currents, and complex ecosystems in almost every niche on the planet. Nature has evolved to utilize this energy in a very efficient and also robust manner, resulting in highly complex, diverse, well-functioning systems that arose and are maintained all within the real-time bio-physical constraints.

The message to learn from ecosystems is elaborated in Jørgensen et al. (2015), *Flourishing within Limits: Following Nature's Way.* The authors, all systems ecologists, have drawn upon ecological theory of growth and development to identify a number of attributes that are evident in ecosystems. Using these properties and approaches—within the known bio-physical constraints—ecosystems are able to thrive and flourish, as the most diverse, complex, integrated, and sustainable systems on the Earth.

In a nutshell, the idea is that ecosystems are constrained by both the available resources as inputs and outputs as they reside as gradient-enhancing conduits between these two flows (see Figure 2, Fath 2017). In addition to solar energy, the rate of material availability is controlled by the bio-geochemical cycles at local and global scales. The rates of water, carbon, nitrogen, phosphorus, etc. all have local and global

TABLE 2   Fourteen Properties Observed in Ecosystems

1. Ecosystems conserve matter and energy
2. There are no trashcans in nature
3. All processes (in nature and society) are irreversible
4. All life uses largely the same biochemical processes
5. Ecosystems are open systems and require an input of work energy to maintain their function
6. An ecosystem uses surplus energy to move further away from thermodynamic equilibrium
7. Ecosystems use three growth and development forms: (1) biomass, (2) network, (3) information
8. Ecosystems select the pathways that move it most away from thermodynamic equilibrium
9. Ecosystems are organized hierarchically
10. Ecosystems have a high diversity in all levels of the hierarchy
11. Ecosystems resist (destructive) changes
12. Ecosystems work together in networks that improve the resource use efficiency
13. Ecosystems contain an enormous amount of information
14. Ecosystems have emergent system properties

Source: After Jørgensen et al. (2015).

Ecosystem Input Constraints
- Solar radiation
- Global carbon cycle
- Rate of nutrient cycling
- Rate of hydrological cycle

Ecosystem Output Constraints
- Rate of decomposition
- Rate of accumulation of unwanted byproducts
- Finding others to take your waste

FIGURE 2   Some basic constraints that ecosystems have evolved to thrive within.

constituents that influence what is available and these constraints then lead to the type of ecosystem observed whether it is a tropical rain forest, deciduous forest, steppe grassland, or Arctic tundra, for example. On the output side, the main constraint is that the products and flows out cannot exceed the capacity of the environment to assimilate them. Often this occurs by linking processes such that the output of one process is the input to another in an integrated, coupled fashion. The most obvious example is photosynthesis and respiration in that the former takes in CO_2 and gives off O_2, while the latter does the opposite. Other wastes, such as organic ones, need to be decomposed and returned back to elemental constituents. Therefore, the rate of decomposition is a main factor for the uptake of outflows, which are biologically mediated and influenced by temperature and water (warm, wet environments have faster rates of decomposition). The biomass foundation for fossil fuels occurred during a time period when for various reasons, the material was unable to be decomposed and therefore was buried under conditions that promoted the conversion to fossil fuels. Today's society is generating plastic pollution faster than it can be decomposed, and will likely also leave a recognizable and discernable (and perhaps useful) layer for future generations, and also likely that some organisms will eventually evolve better processes to decompose plastics thus reaching a new balance between production and consumption.

There are of course key differences between nature and society, namely a temporal aspect that ecosystems happen in the present, in real time—although the consequences, building biomass, building soils, establishing drainage patterns, and biogeochemical cycles, etc. have lasting effects. Socio-economic systems operate with a more future-oriented perspective, in that while the actions are short term, there is thought for a future pay-off. For example, a lion takes only what is needed to satisfy the immediate hunger. There is no utility in killing more prey at that moment as they would be absconded by competitors or would spoil; there is no profit motive or line of credit. A hunter today has in mind to take as many prey (or fish or trees or minerals, etc.) as possible since the surplus can be exchanged for other goods or services or turned into a storable currency commodity. It is not clear how to reconcile this basic difference. Humans cannot and should not give up the intellectual foresight that has evolved, but rather it could be used to see the systemic consequences of actions that exceed the limits, that exceed the regenerative rates provided by nature's processes. The remaining question is both philosophical and one of management, which is: How to align our activities within these available flows and constraints?

Conclusion

If humanity desires to stay within limits, then a first consideration is to know where those limits are. How much resources can humans use from the environment and still be within the limits? A standard ecological concept is one of carry capacity, the maximum number of individuals that can be maintained in an area without degrading that area. Carry capacity is a dynamic concept—it can move higher as we innovate new efficient methods and technologies as observed in agriculture since the time of Malthus; yet, it can also move lower as we despoil forests and fields with logging, erosion, and desertification as observed by Marsh and other witnesses to the collapse of civilizations. Metrics such as Ecological Footprint and Ecological Biocapacity that try to make estimates of both the consumption (footprint) and the resource (Biocapacity) have shown that we have outstripped our resource base. Ecological Footprint exceeds Biocapacity, and at a global level, attention is drawn to a notion called overshoot day, which for 2019 occurred on July 29. Meaning on that date, humanity had already used all the resources "sustainably available" to it for that year. After this date, the remainder of the year represents overshoot or dipping into future reserves and leaving future generations with reduced resources. Given this temporal aspect, we see that limits are not simply bio-physical but also ethical as Leopold instructed. Yet, a rigorous debate continues to this day as to whether there are limits of any kind or not. This creates tension between perspectives on policies and approaches, as some see limits as absent or challenges to overcome, while others want to recast the human scale to accommodate them. The planetary boundaries highlighted through the text give a clear indication of one scale of limits. If perhaps we break beyond the Blue Marble and our progeny lives elsewhere than within the Earth's gravitational pull, then this will shift and delay the ultimate confrontation with those concrete limits. Can this line always be receding into the future? I prefer to take care of the home that we do have on which we have co-evolved rather than rely on one that is only in our imagination.

References

Cline, E.H. 2014. *1177 B.C.: The Year Civilization Collapsed*. Princeton, NJ: Princeton University Press, 237 pp.

Diamond, J.M. 2005. *Collapse: How Societies Choose to Fail or Succeed*. London: Penguin Books.

Endangered Species Act, 16 U.S.C. §§1531-1544 (1973).

Fath, B.D. 2017. Systems ecology, energy networks, and a path to sustainability. Prigogine lecture. *International Journal of Ecodynamics* 12(1), 1–15.

Fiscus, D.A., Fath, B.D. 2018. *Foundations for Sustainability: A Coherent Framework of Life–Environment Relations*. London: Academic Press.

Jacobs, J., 2000. *The Nature of Economies*. New York: Vintage Books, 208 pp.

Jørgensen, S.E. Fath, B.D., Nielsen, S.N., Pulselli, F., Fiscus, D., Bastianoni, S. 2015. *Flourishing within Limits to Growth: Following Nature's Way*. New York: Earthscan.

Kriwaczek, P. 2010. *Babylon: Mesopotamia and the Birth of Civilization*. London: Atlantic Books, 310 pp.

Leopold, A. 1949. *A Sand County Almanac, and Sketches Here and There*. New York: Oxford University Press.

Marsh, G.P. 1864. *Man and Nature; or, Physical Geography as Modified by Human Action*. Cambridge, MA: Belknap Press.

Meadows, D.H., Meadows, D.L., Randers, J., Behrens III, W.W. 1972. The Limits to Growth; A Report for the Club of Rome's Project on the Predicament of Mankind. New York: Universe Books. 205 pp.

Tainter, J.A. 1988. *The Collapse of Complex Societies*. Cambridge: Cambridge University Press.

Turner, F.J. 1893. The Significance of the Frontier in American History. Annual Report of the American Historical Association, *pp. 197–227.*

14

Nuclear Energy: Economics

James G. Hewlett

Introduction

In an entry published in the *Encyclopedia of Energy Engineering and Technology,* I attempted to answer the following question: Is nuclear power economic? Unfortunately, I was unable to give a definitive answer to this question. Most of the work on that entry was done in the 2004–2005 time frame, and at that time, there was very little information on the realized costs of building nuclear power plants in the United States. There was some information on the realized cost of building nuclear power plants abroad (mainly in the Far East). However, that information was at best "sketchy" and difficult to use to estimate costs in the United States. Additionally, in the United States, a number of cost estimates of building hypothetical ("paper") plants at hypothetical ("Middletown USA") sites were published. History has shown that such estimates are always too low.[1] In short, at that time, there was a large amount of uncertainty about nuclear power plant construction costs. Thus, the best that could be done was to conclude that nuclear capital costs would have to be less than $1500 per kW before nuclear power would be economic. (This was substantially less than the cost of building nuclear power plants in the Far East.)

Over the last few years, the economic and political environment has changed, and thus, the question of whether nuclear power is economic needs to be reevaluated. First, a number of estimates of building actual nuclear power plants at actual sites have recently become available. This, by itself, would reduce some of the uncertainty about nuclear power plant construction costs. However, these estimates are two to three times higher than the ones made in the mid-2000s. Additionally, power plant construction costs in general have increased substantially. For example, the realized costs of building coal-fired power plants and wind farms have increased by about 80%–100%.[2] Such increases in construction costs would have major effects on the economics of nuclear power.

Second, unlike when the original entry was written, for a variety of reasons, utilities are now far more interested in nuclear power. In 2006, licensing activity at the U.S. Nuclear Regulatory Commission (NRC) was limited to four early site permit approvals. By issuing an early site permit (ESP), the U.S. NRC approves one or more sites for a nuclear power plant, independent of an application for a combined license to build and operate a power plant. This ESP can be valid for up to 40 years. Since then, 16 utilities have filed applications with the NRC to build and operate a total of 28 nuclear units. Given this increased licensing activity, a reexamination of the underlying economics of nuclear power is certainly in order.

Third, in the 2004–2005 time frame, most forecasts of coal and natural gas prices in 2015–2020 time frame were about $1.45 and $6 ($2009) per mmBtu, respectively.[3] Since then, fossil fuel prices increased substantially and then fell. As of 2010, most forecasts of fossil fuel prices are now greater than the ones made 5–6 years ago. Increased fossil fuel prices will clearly affect the economics of nuclear power.

The purpose of this entry is to reexamine the economies of nuclear power in light of the changed economic environment. Before proceeding with the analysis, two comments about the scope of the analysis will be made. First, in this entry, it is assumed that the investment decision—e.g., the decision to build a nuclear unit or coal-fired power plant—will be made by the owner of a traditional utility. This utility owns and operates other power plants, and the operation of all of them is interrelated. That is, if the utility chooses to operate power plant X less, it would have to operate power plant Y more to meet demand. Otherwise, "the lights would go out." In such cases, the decision maker's objective is to build the unit that minimizes the total cost of building the plant in question and operating all power plants. Suppose, for example, that the decision maker has the choice of building a nuclear or coal-fired power plant. The decision maker will calculate the total system costs if the nuclear plant is built and compare that estimate with the total system costs if the coal plant is built. The decision maker would choose to build the plant that yields the lower total system costs.

The available software to estimate total system costs is complex and expensive and requires many assumptions, and thus, using this approach is beyond the scope of this entry. However, if the alternative to the nuclear power plant is another baseload plant type operating in the same portion of the merit order (baseload demand), total system costs would be minimized by choosing the plant type that has the lower "stand-alone" or levelized cost. (Levelized costs will be defined below. Additionally, because electricity is costly to store, demand will vary over the day, month, and/or year. The portion of total demand that does not vary is called baseload, and the units that are used to meet this demand will run at close to full capacity over the entire year.) This is because the operation of both units under consideration will have the same effect on the operation of the other units. For example, suppose that the levelized cost of building and operating a nuclear power plant for 40 years is 6 cents per kWh, and the levelized cost of building and operating a combined-cycle natural gas–fired power plant is 8 cents per kWh. If both units are assumed to operate in the baseload mode, then the operation of both units will have the same effect on the operation of the other units. In such a case, total system costs would be minimized by building the nuclear plant.

Thus, in this entry, the alternatives to the nuclear power plant are two other baseload plant types—namely, coalfired and combined-cycle natural gas–fired units. By limiting the comparisons to other baseload plant types, the analysis becomes much more tractable and transparent. Unfortunately, by just computing levelized stand-alone costs, many of the renewable technologies must be excluded from the analysis. The stand-alone cost of building and operating a wind farm, for example, can be computed. However, total system costs may not be minimized by building the wind farm even though that plant type has lower stand-alone costs. This is because the effects of the operation of the wind farm and the nuclear unit on the operation of the other units will not be the same.

Second, since fossil fuel prices will probably increase over time, there is a time dimension to the question of whether nuclear power is economic. Because of the recession and utility conservation programs, additional baseload capacity will probably not be needed until around 2020. Thus, in this analysis, the first year of a unit's operation is assumed to be 2020. By focusing on the mid-term, the carbon capture and storage (CCS) technologies will not be considered. Recently, a CCS task force was formed with a goal of bringing 5 to 10 commercial-size CCS units online by 2016.[4] Even if this goal, which is very ambitious, is met, to demonstrate that the technology works, the units would have to operate for probably 5–10 years. In all probably, it would take 15–20 years before the CCS technologies would be commercially available on a widespread basis.

Capital Costs for Nuclear and Coal-Fired Powerplants

One of the major uncertainties in any analysis of the economics of nuclear power deals with the construction cost estimates. This section begins with a discussion about nuclear and coal-fired powerplants' overnight capital cost estimates. Overnight cost is defined as the cost of building a power plant instantaneously at some point in time. It is also a direct measure of the value of the land, labor, and materials needed to build a nuclear power plant. Thus, differences in overnight costs reflect differences in the values of the land, labor, and material needed to build the same unit.

It is obviously impossible to build a plant overnight, so the second part of this section describes how the total project costs are derived from the overnight costs. To do this, a number of important assumptions are needed, and in many studies, they are not articulated. This section will also show why comparisons of total project costs must be made with great care. The fuel costs for coal and natural gas–fired power plants are discussed in Section 4. The other assumptions will be briefly discussed in Section 6. These include nonfuel operating costs and nuclear fuel costs.

Nuclear and Coal-Fired Power Plant Overnight Costs

Prior to about 2007, most analyses/discussions about nuclear power plant overnight capital costs tended to focus on either realized costs of units built in the Far East (mainly Japan) or on the estimated costs of building generic units at generic sites in the United States.[5] Each of these sources had their own set of problems. There are always problems with transferring the experience of reactors built in foreign countries to the United States. Also, publicly available foreign overnight capital cost data are not well documented, so all the costs may not be included in the reported figures. Additionally, research has indicated that cost estimates of generic units built on generic sites were always too low, so there were problems with using the resulting estimates. Nonetheless, the analyses/discussions that based their cost estimates on foreign reactors tended to use overnight nuclear capital costs of about $2700 per kW (2009 dollars).[6] The ones that used the cost estimates of generic units tended to use overnight nuclear construction costs of about $1500–$2000 per kW (2009 dollars).

Over the last few years, as part of the process of getting approval from the state public utility commissions (PUCs) to proceed with their NRC licensing activities, utilities filed cost estimates of building actual powerplants at actual sites. In some cases, it was possible to determine what cost items were and were not included in the estimates. More important, if the overnight construction costs were not directly reported on the filings, they could be directly estimated. On average, these overnight nuclear construction cost estimates, shown in Table 1, were about $4000 per kW.

These estimates are clearly much better than the ones based on generic units built at generic sites. Because problems with the transfer of foreign cost information to the United States are avoided, they are much better than the estimates based on realized overnight costs of reactors built in Japan. However, they are also much higher than the ones based on generic designs at generic sites and were also much greater than the realized overnight costs of units built in Japan.

There are probably at least four reasons why the recent U.S. overnight cost estimates are greater than the realized costs of reactors built the Far East (mainly Japan). First, there are clearly cultural factors at play. Second, over the last 15 years, there has been a slow but relatively constant expansion of nuclear power in Japan, so Japanese builders are further down their learning curves than their counterparts in the United States. (The South Texas project is being built by Toshiba, a large Japanese firm that has built

TABLE 1   Estimated Overnight Costs and Lead Times of Selected Proposed Nuclear Plants

Owner	Plant Type	Plant	Capacity (mWe)	Costs (2009 Dollars per kW of Capacity)	Lead Times (Years)
Tennessee Valley Authority	ABWR	Bellefonte	1371	$3164	NA
Florida Power and Light	ESBWR	Turkey Point	3040	$3811	5–6
Progress Energy (Florida)	AP1000	Levy County	2212	$4541	5
South Carolina Electric and Gas	AP1000	Summer	2234	$4089	5 (unit 2) and 8 (unit 3)
Southern	AP1000	Vogtle	2200	$4535	6
NRG	ABWR	South Texas	2700	$3758	NA
Average				$3983	

Source: Adapted from Du and Parsons.[7]

Note: A correction was made to the Vogtle Estimate reported in Du and Parsons.[7]

ABWR, advanced boiling water reactor; ESBWR, economically simplified boiling water reactor; AP1000, advanced pressurized water reactor.

some nuclear power plants in Japan to another. The cost estimates for that project are not that much different from the others. Thus, the learning may not be transferred from one country to another). Third, it is not clear whether all the costs are being reported in the Japanese figures. This is especially true for the so-called owners' costs. Costs which are the ones incurred by the utility over and above the ones paid to the firms building the unit. In fact, it was never clear if the reported Japanese costs even included any owners' costs. Lastly, many of the major components for the proposed U.S. reactors, such as the reactor vessel, are being manufactured in Asia, so some exchange rate issues could exist.

There are also a number of reasons why the overnight nuclear construction cost estimates obtained from the PUC filings are higher than the ones for generic units. As was just noted, research has shown that generic cost estimates at generic sites are always too low. Additionally, a number of analysts have argued that the growth in overnight costs was due to increases in commodity prices—the prices of iron, steel, cement, and so on—that occurred from around 2005 to 2008.[8] However, about 40%–50% of the overnight cost of building a nuclear power plant is labor related, and therefore, the effects of increases in commodity prices on total construction costs are probably modest. Moreover, the cost estimates based on generic units might have assumed that all of the components were manufactured in the United States. Over the last 10 years, the value of the dollar relative to the yen has fallen. Since many of the major nuclear components for proposed U.S. plants will be imported from Japan, some of the increases in overnight costs could be due to the fall of the dollar. This, however, would depend upon how the contracts with the Japanese firms are structured, and there is very little public information about this. Lastly, with the rapid expansion of nuclear power in China, some bottlenecks in the production of the major components that would increase costs appeared.

The increases in nuclear power plant overnight construction cost estimates raise the question of whether the current cost levels will be permanent. The effects of commodity price increases and/or decreases in the value of the dollar on the cost estimates are very unclear. Thus, the effects of decreases in commodity prices (or increases in the value of the dollar relative to the yen) on costs would also be unclear. Additionally, even with the rapid expansion of nuclear power in the Far East, in the long run, markets would adjust, and the bottlenecks would no longer exist. Because of the complexity of the technology, it would take a number of years for existing firms to build new production facilities and for new firms to enter the market. Thus, the bottlenecks may not be removed for a number of years.

Lastly, currently, only a few firms produce the large nuclear components and build the power plants. In such cases, the behavior of one firm (e.g., Westinghouse) could possibly affect the behavior of its competitor (e.g., General Electric). In economist's jargon, such a market structure is called an oligopoly.

TABLE 2 Estimated Overnight Construction Costs of Selected Proposed Coal-Fired Power Plants

Owner	Plant	Capacity (mWe)	Costs (2009 Dollars per kW of Capacity)
Florida Power and Light	Glades	1960	$2130
Duke Power	Cliffside	800	$2124
AMP Ohio	Megis Co.	960	$3277
AEP Swepco	John W Turk Jr	600	$2508
Average			$2510

Source: Adapted from Du ans Parsons.[7]

One major characteristic of an oligopolistic market is price "stickiness"—i.e., when the firm's costs fall, prices will fall, but with a lag. Thus, even if the costs of the firms that produce the components and build the plants fall, it would take some time before reductions in costs would be reflected in reductions in the prices charged to the utilities.[9,10]

The availability of some actual overnight cost estimates from the PUC filing reduces some of the uncertainty in nuclear power plant construction costs. Unfortunately, the large increases in the estimates introduce another source of uncertainty—namely, whether the current cost levels will be permanent. Given current publicly available information, one can only speculate why overnight capital costs increased and whether they will fall in the future. Moreover, no nuclear power plants in the United States have been built on time and on budget, and cost overruns in the two units under construction in Europe have occurred. Thus, even with a number of detailed cost estimates of building actual units at actual sites, there is still a considerable amount of uncertainty about nuclear overnight construction costs. Indeed, until a few units are actually built, nuclear construction costs are essentially unknown.

Table 1 also shows estimated date of commercial operation leadtimes for a number of proposed nuclear units. As can be seen from Table 1, with the exception of one unit, the estimated leadtimes ranged from 6 to 7 years. It should be noted that these lead time estimates were made just before or at the beginning of the recession. Recently, the date of commercial operation for a number of these units was moved back simply because the capacity was not needed. In all probability, the utilities will also move back the construction start date and keep leadtimes the same. If, however, they begin construction and then revise their estimated date of commercial operation, lead times could actually increase. In the 1970s and 1980s, many utilities building nuclear power plants also increased leadtimes because of the lack of need for capacity. I have shown elsewhere that these actions also affected overnight costs.[11] Thus, this issue is important.

Lastly, Table 2 shows that the average estimated overnight cost of building a number of recent coal-fired power plants was about $2500 per kW. A 2007 Massachusetts Institute of Technology (MIT) study estimated that the costs of building a wide range of coal-fired powerplants were about $1280 to $1360 per kW (2005 dollars).[12] Thus, the cost of building coal-fired power plants also increased substantially, and therefore, the cost growth was not limited to nuclear power.

Derivation of Total Project Costs

The focus of the preceding section was on overnight costs, because they are a direct measure of the values of the land, labor, and materials needed to build a power plant. However, the analysis will use total project (capital) costs. The best way to explain how total capital costs are computed is to outline the steps taken to derive an estimate of the total cost of building any power plant. The first step is to prepare a detailed "bottom-up" estimate using current commodity prices, labor wage rates, and so on. This bottom-up estimate is the product of the estimated quantities of the land, labor, and materials needed to build the unit times the prices of these inputs. Suppose, for example, that the firm is making an estimate in 2010 of the cost of building a nuclear unit that will become operational in 2020. Given a 6-yr lead time, construction must begin in 2015. Thus, the firm would first make a bottom-up estimate using 2010 prices and wage rates. This is the estimate of the overnight costs using 2010 prices. In the example shown in Table 3, the 2010 estimated overnight cost is $4000 per kW of capacity.

Next, the cost of building the unit overnight in 2015—the year that construction of the unit begins—must be estimated. To do this, the firm would make assumptions about how commodity prices, labor wage rates, and so on would change from 2010 to 2015. In the example shown in Table 3, commodity prices, labor wage rates, etc., are assumed to increase at a rate equal to the general inflation rate of 3% per year. Given this assumption, the 2015 overnight cost would be about $4600 per kW ($4000×1.03⁵). Then, assumptions about how the funds are expended over each year of the construction period and how prices change over the construction period would be used to compute total costs in the dollars of the year the funds are expended. In the example shown in column 3 in Table 3, total construction costs excluding financing charges would be about $5000 per kW of capacity. Note that simply because of increases in prices and wages over the 2010–2020 period, costs increased by more than $1000 per kW.

Lastly, there is the issue of how to include the financing costs in total capital costs and analyses of the economics of nuclear power. Because of the complexity of this issue, there are wide variations in both the financing rates and the costs that are included in the total capital cost estimates. In some analyses, including the present one, for a variety of reasons, the total capital cost estimates that are computed do not include any financing costs. In such cases, financing issues are accounted for elsewhere.[7] In other analyses, the total capital cost estimates do include financing costs. Sometimes, the total capital cost estimates include just interest charges on the monies that are actually borrowed.[6] If the analysis is used by a utility that is subject to state-level rate-of-return regulation, the financing charges included in the total capital cost estimate often consist of interest on the funds that are borrowed (debt financing) and an implied charge for the funds that are internally generated (equity financing).

It is always tempting to compare total capital cost estimates that are reported in the media. Columns 4–7 in Table 3 illustrate that this must be done with great care. In particular, these columns show total capital cost estimates using four different assumptions about the financing costs that are included

TABLE 3 Derivation of Total Capital Costs

	Total Capital Costs						
	(1)	(2)	(3)	(4)	(5)	(6)	(7)
Year	Percent of Total Overnight Costs Spent in Each Year of Construction Period	Expenditures in 2015 Dollars (per kW)[a]	Expenditures in Dollars of Year Funds Expended (per kW)[a]	Financial Costs Not Included (per kW)	Column 4 Plus Financing Charges—Lower Financing Rate (per kW)[b]	Column 4 Plus Financing Charges—Higher Financing Rate (per kW)[c]	Column 4 Plus Just Debt Component (per kW)[d]
2015	10.00%	463.7	$463.7	$463.7	$695.9	$925.1	$560.2
2016	15.00%	$695.6	$716.4	$716.4	$1004.8	$1273.9	$838.6
2017	20.00%	$927.4	$983.9	$983.9	$1289.7	$1559.3	$1116.0
2018	30.00%	$1391.1	$1520.1 ˴	$1520.1	$1862.2	$2147.1	$1670.8
2019	15.00%	$695.6	$782.9	$782.9	$896.3	$985.5	$833.8
2020	10.00%	$463.7	$537.6	$537.6	$575.2	$603.1	$554.8
Total Capital Costs per kW		$4637.1	$5004.6	$5004.6	$6324.1	$7494.1	$5574.1
Total Capital Costs Billions of Dollars		$ 10.2	$ 11.0	$ 11.0	$ 13.9	$ 16.5	$ 12.3

[a] Overnight capital costs in $2010 of $4000 per kW of capacity and a 3% annual escalation rate in costs were used to derive the data in columns 2 and 3

[b] A 7% financing rate was used.

[c] A 12 % financing rate was used.

[d] An 8 % debt rate was used.

in them. As can be seen from this table, the total capital costs range from $5000 to $7500 per kW, depending upon how the financing costs are reported. Again, comparisons of cost estimates without detailed knowledge about what is included in them should not be done.

Environmental Costs and Regulations

In the long run, another important factor is the environmental costs of generating electricity from fossil fuel–fired and nuclear power plants. The fossil fuel–fired power plant's capital cost estimates shown in Table 2 include the expenses needed to meet sulfur dioxide and nitrogen oxide limits imposed by the Environmental Protection Agency. The estimates also include the costs of meeting current federal and state water discharge regulations. Additionally, in the analysis reported in Section 6, an explicit fee on carbon dioxide emissions will be included. The costs of all existing laws and regulations affecting fossil fuel–fired power plants, and an important proposed one are, therefore, included in the analysis.

Nuclear power has its own set of environmental costs—namely, the possibility of exposing the public to radiation, decommissioning, and radioactive waste disposal. The cost estimates of the units shown in Table 1 reflect designs that met NRC requirements as of about 2007 and 2008. These designs may, however, have to be changed because of additional NRC requirements, which could have cost implications. The design of the AP1000, a 1100-MW pressurized water reactor, was approved by the NRC in 2006. However, a number of changes to that design have been approved, and one additional change is currently under review by the NRC. Similarly, the design of the ABWR, a 1600-MW boiling water reactor developed by GE, was approved by the NRC in 1997. Currently, GE is in the process of renewing the NRC approval of that design. Additionally, in 2009, a number of design changes were submitted by GE to the NRC. The NRC is currently reviewing these changes.

Nuclear power decommissioning deals with the dismantlement of the plant and the decontamination of the site so it can be used for other purposes. The NRC must approve the utilities' plans, set residual radiation standards, and oversee the actual dismantlement of the plant. To insure that funds will be available when units are decommissioned, the state PUCs require that monies be placed in trusts. In the analysis presented in Section 6, it was assumed that decommissioning would cost $600 million and would occur 40 years after the plant's date of commercial operation.

The spent fuel from a light water reactor will be radioactive for millions of years, and the disposal of that waste is a major economic and political issue. There are two basic methods of disposing of the spent fuel from nuclear reactors: geological disposal and reprocessing/recycling. Reprocessing/recycling consists of extracting usable fuel from the waste and using it in other reactors. Currently, this is done in a few countries, most notably France and Japan. Needless to say, reprocessing/recycling is very controversial. It is very expensive and, at least in its current form, the reprocessed spent fuel could be used for military purposes. Thus, there are major proliferation concerns with reprocessing. Lastly, reprocessing has its own set of waste disposal problems.

With the passage of the Nuclear Waste Policy Act (NWPA) of 1982, the United States formally chose direct geological disposal of the waste. The initial act directed that the Department of Energy (DOE) study the feasibility of burying the waste at a number of sites. In 1988, however, Congress directed DOE to focus on just one site—Yucca Mountain, Nevada—and in 2003, the President formally chose Yucca Mountain as the country's high-level waste repository site. Under the NWPA, the state of Nevada could veto the President's decision, which they did. This veto was then overturned by Congress, and in 2008, the DOE submitted an application to the NRC for their approval to build the repository.

In 2010, the Obama Administration decided to stop work on Yucca Mountain, and DOE formally requested the NRC to permit them to withdraw the application. As of early 2011, the NRC has yet to publicly announce their decision. Additionally, the Administration abolished the Office of Civilian Radioactive Waste Management—the office within DOE that was managing the Yucca Mountain project. In response to the Obama Administration's actions, a number of lawsuits were filed by various states and localities in Federal Court. These lawsuits claim that DOE does not have the authority to unilaterally stop work on Yucca Mountain. As of early 2011, the ultimate outcomes of these lawsuits are unknown, and it is unclear what would happen if the courts ruled against DOE.

The NWPA requires that the government collect a fee of one mill (one-tenth of a cent) per kilowatt hour of electricity generated from nuclear power, and this charge is included in the analysis. Given the state of the U.S. spent fuel disposal policy, it is impossible to say anything about ultimate waste disposal costs. Regardless of what the ultimate cost is, at least for geological disposal, many of the costs will be incurred by future generations, and because of the "magic of compound interest," the total costs "today" will be small. To illustrate the effects of discounting, the yearly estimated costs of building and operating Yucca Mountain will be used. These costs were derived from a 2008 estimate prepared by DOE and the total undiscounted costs were about $100 billion. (Since Yucca Mountain has been abandoned, these estimates should be viewed as the cost of some hypothetical repository.) The cumulative costs are shown in Figure 1. As can be seen from this figure, roughly 50% of the expenditures will be made from year 50 to year 150, and the bulk of the expenditures will be made from year 30 to year 80. The point here is that the bulk of the costs of a geological repository will probably be incurred many years in the future.

Because of the complexities involved in geological disposal of the waste, it is quite possible that the ultimate cost of such a repository will be much greater than $100 billion. For argument's sake, suppose that the yearly expenditures are five times those shown in Figure 1. Figure 2 shows the present value "today" of the yearly expenditures using various discount rates. Using a 2% discount rate, which is very low, the present value "today" of the yearly expenditures is about $200 billion. Many private sector discount rates range from 7.5% to 10%, and if these rates were used, the present value "today" of $500 billion s incurred over a 150 years period would be less than $50 billion—roughly one tenth of the undiscounted costs.

Because of the discounting process, very large costs imposed on future generations will appear to be very small "today." Consequently, some intergenerational equity issues dealing with evaluating the back-end costs exist. Some economists have attempted to include intergenerational fairness considerations in discounting, but unfortunately, there is no consensus about how to incorporate equity issues in discounting.[14,15] Thus, there are some equity issues dealing with nuclear power that cannot be resolved with economic analysis.

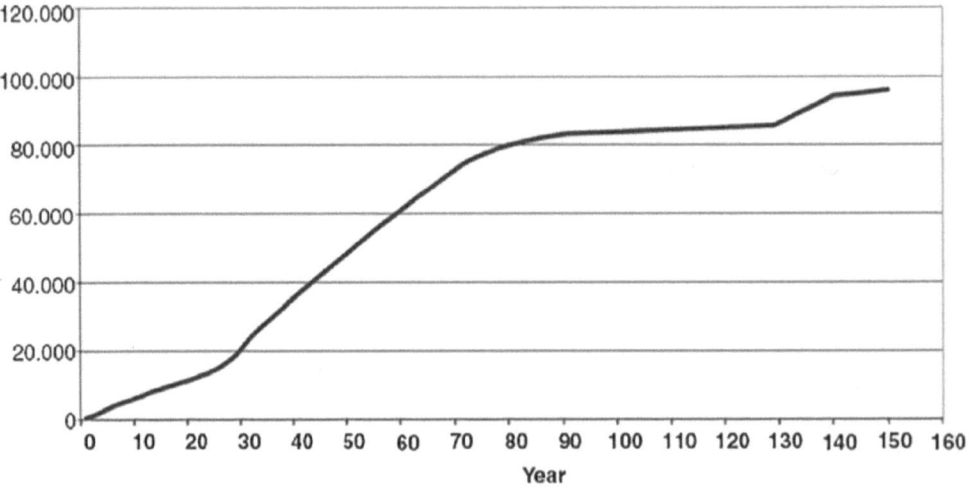

FIGURE 1　Cumulative cost of building and operating a hypothetical geological nuclear waste repository (millions of 2007 dollars).
Source: Adapted from *Analysis of the Total System Life Cycle Cost of the Civilian Radioactive Waste Management Program, Fiscal Year 2007.*[13]

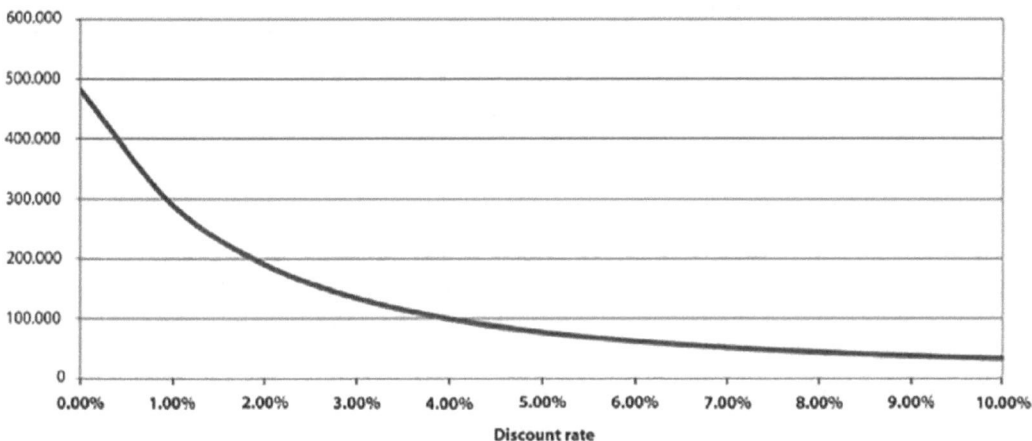

FIGURE 2   The effects of discounting on the cost of building and operating a hypothetical nuclear waste repository (present value of costs in millions).
Note: The costs are assumed to be 5 times the ones shown in Figure 1.
Source: Adapted from Analysis of the Total System Life Cycle Cost of the Civilian Radioactive Waste Management Program, Fiscal Year 2007.[13]

Discount Rates

The choice of the appropriate discount rate is another very important assumption in any analysis of the economics of nuclear power. To compare the economics of different plant types, costs that are incurred in the future must be discounted back of the present ("today"). This discounting process will account for the fact that "a dollar today is worth more than a dollar tomorrow." Nuclear capital costs are much greater and the operating costs are much less when compared with fossil fuel–fired power plants, especially natural gas–fired units. Thus, a larger percentage of the total cost of building and operating a nuclear plant is incurred "upfront" when compared with a fossil–fired power plant. The higher the discount rate, the greater would be the weight placed on the "up-front" costs, and thus, higher discount rates would tend to favor natural gas and, to a lesser extent, coal-fired power plants relative to a nuclear unit.

Since most spreadsheets have routines that calculate present values, the mechanics of discounting are trivial. However, the same cannot be said for the choice of the discount rate. The discount rate should reflect the risk of the project—i.e., building and operating a power plant. There is a long literature on estimating discount rates, but unfortunately, in most cases, the necessary financial data (stock prices and bond yields) are at the utility and not at the project level.[16] This presents some problems when utilities own assets that have different risks. In particular, most but not all utilities own transmission and distribution facilities in addition to generating plants. The risks associated with building and operating the transmission/distribution (TD) system are very different from the risks of building and operating powerplants. Thus, the risks reflected in observed utility level data are some type of weighed average of the risks of the TD and generating plant assets. Additionally, the risks reflected in recent utility level data largely deal with operating factors because very few baseload power plants are currently under construction. Thus, project specific risks (i.e., discount rates) for large-baseload power plants cannot be estimated with actual data. Indeed, this is one of the classic problems in finance—the choice of a discount rate when a low-risk firm undertakes a high-risk investment.

Since the choice of a discount rate is largely judgmental, the analysis described in Section 6 will use two of them. (See Table 4.) Financial theory states that investors will require higher returns for bearing risks that are nonrandom.[16] (Financial economists often refer to nonrandom risk as nondiversifiable or systematic risk. Even when observable data can be used, the estimation of such risks is very difficult).

TABLE 4   Assumption Used in the Analysis

	Plant Type		
Assumption	Nuclear	Coal	Combined-Cycle Natural Gas
Unit size (mWe)	1,000	1,000	1,000
Capacity factor	85.00%	85.00%	85.00%
Heat rate	10,400	8,870	6,800
Overnight costs (dollars per kW of capacity)	NA	2300	850
Lead times (years)	6	4	3
Fixed O&M costs (dollars per kW of capacity)	96	51	23
Variable O&M (dollars per kwh)	0.0004	0.00357	
Fuel costs (mills per kwh)	7	NA	NA
Waste fee (mills per kwh)	1	NA	NA
Decommissioning (million $s)	600	NA	NA
CO2 fee for each dollar per metric ton CO2 (2009 $/mmBtu)[a]	0	0.095	0.053
Escalation/Inflation Rates:			
Inflation rate	2.00%	2.00%	2.00%
Annual O&M real escalation rates	1.00%	1.00%	1.00%
Annual fuel cost real escalation rate	0.50%	0.20%	1.70%
Annual real capital cost escalation rate	0.00%	0.00%	0.00%
Financial:			
Tax rate	37.00%	37.00%	37.00%
Debt fraction—higher rate	40.00%	40.00%	40.00%
Debt fraction—lower rate	60.00%	60.00%	60.00%
Cost of debt capital—higher rate	8.00%	8.00%	8.00%
Cost of debt capital—lower rate	6.50%	6.50%	6.50%
Cost of equity capital—higher rate	15.00%	15.00%	15.00%
Cost of equity capital—lower rate	11.00%	11.00%	11.00%
Weighted after-tax cost of capital—higherrate[b]	11.02%	11.02%	11.02%
Weighted after-tax cost of capital—lower rate[b]	6.04%	6.04%	6.04%

Source: Adapted from Du and Parsons.[7]

Note: All costs are in 2009 dollars. Except for the financial assumptions, the others were generally obtained from Du and Parsons.[7]

[a] These values were used to compute the increase in fuel costs per kilowatt hour because of a carbon fee. For example, the increase in fuel costs for a coal plant caused by a fee of $20 per metric ton carbon was computed as follows: 20˙.095˙(8870/1000)=16.86 mills per kwh.

[b] These values were used as the discount rates. These rates cannot be compared with the ones used in any study that excluded corporate income taxes.

This is because random risks can be eliminated by constructing portfolios of diversified assets. Thus, the discount rate will be a direct function of the level of nonrandom risk. The higher discount rate assumes that the nonrandom risks of building and operating any power plant are 50% greater than the risk for the average investment. The lower discount rate assumes that the nonrandom risks are about 20% less than the risk for the average investment. The higher and lower discount rates are consistent with the nonrandom risks observed in the airlines/telecommunications and manufacturing industries, respectively. The higher discount rate is also slightly greater than the one used for nuclear power in the MIT study, and the lower rate is slightly less than the one used for fossil fuel–fired power plants in the MIT study.[6]

To properly interpret the differences in the two rates, the distinction between risk reduction and risk shifting becomes important. One example of risk shifting, as opposed to risk reduction, is the cost-based rate-of-return regulation of utilities. Under this form of regulation, utilities can recover all the costs for projects that are ex ante prudently expended but ex post uneconomic. In such cases, some of the risks are being shifted from utility shareholders to electricity consumers. State-level rate-of-return regulation, therefore, does not reduce risks but rather shifts some of the risks to consumers. Another example is long-term fixed-price purchase power contracts between deregulated generating companies and the regulated transmission and distribution firms. Again, such contracts do not reduce risks but instead shift the risks associated with volatile wholesale electricity prices from utility shareholders to consumers. In short, the difference in the two discount rates shown in Table 4 reflects differences in the underlying risks of building and operating power plants and not who bears that risk.

As Table 3 shows, the total cost of building a 2200-MW-nuclear power plant could be in the range of $10 to $15 billion. Given the size of many U.S. utilities, the failure of a $10-billion project may result in the firm's bankruptcy or insolvency. Thus, many utilities who are planning to build nuclear power plants are attempting to dilute bankruptcy possibilities by forming joint ventures or, in one case, a merger. The discussion above abstracts from bankruptcy risk per se because it a function of the size of the project relative to the size of the firm along with the underlying risk of the project. In other words, the risk of a project is not a function of the firm's size but instead is determined by the variability in the underlying costs and revenues.

Cost of the Alternative

A fourth factor affecting the decision to build a nuclear power plant is the cost of the alternative. As was noted in Section 2, coal-fired power plant construction costs have escalated over the last few years. This cost growth has introduced some uncertainties related to the permanency of the increases. However, environmental considerations aside, most of the uncertainty related to the cost of generating electricity from fossil fuel–fired power plants deals with fuel prices. Historical and projected coal prices are shown in Figure 3. As this figure shows, after adjusting for inflation, coal prices fell until about the year 2000 and then increased by about 30% from 2005 to 2010. Two projections suggest that coal prices will remain relatively constant at about $2.00 per mmBtu over the 2010–2020 period, and the other one shows coal prices increasing to their 2009 levels. (In this section, all the prices are in 2009 dollar. Additionally, the first year of the unit's commercial operation is assumed to be 2020.)

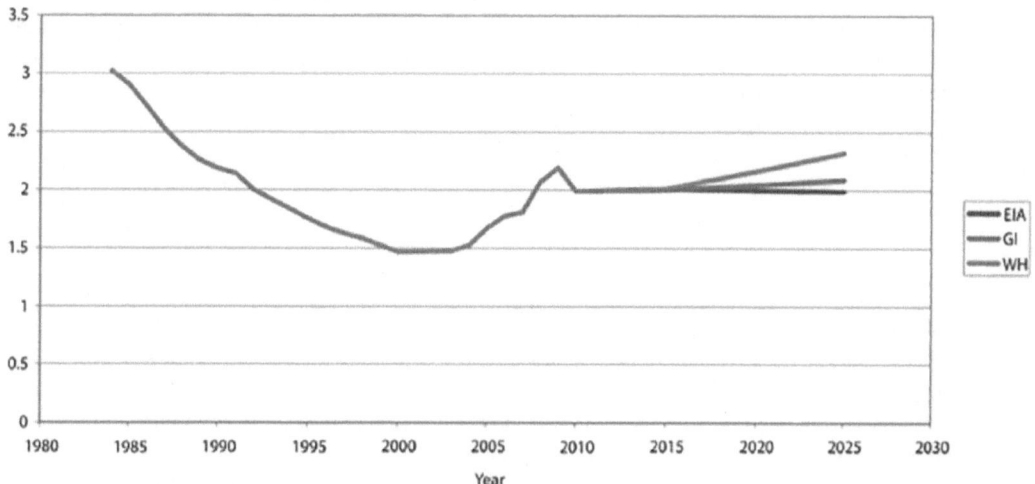

FIGURE 3   Delivered price of coal to electric utilities, 1984–2025 (2009 dollars per mmBtu).
Note: EIA, Energy Information Administration; GI, IHS Global Insights; WH, Wood MacKenzie Company. For the GI and WH projections, only the years 2015 and 2025 were reported.
Source: Annual Energy Outlook, 2010.[17]

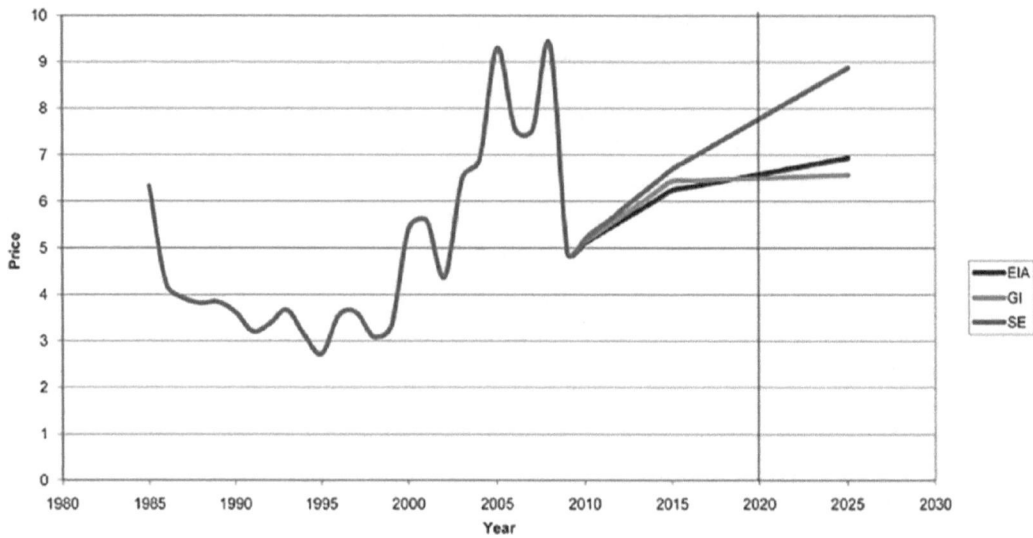

FIGURE 4   Delivered price of natural gas to electric utilities, 1985–2025 (2009 dollars per mmBtu).
Note: EIA, Energy Information Administration; GI, IHS Global Insights; SE, Strategic Energy and Economic Research. For GI and SE projections, only the years 2015 and 2025 were used.
Source: Annual Energy Outlook, 2010.[17]

Historically, natural gas prices have been much more volatile than coal prices (see Figure 4). After natural gas prices were deregulated in the early 1980s, they fell from more than $6 to about $3–4 mmBtu. In about 1998, natural gas prices began to increase, and by 2008, they increased to more than $9.00 mmBtu. Then, partly because of the recession, they fell by more than 100% to about $5 mmBtu. All three projections shown in Figure 4 have natural gas prices increasing to more than $6 mmBtu by about 2014. Over the 2015–2020 period, two of the three projections have natural gas prices increasing by relatively small amounts, whereas the third one has prices continuing to increase to more than $8.00 mmBtu by 2020.

As Table 5 shows, there are also wide regional variations in prices of coal and, to a lesser extent, natural gas. These variations are largely due to the cost of getting the fuel from the source of supply to the end users. Regions of the country that are far away from the source of supply tend to have higher costs, and the ones located close to the coal mines or gas fields have lower costs. Thus, the economics of nuclear power have a regional dimension.

TABLE 5   2010 Delivered Price of Coal and Natural Gas to Electricutilities (2009 Dollars per mmBtu)

Census Region	Coal	Natural Gas
New England	3.14	5.70
Mid Atlantic	2.23	5.03
South Atlantic	2.86	5.87
East North Central	1.88	4.30
East South Central	2.24	4.66
West North Central	1.25	4.62
West South Central	1.55	4.40
Mountain	1.55	5.00
Pacific	2.24	4.89
US average	1.99	4.85

Source: Annual Energy Outlook, 2010.[17]

Results

As was noted in Section 2, the cost of building nuclear power plants is highly uncertain. There are issues dealing with the permanency of current construction cost levels and also their accuracy. Wide regional variations in coal prices and uncertainty about future natural gas prices also exist. Because of these factors, point estimates of the levelized cost of generating electricity from coal, natural gas, and/ or nuclear power plants have little value. The general approach used here is, therefore, to derive various combinations of nuclear overnight capital costs and "current" coal (natural gas) prices that would result in nuclear power and coal (natural gas) being equally economic—i.e., the levelized costs of coal (natural gas) and nuclear power are the same. The levelized cost is defined as the constant real price of electricity that would result in the net present value of the project (discounted revenues less discounted costs) equal to zero. The method used to compute the levelized costs can be found in the work of Du and Parsons.[7] Then, the combinations of fossil fuel prices and overnight nuclear capital costs that would result in nuclear power being economic (or uneconomic) could be determined. While this is not ideal, it is the best that can be done, given all the uncertainty.

To implement this approach, it was necessary to "fix" all the other variables affecting the economics of nuclear power, and unfortunately, there is some uncertainty in all of them. Some unreported sensitivity analyses suggested that variations in leadtimes, heatrates, and nonfuel Operations and Maintenance (O&M) costs had relatively minor effects on the basic conclusions of this analysis. Additionally, as was discussed in Section 2, there is uncertainty about the permanency of the increases in the coalfired power plant overnight construction costs. As will be seen shortly, fixing the coal capital costs will not have any major impact on the basic conclusions of this analysis.

The results of a comparison of the economics of nuclear power relative to coal-fired powerplants using the higher discount rate are shown in Figure 5a. (In this section, except where noted, all costs and prices are in 2009 dollars.) The solid line shows the combinations of overnight nuclear capital expenses and current coal prices that would result in both plant types having the same levelized cost. Thus, any combination of overnight nuclear capital costs and coal prices that fall in the region denoted as E would result in nuclear power being economic. Similarly, any combination of these two factors that fall in the region U would result in nuclear power being uneconomic relative to coalfired power plants.

As was noted above, overnight nuclear capital costs derived from a small number of PUC filings averaged about $4000 per kW. Given overnight capital costs of $4000 per kW, projected coal prices would have to be about $6 per mmBtu in 2020 before nuclear power would be competitive with coal-fired power plants (point A in Figure 5a). These coal prices are about two and a half to three times their 2010 levels and are also much greater than the projected values. Coal prices of $6 per mmBtu are also much greater than their 2010 levels in regions of the United States that are not close to the coal reserves.

Similarly, given projected coal prices in the midterm of about $2 mmBtu, nuclear overnight capital costs would have to fall to levels roughly comparable with the ones for the coal-fired power plants, before nuclear power would be economic. Given $2 per mmBtu coal prices and the nuclear operating cost assumptions shown in Table 4, the operating costs of both plant types are about the same. Thus, for nuclear power to be competitive with coal-fired powerplants, the overnight capital costs of the two technologies would have to be similar (point B in Figure 5a).

The results of the comparison of nuclear power to coalfired power plants using the lower discount rate are shown in Figure 5b. These results are qualitatively similar to the ones using the higher discount rate. That is, if the overnight nuclear capital costs would be about $4000 per kW, coal prices in 2020 would have to be much greater than the ones reported in a number of recent studies. Additionally, using projected coal prices in 2020 of about $2 per mmBtu, overnight nuclear capital costs would have to fall to about $2500 per kW before nuclear power would be competitive with coal-fired power plants.

To summarize, if nuclear overnight capital cost estimates found in a number of recent PUC filing are at all indicative of what it would actually cost to build a nuclear power plant, and if the external costs of carbon dioxide (CO_2) emissions are ignored, nuclear power is not competitive with efficient coal-fired

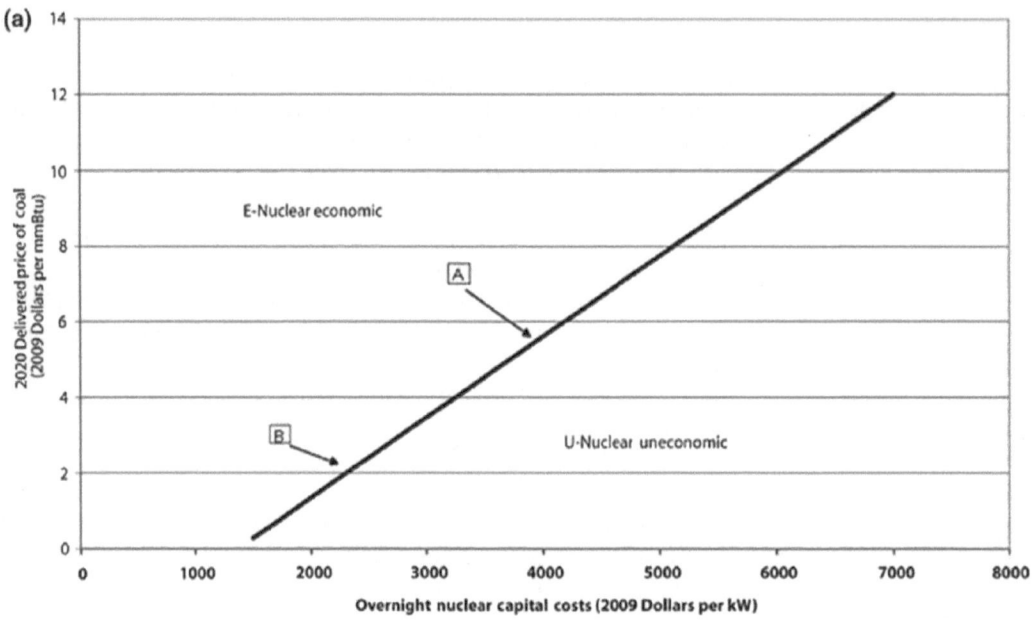

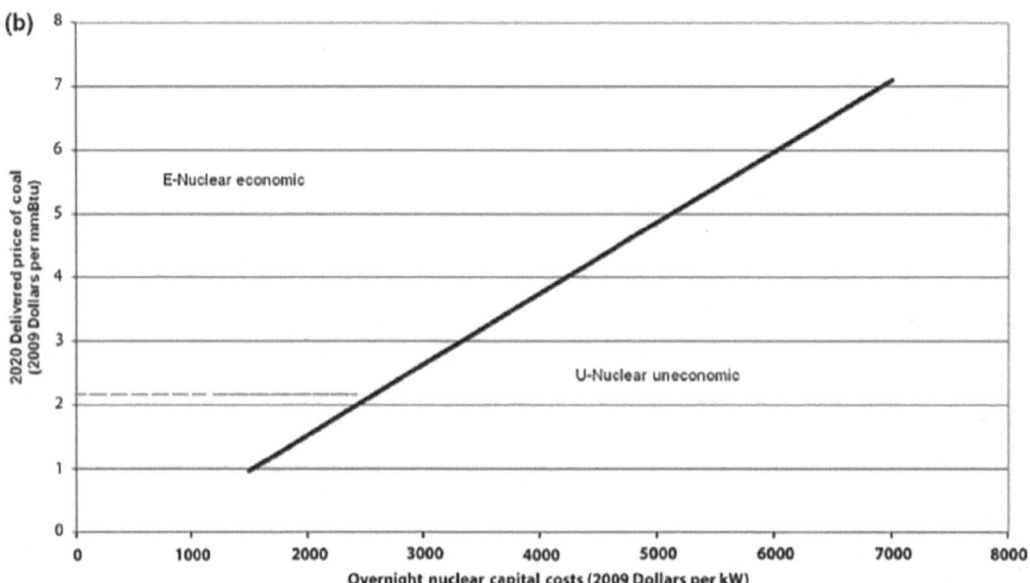

FIGURE 5 The economics of a nuclear power plant relative to a coal-fired power plant: (a) higher discount rate (b) lower discount rate.

Note: The solid line shows the combinations of 2020 coal prices and nuclear capital costs that would produce the same levelized cost for both plants. The combinations of coal prices and nuclear capital costs that would fall in the region E (U) would produce levelized costs that are lower (higher) for the nuclear plant than for the coal plant.

power plants. Given nuclear capital costs of about $4000 per kW, coal prices would have to increase substantially before nuclear power would be competitive with coal-fired power plants. Also, given the assumptions shown in Table 4, these conclusions do not depend upon the assumed discount rate. As will be seen shortly, the same is not true for comparisons of nuclear power with natural gas–fired power plants.

The economics of nuclear power relative to an efficient combined-cycle natural gas–fired power plant using the higher discount rate is shown in Figure 6a. Given overnight nuclear capital costs of $4000 per kW, natural gas prices would have to increase to about $9.50 per mmBtu before nuclear power would be economic relative to gas–fired power plants (point A in Figure 6a). Natural gas prices of over $9 per mmBtu in 2020 are much higher than their 2010 levels and are also higher than their projected levels shown in Figure 4. Similarly, using the higher discount rate and 2020 natural gas prices

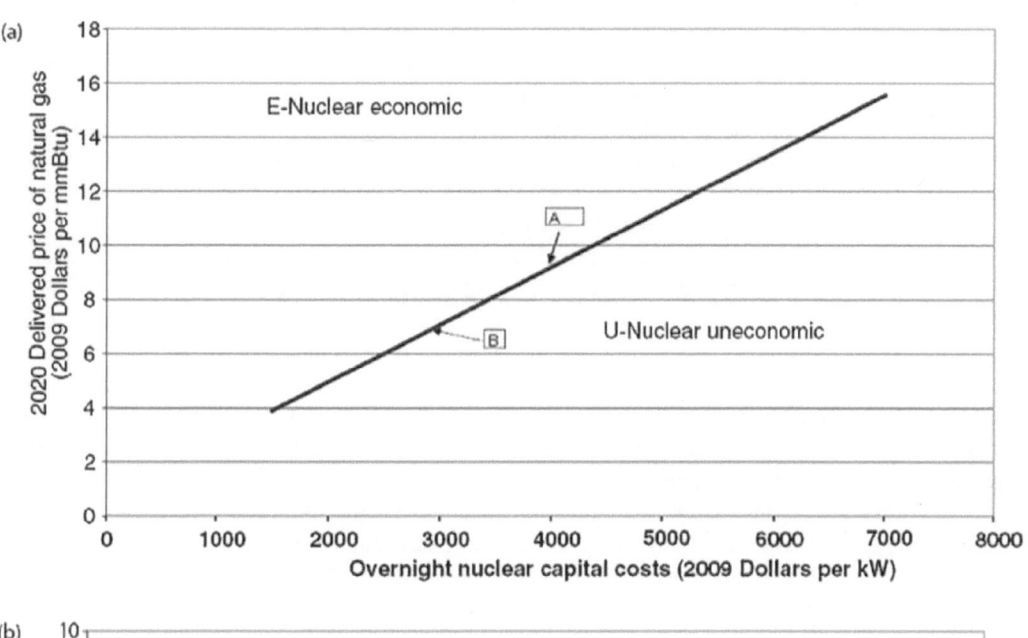

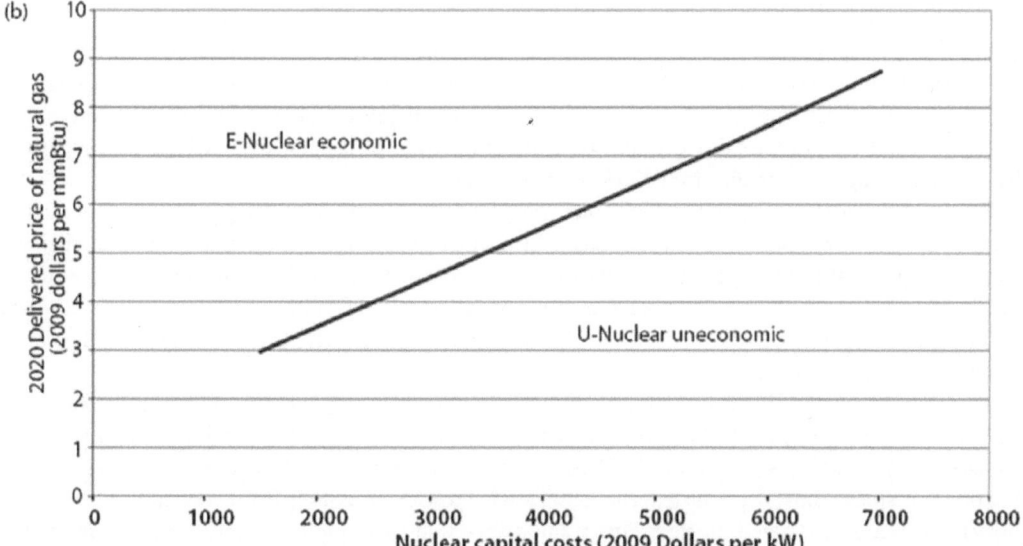

FIGURE 6 The economics of a nuclear power plant relative to a combined cycle natural gas-fired power plant: (a) higher discount rate (b) lower discount rate.

Note: The solid line shows the combinations of 2020 natural–gas prices and nuclear capital costs that would produce the same levelized cost for both plants. The combinations of gas prices and nuclear capital costs that would fall in the region E (U) would produce levelized costs that are lower (higher) for the nuclear plant than for the gas–fired plant.

of about $7 per mmBtu, overnight nuclear capital costs would have to be about $3000 per kW before nuclear power would be competitive with natural gas-fired power plants.

Given overnight nuclear capital costs of $4000 per kW, roughly 75% of the total cost of generating electricity from the nuclear unit is incurred upfront in terms of construction costs. However, about 75% of the total cost of generating electricity from a combined-cycle natural gas–fired plant is incurred in future years. Again, the lower the discount rate, the lesser will be the weight placed on the up-front costs, and thus, lower discount rates would favor nuclear power. The comparison of the economics of nuclear power plants relative to gas–fired power plants using the lower discount rate, shown in Figure 6b, suggests that this is the case. Using the lower discount rate and nuclear capital costs of about $4000 per kW, natural gas prices would have to increase from their 2010 levels of about $5.00 to about $6.00 per mmBtu by 2020 before the nuclear power plant would be economic. Natural gas prices of about $6 per mmBtu in 2020 are slightly lower than two of the three projections shown in Figure 4. Thus, if building and operating any power plant is perceived to be a relatively low-risk endeavor, even if the external costs related to CO_2 emissions are ignored, nuclear power is marginally economic relative to combined-cycle natural gas–fired power plants.

This result was based on the assumption that real natural gas prices would increase at an annual rate of .75% from 2021 to 2061. At least through 2035, this escalation rate is relatively high. If a lower escalation rate in annual real natural gas prices of 0.3% were assumed, and given nuclear overnight capital costs of $4000 per kW, 2020 real natural gas prices would have to be about $7 per mmBtu before a nuclear power plant would be competitive with a combined-cycle natural gas–fired power plant. Natural gas prices of $7 per mmBtu in 2020 are greater than two of the three projection shown in Figure 4. Thus, using lower discount rate, nuclear power would be extremely marginally competitive with combined-cycle natural gas–fired power plants.

As was noted above, the fossil fuel capital costs include the expenses needed to meet current sulfur dioxide and nitrogen oxide emission levels. However, the external costs related to CO_2 emissions have not been included. The size of these costs is, however, highly uncertain. Indeed, the extent to which CO_2 emissions have resulted in global warming and the associated costs of the warming of the atmosphere are still being debated. However, one recent study recommended using external costs in 2020 of about $7 to $42 or perhaps about $80 per metric ton of CO_2 emissions if the Earth warms faster than expected.[18] Additionally, some recent analyses of the effects of recently proposed U.S. "cap-and-trade" programs estimated that in 2020, the "price" of CO_2 emissions would range from about $20 to $100 per metric ton in 2020.[19] Lastly, in Europe's CO_2 cap-and-trade program, over the last few years, CO_2 emission permits have been trading in the range of 15€–20€ per metric ton.

Because of these uncertainties, the best that could be done is to compute the combinations of nuclear capital costs and CO_2 prices that would result in nuclear power being economic relative to coal and combined-cycle natural gas–fired power plants. This, of course, requires that coal and natural gas prices in 2020 be fixed at $2.00 and $7.00 per mmBtu, respectively. The results of this exercise for coal and natural gas–fired power plants are shown in Figures 7 and 8, respectively. As before, both the higher and lower discount rates were used.

These results suggest that if nuclear power construction costs were about $4000 per kW, CO_2 prices in 2020 would have to exceed $20–$40 per metric ton before nuclear power would be economic relative to coal-fired power plants (see Figure 7). These CO_2 prices are in the same range as ones used in a number of discussions. Thus, once the external costs related to CO_2 emissions from very efficient coal-fired power plants are considered, nuclear power could very well be economic relative to coal-fired power plants. However, using the higher discount rate, the price of CO_2 emissions would have to be somewhat higher (about $50–$60 per ton) before nuclear power would be competitive with combined-cycle natural gas–fired power plants. As was just noted, using the lower discount rate, nuclear power would be very marginally economic relative to gas–fired power plants even if the price of carbon were zero.

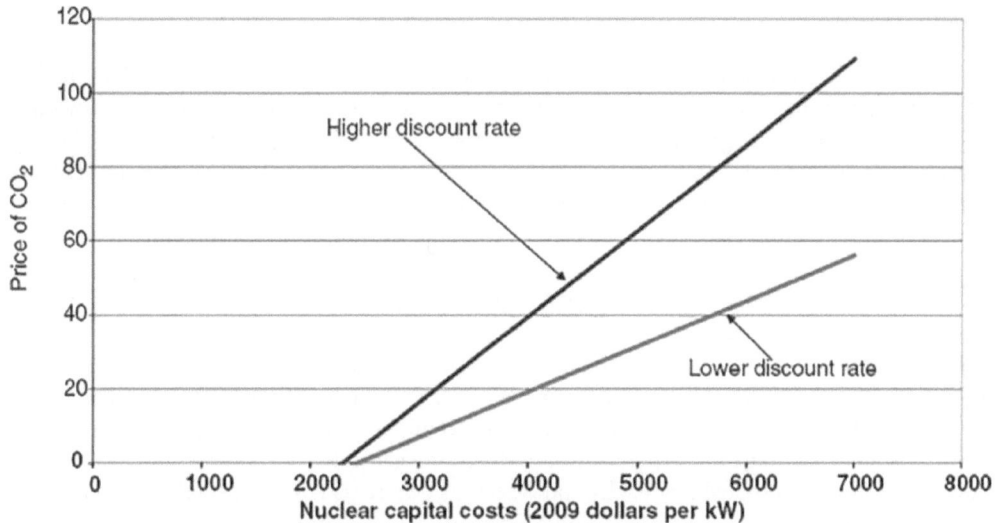

FIGURE 7 The economics of a nuclear power plant relative to a coal-fired power plant: costs of CO_2 emissions included.

Note: The solid line shows the combinations of CO_2 prices and nuclear capital costs that would produce the same levelized cost for both plants. The combinations of coal prices and nuclear capital costs that would fall in the region above (below) or to the left (right) would produce levelized costs that are lower (higher) for the nuclear plant than for the coal plant. Real coal prices in 2020 were assumed to be about $2.10 per mmBtu. The carbon prices are in 2009 dollars per metric ton.

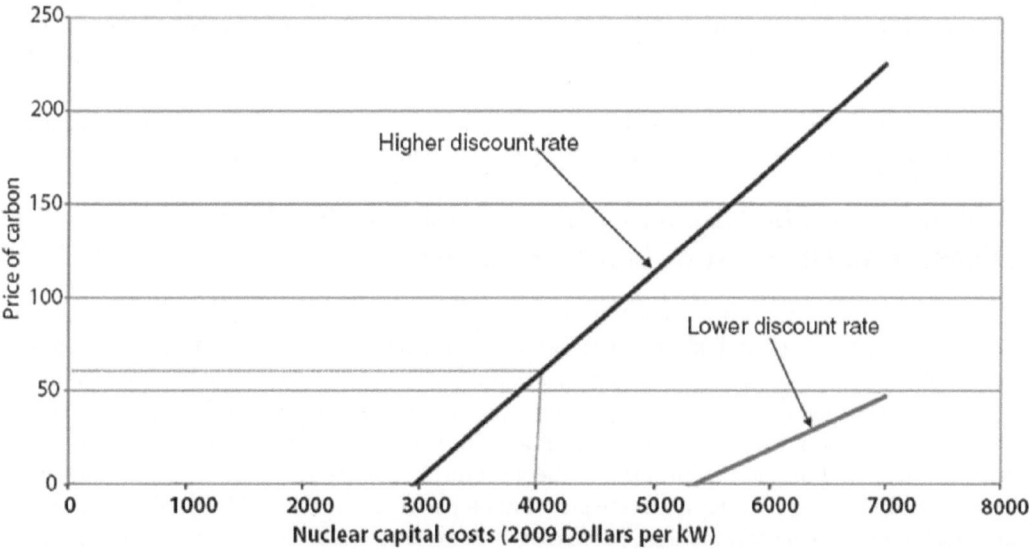

FIGURE 8 The economics of a nuclear power plant relative to a combined-cycle gas–fired power plant: cost of CO_2 emissions included.

Note: The solid line shows the combinations of CO_2 prices and nuclear capital costs that would produce the same levelized cost for both plants. The combinations of natural gas prices and nuclear capital costs that would fall in the region above (below) or to the left (right) would produce levelized costs that are lower (higher) for the nuclear plant than for the gas plant. Real natural prices in 2020 were assumed to be about $7.00 per mmBtu. The carbon price is in 2009 dollars per metric ton.

Conclusions

This entry attempted to answer the following question: Is nuclear power economic? Two major factors will influence the economics of nuclear power, and since there are major uncertainties with both of them, it is impossible to give an unqualified answer. The most obvious one is the cost of building a nuclear power plant. In a number of recent PUC rate cases, overnight nuclear power plant capital costs of around $4000 per kW were used. If these estimates are indicative of realized overnight costs, environmental considerations aside, then nuclear power would generally not be economic. The analysis in this entry suggests that overnight nuclear power plant capital costs would have to fall to between $2500 and $3000 per kW before nuclear power would be economic. The one exception to this conclusion is if a very low discount rate is used.

The second factor is the cost of the alternative, which would be either coal-fired power plants or combined-cycle natural gas–fired units. The major uncertainty here is the enactment of policies limiting CO_2 emissions. The economics of nuclear power would be greatly improved if all of the external costs related to global warming were included in the cost of generating electricity from fossil fuel–fired power plants. It must be noted that nuclear power has its own set of environmental costs in the form of nuclear waste disposal and decommissioning. Since these costs will be incurred over hundreds of years, because of discounting, the back-end costs are very small "today." Some have objected to discounting expenses incurred over very long time periods because this procedure represents a strong incentive to impose large costs on future generations. However, such equity considerations are outside the realm of economic analysis.

Lastly, there are probably at least two reasons why some utilities are interested in nuclear power even though the economics appears to be unfavorable. First, recent research using data from the United Kingdom found that nuclear power can be used as a hedge against volatile natural gas prices if the utility was subject to price regulation.[20] It is possible that utilities are using nuclear power as a hedge. This might partially explain why most of the utilities interested in nuclear power are located in states that have not been deregulated. Second, a number of utilities that are interested in nuclear power and their state-level regulators are assuming that eventually some type of explicit carbon price will be enacted. The analysis in the present entry does show that the economics of nuclear power is improved considerably when the cost of CO_2 emissions from fossil fuel–fired power plants is included.

Addendum—The Impact of the Fukushima Nuclear Disaster on the Cost of Nuclear Power

On March 11, 2011, as a result of a major earthquake and tsunami in Japan, the Fukushima nuclear power plant lost all of its on-site and off-site power, which is needed to operate the plant's safety systems. At least as of May 2011, it appears that the earthquake damaged the transmission facilities needed to supply off-site power, and the tsunami damaged all of the on-site power sources, including the emergency on-site backup diesel generators. As a result, without any source of power, the safety systems could not cool the reactor, and the fuel literally melted. The spent fuel is submerged in water in large storage pools, which cool the high-level waste. When the plant lost all power, and water could not cool the spent fuel, this resulted in a number of explosions. Because of these explosions and attempts to cool the reactor and spent fuel from off-site water sources, significant amounts of radiation were discharged into the air and water surrounding the plant.

As of May 2011, the owner of Fukushima and the Japanese government estimates that the plant will not be stabilized (cooled) until January 2012. Thus, it could take months or even years before authorities can determine what exactly happened to the plant. Indeed, it took 5 years before workers could enter the Three Mile Island plant to determine the exact nature of that accident. Consequently, at this point, it is impossible to determine with any degree of confidence what the impact of the Fukushima disaster will be on the cost of nuclear power. Nevertheless, it is possible that the disaster could affect the economics of nuclear power in a number of ways. First, there could be design changes to existing and new power plants that will increase capital costs, especially in the areas of backup power and spent fuel storage. Second, the disaster will clearly affect public acceptance of nuclear power, and the licensing hearings at the NRC could become more controversial, which would increase leadtimes. Third, in other research, I found evidence of a very small (1 to 2 percentage points) risk premium on the common stock of nuclear U.S. utilities resulting from the accident at Three Mile Island.[21] Over time, capital markets could have similar reactions to the Fukushima disaster. This would increase the cost of financing the construction and operation of a nuclear plant.

Acknowledgments

I would like to thank my past and present colleagues at the Energy Information Administration for their many useful discussions with me about the economics of nuclear power and their assistance in my research in studying this subject. However, the views and opinions stated in this entry are the author's alone and do not represent the official position of the Energy Information Administration or the United States DOE. All errors are the sole responsibility of the author.

References

1. Merrow, E.; Phillips, K.; Myers, C. *Understanding Cost Growth and Performance Shortfalls in Pioneer Process Plants;* The Rand Corporation: Santa Monica, CA, 1981.
2. Wiser, R.; Bolinger, M. *2008 Wind Technologies Market Report;* Lawrence Berkley National Laboratory: Berkley CA, 2009, available at http://www.nrel.gov/analysis/pdfs/46026.pdf (accessed April 2011).
3. *Annual Energy Outlook: 2004,* DOE/EIA-0383 (2004); U.S. Department of Energy, Energy Information Administration: Washington, DC, 2004.
4. *Report of the Interagency Task Force on Carbon Capture and Storage,* DOE/FE-0001 (2010); U.S. Department of Energy, Assistant Secretary for Fossil Energy: Washington, DC, 2010, available at http://fossil.energy.gov/programs/sequestration/ccstf/CCSTaskForceReport2010.pdf. (accessed April 2011).
5. *Projected Costs of Generating Electricity: 2010 Update;* Nuclear Energy Agency-International Energy Agency, Organization of Economic Co-operation and Development: Paris, France, 2010.
6. Deutch, J.; Moniz, E.; Joskow, P. *The Future of Nuclear Power*; Massachusetts Institute of Technology: Cambridge, MA, 2003, available at http://web.mit.edu/nuclearpower/pdf/nuclearpowerfull.pdf (accessed April 2011).
7. *World Energy Outlook: 2008;* International Energy Agency: Paris, France, 2008.
8. Du, Y.; Parsons, J. *Update of the Cost of Nuclear Power,* Working Paper 09-004; Center for Energy and Environmental Research, Massachusetts Institute of Technology: Cambridge, MA, 2009.
9. McCabe, M. Principals, agents, and the learning curve: The case of steam electric power plant construction. *J.Ind. Econ.* **1996**, *20* (4), 240–270.
10. Cantor, R.; Hewlett, J. The economics of nuclear power: Further evidence of learning, economies of scale and regulatory effects. *Resour. Energy* **1988**, *10* (4), 315–335.
11. Hewlett, J. Why were the nuclear power plant cost and leadtime estimates so wrong?. In *Nuclear Power at the Crossroads;* Lowinger, T., Hinman, G., Eds.; University of Colorado Press: Bolder CO, 1994; 121–148.
12. Deutch, J.; Moniz, E. *The Future of Coal;* Massachusetts Institute of Technology: Cambridge, MA, 2007, available at http://web.mit.edu/coal/ (accessed April 2011).
13. *Analysis of the Total System Life Cycle Cost of the Civilian Radioactive Waste Management Program, Fiscal Year 2007,* DOE/RW-0591; U.S. Department of Energy, Office of Civilian Radioactive Waste Management: Washington, DC, 2008.

14. Portney. P.; Weyant, J. *Discounting and Intergenerational Equity;* Resources for the Future: Washington, DC, 1999.
15. Sumaila, U.; Walters, C. Intergenerational discounting: A new intuitive approach. *Ecol. Econ.* **2005,** *10* (2), 135–142.
16. Copeland, T.; Weston, F. *Financial Theory and Corporation Policy,* 3rd Ed.; Addison-Wesley: New York, 1988.
17. *Annual Energy Outlook, 2010,* DOE/EIA-0383 (2010); Energy Information Administration: Washington, DC, 2010.
18. Parry, I.; Williams, R. Is a carbon tax the only good climate policy. In *Resources;* Resources for the Future: Washington DC, Fall 2010, 176; 38–41, available at http://www.rff.org/resourcesno176/parry (accessed April 2011).
19. *Energy Market and Economic Impacts of the American Power Act of 2010,* SR/OIAF/2010-01; U.S. Department of Energy, Energy Information Administration: Washington, DC, 2010.
20. Roques, F.; Newbery, D. Nuclear power: A hedge against uncertain gas prices and carbon prices. *Energy J.* **2006,** *27* (4), 72–95.
21. Hewlett, J. *Investor Perceptions of Nuclear Power,* DOE/ EIA-0446; Energy Information Administration, U.S. Department of Energy: Washington, DC, 1984.

15

Remote Sensing and GIS Integration

Egide Nizeyimana

Compatibility Issues between Remote Sensing and GIS

A GIS can be defined as a set of computer tools for capturing, storing, analyzing, and displaying spatially referenced data. The data within a GIS consists of two elements: spatial entities represented by points (e.g., well locations), lines (e.g., streams, road networks), and polygons (e.g., soil delineations) and attribute data or information that describes characteristics of the spatial features. The spatial entity is referenced to a geographic coordinate system and is stored in either a vector or raster model. GIS is primarily a platform that integrates spatial information from variable data sources and provides tools to overlay and analyze it.

RS is compatible with GIS because the information acquired by imaging sensors carried aboard satellites and airplanes is geographic in nature, referenced to a known coordinate system, and in a grid layout, a raster data model commonly found in most popular GIS software packages. In a raster model environment, the analysis is performed pixel by pixel. Current commercial GIS-based software packages have vector and raster capability analyses and allow data conversion from one data model to the other. Nonimaging systems are also important to RS/GIS integration because they provide point data, which at georeferenced locations, are often combined with other environmental data for site-specific assessments.

Linking Remote Sensing and GIS

Remote Sensing as a Source of Spatial Data

RS is one of the most important sources of land use/cover information used in GIS analyses. It provides information on the location and spatial and temporal distribution of land cover on the Earth's surface. Land use/cover distributions may be derived from aerial photographs after they are corrected for relief displacement and distortions caused by camera angle, and registered to a coordinate system. Delineations of different land use/cover types on photographs are made from visual interpretations of characteristics (e.g., tone, texture and color) aided by optical devices. Digital maps of these classes are created by digitizing and processing boundaries between land uses or by scanning photographs covering the area of interest and screen-digitizing their boundaries.

Land use distributions on the Earth's surface may also be obtained using RS imaging systems. In this case, RS provides opportunities in the area of land use planning which would not be otherwise available. While aerial photographs are effective and appropriate for analyses of small areas, RS offers tremendous advantages when planning for large areas such as river basins and regions. The fact that the land surface is observed from reflected/emitted energy for large areas and over a wide range of wavelengths allows for easy differentiation of existing land uses (e.g., wetlands, degraded landscapes).[1] Earth observation satellites also offer a repetitive coverage of the land, thus providing the possibility of monitoring land use pattern changes over time. Gathering information on land uses at several time intervals is particularly important in the monitoring and land evaluation stages of the land use planning process. Factors such as plant stress, crop growth and yields that serve as measures of agricultural productivity can be rapidly estimated following digital processing and analysis of RS imagery. RS has been used in many instances to monitor land surface conditions such as soil degradation and soil salinity.[2] It is true that RS cannot replace field mapping of land use/cover. However, RS supplements provide information that would not otherwise be available to land use planners and managers. There is no doubt that the development of sensors of high spatial resolution (1 m and higher) presently in orbits would increase the use of RS in land use planning and management.

As indicated before, RS and GIS technologies are highly compatible primarily because of the nature of RS as a source of spatial land use/land cover used in various environmental applications. Remote sensed imagery is a grid cell layer, a data format easily handled by GISs. In the RS/GIS integration, GIS appears as a platform that stores and integrates spatial data from different sources including remote sensing and output information needed for environmental analyses depending on the type of analysis sought. Land use/cover distributions derived from RS are input to GIS, GIS-based systems (SDSS), and models (e.g., hydrologic/water quality, crop yields, primary productivity) commonly used in land evaluation for land use planning and management. GIS has been proven to be a valuable tool in integrating RS-derived data with climate and land surface parameters (soils, terrain, etc.) to generate digital maps of ecosystem productivity or vulnerability to environmental factors.[3] The parameterization of the Terrestrial Ecology Model (TEM) and a forest ecosystem model using GIS- and RS-derived data at regional scales has also been accomplished.[4,5] The RS/GIS approach has been adopted by many state and federal agencies involved in environmental assessments. For example, in the GAP Analysis Program, the U.S. Geological Survey (USGS) and collaborative institutions map and/or model potential natural habitats of native vertebrate species from remotely sensed data across the country, and use GIS map overlay tools to determine the degree of richness of habitats in these species.[6]

Finally, GIS and RS applications have promoted the development of other high-resolution spatial technologies that enhance the land use assessment and planning. Some of these are the global positioning system (GPS) and digital orthophoto quadrangles (DOQ). GPS allows the user to record accurately and rapidly geographic coordinates of any location in the field with precisions ranging from several meters to a centimeter. Soil, terrain and land use attributes observed or measured can then be input to a GIS along with their precise locations and extent. GPS coupled with a GIS can improve the accuracy of land quality mapping by increasing the spatial variability of soil and landscape attributes. DOQs, on the other hand, are digital images of aerial photographs that were corrected to remove relief displacement and distortion caused by the camera angle. The USGS distributes single-band, 256-scale, gray-scaled DOQs at 1 m grid resolution. Although DOQs have not been used extensively in land use planning in the past, their high resolution and photograph-like characteristics make them a potential source of data in this area. Furthermore, repetitive DOQs acquired at different times can be used to monitor the magnitude of changes in land use over time. The relationship between GIS, RS, and other data sources is shown in Figure 1.

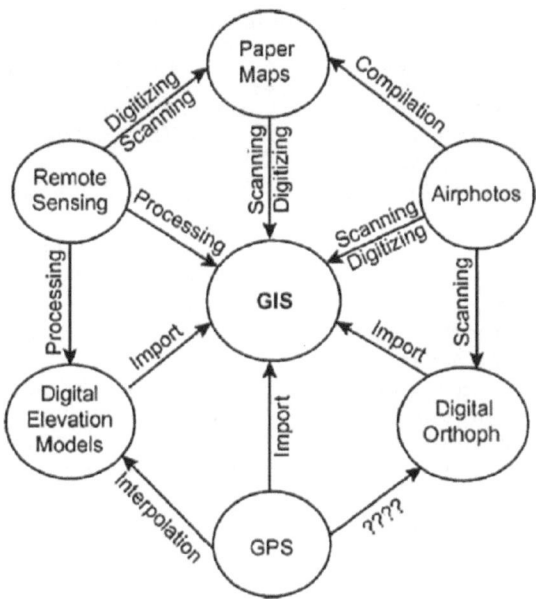

FIGURE 1 Relationships between remote sensing and GIS.
Source: Adapted from Nizeyimana and Petersen.[7]

Applications of Remote Sensing/GIS in Soil Science

The need for geospatial data for use in various spatial database development and analyses in industry, government, and universities have increased the demand for remotely sensed data. In the area of soil science, the remote sensing technology is used to acquire data, and digitally process and analyze it. Subsequent analyses may involve error assessment, and data conversion from raster to vector format and vice versa before the data is merged with other datasets in GIS for specific analyses. Table 1 summarizes remote sensing techniques used in various aspects of soil science and potential limitations for use in GIS and advantages of each. These are laboratory approaches, field methods, and aircraft-/satellite-based methods. The latter involves interpretation and analyses of digital images, color composites, or radiances.

Remote Sensing/GIS integration for Site-Specific Farming

The goal of site-specific farming (SSF) or precision farming is to optimize the profitability of a farm practicing variable management according to soil conditions found at each site. The concept is based primarily on the fact that soil chemical and physical properties that affect crop production (e.g., pH, nutrients, available water, impeding layers) vary spatially across agricultural landscapes Management practices such as fertilizer and pesticide applications, irrigation water, and crop varieties should, therefore, be prescribed according to this soil variability. The profitability associated with variable rate applications of SSTs should avoid or reduce waste and the risk for environmental pollution because these agrochemicals are applied to the field only in amounts needed for optimal crop growth.

In an SST, the real-time detection of continuous soil variable is made possible by mobile devices including sensors and differential GPS units while agrochemicals are applied using variable rate field applicators mounted on farm equipments. These sensors are based on the same principles as RS and have been developed to detect, directly or indirectly, soil properties such as soil moisture, soil texture, nitrates, etc. The data acquired, along with their respective locations, are often integrated with spatial information from other sources such as soil databases, climate, landscape, and satellite/airborne remote sensing to delineate meaningful management zones within the farm.

TABLE 1   Documented Uses of Remote Sensing in Various Areas of Soil Science

Application	Advantages	Potential Limitations	References
Laboratory methods	Provides accurate measurements of reflectance values	Provide point data rather than areal extent of soil properties	[8]
Field methods	Easily related to on-site conditions	Provide limited data coverage; measurements are affected by soil conditions (soil roughness, moisture), sun angle, etc.	[9]
Airborne -/ satellite-based methods	Provides areal extent and temporal coverage of soil properties	Radiance measurements affected by atmospheric and soil-surface conditions	(–)
a) Interpretation of digital images	Good results when data are well analyzed and interpreted	Image analysis can be costly; requires experienced technician for good results; field verification or prior knowledge of area required for best results	[10]
b) Interpretation of color composites	Easy to use and rapid interpretations	Intensive field verification is needed for good results due to the fact that interpretation is based on differences in tone and physical characteristics of objects; requires experienced interpreter for good results	[11]
c) Interpretation of radiances	Relatively easy, cheap, quantitative; data can be normalized to remove environmental effects	Results often unreliable	[12]

Source: Adapted from Nizeyimana and Petersen.[13]

References

1. Petersen, G.W.; Nizeyimana, E.; Evans, B.M. Applications of geographic information systems in soil degradation assessments. In *Methods for Assessment of Soil Degradation*; Lal, R., Blum, W.H., Valentine, C., Rose, B.A., Eds.; Advances in Soil Science; CRC Press: Boca Raton, FL, 1997; 377–391.
2. Raina, P.; Joshi, D.C.; Kolarkar, A.S. Mapping of soil degradation by remote sensing on Alluvial plan, Rajasthan, India. Arid Soil Res. Rehab. **1993**, *7*, 145–161.
3. Parrish, D.A.; Townsend, L.; Saunders, J.; Carney, G.; Langston, C. USEPA Region 6 Comparative Risk Project: Evaluating Ecological Risk. EPA Unpublished Report, 1993.
4. Pan, Y.; McGuire, A.D.; Kicklighter, D.W.; Melillo, J.M. The importance of climate and soils for estimates of net primary production: a sensitivity analysis with the terrestrial ecosystem model. Global Change Biol. **1996**, *2*, 5–23.
5. Lathrop, R.G., Jr.; Abler, J.D., Jr.; Bognar, J.A., Jr. Spatial variability of digital soil maps and its impact on regional ecosystem modeling. Ecol. Model. **1995**, *82*, 1–10.
6. Scott, J.M.; Davis, F.; Csuti, B.; Noss, R.; Butterfield, B.; Groves, C.; Anderson, H.; Caicco, S.; D'Erchia, F.; Edwards, T.C., Jr.; Ulliman, J.; Wright, R.G. GAP Analysis: a geographic approach to protection of biological diversity. Wildlife Monogr. **1993**, *123*, 1–41.
7. Nizeyimana, E.; Petersen, G.W. Land use planning and environmental impact assessment using GIS. Environmental Modeling Using GIS and Remote Sensing. 2001.
8. Stoner, E.R.; Baumgardner, M.F. Characteristic variations in reflectance surface of soils. Soil Sci. Soc. Am. J. **1981**, *45*, 1161–1165.

9. Gausman, H.W.; Leamer, R.W.; Noriega, J.R.; Rodriguez, R.R.; Wiegand, C.L. Field-measured spectrometric reflectance of disked and non-disked soil with and without wheat straw. Soil Sci. Soc. Am. J. **1977**, *41,* 493–496.

10. Connors, K.F.; Gardner, T.W.; Petersen, G.W. Digital analysis of the hydrologic components of watersheds using simulated SPOT imagery. Proceedings of Workshop on Hydrologic Applications of Space Technology; IAHS: Cocoa Beach, FL, 1985; Vol. 160, 355–365.

11. Bocco, G.; Palacio, J.; Valenzuela, C.R. Gully erosion modeling using GIS and geomorphologic knowledge. ITC J. **1990**, *3,* 253–261.

12. Pickup, G.; Nelson, D.J. Use of landsat radiance parameters to distinguish soil erosion, stability, and deposition in Arid Central Australia. Remote Sens. Environ. **1984**, *16,* 195–209.

13. Nizeyimana, E.; Petersen, G.W. Remote sensing applications to soil degradation assessments. In *Methods for Assessment of Soil Degradation;* Lal, R., Blum, W.H., Valentine, C., Rose, B.A., Eds.; CRC Press: Boca Raton, FL, 1997; 393–405.

16

Solid Waste: Municipal

Angelique
Chettiparamb

Introduction

Waste is normally understood as something that has no value. The *Oxford Dictionary* defines waste as "unwanted or unusable material, substance, or by-product" (http://oxforddictionaries.com/definition/waste). The United Nations Statistical Division defines it as

> Wastes are materials that are not primary products (produced for the market) and for which the generator has no further use in terms of production, transformation or consumption and therefore wants to dispose of [cited in Basel Convention, 2006][1] (p. 140).

The centrality of "value" and "use" in defining waste draws attention to ways and means by which these come to be attributed to waste. Since this attribution can vary greatly with users, with time, and across societies, it becomes difficult to objectively define waste. As Davoudi[2] (p. 131) says

> ... it is difficult to determine the boundaries between what *is* and what is *not* waste. In other words it is difficult to determine exactly when a material ceases to be a resource (with social, economic and environmental values) and *becomes* a waste [original italics].

Understanding waste and its value or lack of it as an attribute bestowed upon it by society allows for an examination of societal values and practices that lead to the formation of waste. Waste can thus be understood in two distinct but complementary ways: first, in terms of its intrinsic qualities, which then suggest particular use and nuisance values, and second, its socially constructed qualities that draw attention to the political economy and the socio-cultural practices that reproduce the worth or worthlessness of waste.

This entry is concerned with "municipal solid waste" (MSW). It includes waste produced by households, small businesses, and industry, which are of a similar nature and therefore collected by the municipal authorities. Wastes such as construction and demolition wastes, heavy industrial wastes, etc., are excluded from this definition.[3] However, it is not uncommon to find contamination of municipal waste by small quantities of other types of waste, and often these can be made up of harmful substances such as pathological hospital waste, abattoir waste, harmful industrial waste, or even feces.[4]

The aim of this entry is to provide a broad-ranging overview of the nature of MSW and managerial practices associated with it. Particularly stressed are the historical growth of MSW as an environmental problem and the geographical variations—the global commonalities and differences—in the nature of MSW. The historical trajectories and geographical constancies in the nature of solid waste in turn have given rise to particular normative principles for the management of MSW. These are reviewed. These principles emanate from an understanding of the intrinsic nature of solid waste, but are, however, mediated by place- and time-specific ways of organizing governance. The patterning of these ways of organizing governance is finally reviewed.

Waste: A Historical Overview

Waste has always been an issue since human beings started living in societies. The severity of waste management has, however, varied, culminating in it becoming a serious environmental issue today. Compared with present times, in pre-industrial times, material possessions and artifacts were limited and consequently valued more highly. Everything that could be repaired and reused would be repaired and reused. Also, with limited population numbers and densities, more space for disposal and limited waste generation, waste was, relatively speaking, manageable and less polluting. In terms of waste composition, in most societies during these times, domestic waste was made up of ash from fires (for cooking and heating), wood, bones, and vegetable waste.[5] These were allowed to either compost or decay aerobically (i.e., decomposition that takes place in the presence of abundant oxygen, which then do not result in foul odors). This is a viable form of managing organic waste if the quantity of waste produced is less and if the waste is spread out to decay.

The above situation changed with industrialization and consequent urbanization, first, in the Western, now developed world. Rising population numbers in cities led to high densities and less space for aerobic waste disposal. Besides, waste generation itself increased with the availability of cheaper goods through mass production and the rise of a consumer culture. The nature of waste generated also saw a diversification with new waste materials such as plastics entering into the waste stream in significant quantities and materials such as ash declining due to the emergence of water-, gas-, oil-, and electricity-based central heating systems and cooking appliances. Roberts[6] reports that during this period, waste was being dumped in streets, alleys, open land, or common water sources, where pigs and other animals could devour them. Thus, waste management during the early periods of industrialization was almost non-existent and very similar in both England and the United States.

The history of institutionalization of waste management practices in the United Kingdom begins with the Public Health Act of 1848. This is when management of waste was directly linked to public health outcomes. The subsequent Public Health Act of 1875 required all local authorities to arrange for the removal and disposal of domestic solid waste.[5] In the United States, the first public health code was enacted in 1866 in New York City. With the establishment of a correlation between sanitation and public health and the occurrences of a series of epidemics in different states, the federal government of the United States came to establish the National Board of Health in 1879. Waste handling and disposal systems then came to be gradually set up in the United States. However, dumping in landfill sites (mostly near water sources) was the most common disposal method adopted.[6] In Britain, additionally, incineration plants (destructors) were constructed in around 250 locations in an attempt to recover energy.[5]

The environmental impacts of unsanitary disposal methods were soon felt. Thus, in 1929, the federal government in the United States first issued restrictions on the location of waste dumps by requiring these to be away from river banks. Waste disposal at sea was prohibited in 1934 and restrictions on disposing waste 20 miles within the shoreline came into existence in the 1930s.[6] During this time, the concept of a sanitary landfill (waste alternated with layers of sand or other inert material) was developed in Britain. The sanitary landfills were thought to be less polluting and less dangerous than the previous waste dumps. Further, explicit planning powers over waste management sites were granted to local authorities by the Town and Country Planning Act of 1947. The concept of sanitary landfills subsequently spread to the United States, and the 1930s and 1940s saw the establishment of a number of sanitary landfills in the country. However, in overall terms, these sanitary landfills were still just around 37% of the total landfills.[5,6]

The post-war years saw a rise in consumer culture. Easy access to mass-produced material goods soon saw an increase in the quantity of waste generation. Also, new nonbiodegradable products started entering the waste stream. Although the practice of sanitary landfills grew both in the United States and the United Kingdom, the unseen environmental effects of landfill sites started to surface. This, together with increasing environmental awareness, led to more proactive intervention in waste management. For instance, the Resource Conservation and Recovery Act, which aimed at conserving resources and reducing waste, thereby protecting the environment, was enacted in the United States in 1976. Similarly, in the United Kingdom, the Control of Pollutions Act was enacted in 1974, which then later evolved into the Environmental Protection Act 1990 wherein waste minimization and recycling are priorities in waste management.[5,6]

The above broad overview of the history of waste management in the West, particularly the United States and United Kingdom, show that differing socio-economic conditions through time have led to differences in waste characteristics and notions of how it should be handled and treated. Overall economic growth and development have, in the past, resulted in a growth in the quantity of solid domestic waste generated, and the very composition of waste. Historic differences in solid waste management (SWM) practices can today be seen in geographical differences in SWM practices. Thus, in developing countries, the quantity of waste generated tends to be less and the composition of waste tends to be distinctly different from the composition of waste in developed countries. However, waste characteristics can also differ with climate, seasons, the location, and the developmental status of specific locations. The next section further explores the nature and quantity of domestic solid waste and geographical differences in the same.

Waste Characteristics and Waste Quantities

There are many ways of classifying MSW and different degrees to which component streams can be differentiated. The forms of classification used in a particular instance will normally depend on the purpose of classification. In this entry, to understand the environmental pollution associated with domestic solid waste in broad terms, and the geographical variations in the type and severity of environmental pollution, a broad classification adopted by the United Nations Environmental Programme (UNEP)[4] is used. This classification is useful to understand some of the major differences in the nature of waste produced in developed and developing regions.

The UNEP[4] classifies MSW into two: organic wastes and inorganic wastes. Organic wastes can be further classified into putrescible, fermentable, and non-fermentable organic wastes. The putrescible components within domestic waste can decompose very rapidly with the production of unpleasant odors (e.g., food). The fermentable components also decompose rapidly, but do not give rise to the unpleasant odors associated with putrescible waste (e.g., crop). Non-fermentable wastes decompose very slowly (e.g., wood).

One of the major differences in waste characteristics between the developed and the developing world is in the higher proportionate amount of organic material present in the municipal waste generated in developing countries. This can be seen in Table 1 below.

The proportionate weight of putrescibles in waste in India is around 4.5 times more than that of Paris (75.2% vs. 16.3%). Significant difference can also be seen in the quantity of paper present in waste. Generally, the quantity of paper in the waste varies with the level of economic development of an area. Therefore, from Table 1, it can be seen that the quantity of waste in Paris and California is 40.9% and 40.8%, respectively, while the quantity of waste present in India is just 1.5%. This difference is mainly due to low levels of packaging combined with high levels of formal and informal reuse, recycling, and recovery practices widely prevalent in developing countries. Further, waste from warm and humid tropical

TABLE 1   Comparison of Solid Waste Characteristics Worldwide (% wet wt)

Location	Putrescibles	Paper	Metals	Glass	Plastics, Rubber, Leather	Textiles	Ceramics, Dust, Stones	Wt (g)/ cap/ day
Bangalore, India[1]	75.2	1.5	0.1	0.2	0.9	3.1	19.0	400
Manila, Philippines[2]	45.5	14.5	4.9	2.7	8.6	1.3	27.5	400
Asunción, Paraguay[2]	60.S	12.2	2.3	4.6	4.4	2.5	13.2	460
Seoul, Korea[3]	22.3	16.2	4.1	10.6	9.6	3.8	33.4"	2000[a]
Vienna, Austria[4]	23.3	33.6	3.7	10.4	7.0	3.1	18.9[b]	1180
Mexico City, Mexico[5]	59.8[c]	11.9	1.1	3.3	3.5	0.4	20.0	680
Paris, France[4]	16.3	40.9	3.2	9.4	8.4	4.4	17.4	1.430
Australia[7]	23.6	39.1	6.6	10.2	9.9		9.0	1.870
Sunnyvale, California, USA[6]	39.4[d]	40.8	3.5	4.4	9.6	1.(1)	1.3	2000
Bexar Comity, Texas, USA[6]	43.8"	34.0	4.3	5.5	7.5	2.0	2.9	1816

[a]Includes briquette ash (average).
[b]Includes "all others."
[c]Includes small amounts of wood, hay, and straw.
[d]Includes garden waste.
Source: Data from UNEP (2005, p. 2).

and subtropical countries (where most of the developing world in located) also tend to be high in plant content, while waste from colder climates may have higher levels of ash (if coal or wood is used).

Besides differences in waste characteristics, a significant observation from Table 1 is the differences in overall quantities of waste production. Lower levels of consumption, lower levels of packaging in consumption practices, as well as higher levels of formal and informal reuse and recycling mentioned earlier, result in significantly low levels of waste production and accumulation. Thus, Table 1 shows that California produces 5 times more waste per capita per day than does Bangalore in India. However, it must be remembered that the management of solid waste, especially collection and disposal, continues to remain a significant municipal challenge in most developing countries and can thus result in less overall quantities of waste collected and disposed.[7]

The above account draws a sharp distinction between waste characteristics found in the developed world and the developing worlds. It is, however, useful to remember that the boundaries of this distinction may be blurred in practice as particular locations in developed countries may reveal character-istic very similar to those found in developing countries and vice versa. Therefore in many ways, the above distinction provides a caricature. It nevertheless performs a pragmatic function in that it alerts an observer to potential differences based on level of development that may be found in municipal waste characteristics and quantities. The actual use of this distinction as applied to particular locations must, in the end, be necessarily informed by concrete realities.

While the quantity and composition of waste may vary with time and location, two significant attri-butes remain intrinsic to municipal waste regardless of the context: first, the site of production of waste is diffused and variable, and, second, municipal waste has potential environmental nuisance values. Solid waste management practices instituted by municipalities worldwide reflect the challenges posed by these intrinsic characteristics and therefore certain managerial practices have come to be identified as constitutive of municipal waste management worldwide. The next section provides an overview of such managerial practices.

Municipal Solid Waste Management Practices

Municipal solid waste management practices can be generally said to comprise of storage and collection practices (including practices of collection for reuse, recycle, and recovery), transportation practices (sometimes including practices of transferring waste from one form of transport to the other), and disposal practices. All, or some of the above, may occur through formal practices designed and imple-mented by the municipal authority or through informal voluntary initiatives organized individually or in community groups.

Storage and collection practices include various types of storage receptacles designed for waste pro-duction units such as households/commercial establishments and community storage receptacles for storage of collected waste. Organization of proper storage of waste is essential as in its absence waste can be dumped in public spaces/drains. Storage units may also be custom made to suit particular waste management technologies such as the use of receptacles of particular design to suit waste transportation vehicles. Further, storage systems may include practices to encourage reuse/recycle/recovery by using specific color-coded waste receptacles that enable the streamlining of various streams.

Transportation practices consist of the movement of collected waste to disposal sites or transfer sites (where waste is collected and then transferred to vehicles suitable for transporting to more distant final disposal sites). Collection and transportation practices may also include managerial contracts forged with the private sector or voluntary sector for the collection and transportation of waste produced in dispersed locations.

Disposal practices include the many options through which waste is disposed, such as incineration plants, compost plants, landfills sites, and anaerobic biogasification plants, for the end disposal of wastes that are not reused and recycled. Disposal options can include material recovery options (such as com-posting) or energy recovery options (such as anaerobic biogasification plants or energy recovery from landfill sites). Disposal options may also aim to just get the waste out of sight by either dumping it in waterways, quarries, and other similar areas, or by transporting it afar (often to less developed areas). Further, disposal practices need not always be organized by municipal authorities, but can also be found at individual, household/establishment, or community level.

The above SWM practices are typically organized in accordance with generally acknowledged prin-ciples. Some of the principles that are widely adopted in SWM practices today are reviewed below.

Normative Principles in SWM

The "waste hierarchy" is a normative principle that has had a significant impact on practices of SWM in recent times (for instance, the principle has been very influential in the formulation of European waste management practices).[8] A diagrammatic representation of this is given in Figure 1.

The waste hierarchy advocates a waste management system oriented toward a hierarchy of mana-gerial options that target waste reduction, reuse, recycling, and recovery, in that order of preference, thus reducing the amount of waste that needs to be finally disposed in a landfill. Strategic aims such as "zero waste" initiatives' aimed at minimizing waste toward zero by reducing consumption, arranging for reuse by repair or modification, and maximizing recycling, thereby bringing materials back into nature or the marketplace, are based on the normative principle promoted by the "waste hierarchy".

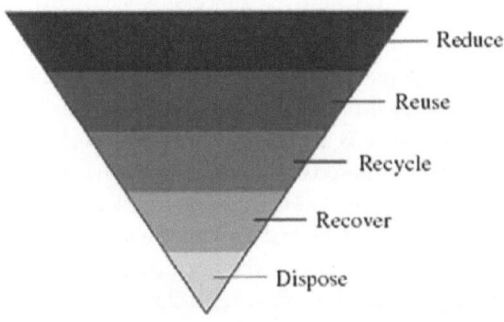

FIGURE 1 Waste hierarchy.
Source: Chettiparamb et al. (p. 327).[7]

Such initiatives are typically aimed toward instilling a re-valuation of waste as a resource rather than as something to be disposed off.[9]

Life cycle assessment (LCA) in waste has influenced waste reduction practices since around the 1990s.[10] It focuses on the environmental burdens associated with a product, a process, or an activity. LCA encompasses the entire life cycle of the system studied, including material and energy flows in raw materials, their acquisition, manufacturing processes, usage, and waste disposal/treatment, leading to what is known as a "cradle to grave approach." Thus, LCA helps expand the framework of analysis to beyond the relatively limited confines of the actual SWM system, thereby bringing in wider environmental impacts, some of which may have a greater impact on the environment than just the analysis of the SWM system itself. The LCA can thus lead to an improvement analysis that suggests ways in which the impacts on the environment can be reduced.[9] In practice, however, the LCA, although normatively influential, has been difficult to implement mainly because of the enormity of data demanded, problems in the reliability of the data, and other pragmatic difficulties in employing the technique.[11]

Integrated solid waste management (ISWM) is yet another principle that has been influential in organizing whole SWM systems. ISWM is based on the idea that any one method of SWM may not be adequate in addressing the challenges that the various streams of waste present. Instead of viewing the many streams of solid waste as distinct and separate, ISWM proposes that the most appropriate method must be adopted for each stream in a way that the sum total of the environmental impact is kept to a minimum. Thus, paper and other recyclable material must be recycled, while compostable material must be composted or digested anaerobically, and so on. The overall key aim is to minimize landfill while maximizing recycling and recovery.

The above account suggests that there is relative stability surrounding understandings of ideal ways of managing MSW. Studies of the governance of MSW, however, show wide variations, including shifting relative emphasis on these normative principles occurring both historically and within and across particular contexts. These variations are reviewed in the next section.

Policy Drivers, Policy Regimes, and Modes of Governance in SWM

The nature of the "political economy" of a region has been identified as a key determinant of MSW. The term "political economy" is defined by O'Brien[12] (p. 270) as "a regulated social framework for transacting values, comprising an arrangement of practices, relationships and institutions." He argues that industrialized societies are primarily "rubbish societies" in that these societies are infused with a relationship to waste and wasting that allows their social, political, and cultural systems and their own self-understanding to develop and change. If the centrality of wasting processes in industrialized societies is acknowledged, a political economy around wasting needs to develop, what O'Brien terms as "rubbish relationships"[12] (p. 287). Further, it then becomes necessary to valorize and promote particular institutional relations that organize and stabilize rubbish flows in ways that allow the reproduction and survival of society. This valorization and promotion reconfigures SWM practices, thus lending more credibility to particular normative principles resulting in patterns in what has been termed as "drivers,"[13] "policy regimes,"[14] or "modes of governance"[15] in SWM. These then give rise to new terminologies such as "waste citizenship" (how waste is accessed),[16] waste networks and waste flows (how waste is distributed),[17] and "waste commons" (hard rubbish meant for scavenging),[18] to name a few.

The study by Wilson[13] sought to identify key drivers that have led to the adoption of particular mechanisms or strategies in SWM both historically and around the world. A survey conducted by Wilson[13] among "colleagues" identified six key drivers for the organization of SWM practices: 1) public health; 2) environmental protection; 3) resource value of waste; 4) closing the loop; 5) institutional and responsibility issues; and 6) public awareness. The relative emphasis among these varies over time and differs with local circumstances. Wilson[13] argues that public health emerged as a key driver in the 19th century and has been more or less taken for granted in developed countries. It, however, still persists as a key driver in emerging and developing countries with hot climates. The environmental protection driver came to the fore in the 1970s and still remains a key driver in the developed world. Energy/climate change has now become central to this agenda. In developing countries, on the other hand, the focus remains on initial steps such as the phasing out of uncontrolled disposal and the promotion of clean development mechanisms. The resource value of waste was a driver until the early 20th century underpinning major industrial economies in 19th century London, and 20th century China, Soviet Union, and Eastern Europe. In developed countries, however, this driver was soon replaced by a more holistic approach to waste management. The resource value of waste remains a key driver in developing countries as large numbers of the poor survive on reuse, recycling, and recovery practices. Further, the import of waste as an industrial raw material continues in some developing countries, thus making waste integral to those economies. The waste hierarchy came to be advocated in the late 1970s and triggered a more holistic approach to resource management. Wilson[13] argues that this now means that waste prevention and recycling are key priorities in the developed world with sustainable production and consumption, integrated production policies, and zero waste policies emerging as key drivers. Notions of institutional responsibility emerged as a key driver when municipalities were charged with waste collection responsibilities in the late 19th century. While in the developed world, this responsibility is more or less taken for granted, with even an extended notion involving the private sector emerging, in developing countries, the ability to discharge this function is still limited with notions of capacity building and good governance emerging as key drivers. Finally, Wilson[13] argues that public awareness has become a key driver with waste management moving up in people's priorities as living standards increase. The environment, climate change, and resource management is informing this increase in public awareness in the developed world. In the developing world, however, public awareness remains focused on food, shelter, security, and livelihoods, with waste drawing attention only when these are threatened. The study concludes that

> "There is no one, single driver that can be seen as 'dominant'; rather, all of the six groups of drivers are important, and the balance between them will vary between countries depending on local circumstances, and indeed between stakeholder groups depending on their particular perspective"—Wilson (pp. 205–206).[13]

The "development drivers" described above therefore have a key role in organizing practices and selecting particular normative orientations and stabilizing these within institutional relations and a variety of governing instruments.

Gille[14] extends Young's[19] concept of "resource regimes" to waste. He defines it as follows:

> At their [resource regime's] core is a structure of rights and rules, which implies a certain distribution of advantages and disadvantages. Social institutions determine what wastes, and not just what resources are considered valuable by society, and these institutions regulate the production and distribution of waste in empirically tangible ways (p. 1056).

Gille explores the history of waste management in Hungary and identifies three distinct regimes in the history. He calls these "the metallic regime (1948–1974), the efficiency regime (1975–1984), and the chemical regime (1985–present)" (p. 1056). In the metallic regime, Gille argues that waste was seen as a particular kind of material—something that was like metal scrap and thus discrete, non-toxic, and infinitely recyclable and reusable. Under this benevolent perception of waste, the key actors were class-conscious workers and citizens who would collect and find new uses for all sorts of wastes. The efficiency regime was marked by a "monetized concept of waste: [where] waste was seen as a cost of production, and waste reduction and reuse were seen as steps to increase efficiency" (p. 1057). Policy tools thus included financial incentives and professionals with economic and technical expertise came to be valued. In the chemical regime, waste came to be seen as useless and even harmful. Waste liquidation came to be very important, and scientists, engineers, and the chemical industry became key players. The waste regime concept thus traces the broad patterns that structure certain normative values in particular periods and locations. **Table 2** Modes of governing MSW.

TABLE 2   Modes of Governing MSW

	Mode			Components	
Disposal	Governmental rationality (policies and programmes)	Governing agencies	Institutional relations	Governmental technologies (examples)	Governed entities
	Economic efficiency Public health Environmental efficiency	Local authorities Regulator	Devolved hierarchy	Dustbins Weekly collections Landfill sites Contracts Best Practicable Environmental Option assessment	Municipal waste Ratepayers
Diversion	Reducing the (global) environmental impact of landfill (EU Landfill Directive 1999 *Waste Strategy WOO*)	European Union DEFRA Local authorities	Multilevel Strongly hierarchical	Performance targets and auditing New policy instruments Funding mechanisms and criteria Education campaigns	Successively lower government tiers Individuals as passive citizens Differentiated waste streams
Eco-efficiency	Reducing the environmental impacts of waste; recovering value (waste hierarchy; meeting targets)	Local authorities, waste contractors, community-waste-sector organizations	Heterarchy Networks	Kerbside collections New technologies Reuse and reduction practices (e.g., nappies, compost)	Individuals as active citizens Differentiated waste streams
Waste as resource	Reducing the environmental impacts of waste; social and economic benefits	Nongovernmental organizations and networks	Solidarity Community	Provision of alternative infrastructures and collections	Individuals as community members Waste as a resource

Source: Data from Bulkeley and Watson (p. 2740).

Bulkeley and Watson[15] develop the notion of "modes of governance" in waste in order to capture both structures and processes of governing while also recognizing the plurality of modes through which governance is established. They define a mode of governing as

> A set of governmental technologies deployed through particular institutional relations through which agents seek to act on the world/other people in order to attain distinctive objectives in line with particular kinds of governmental rationality (p. 2739).

The authors analyze SWM in the United Kingdom and identify four distinctive, but coexisting, modes of governance. These are the disposal mode, the diversion mode, the ecoefficiency mode, and the waste as resource mode. The various components of waste management associated with these different modes are reproduced in Table 2 above.

The three examples of development drivers, regimes, and modes of governance reviewed above illustrate how the management of solid waste can be very different across geographical regions even if the fundamental building blocks of managing waste—those that emanate from its intrinsic qualities—remain the same.

Conclusions

This entry has provided a broad overview of the nature of MSW and the management of it. It has highlighted that the definition of what waste is, will depend on particular contexts as the attribution of value and non-value takes place subjectively and contingently. Further, the history of waste management practices shows that the nature of municipal waste has varied with the level of industrialization and climatic conditions of a location. This variation can also be seen today in the difference in waste characteristics of developed and developing countries.

The dispersed nature of production sites of MSW and the potential environmental nuisance value of MSW can, however, be said to be intrinsic to waste. Fairly consistent management practices have arisen from these intrinsic attributes. Further, certain normative principles have also arisen to guide the organization of these managerial practices. The selection and mobilization of the normative principles, however, take place within broader governance frameworks that have been variously termed as development drivers, waste regimes, or modes of governance in waste. These have also been reviewed. In conclusion, this entry has shown that while the intrinsic nature of SWM has led to selective widely stabilized managerial practices, the organization of these practices and the governance of the same may be embedded in very different governance frameworks, thereby giving rise to a myriad of permutations in the actual practices of SWM.

References

1. Basel Convention, 2006. *Vital Waste Graphics 2.* Basel Convention: Geneva. Available at http://www.grida.no/files/publications/vital-waste2/VWG2_p3to46.pdf (accessed January 2012).
2. Davoudi, S. Governing waste: Introduction to special issue. J. Environ. Plan. Manage. **2009**, *52* (2), 131–136.
3. DEFRA. *Local Authority Collected Waste—Definition of Terms,* 2011. Available at http://www.defra.gov.uk/statistics/environment/waste/la-definition/ (accessed February 2012).
4. UNEP. *Solid Waste Management,* 2005; Vol. 1. Available at http://www.unep.org/publications/contents/pub_details_search.asp?ID=3799 (accessed February 2012).
5. Waste Online. *History of Waste and Recycling Information Sheet;* 2004. Available at http://dl.dropbox.com/u/21130258/resources/InformationSheets/HistoryofWaste.htm (accessed January 2012).
6. Roberts, J. *GARBAGE: The Black Sheep of the Family: A Brief History of Waste Regulation in the United States and Oklahoma;* Office of the Secretary of Environment: State of Oklahoma, 2011. Available at http://www.deq.state.ok.us/lpdnew/wastehistory/wastehistory.htm (accessed January 2012).
7. Chettiparamb, A.; Chakkalakkal, M.; Chedambath, R. In my backyard!: An alternate model of solid waste management. Int. Plan. Stud. **2011**, *16* (4), 313–331.
8. Schmidt, J.H.; Holm, P.; Merrild, A.; Christense, P. Life cycle assessment of the waste hierarchy—A Danish case study on waste paper. Waste Manage. **2007**, *27* (11), 1519–1630.
9. Glavic, P.; Lukman, R. Review of sustainability terms and their definitions. J. Cleaner Prod. **2007**, *15* (18), 1875–1885.
10. Sundqvist, J.O. System analysis of organic waste management schemes—Experiences of the ORWARE model. In *Resource Recovery and Reuse in Organic Waste Management;* Lens, P., Hamelers, B., Hoitink, H., Bidlingmaier, W., Eds.; IWA Publishing: London, 2004.
11. Ekvall, T.; Aseefa, G.; Bjorklund, A.; Eriksson, O.; Finnveden, G. What life-cycle assessment does and does not do in assessments of waste management. Waste Manage. **2007**, *27*, 989–996.
12. O'Brien, M. Rubbish values: Reflections on the political economy of waste. Sci. Cult. **1999**, *8* (3), 269–295.
13. Wilson, D.C. Development drivers for waste management. Waste Manage. Res. **2007**, *25* (3), 198–207.

14. Gille, Z. Actor networks, modes of production, and waste regimes: Reassembling the macro-social. Environ. Plan. A **2010**, *42*, 1049–1064.
15. Bulkeley, H.; Watson, M. Modes of governing municipal waste. Environ. Plan. A **2007**, *39*, 2733–2753.
16. *Gutberlet, J. Recovering Resources—Recycling Citizenship: Urban Poverty Reduction in Latin America; Ashgate: Aldershot, U.K., 2008.*
17. Davies, A.R. *The Geographies of Garbage Governance;* Ashgate Publishing: Aldershot, U.K., 2008.
18. Lane, R. The waste commons in an emerging resource recovery regime: Contesting property and value in Melbourne's hard rubbish collections. Geogr. Res. **2011**, *49* (4), 395–407.
19. Young, O. *Resource Regimes: Natural Resources and Social Institutions;* University of California Press: Berkeley, CA, 1982.

Bibliography

1. *Baud, I.; Post, J.; Furedy, C. Solid Waste Management and Recycling: Actors, Partnerships and Policies in Hyderabad, India and Nairobi, Kenya;Kluwer Academic Publishers: Dordrecht, the Netherlands, 2010.*
2. Bulkeley, H.; Watson, M.; Hudson, R.; Weaver, P. Governing municipal waste: Towards a new analytical framework. J.Environ.Policy Plan. **2005**, *7* (1), 1–23.
3. Davoudi, S. Planning for waste management: Changing discourses and institutional relationships. Progr. Plan. **2000**, *53*, 165–216.
4. Davoudi, S. Scalar tensions in the governance of waste: The resilience of state spatial Keynesianism. J. Environ. Plan. Manage. **2009**, *52* (2), 137–156.
5. Tchobanoglous, G.; Kreith, F. *Handbook of Solid Waste Management*, 2nd Ed.; McGraw Hill: New York, 2002.
6. UN-HABITAT. *Solid Waste Management in the World's Cities: Water and Sanitation in the World's Cities 2010.* UN-HABITAT, 2010. Available at http://www.unhabitat.org/pmss/listItemDetails.aspx?publicationID=2918 (accessed October 2012).
7. Zhu, D.; Asnani, P.U.; Zurbrugg, C.; Anapolsky, S.; Mani, S. *Improving Municipal Solid Waste Management in India: A Sourcebook for Policy Makers and Practitioners;* The World Bank: Washington, D.C., 2008.

17

Sustainability and Planning

Richard Cowell

Introduction

This chapter offers a critical assessment of the role of planning in promoting sustainability. "Planning" can be defined broadly as an activity by which humans seek consciously to shape their collective future,[1] but it embraces activities with enormous variation in scope and scale. Planning can be undertaken from the local through to the national scale, and it entails the detailed regulation of land use and management as well as the setting of strategic direction for an array of public and private sector actions.

The first half of this entry charts the emergence of a worldwide interest in "planning for sustainability" and outlines key features of the dominant, conventional approach. This typically involves the following components: identifying sustainability goals, allocating actions, seeking integration between governmental sectors, engagement of different stakeholder groups, and appraisal and monitoring.

The second half of the entry assesses the extent to which planning for sustainability has achieved results and offers explanations for these outcomes. It notes that there have been significant shortfalls in implementation, partly attributable to the deficiencies of planning as a force for change but also because of the fundamentally contested nature of sustainable development as an objective. In response, two alternative perspectives to the conventional approach to planning for sustainability are discussed: (1) adaptive planning and transition management, which emphasizes the complexity of the task and the need to respond to uncertainty, and (2) a more critical perspective, which highlights how planning systems have actually exerted leverage over business as usual.

Some preliminary points of definition are required. A distinction is often made between "sustainability," a concept with an environmental focus that describes the long-term continuation of a system, and "sustainable development," a concept that is more overtly concerned with social progress and distributive equity as well as environmental quality. However, there is much debate about the meaning of either term and little consistency in the way that they are used.[2] This handbook entry adopts the term

TABLE 1   The Two Main Categories of Planning for Sustainability

Type of Approach	Scale	Focus	Power
National sustainability planning	National, with focused attention on individual sectors	Achieving environmental targets consistent with sustainability	Overarching coordination, target-setting, backed by monitoring, oversight arrangements and sector-specific measures
Land use or spatial planning	Mainly local, but subject to national policy guidance	Land use implications arising from array of environmental, social, and economic demands	Regulation of land use change through zoning procedures and spatial steering; wider coordinating role

"sustainability" and argues that rather than focusing on *ex ante* definition, practitioners must be alert to the particular interpretations of the concept that get used in any given planning process, and to what is actually being sustained.

"Planning," too, requires some further explanation. Although planning can be undertaken by any kind of organization—public, private, or voluntary sector—here, we are primarily concerned with planning as a collective, public activity, undertaken by governments at all levels (local, regional, national). Planning can thus be seen as one of a number of tools available to government to promote sustainability, working alongside, or sometimes in conjunction with, regulations, market-based instruments, voluntary measures, and so on.

There are a number of categories of approach to "planning for sustainability" that are covered within this entry. The first category is the creation of "national sustainable development plans" designed specifically to promote the achievement of sustainable development (as encouraged by Agenda 21), which tends to provide high-level strategic direction across an array of activities.[3] An exemplar is the National Environmental Policy Plans (NEPPs) that were drawn up for the Netherlands from the 1980s to the early years of the 21st century. The second category of approaches is the incorporation of sustainability objectives into existing planning processes for coordinating the use of land ("land use planning" or "spatial planning"). Many countries have planning systems that seek to steer the use and development of land, and provide a framework for making decisions about development projects. Key features of the two categories are set out in Table 1 (above).

History and Context

Ever since "sustainable development" first arrived on global policy agendas, planning has been identified as a means of achieving it. A commitment to producing national sustainability plans was a key outcome of the United Nations Conference on Environment and Development, held in Rio in 1992, with its main statement of principles—Agenda 21—regarding "environmentally sound physical planning" as essential to sustainable development.[4] The European Union, too, has given planning a significant role in shifting from reactive responses to environmental problems toward anticipatory and strategic solutions.[5]

Over the same period, sustainable development took hold as a powerful ethical and intellectual framework for existing planning activities, notably land use planning or spatial planning, but also planning across a range of sectors—transport, economic development, tourism, and so on. By way of definition, land use planning usually concerns the management of competing uses of land, typically through regulatory means. This differs somewhat from "spatial planning," which—in the form adopted in the Netherlands or France, for example—has a wider role in coordinating the actions of all sectors that have spatial consequences (transport, economic development, etc.). For convenience, "land use planning" is the term used here, but the analysis also applies to forms of spatial planning. Many countries around the world have instituted requirements that their land use or spatial planning procedures promote sustainability.[6-11]

For many commentators, the connections between sustainability and planning are self-evident. The planning profession has long identified itself with shaping the future[12] and sustainability is a long-term goal, bound up with moral responsibilities to future generations. Thus, if sustainability can be regarded as a preferred future state of affairs—whether that is applied to an ecosystem, a nation, or a city—planning provides a vehicle for steering society toward it. Planning offers the prospect that the transition to sustainability can be orderly, rather than precipitated by ecological, social, or economic crises.

Planning also provides important coordinating and integrating functions. Planning offers mechanisms for identifying and fostering "win-win-win" solutions between economic development, environmental protection, and social equity, the prospect of which has always been central to the political appeal of sustainable development. Planning has also been identified as having procedural virtues for promoting sustainability, in that the processes of formulating plans can allow for the engagement of civil society, diverse stakeholders, and foster wider learning.[13]

However, if it is widely asserted that we should "plan for sustainability," the reality is more complex. At a fundamental level, some would contest the possibility of steering complex societies and unpredictable ecosystems toward pre-given goals; for Lindblom, the best societies can hope for is "muddling through."[14] Governments and societies vary in the faith that they place in (state) planning compared with markets as means of coordination. States also vary in the wider effectiveness of their governance institutions, with developing countries especially often lacking capacity. Consequently (as discussed further below), many efforts to plan for sustainability fail to fully achieve their goals, for all sorts of reasons—some economic, some political, and some practical.

Thus, to understand the prospects of planning for sustainability, one needs to consider both ideals and what happens in practice, as this entry seeks to do.

Planning for Sustainability—A Conventional Perspective

This account of planning for sustainability is organized around what can be regarded as the conventional perspective, as depicted in Figure 1 below. It echoes core features of long-standing rational planning models: exhibiting the procedural stages of "survey-analyze-plan-monitor," and with a clear relationship between means and ends. Although this traditional, instrumental view of planning has come under criticism from various directions,[15] it still provides a useful structuring device for this review. Moreover, this perspective remains deeply entrenched in planning practice in many arenas, and concern for sustainability has arguably reinforced its position.[16] Problems with this perspective are discussed later in this entry.

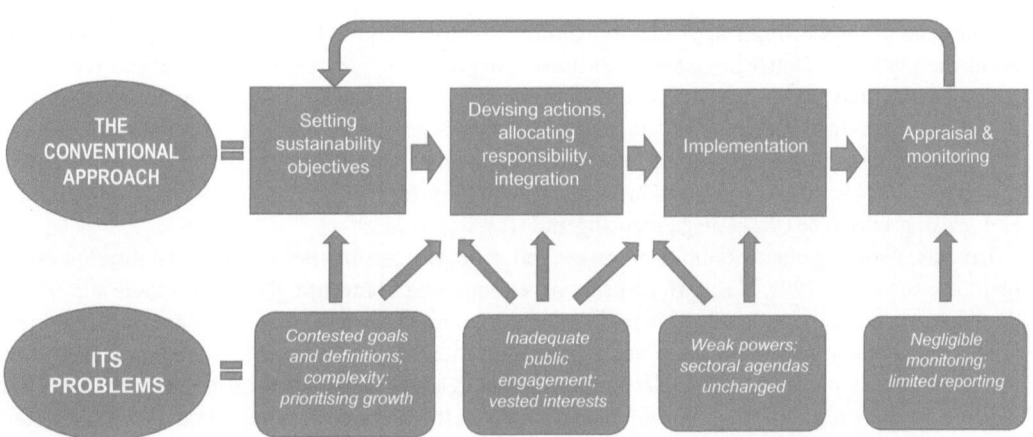

FIGURE 1 The conventional approach to planning for sustainability and its problems.

Setting Objectives and Interpreting Sustainability

At the core of the conventional view of planning for sustainability is the assumption that one can specify what makes a society more sustainable and express this in objectives toward which development can be steered. These objectives might be environmental constraints, trends or targets, or some vision of a more sustainable urban form. Objectives might be designed to avoid losses of biodiversity; to reduce the depletion of primary resources and/or replace them with renewable or less polluting substitutes; to limit the release of certain wastes or pollutants; or to express social goals for literacy, health, poverty, access to services, and so on.

In effect, this part of the process is an exercise in defining outcomes, which seek to capture "the overall quality or sustainability of human well-being and the eco-systems on which it ultimately depends."[17] Sometimes the process of determining objectives takes place within the planning process, but very often objectives are "imported" from other policy arenas. A pre-eminent example is climate change, where targets for greenhouse gas emission reductions cascade down from international agreements into national action plans and then into planning. Other examples include air quality standards, which may emanate from debates in public health. Where the objectives relate closely to land use, then they may be expressed in vision statements, diagrams, or maps.

Of course, there is no singular, neutral definition of sustainability, but a range of interpretations with quite different developmental consequences. Planning for sustainability is often therefore more than an exercise in applying pre-given goals; it forms part of the process of working out how sustainability should be interpreted. Planning processes may exhibit "weak" forms of sustainability,[18] so-called either because they explicitly prioritize economic growth or because they give no clear priority between economic, social, and environmental goals. The goal of planning for sustainability may thus be expressed as finding a "balance" between these goals, allowing for environmental quality to be traded off for economic development.

Alternatively, planning approaches may pursue "stronger" conceptions of sustainability.[18] Here, the guiding assumption is that society should be precautionary in the face of significant, irreversible losses of valued environmental assets—be they planetary life support systems or treasured landscapes—and give priority to maintaining their value over time. The implications for planning are that environmental constraints are pre-eminent, goals should be set to reflect this, and that development trajectories must be adjusted to observe these constraints and avoid degrading the ability of the environment to support human welfare. Stronger conceptions of sustainability draw support from justice to future generations, and the idea that they should inherit an equivalent stock of "environmental capital"[19]; or from biological notions of observing the "carrying capacity" of an environment to sustain a particular activity without deterioration.[20]

One example of a stronger approach to planning for sustainability is the NEPPs of the Netherlands. From the 1980s, the Dutch Environment Ministry began developing plans that sought to respond to the worsening and interconnected nature of environmental problems through what was regarded as a "holistic" planning process. A key starting point was analysis of the level of pollution reduction and resource efficiency that was required "to achieve sustainable development in one generation."[21] From this, the NEPPs identified targets and brought together steps for reducing pressures on the environment and "closing substance cycles" (e.g., reducing and recycling wastes).

Land use planning offers examples of weaker and stronger interpretations of sustainability. For example, since the early 1990s, U.K. governments have required all land use plans to promote sustainable development; however, national guidance has often interpreted this in weak terms—as achieving simultaneously economic growth, inclusive social progress, protecting the environment, and making prudent use of natural resources.[22] Nevertheless, one can point to land use plans at local and regional levels that have sought to determine the environmental capacity of their area to accommodate development, and introduce policies to ensure that development does not exceed it.[23,24] Observing environmental limits remains an important idea in planning for sustainability.[25]

Devising Actions and Allocating Responsibility

The creation of a plan can itself exercise persuasive effects. There is a long history of planning processes pursuing some form of "visionary idealism" to stimulate public, stakeholder and political support and promote change, many of them connected to aspects of sustainability.[26] However, the second stage of the conventional approach to planning for sustainability usually entails specifying the particular actions required to achieve the objectives of the plan.

Various approaches can be used to identify those activities most responsible for placing unsustainable pressures on the environment—be they industry, government or the public, or particular sectors of activity (transport, housing, agriculture, etc.)—and thus requiring action. One such approach is "backcasting," a form of futures study. The backcasting process begins by envisioning a desirable end state—in this context, one deemed to be sustainable—then works backward from that to identify actions that will move society toward it.[27]

Depending on the political culture of the country concerned, the process of preparing a sustainability plan may involve efforts to secure the participation of different sectors of society. This participation can take place at the objectives-forming stage as well as in the selection of actions. The rationale for participation is at least fourfold:

- To try to achieve some "buy-in" from those stakeholders with the capacity to affect outcomes, and reduce the chance of conflict at the implementation stage
- To draw in knowledge and ideas beyond that possessed by government alone
- To create potentially more just outcomes by conferring recognition and opportunities for participation on a range of social groups[28]
- To create greater democratic legitimacy and societal support for the transitions that sustainability requires.[13]

Some theories of planning see the participatory processes of planning as key to its role in achieving change, notably collaborative planning theory.[15] The interactions and dialogue that takes place in planning processes—where they require parties to make and defend arguments in relatively open arenas—can help shift the attitudes and beliefs of those involved. Planning exercises thus help foster wider learning, develop new knowledge and relations between groups of people, and assist in finding common ground between contending parties. This, in turn, builds up the capacity of society for resolving future problems.

Collaboration between stakeholders was a key element of producing the NEPPs in the Netherlands. Government and business constructed sectoral plans for reducing environmental impacts and technological change bound together by voluntary agreements called "covenants" (albeit often with the threat of tighter regulation should voluntary compliance fail). The Dutch NEPPs also incorporated environmental non-governmental organizations (NGOs) into the planning process. In many countries, the existing land use or spatial planning system has long-provided opportunities for the public and environmental groups to contribute to the decision-making process,[29] and planning for sustainability builds on these traditions.

If planning is to have some strategic role in promoting sustainability across society, then the process of allocating responsibility and devising actions may entail "environmental policy integration."[30] This is a process by which key policy sectors coordinate their actions toward shared sustainability objectives, cooperate on implementation, and reduce conflicting objectives. Examples would be agricultural policy and transport policy being attuned to the delivery of biodiversity or air quality objectives, rather than just their own developmental goals. In procedural terms, integration may be achieved by various innovative governance arrangements, including altering the objectives of sectors, or through various forms of deliberation (i.e., getting key departments round the table) and policy appraisal, as discussed below.

Closely connected with the assignment of responsibility is the identification of actions and policy instruments—the means by which the plan will actually be implemented. Where sustainability goals are being integrated into existing land use planning processes, then the powers of those processes will dictate how the goals can be achieved. In many countries, this entails the incorporation of policies into a plan, with some spatial expression of future goals, which then informs zoning controls or some other regulatory process to achieve compliance. Where sustainability is concerned, these regulations can embrace[31]:

- Promoting sustainable urban form (by controlling building density, or directing development to locations accessible by walking, cycling or public transport)
- Protecting sensitive, valued spaces from damaging development (e.g. for biodiversity)
- Requiring that new developments are designed and built in ways that reduce their adverse impacts.

With national sustainability plans, the plan may provide a long-term framework for shorter- and medium-term actions to be undertaken by a range of actors—government departments, industrial sectors, and the community and voluntary sector. Goals may therefore ultimately be implemented by an array of measures: standards and regulations, market-based instruments like green taxes, channeling investment or research funding, and various forms of voluntary mechanisms (like information campaigns, promoting best practice, etc.).

Appraisal and Monitoring

Forms of appraisal and monitoring are integral to conventional approaches to planning for sustainability and may be used either ex post or ex ante to guide the planning process. Ex post entails monitoring the effects of the plan after implementation (either progress toward goals or of wider unintended effects), and using the information to make adjustments or revisions in cyclical, learning processes. Ex ante appraisal entails assessing the likely effects of the plan while it is being formulated, perhaps to ensure that it is contributing sufficiently to sustainability.

The use of appraisal techniques has become a significant part of planning for sustainability and links to wider suites of tools used in environmental management, many drawing on international principles and methodologies.[10] Strategic Environmental Assessment (SEA) is required in many countries for plans, programs, and policies likely to affect the environment (e.g., under European Union Directive 2001/42/EC). Through SEA, the draft policies of plans are assessed for their environmental impacts and, in many formulations, this entails assessment against sustainability criteria and objectives.[32] SEA can also enable the comparison of different options and, potentially, facilitate debate over alternatives, as well as fostering environmental policy integration by identifying where sectoral policies are poorly aligned with sustainability goals.

Environmental indicators may be used to assess changes in the economy, society, and environment, and signal whether those changes lead us toward or away from sustainability. Some indicators reflect a particular environmental parameter. For example, in the United Kingdom, the percentage of new development located on brownfield land (i.e., land that has been built on before) is an important sustainability indicator for the land use planning system, with the goal being that increasing development on brownfield sites will reduce sprawl, slow down the consumption of "greenfield" land, and support compact, less car-dependent urban forms.

The search for better appraisal and measurement systems has been a constant feature of planning for sustainability debates: where "better" can be taken as more comprehensive, more integrated and quantifiable (i.e., combining different effects into single units), more comprehensible to diverse audiences or decision-makers, or better linked to core principles of sustainability like environmental limits. One example is the "ecological footprint," which seeks to provide a proxy measure of human demands on the environment by assessing how much biologically productive land and sea is appropriated to maintain a given consumption pattern and assimilate the waste produced.[33] There are other approaches also seeking to capture the totality of material and ecological flows associated with "urban metabolisms,"[34] including "exergy," a concept that derives from thermodynamics and ecosystem health analysis.[35]

Rising in profile from 2010 onwards has been interest in assessing the effects on ecosystem services; an approach that reflects a long-standing belief in environmental appraisal circles that for decision-makers to take environmental effects seriously they need to be expressed in economic terms.[36]

Has Planning for Sustainability Had Any Effect?

Across the world, numerous plans have been produced that would claim to be promoting sustainability. In many countries, sustainable development has become "mainstreamed" as a core goal of land use or spatial planning—with almost all such plans, and much national planning policy too, now making reference to the importance of sustainability.[7,8,10,11] What is less clear is how far these planning *outputs* have promoted more sustainable *outcomes*. This section reviews the effects of planning for sustainability and then explains why outcomes rarely match aspirations.

One can point to beneficial outcomes in a number of spheres. Sustainability-oriented land use plans have contributed to an incremental greening of urban development. By providing locational guidance, plans have helped steer development to sites where its impacts have been lower than they might otherwise have been, reducing inappropriate development in areas of high existing environmental value, and focused development on advantageous locations—such as sites that encourage access by public transport or foot, or face reduced risks from flooding. The planning system also confers greater protection to environmental assets, such as valued wildlife sites, and requires more exacting mitigation and compensation measures when development has damaging effects. Over the same period, plans have mobilized policies to enhance the environmental performance of built development, for example, by promoting better design, or as one of a suite of tools for ensuring development is adapted to the risks of climate change. Plans have developed innovative regulations,[31,37] such as the "One Planet Development" policy in Wales, allowing development in the countryside where it demonstrably attains a low Ecological Footprint.[38]

Nevertheless, the overall effects of planning on the sustainability of development have been modest, and uneven between issues and places. Unsurprisingly perhaps, land use planning has made the greatest progress where sustainability agendas link to traditional planning concerns, such as land allocation and siting policies, or issues like transportation where technical fixes are available. Where particular environmental issues are of recognized importance, then planning has addressed them, such as water scarcity in parts of the United States and Australia.[10,39] The contribution of planning is often less clear in relation to global, systemic issues, like reducing greenhouse gas emissions, poverty or resource depletion, where wider societal transformation is required.

With national sustainability planning, too, most strategies have gravitated "toward the cosmetic rather than the ideal."[40] The NEPPs from the Netherlands were most successful in reducing wastes, pollution, and the use of toxic materials; however, progress has been greatest in areas where industry saw the potential for economic gains from improving environmental performance, and where technological solutions were readily available.[41] Moreover, the NEPP strategies have struggled to address issues caused by mass consumption and growing personal mobility, such as carbon emissions from transport. If there has been any decoupling of economic growth from environmental degradation, it is mostly relative decoupling (falling environmental impact per unit of growth) rather than absolute decoupling (environmental impacts falling in absolute terms). Many countries have still not produced national sustainability plans, and those that have done so too rarely institute cycles of reporting and updating.[3]

Thus, despite 30 years of "planning for sustainability" around the globe, progress in fostering more sustainable patterns of development has been patchy at best, and often disappointing. As noted above, nations differ greatly in their capacity for organized and effective planning.[42] That said, success is partial in many developed countries, including those with strong traditions of strategic plan-making and reputations for environmental concern, for example, the Netherlands[43] and Sweden.[44] Despite some modest gains, the overwhelming perception is one of a significant "implementation deficit," in which the outcomes achieved through planning fail to match the rhetoric of sustainability or meet the more challenging targets. How might this "deficit" be explained?

Some see it as a simple time lag between aspiration and realization. Plans take time to prepare and implement; consequently, they can omit newer, emerging issues, like sustainable urban food.[45]

Ambitious planning for sustainability also confronts the limited power of planning mechanisms to effect change. The regulatory powers of land use planning do not apply to every sphere of social or technical change. It can be used to regulate new built development—indeed, new "eco-cities" or "sustainable communities" with higher sustainability credentials than the norm are frequently held up as exemplars of what can be achieved.[46] However, planning has limited scope to reshape existing buildings or economic systems, especially those that unfold at wider spatial scales. Thus, land use planning might be successful in steering the location of supermarkets, even influence the environmental performance of their built form (perhaps by requiring certain energy efficiency standards), but exercises little control over the systems of food production and consumption channeled through them. Research suggests that planning professionals have sometimes molded planning for sustainability toward orthodox planning tools and approaches rather than looking at problems afresh.[11,45]

However, there are fundamental problems in conceiving of this underachievement as an "implementation gap," with the expectation that this will be closed over time with "more knowledge," "better practices," or "improved participation."

One problem with the conventional, linear view of planning for sustainability is the assumption that sustainability has an agreed meaning that can be translated into clear consensual goals, which can inform singular blueprints for progress. In reality, although there may be broad agreement on the principles of sustainability, translating this into more specific interpretations in the planning process can cause dispute, as fundamental questions are raised about human wants and needs, the relationship between state and market, and environmental values. Planning for sustainable development cannot therefore be a value-free, technical exercise.[2] Even within the environmental sphere, there is often significant scientific uncertainty about the causes of change, and the position (or existence) of ecological thresholds or "capacities," which affects the definition of targets or indicators.

Planning for sustainable development has also exposed the difficulties of achieving outcomes that are win-win for social, economic, and environmental goals. Synergies between goals can be found: for example, planning for more walking-friendly cities has benefits for human health, environmental protection, and local economies where congestion is reduced. In many other circumstances, goals cannot readily be reconciled without winners and losers. Protecting valued landscapes or habitats can displace local livelihoods, including those which are highly sustainable.[9,45] Displacement can also be caused by the expansion of renewable energy like solar power.[47] Newly enhanced or created sustainable communities can make property too expensive for poorer existing residents.[39] Environmental constraints may challenge particular directions of economic growth. Planning for sustainability may thus struggle to align competing conceptions of what it means for development to be sustainable.[7]

One also needs to understand the political and economic context in which planning for sustainability takes place, particularly macro-economic shifts that have made communities and nations dependent on the private sector to deliver development, and public sector austerity programs that reduce the resources for long-term environmental activity.[48] The vulnerabilities this creates are illustrated by particular projects, for example, Masdar Eco-City in Abu Dhabi. This new settlement was planned with very high sustainability ambitions, but the 2008 financial crash undermined the economics of the project, leading to the most innovative environmental dimensions being abandoned.[49] More widely, economic concerns may explain why planning authorities may be reluctant to exercise stringently the regulatory powers they do possess, where that could threaten jobs and growth. Plans that have sought to promote "strong" interpretations of sustainability, by seeking to keep development within environmental limits, have faced resistance when confronting what "seem to be inexorable upward trends in production, consumption, and mobility."[7]

This also explains why many exercises in planning for sustainability have adopted "weaker" interpretations of the concept, seeking only to achieve some "balance" between economic, environmental, and social goals. Sometimes the interpretation of sustainability has been weaker still, with planning goals framed in ways that accommodate rather than question unsustainable patterns of production and demand, such as for road space, mineral resources, or speculative property assets, causing contradictions. For example, although Dutch spatial plans sought to encourage greener forms of travel, they still promoted the Netherlands' international role as a transport hub, incorporating goals for expanding seaports and airports, with deleterious environmental consequences.[50] Here, one can see that the problem is not a straightforward "implementation deficit," but that sustainable development has been defined in ways that accommodate the status quo.

These difficulties have raised further questions about the procedural dimensions of planning for sustainability.

- *Participation and expertise*: There is evidence to support the argument that wider societal participation helps support plans that emphasize sustainability.[8] However, securing "adequate" public engagement in planning processes encounters a number of problems. The range of complex ecological issues that sustainability has brought into planning processes raises questions about how different forms of knowledge are handled and tested. A common risk is that technical elites and industry representatives dominate the process, to the exclusion of other groups. These dilemmas, in turn, highlight the risks of collaborative planning processes that give privileged access to key economic players, which act to protect their immediate interests; Van der Straaten[51] suggests that the NEPPs underperformed because powerful actors were able to weaken targets.
- *Futurity and uncertainty*: For all that planning and sustainability both seek to respond to long time horizons, the realities on which present actions will be based are constantly shifting in the light of new knowledge and social changes. This creates tensions between instituting long-term plans and remaining responsive to change[52] without caving in to expedient short-termism.[53]
- *Environmental policy integration*: The idea that planning for sustainability requires "integration" has also been easier to say than to achieve in practice.[30] The kind of voluntarist, collaborative approaches to integration that planning processes tend to foster has proved insufficient in shifting dominant sectoral objectives toward sustainability goals.[54] National sustainability strategies are still often led by environment ministries, which then struggle to exercise much influence over other sections of government. Ironically, land use planning is often left managing the side effects of sectoral policies that national sustainability strategies have failed to integrate with sustainability goals.

The overall evidence of the years since Agenda 21 is that planning is a desirable but insufficient ingredient in progressing sustainable development. For some, the problem is the concept of sustainability itself, which has been accused of being vacuous, vague, too contradictory and too readily captured by powerful interests. Some have shifted their attention to goals that seem to offer firmer anchors for progress, like resilience or decarbonization,[26] but such concepts may be no less prone to interpretive and political challenges when used in actual planning.

Perhaps the problems lie partly in the planning approach? The implementation problems discussed above expose how conventional approaches to planning for sustainability, depicted in Figure 1, tend to adopt an "ineffective engineering ideal,"[21] seeing society as a technical system that can be steered from the center to deliver targets. This strategy can work for narrow issues like pollution control but performs less well for complex social problems. However, if the above analysis shows planning for sustainability to be a more complex and challenging task than it might at first appear, it also points the way to two alternative perspectives.

Rethinking Planning for Sustainability

Adaptive Planning and Transition Management

If initial interest in using planning to promote sustainability was often guilty of naive optimism in what could be achieved, one by-product of these implementation difficulties has been a better understanding of the problem.

Sustainable development is pursued in a world of multidimensional, intersecting and dynamic complex systems. We cannot expect to describe them fully, much less predict future effects. We may lack even suggestive evidence about many emerging problems, whose influences will ripple unpredictably through complex socio-ecological systems.[55]

One response has been to advocate approaches to promoting sustainability that are better suited to this reality, by showing "explicit appreciation of complexity and uncertainty, likelihood of surprise, and need for flexibility and adaptive capacity."[55] Some would describe these approaches as "adaptive planning"[56]; others have developed concepts of "transition management."

In practice, adaptive planning and transition management embrace elements of the conventional approach to planning for sustainability, but with different emphases. They incorporate an interest in developing shared long-term objectives, sustainability visions, and being anticipatory, as well as the use of indicators and monitoring to assess progress. However, there is less emphasis on "the old plan-and-implement model aimed at achieving particular outcomes"[55] in favor of "a more open-ended, process oriented philosophy." Rather than pursuing a single vision of sustainability, there is a concern to explore a diversity of different routes in a more iterative way, using a portfolio of ideas including "transition experiments," to develop circles of learning and adaptation. Adaptive planning takes an iterative approach with a greater emphasis on "monitoring, researching, and adjusting," allowing the redirection of previous management goals and activities "in light of new information and surprise,"[57] which could be an environmental or social change. There is less accent on achieving specific goals and more on institutional designs that prove resilient in the event of failure.

Some transition management approaches also take a much wider view of what needs to be changed if sustainability is to become a reality; from more targeted, incremental improvements in existing development forms to "promoting the sustainable reconfiguration of entire socio-technical systems."[58] This reflects the fact that "the power to shape structural change in society and technology is distributed across a multitude of actors and societal subsystems."[59] Transition management also emphasizes fostering innovation.

There is limited evidence for the actual effects of transition management approaches,[60] but the ideas have been applied in the Netherlands. Dissatisfied by the poor performance of the NEPPs, and keen to "revitalize" the process,[58] policy makers incorporated transition management in the fourth NEPP, released in 2001, and committed the government to sustainability transitions in energy, transport, water management, natural resource use, and agriculture.[55] The process involved creating a series of "platforms" in which stakeholders came together, to develop "transition themes" and pathways to those goals, including "transition experiments" to explore in practical terms how to progress along those pathways.[58]

Whatever its theoretical merits, the Netherlands' initial experiences raise questions about how far, in practice, elements of transition management have actually exerted pressure on business-as-usual. Smith and Kern[58] show how, when applied to the energy sector, the desire to incorporate relevant, powerful actors from industry and government departments into the process left it open to being captured by them; a wider risk of stakeholder engagement, as noted above. The result was that the transitions management approach failed to displace a narrow focus on technological costs and benefits or to exert much influence on wider energy policy. Absent also was any wider public debate about the transition goals or pathways. Other commentators have expressed a concern that transition management does not avoid the problems of conventional, linear approaches to planning for sustainability in assuming that there could be a consensus among diverse stakeholders about what constitutes a sustainable development path.[61]

Arguably, transitions management approaches also display another problem of much planning for sustainability: a tendency to focus excessively on new, more innovative forms of development, and give insufficient attention to how societies challenge and displace existing, unsustainable development forms.[62] However, planning can have a role in this, as explained below.

Planning and the Politics of Sustainability Transitions

The second alternative perspective considered here requires us to look beyond the intended goals of planning, to understand how planning processes have actually served to promote more sustainable forms of development. This requires us to view planning less as an activity on its own and more in relation to wider systems of public policy, and the politics of decision making. Viewed in this way, the impact of planning on sustainability may be messier, and less obvious in the short term, but more profound over the long term.[7,63]

One important feature of planning processes is that they provide "opportunity structures"[64]—including consultation procedures over plans and projects, legal review mechanisms, and various forms of environmental assessment—that allow the involvement of the public and interest groups, and through which argument about social purpose can take place. A distinctive feature of the opportunity structures provided by planning is the scope they offer for cross-scalar discussions, that is, through engagement about future development projects or land uses participants may also come to question the wider policies and values that lie behind them. Participants may raise fundamental concerns about the environmental sustainability of certain forms of development, and press for lower-impact alternatives.[45,65] The arguments generated can also exert pressure on environmentally damaging forms of development through extra costs, delays, and the ever-present threat of veto. Over time, the cumulative effect of these discursive and material impacts has sometimes undermined previously dominant policy approaches, loosening the grip of powerful interests, fostering acceptance of tightening environmental conditions, and opening up long-held policy norms.[63]

One can see these dynamics in a number of contexts. Challenges made through (and beyond) planning processes to new road schemes and their environmental impacts have helped reinforce arguments for a change in transport policy, away from predicting travel demand and building road space to accommodate it, toward greater emphasis on public transport and managing travel demand. Airport expansion planning has also become an important venue for debates about the environmental risks and societal worth of unfettered aviation growth.[7,66] Hudalah et al.[67] chart the way in which environmental NGOs in Indonesia exploited moments of opportunity in project and plan decision making to form new coalitions, and increase the weight given to wider development agendas based on conserving ecological functions. Planning is important because it is where broad policies and otherwise abstract sustainability concepts meet real, concrete environmental changes, stimulating public engagement and demanding solutions.

For all that it might be messy, conflictual, and unintended, one can see how existing planning systems have generated the kind of debate and policy learning that transitions to sustainability are often deemed to demand.[55] It is often conflict as much as cooperation that forces new perspectives onto the agenda. Moreover, this function of planning has reinforced sustainability transitions by subjecting unsustainable development to intensive questioning and, in some cases, presaging its demise—a dimension of change that the technology/innovation-centered approaches of transitions management tend to neglect.[62]

The challenge is to maintain openness of planning processes to wide-ranging debate, often in face of efforts by governments to keep fundamental questions of need and risk off the agenda.[7]

Conclusion

A central problem with the proposition that one can "plan for sustainability" is that sustainability is not a concept that can be defined precisely and then implemented.[68] Rather, planning is one of a number of arenas in which the meaning of sustainability is debated and constructed, often in the face of significant challenges. Indeed, the experiences of planning have been important for showing that formulating policies for sustainability can open up quite fundamental questions about needs, demands, and environmental value. Such questions are implicit in asking—"what is it that we should sustain?" Because of the disagreements surrounding such fundamental questions, progressing toward sustainable development is inevitably a process of "muddling through,"[14] even when plans look clear on paper, and likely to be contested—especially perhaps if plans are seeking to make a real difference to unsustainable trends.

However, although the conventional, linear view of planning for sustainability displays a number of weaknesses, planning is a vital component of social progress toward sustainable development. Through the intended goals and actions of plans, more sustainable forms of development can be delivered; at the same time, the spaces for deliberation and challenge that planning creates form a key part of the "ongoing territorial struggle"[69] to understand and institutionalize more sustainable forms of growth.

This experience suggests that when examining processes of planning for sustainability, one needs to do more than look at whether the core features of the conventional approach are in place: clear goals and targets, well-chosen actions, mechanisms for public and interest group engagement, and systems for monitoring and review. One should also ask the following questions: what are the opportunity structures for influencing the plan and who is able to access and exploit them? What scope is there for raising alternative development strategies and challenging dominant but unsustainable development priorities?[70]

References

1. Rose, E. Philosophy and purpose in planning. In *The Spirit and Purpose of Planning*, 2nd Ed.; Bruton, M.J., Ed.; Hutchinson: London, 1984; 31–65.
2. Springett, D.; Redclift, M. Sustainable development: History and evolution of the concept. In *Routledge International Handbook of Sustainable Development*; Redclift, M., Springett, D., Eds.; Routledge: London, 2015; 3–38.
3. Steurer, R.; Martinuzzi, A. (2007) 'Editorial: From environmental plans to sustainable development strategies', *Eur. Environ.* **2007**, *17*, 147–151.
4. United National Conference on Environment and Development (UNCED). *Agenda 21: The United Nations Programme of Action from Rio*; United Nations Department of Public Information: New York, 1992; para 7.28.
5. CEC. *Towards Balanced and Sustainable Development of the European Union*; Office for Official Publications of the European Commission: Luxembourg, 1999.
6. Bishop, K. Planning to save the planet? Planning's green paradigm. In *British Planning Policy in Transition*; Tewdwr-Jones, M., Ed.; UCL Press: London, 1996; 205–219.
7. Owens, S.; Cowell, R. *Land and Limits: Interpreting Sustainability in the Planning Process*; Routledge: London, 2002.
8. Conroy, M.; Berke, P. What makes a good sustainable development plan? An analysis of factors that influence principles of sustainable development. *Environ. Plan. A* **2004**, *36*, 1381–1396.
9. Lestrelin, G.; Castella, J.; Bourgoin, J. (2012) Territorialising sustainable development: The politics of land-use planning in Laos. *J. Contemp. Asia 42* (4), 581–602.
10. Davidson, K.; Arman, M. (2014) Planning for sustainability: An assessment of recent metropolitan strategies and urban policy in Australia. *Aust. Plan.* **2014**, *51* (4), 296–306.
11. Persson, C. Deliberation and doctrine? Land use and spatial planning for sustainable development in Sweden. *Land Use Policy* **2013**, *34*, 301–313.
12. Davies, J.G. *The Evangelistic Bureaucrat: A Study of a Planning Exercise in Newcastle upon Tyne*; Tavistock: London, 1972.
13. Blowers, A. *Planning for a Sustainable Environment*; Town and Country Planning Association: London, 1993.
14. Lindblom, C. The science of 'muddling through'. *Public Adm. Rev.* **1959**, *19*, 79–88.
15. Healey, P. *Collaborative Planning: Shaping Places in Fragmented Societies*; Macmillan: Basingstoke, 1997.
16. Rydin, Y. Sustainable development and the role of land use planning. *Area* **1995**, *27* (4), 369–377.

17. Adger, W.N.; Jordan, A. Sustainability: Exploring the processes and outcomes of governance. In *Governing Sustainability*; Adger, W.N., Jordan, A., Eds.; Cambridge University Press: Cambridge, 2009; 5.

18. Jacobs, M. *The Green Economy*; Pluto Press: London, 1991.

19. Pearce, D.W.; Turner, R.K. *Economics of Natural Resources and the Environment*; Harvester Wheatsheaf: Hemel Hempstead, 1988.

20. Meadows, D.H.; Randers, J.; Meadows, D.L. *Limits to Growth: The Thirty Year Update*; Earthscan: London, 2005.

21. Van der Straaten, J.; Ugelow, J. Environmental policy in the Netherlands: Change and effectiveness. In *Rhetoric and Reality in Environmental Policy. The Case of the Netherlands in Comparison with Britain*; Wintle, M.; Reeve, R., Eds.; Ashgate: Aldershot, 1994; 127.

22. Department of Transport, Environment and the Regions. *A Better Quality of Life: A Strategy for Sustainable Development*; Cmnd 4345; The Stationery Office: London, 1999.

23. Cowell, R.; Owens, S. Sustainability: The new challenge. In *Town Planning into the 21st Century*; Blowers, A., Evans, B., Eds.; Routledge: London, 1997; 15–32.

24. Counsell, D. Attitudes to sustainable development in the housing capacity debate: A case study of the West Sussex structure plan. *Town Plan. Rev.* **1999**, *70* (2), 213–229.

25. Haines-Young, R.; Potschin, M.; Cheshire, D. *Defining and Identifying Environmental Limits for Sustainable Development. A Scoping Study*; Final Overview Report to Defra, Project Code NR0102; Defra: London, 2006.

26. Gunder, M. Sustainability: Planning's saving grace or road to perdition? *J. Plan. Educ. Res.* **2006**, *26* (2), 208–221.

27. Miola, A., Ed. *Backcasting Approach for Sustainable Mobility*; European Commission Joint Research Centre: Luxembourg, 2008.

28. Schlosberg, D. *Defining Environmental Justice. Theories, Movements, and Nature*; Oxford University Press: Oxford, 2007.

29. Lowe, P.; Goyder, J. *Environmental Groups in Politics*; Allen and Unwin: London, 1983.

30. Lafferty, W.M. From environmental protection to sustainable development: The challenge of decoupling through sectoral integration. In *Governance for Sustainable Development. The Challenging of Adapting Form to Function*; Lafferty, W.M., Ed.; Edward Elgar: Cheltenham, 2004; 191–220.

31. Gurran, N.; Gilbert, C.; Phibbs, P. (2015) 'Sustainable development control? Zoning and land use regulations for urban form, biodiversity and green design in Australia', *J. Environ. Plan. Manage.* **2015**, *58*(11), 1877–1902.

32. Therivel, R.; Partidario, M. *The Practice of Strategic Environmental Assessment*; Earthscan: London, 1996.

33. Wackernagel, M.; Rees, W. *Our Ecological Footprint: Reducing Human Impact on the Earth*; New Society Publishers: Gabriola Island, 1996.

34. Curry, R.; Ellis, G. Material flow analysis. In *Routledge Companion to Environmental Planning*; Davoudi, S., Blanco, H., Cowell, R., White, I., Eds., 2019, 346-357, Routledge, Abingdon UK and New York.

35. Jørgensen, S.E. *Eco-Exergy as Sustainability*; WIT Press: Ashurst, 2006.

36. Cowell, R.; Lennon, M. The utilization of environmental knowledge in land use planning: Drawing lessons for an ecosystem services approach. *Environ. Plan. 'C': Gov. Policy* **2014**, *32*, 263–282.

37. Harris, N. Exceptional spaces for sustainable living: The regulation of one planet developments in the open countryside. *Plan. Theory Pract.* 2019 *20*(1), 11–36.

38. Jepson, E.J.; Haines, A.L. Zoning for sustainability: A review and analysis of the zoning ordinances of 32 cities in the United States. *J. Am. Plan. Assoc.* **2014**, *80* (3), 239–252.

39. Long, J. Constructing the narrative of the sustainability fix; sustainability, social justice and representation in Austin TX. *Urban Stud.* **2016**, *53* (1), 149–172.

40. Meadowcroft, J. National sustainable development strategies: Features, challenges and reflexivity. *Eur. Environ.* **2007**, *17*, 152–163.

41. Van Muijen, M. The Netherlands: Ambitious on goals—Ambivalent on action. In *Implementing Sustainable Development: Strategies and Initiatives in High Consumption Societies*; Lafferty, W., Meadowcroft, J., Eds.; Oxford University Press: Oxford, 2000; 142–173.

42. Jänicke, M.; Weidner, H., Eds. *National Environmental Politics: A Comparative Study of Capacity-Building*; Springer: Berlin, 1997.

43. Weale, A. *The New Politics of Pollution*; Manchester University Press: Manchester, 1992.

44. Fudge, C.; Rowe, J. Ecological modernisation as a framework for sustainable development: A case study in Sweden. *Environ. Plan. A* **2001**, *33*, 1527–1546.

45. Adams, D., Scott, A.J.; Hardman, M. Guerrilla warfare in the planning system: Revolutionary progress towards sustainability? *Geogr. Ann. B* **2013**, *95* (4), 375–387.

46. Pandis Iverot, S.; Brandt, N. The development of a sustainable urban district in Hammarby Sjöstad, Stockholm, Sweden? *Environ. Dev. Sustain.* **2011**, *13*, 1043–1064.

47. Yenneti, K.; Day, R.; Golubchikov, O. Spatial justice and the land politics of renewables: Dispossessing vulnerable communities through solar energy mega-projects. *Geoforum* **2016**, *76*, 90–99.

48. Flint, J.; Raco, M., Eds. *The Future of Sustainable Cities. Critical Reflections*; Policy Press: Bristol, 2012.

49. Jensen, B. Masdar City: A critical retrospective. In *Under Construction: Logics of Urbanism in the Gulf Region*; Wippel, S., Bromber, K., Steiner, C., Krawietz, B., Eds.; Routledge: London, 2016; 45–54.

50. Woltjer, J. *Consensus Planning: The Relevance of Communicative Planning Theory in Dutch Infrastructure Development*; Ashgate: Aldershot, 2000.

51. Van der Straaten, J. The Dutch NEPP: To choose or to lose. *Environ. Polit.* **1992**, *1* (1), 45–71.

52. Smith, A.; Stirling, A. Moving outside or inside? Objectification and reflexivity in the governance of sociotechnical systems. *J. Environ. Policy Plan.* **2007**, *9* (3/4), 351–373.

53. Boston, J. *Governing for the Future. Designing Democratic Institutions for a Better Tomorrow*; Emerald Group Publishing: Bingley, 2016.

54. Degeling, P. The significance of 'sectors' in calls for urban publish health intersectoralism: An Australian perspective. *Policy Polit.* **1995**, *23* (4), 289–301.

55. Kemp, R.; Parto, S.; Gibson, R. Governance for sustainable development: Moving from theory to practice. *Int. J. Sustain. Dev.* **2005**, *8* (1/2), 12–30.

56. Holling, C.S. *Adaptive Environmental Assessment and Management*; Wiley: Chichester, 1978.

57. Lessard, G. An adaptive approach to planning and decision making. *Landsc. Urban Plan.* **1998**, *40*, 81–87.

58. Smith, A.; Kern, F. The transitions storyline in Dutch environmental policy. *Environ. Polit.* **2009**, *18* (1), 79–98.

59. Newig, J.; Voss, J.; Monstadt, J. Editorial. Governance for sustainable development in the face of ambivalence, uncertainty and distributed power. *J. Environ. Policy Plan.* **2007**, *9* (3/4), 185–192.

60. Jordan, A. The governance of sustainable development: Taking stock and looking forwards. *Environ. Plan. C Gov. Policy* **2008**, *26*, 17–33.

61. Walker, G.; Shove, E. Ambivalence, sustainability and the governance of socio-technical transitions. *J. Environ. Policy Plan.* **2007**, *9* (3/4), 213–225.

62. Shove, E.; Walker, G. Transitions ahead: Politics, practice, and sustainable transition management. *Environ. Plan. A* **2007**, *39* (4), 763–770.

63. Cowell, R.; Owens, S. Governing space: Planning reform and the politics of sustainability. *Environ. Plan. C Gov. Policy* **2006**, *24* (3), 403–421.

64. Kitschelt, H. Political opportunity structures and political protest: Anti-nuclear movements in four democracies. *Br. J. Polit. Sci.* **1986**, *16*, 58–95.

65. Freudenberg, N.; Steinsapir, C. Not in our backyards: The grassroots environmental movement. *Soc. Nat. Res.* **1991**, *4* (3), 235–245.
66. Hayden, A. Stopping Heathrow Airport expansion (for now) lessons from a victory for the politics of sufficiency. *J. Environ. Policy Plan.* **2014**, *16* (4), 539–558.
67. Hudalah, D.; Winarso, H.; Woltjer, J. Planning by opportunity: An analysis of peri-urban environmental conflicts in Indonesia. *Environ. Plan. A* **2010**, *42* (9), 2254–2269.
68. Owens, S. Land, limits and sustainability: A conceptual framework and some dilemmas for the planning system. *Trans. Inst. Br. Geogr.* **1994**, *19*, 439–456.
69. Dierwechter, Y. *Urban Growth Management and Its Discontents. Promises, Practices and Geopolitics in U.S. City Regions*; Palgrave Macmillan: New York, 2008.
70. Owens, S.; Cowell, R. *Land and Limits: Interpreting Sustainability in the Planning Process*, 2nd Ed.; Routledge: London, 2010.

18

Sustainable Development

Mark A. Peterson

Introduction

This entry defines sustainable development and its three basic aspects. Because sustainable development is a relatively new concept, a short history and description of the drivers that lead to sustainable development are described. Then, a sustainable energy future is presented. To be sustainable as a society requires cooperation and collaboration rather than command and control management. A very key aspect of sustainable development, social synergy, is covered. Also, because sustainable development is relatively new but essential to build a better and viable future, children from the earliest age through college need to learn to understand and apply the concepts of sustainable development. A paragraph is included that describes current efforts in the United States to incorporate education in sustainable development (ESD) in K–12 and college curricula.

Sustainable development encompasses stewardship of many areas of human and planetary life. In business, one of the key motivators is to implement sustainable development measures to be profitably successful indefinitely. To do this requires that businesses show their due diligence to both society and the environment while maximizing profits. Sustainability reporting assists businesses in assessing their efforts. Sustainability reports are both management and public relations tools. An innovative, very effective, time-tested form of sustainable development, "renting a service" rather than "selling a product," is covered. Then, sustainable development for community vitality is described. As the sizes, types, and socioeconomics of communities vary considerably, so do all of the related aspects of evolving them to be sustainable. Reference is made to a web site that thoroughly describes all aspects. Some of the subjects covered on the web site are briefly summarized in this entry. The final section covers the vast, opening market of sustainable development in developing countries.

What Is Sustainable Development?

Sustainable development is development that meets the needs of the present without compromising the ability of future generations to meet their needs. Sustainable development has three aspects:

1. Social (people)
2. Environmental (planet)
3. Economic (profits)/prosperity

All development affects all three aspects. All three aspects are interdependent. Thus, being mindful of these interdependencies in management and leadership decisions will result in the best overall solution—a win-win-win solution that maximizes success and minimizes any negative social, environmental, and economic costs. This is called managing the triple bottom line of people, planet, and profits. This is also called whole systems thinking[1,2] because all relevant factors are considered as a whole. The role of engineers is to help their clients be successful. This requires integrated whole systems thinking that covers all related liabilities that a company or community (their client) may have and provides the most efficient and profitable solution to the challenge. Often, the best whole system solution is also the most efficient and most sustainably profitable.

The environmental (planet) aspect is significantly affected by energy consumption and management, including: the entire national power infrastructure and distribution, transportation, plus the construction and renovation of all residential, commercial, and industrial facilities.

History, Environmental Degradation, and National Security

During the last century, while fossil fuels were abundant and cheap, those fuels fulfilled a majority of our energy conversion needs. The mounting problem is that combustion emissions have fouled the environment in a number of ways, resulting in increases in respiratory illnesses, mercury pollution, and a rise in global temperatures. The quantity of easily retrieved fossil fuels is significantly depleted. Coal is still relatively abundant, but it does not burn cleanly. Technologies need to be developed to both mine the coal safely and to burn it cleanly. Regarding petroleum, many countries that are not friendly to the United States control most of the remaining easily extractable sources. Many national security advisors have indicated the urgency of severing our dependence on foreign oil as a part of an overall strategy for the security of the United States and as a means to prevent oil-related conflicts.[3,4] It is also apparent that there is a need to protect and allow the environment to regenerate. The effect of using fossil fuels extensively and inefficiently is that we are simultaneously poisoning the environment and ourselves. Nuclear power emissions are clean, but the nuclear power industry has significant obstacles such as storage of radioactive wastes for many thousands of years. In addition, there are security concerns to safeguard radioactive material from being stolen for production of atomic weapons.

Governors and mayors are taking action to implement clean energy technologies. In response to clear signs of increased cost from continuing to use fossil fuels and scientific evidence showing that by burning fossil fuels we are initiating a possibly devastating global warming trend[5] that could flood coastal cities, disrupt the food chain, and change climate patterns significantly, many states have taken the initiative and enacted renewable energy portfolios to fund the transition to renewable energy resources. Many remaining states are in the process of developing their own renewable energy portfolios. These renewable energy portfolios provide significant state- and utility-sponsored financial incentives for the commercial, industrial, and residential use of renewable energy systems and fuels. On the city level, many mayors from major cities around the world have made commitments to cut greenhouse gas emissions to slow the rate of global warming.[6] Many of these cities are coastal and could be severely impaired or destroyed from rising sea levels from global warming. So, civic action to switch to cleaner energy options is beginning in earnest.

Energy Future

What this means to energy engineers is that petroleum-derived fuels are on their way out over the next half century. Hydrogen (where the hydrogen is derived from renewable energy sources), ethanol, biodiesel, and other forms of renewable fuels are on their way in. Direct and indirect conversion of solar energy, including wind, biomass, wave/tidal power, and small-scale hydroelectric power will increasingly be part of the energy infrastructure that energy engineers will design and build. The bottom line with energy is that it needs to be relatively nonpolluting and indefinitely available. It is a very dynamic time for energy engineers as the entire, worldwide energy picture transitions to clean renewable technologies. This will eventually add a lot of stability to the world economy, the world political environment and to everyone's lives. The stability will come from the fact that renewable energy technologies can be used to tap the natural energy resources that are available everywhere. Stability will also come, as the environment regenerates, the climate stabilizes and resources remain available for our sustenance.

How We Socially and Professionally Interact with Each Other Determines Our Degree of Success

Another aspect of sustainability has to do with how well we interact and collaborate. In the past, management of most activities was by a top-down hierarchy. Now humanity is evolving and it is driven by high levels of sophistication in technology and communications, which has resulted in individual knowledge and skill level increases. Plus, many families are now structured such that individual adults and children are taking more independent responsibilities for the many aspects of their lives. This has all created a desire by many people to be more intimately involved in solutions rather than just letting someone else do the thinking. In this new economy, teamwork and open communication are important to bring all stakeholders together, whether in a business or community, to manage by consensus and cooperation. Everyone affected should have the opportunity to be involved in the solution, even if just by being informed as the planning and decision-making are in process. The outcome then is one that promotes efficiency, for the simple reason that all persons affected are involved, which promotes enthusiasm and "buy-in." Typically with this process, more work and time are invested up-front such that all aspects are considered and thus everything proceeds more efficiently down-stream.[7]

Creative, Cooperative, Design and Planning Teamwork[2]

The ASHRAE GreenGuide recommends "integrated design teams" that have all of the design, economic, planning, and other related disciplines involved up-front to create better designs. If this process is not used, design typically proceeds in a series of "handoffs" that tend to compound problems, as each succeeding team designs "around" any incompatibilities that the previous designers have already finished. This adds unnecessary complexities and inefficiencies, which increase construction and life-cycle costs. Through coordination and the collaboration of designers, architects, engineers, and key players, an integral design can be created that functions as an efficient system and not as a collection of parts that are force-fit together. This approach has enabled design teams to design very energy efficient, comfortable, aesthetically and environmentally friendly buildings at or less than the conventional price per square foot of traditionally designed buildings. So, working as a team from the beginning is the most efficient way of designing because potential conflicts are resolved up-front rather than later at a higher cost. Effectually, when people are creating and they know that their contributions are respected, superior planning and designs are achieved. The upfront work of coordinating planning and brainstorming sessions with many people of diverse backgrounds can be a challenge. However, the results and life-cycle costs are almost assuredly optimal.

Education

The United Nations has declared the decade from 2005 to 2014 as The Decade of Education for Sustainable Development.[8]

There are many national teams around the world that have taken the lead to work across the educational spectrum—including public and private education, primary and higher education, independent, charter, and home-schooling—to incorporate ESD in their curricula. The U.S. Partnership for The Decade of Education for Sustainable Development[9] was formed to facilitate implementation of ESD in the United States.

Effective education can demonstrate the inter-relationship and interdependence of people, planet, and profits in all life activities. We have the technology to transition to a clean energy future and to manage materials in a cyclic manner. We know that pollution is causing global environmental change. We also know that teamwork and cooperation get better results than working in a hierarchical or isolated manner. Education in sustainable development will show that the current "consume and throw away" economy no longer works for the benefit of humanity and life. Rather, there is a need for cyclic, whole-systems thinking that integrates all relevant factors into the best, longest lasting results.

Sustainability Reporting

Realizing the importance of being "sustainable" and understanding that they are good long-term investments, many companies are developing sustainability reports. These reports are strong management tools that show how well a company is progressing with their sustainability, their corporate social responsibility (CSR), and their goals to continually improve. They provide an openness that stakeholders (employees, stockholders, regulatory agencies, customers, and community leaders) expect so they know if a company is a good place to work, a good investment, or a good neighbor. These reports are available to the public (free download).[10] For engineers, preparation of sustainability reports involves the collection and analysis of energy and environmental data. Demonstration of energy savings plans and associated pollutant emission reductions can have significant public relations and market share value. This can promote higher sales and flow of investor capital as companies prove their social responsibility and long-term viability.

"High 5!—Communicating your Business Success through Sustainability Reporting—A Guide for Small and Not-So-Small Businesses," from the Global Reporting Initiative, describes the benefits of sustainability reporting:

Sustainability reporting has many advantages that benefit different areas of your business. Some benefits are purely financial while others deal with customer or employee satisfaction. Sustainability reporting helps organizations identify and address their current and potential risks, saving time and money in the short and long term. As the public becomes more aware of your efforts, customer loyalty and credibility of your business will greatly increase. When you take a deeper look into your daily business operations through sustainability reporting you will be able to discover new opportunities.

Businesses continuously seek to generate income and acquire a competitive edge by identifying new market opportunities and determining current and potential risks. When your organization embraces a sustainability perspective, i.e., simultaneously addressing social, environmental and economic issues, you can benefit from cost-savings and improvements in product quality and employee performance.

When considering how to improve financial performance, many organizations only look at financial aspects, such as the cost of purchasing goods, personnel costs, or tax payments. However, working on environmental and social issues can also positively affect your financial bottom line.

Sustainability reporting is the way to identify these potential benefits and realize the economic gains. It helps you achieve your business goals by setting up a continuous improvement process based on target setting and progress measurement. All in all, sustainability reporting helps you to acquire that competitive advantage.

Renting vs. Buying (A Sustainability Innovation)[11,12]

Here is an example of systems thinking to ensure that a product has minimal environmental impact plus high social and economic value:

In today's consumer/throw-away economy, typically a product is manufactured and sold. There is producer incentive to minimize the use of labor and material resources put into a product, thus saving on cost. The product is sold as cheaply as possible to maximize sales. The product eventually wears out and is disposed of. A product that breaks and is disposed of soon after its warranty period is best for sales, so that a consumer will go out and buy a replacement. This is obviously a wasteful scenario, which is prevalent in commerce today. However, that is changing.

A more efficient scenario that many companies have successfully deployed for years is to manufacture and rent a product. Then, there is an incentive to maximize the utility and life of the product, thus ensuring an income to the owner/renter for as long as possible into the future. (This is what sustainable development is all about—maintaining economic flow for as long as possible into the future.) When this product is manufactured, due care in manufacturing processes and sufficient material are used to maximize durability, reliability, and longevity. Customer satisfaction is high from having a reliable product, which also builds name recognition and increases desirability. The product is designed to be easily maintainable, again, to maximize longevity, and also to facilitate dismantlement of the product when it has come to the end of its useful life. Thus, the parts can be easily remanufactured and reused or segregated for efficient recycling of the raw materials. So, maximum utility and income is achieved from the product, and most of the resources that went into manufacturing and maintaining the product throughout its life are recycled with minimal impact to the environment.

So, engineers with sustainable development in mind might be thinking of the "rented" product scenario, which is inherently efficient, rather than the "sold" product scenario, which is inherently wasteful. For example, Interface, Inc., the largest carpet manufacturer in the world and a company that has committed itself to be as sustainable as possible, is using this renting concept in its products. They rent carpet tiles. As the carpet eventually wears, tiles are replaced and recycled to make new tiles.

Community Development to Decrease Energy Costs[13]

Over the last three-quarters of a century, there was an assumption that gasoline and diesel fuel would be cheap and plentiful, indefinitely. Thus, urban sprawl developed because of the low cost of owning and operating one or more cars. This is not the case anymore! Now, with the realization that cars are expensive to both purchase and operate, there are efforts by many cities to re-establish neighborhoods that have all of the amenities needed for occupants all within a short distance that can be covered on foot, with a bicycle, or with convenient public transportation. These cities are excluding automobiles from certain areas and some are charging admission fees for cars to enter semirestricted areas. This has helped to revitalize many city commercial districts because of the park-like feeling of being in these areas without the noise and exhaust from cars. With the economic benefits that have been realized by these arrangements, more and more urban and suburban cities are pursuing these types of commercial district renovations in their cities.

Obviously, society has made a major investment that established urban/suburban sprawl. That investment now needs to be made sustainable. Consequently, there will be major investments in clean fuels such as biodiesel, ethanol, and renewably derived hydrogen to power the huge fleet of vehicles in the United States. A remaining economic burden to maintain suburban living is the maintenance of roads. However, that may be relieved as convenient, modern, and cleaner public transit busses, trains, and other guided vehicles are developed into transportation networks, thus decreasing the number of cars on the road. Energy engineers will be integral to this transition in transportation.

So, the paradigm for energy engineers will evolve as community structures around urban and suburban cities change and improve for energy efficiency. There will probably be a tendency to create nodes, where residents have a majority of the amenities nearby. Then, clean, comfortable, energy-efficient public transportation and express lanes for cars will connect each node.

Overview of the Many Aspects of Sustainable Communities Infrastructure and Nature

The web site http://www.conservationeconomy.net/IN-DEX.CFM (courtesy of the Ecotrust) summarizes the many and various aspects of sustainable communities and preserving our natural assets. Many social factors that play a role in the culture and systems changes associated with sustainable development are thoroughly explained on the web site. Engineers should know these social interactions, which are cohesive and essential to sustainable communities. Readers are encouraged to visit the web site to get a flavor of all of the aspects of sustainability or to narrow in on aspects that are of particular interest to them. The following is a narrative summary of some of the entries on the web site:

A conservation economy describes how social capital, natural capital, and economic capital can be synergetic and sustainable.

Social capital covers fundamental needs, which include: the strong need for local sources of food; accessible, healthy shelter; healthy environment and access to healthcare; plus access to knowledge about the interconnectedness of us to our environment and to each other. The section on community discusses collaborative processes that honor: social equity, which promotes prosperity for all; security from fear and violence; recognition of the wealth and strength in our cultural diversity and establishment of a will to preserve it; plus, local celebrations to honor a sense of place with the community and environment. It proceeds to describe the importance of enjoying beauty and play, to relieve stress from our busyness; learning to welcome transitions that improve communities as a whole; and establishment of civic society where all residents can manage their communities collaboratively.

Natural capital includes the atmosphere, biosphere, and earth. To sustain it requires ecological land use, which includes connected wild lands, in which indigenous animals, plant and people can coexist together. Protected core reserves can be set aside for native plants and animals to thrive, without interference. Wildlife corridors can connect reserves such that animals may migrate freely and parks can be established to act as buffer zones between developed and undeveloped areas. Productive rural areas can be re-established through: sustainable agriculture, which does not cause runoff of pollutants into streams and is more in harmony with natural processes; sustainable forestry, which thins rather than clear-cuts stands of trees; sustainable fisheries, which establish quotas, such that species are not depleted; and eco-tourism, to provide natural getaways and education for people that want to learn more about sustainability and to be close to nature. Compact towns and cities, with human scale neighborhoods, green buildings, convenient transit access, ecological infrastructure, and urban growth boundaries will create healthy, vibrant communities.

Economic capital in healthier communities and commerce will tend to expand through synergies with social and environmental capital, thus building prosperity.

Energy Engineering for Developing Countries[13]

Approximately four billion of the six and one half billion people on the planet live in extreme poverty, where a poor sanitation infrastructure results in disease, there is minimal economic productivity, and there is significant economic burden on governments and aid agencies. However, recently, it has been found that given some of the modern necessities such as water wells, electronic communication, and dependable energy, people that are impoverished can quickly and enthusiastically become productive to their communities and not be an economic burden.

The use of small photovoltaic power systems in villages has literally energized villages into minieconomic zones of relative wealth and flow of capital, thus creating self-sufficiency rather than dependence.

Just providing electricity for lighting, water pumps, and small power tools can tremendously boost the productive capabilities of a village.

The establishment of cell phone repeaters and wireless infrastructure is much less expensive than running miles of telephone cable. A limited number of cell phones in villages have also promoted prosperity because farmers and merchants can effectively communicate and market their products.

As developing countries continue to develop, water, energy, and communication will be vital to their success. So, energy engineers will be a key part of this process. Many multinational and small companies and financial institutions are tapping into bringing these four billion people into the world of commercial success and out of the world of poverty. So, engineers will play an important part in this next major step toward a sustainable world.

Conclusion

Though we are just now seeing the start of a transition to sustainable development, it is clearly economically, ecologically, and socially advantageous to choose this means to success, prosperity, well-being, stability, and peace. We have an abundance of all of the material and energy resources we need. We have the ability, knowledge, and conscience to do the best for humanity, all of life, and the planet. We have the inspiration and insight to transition to a sustainable world. Now all it takes is the willingness to accept change and make the transition to sustainability. The outcome will be a world that is more prosperous and stable than it is now. Sustainable development is a goal to embrace and make part of our daily decisions.

References

1. *Waage, S. Ants, Galileo, and Gandhi—Designing the Future of Business Through Nature, Genius, and Compassion*; Greenleaf Publishing: Sheffield S3 8GG, U.K., 2003; [www.greenleaf-publishing.com].
2. Stecky, N. *Introduction to the ASHRAE Greenguide for Leed*, Globalcon 2005 Proceedings CD. Available at https://www.aeecenter.org/store/detail.cfm?idZ895&category_idZ6.
3. Woolsey, R.J.; Hunter, L.; Amory, L. *Energy Security: It Takes More Than Drilling.* Available at http://www.csmonitor.com/2002/0329/pns02-coop.html.
4. Woolsey, R.J.; McFarlane, B. Conference Report: *Renewable Energy in America: The Call for Phase II, Session 3: Using Renewable Energy to Meet Our National Security Needs*; December 2004. Available at http://www.acore.org/pdfs/04policy_report.pdf.
5. Available at http://www.insnet.org/ins_headlines.rxm?custZ2&idZ1234.
6. Available at http://www.insnet.org/ins_headlines.rxm?custZ2&idZ1248.
7. Anderson, C.; Katharine, R. *The Co-Creator's Handbook,* http://www.globalfamily.net.
8. Available at http://portal.unesco.org/education/en/ev.php-URL_IDZ27234&URL_DOZDO_TOPIC&URL_SECTIONZ201.html.
9. Available at http://www.uspartnership.org.
10. Available at http://www.CorporateRegister.com.
11. Williard, B. *The Sustainability Advantage—Seven Business Case Benefits of a Triple Bottom Line*; New Society Publishers: British Columbia, V0R 1X0, Canada, 2002; [http://www.newsociety.com].
12. *Doppelt, B. Leading Change Toward Sustainability—A Change-Management Guide for Business, Government and Civil Society*; Greenleaf Publishing: Sheffield S3 8GG, U.K., 2003; [http://www.greenleaf-publishing.com].
13. Available at http://magma.nationalgeographic.com/ngm/data/2001/07/01/html/ft_20010701.3.html.
14. Available at http://www.csrwire.com/print.cgi?sfArticleIdZ3709.

19

Urban Agriculture

Natalia Fath

How we eat determines, to a considerable extent, how the world is used.

Wendell Berry (2010)

Introduction

The practice of growing food in cities, or urban environments, is known as urban agriculture. Integration of food production into the urban ecosystem is not a new phenomenon; however, the reasons behind this growing practice have changed over the past few centuries. Urban agriculture arose during difficult times, such as during war or depression, when the growth of vegetables on any available plots was encouraged as a response to dwindling food supply and food shortages. In addition, this was considered as an activity that provided jobs and income and boosted spirits and community cohesion during harsh times. The concept of "allotment gardens" was developed in Germany as a response to rapid industrialization in Europe throughout the 19th century (Holmer and Drescher 2005). Growth of cities during that time was accompanied by growing food insecurity and by allowing citizens to use open spaces to grow food, originally called "the gardens of the poor", the administrations of the cities, thus, contributed to the idea of urban farming. The late 19th century witnessed the initiative by Detroit's Mayor, Hazen S. Pingree, when the citizens were asked to utilize any available land to grow vegetables as a response to economic slowdown (Austin 2019) – coincidentally, now that Detroit is again a leader in urban farming due to recent economic collapse with the loss of the automobile industry. These original initiatives were further developed during World War I and World War II (WWII) and received a new meaning – they were called "victory gardens". Misallocation and devastation of agricultural land in Europe and loss of farmers to military recruitment led to the growth in popularity of these fruit and vegetable gardens, specifically in the United States, Canada, the United Kingdom, Germany, and Australia (Drescher 2001). In the United States alone, these gardens produced about 1.45 million quarts of canned fruits and vegetables in 1918 (Schumm 2018). The United States Department of Agriculture (USDA) further continued to encourage expansion of the victory gardens during the hardships of WWII. Thus, the existence of agricultural production in urban environments has evolved to be more systematic and expected.

179

The purpose of establishing those gardens in the cities has shifted since the end of WWII. During times of relative food abundance, these practices may be seen from a different perspective. While still providing food, which is especially crucial for food-insecure districts, and part-time or full-time jobs, urban farms also now looked at as a place for social gathering or as a green area that helps fighting air pollution in cities. Advantages of urban farming include but are not limited to social, economic, and health benefits. However, there are some risks involved as well, which include but are not limited to required high inputs of energy and time, and high probability of polluted urban soil and air. It has been attempted to estimate the proportion of food that is provided by urban agriculture globally. According to van Veenhuizen and Danso (2007), 15%–20% of the world's food supply comes from urban agriculture. A recent publication by Clinton et al. (2018) suggests that the above numbers are highly overestimated, perhaps because of inclusion of peri-urban areas and livestock production. The authors claim that if the space is utilized at its maximum, and the practices in place are intense, then urban agriculture could produce ~5% of the global crop production.

Types of Urban Farming

Food production in cities can take various forms. There are different projects and activities that have been developing and have now firmly established themselves in urban environments. Several major categories should be mentioned in respect to urban farming: (1) community-based, (2) institutional, (3) public, and (4) commercial (Steele 2017). The borderlines between them are not rigid and, in particular cases, may carry the features of two or three categories. Urban agriculture is a flexible concept and adjusts to the local conditions, thus acquiring various forms. There is not a widespread agreement on the classification of types of urban farming, however, there are types that are commonly recognized. Specifically, community and backyard gardens; rooftop and balcony gardening; growing in vacant lots, right-of-ways, and parks; aquaponics; hydroponics; fruit trees and orchards; market farms; raising livestock and beekeeping (Steele 2017). In addition, it can be stated that post-harvest activities, for example, "creating value-added products in community kitchens, farmers' markets and road-side farm stands, marketing crops and products, and addressing food waste" (Steele 2017), should also be considered as overall urban agricultural activities.

Community and backyard gardens involve growing food for the family or the neighborhood; rooftop gardening utilizes available space on roofs while growing in vacant lots puts a productive use on otherwise abandoned land; hydroponics practices soilless food growth; and aquaponics combines the latter with raising fish. Some cities allow raising chickens or keeping beehives in urban environments.

Additionally, the classification system of urban farming types can explore the following categories: location, people involved, products grown, economic activities, product destination, scale, and technology used (RUAF Foundation 2019). However, currently, urban agriculture researchers have been using their own approach to typify this activity, which has led to a significant variety of divisions and subdivisions of local urban farming systems. Thus, it appears to be beneficial to continue developing a consistent viewpoint of types of urban farms (van Veenhuizen and Danso 2007).

Regardless of the classification system used, it is imperative to realize that food production in cities is fully integrated into the urban economic and ecological systems, and thus should be an important part of urban planning policies (RUAF Foundation 2019).

Benefits of Urban Farming

Social and Cultural Benefits

The practice of growing food in cities can be viewed as an excellent opportunity for social integration, cultural exchanges, and education for both adults and youth. Through collective action and advocacy, city dwellers build and strengthen their communities, as urban farms in their various forms mostly appear to be based on a sense of community (Holland 2004). Working as a team on underutilized land plots and transforming them into productive environments that can feed them or can be sold, residents improve the quality of their immediate surroundings and develop bonding (see Figure 1). Of particular importance, relevant to urban food production is the opportunity for people of different ages to interact which promotes intergenerational bond, and the chance for disadvantaged or segregated groups such as people from different socioeconomic classes or different backgrounds, orphans, women, recent immigrants without jobs, elderly and disabled people, to get involved in community building, to become engaged citizens, and to maintain decent livelihoods (Santo et al. 2016).

Urban farming, thus, can be an "agent of change" as it encourages social bonding and activities that have the potential to promote social, economic, and environmental changes at the local level (Holland 2004).

Urban farms can also play an important role in educational functions – participants, and especially youth, learning about ecology and the process of food growth, which in turn, can encourage growth in agricultural literacy and reconnection to nature (Santo et al. 2016). Advancing urban dwellers' knowledge on how their food is grown and distributed, may lead to the appreciation of the urgent need for cities to adopt "circular" food production system rather than current "linear" to ensure their own sustainability. Specifically, urban "outputs", like organic waste production, can and should become an "input" into urban food production system (Deelstra and Girardet 2000). Recycling and composting of organic matter and nutrient recycling on local urban farms reduce dependence on delivered "inputs" and keeps this waste from landfills. By participating in the food-producing process, citizens foster an understanding and appreciation of systems thinking and the cycles of life; and therefore, the importance of urban sustainability as the world is predominantly urban and the pattern is predicted to continue into the future.

Economic Benefits

As some types of urban agriculture utilize vacant lots, it leads to increased property values, which in turn generates increased tax revenues (Steele 2017). This is particularly true in economically disadvantaged neighborhoods (see figure 1). Among other economic benefits, it can be mentioned that this activity

FIGURE 1   Baltimore, MD, USA – low-income neighborhood, underutilized land. Real Food Farm, Perlman Place. Planted fruit trees (fig). (*Photo credit: N. Fath.*)

adds jobs to the local economy and stimulates capital investment and redevelopment (Santo et al. 2016). Guitart et al. (2012) analyzed published research on community gardens. Their review found that all 13 of the studies which examined property values reported increased property values associated with the existence of the community garden. Similar research findings can be found in Voicu and Been (2008).

It can be argued that urban agriculture makes fresh produce more affordable. This is particularly important for low-income neighborhoods and for poorer countries where between 50% and 70% of income on being spent on food. However, based on the currently available studies, it can be concluded that the criteria and methods used to assess specific profits and other economic impacts differ markedly and/or are lacking (van Veenhuizen and Danso 2007). In addition to the direct economic benefits to the participants, urban agriculture also encourages the growth of small enterprises; farming activities require necessary agricultural inputs that can vary from compost and organic pesticides production to tool production and managing of the outputs that can include packaging, marketing, and transportation of products (Homem de Carvalho 2001).

Besides the above discussed economic advantages, urban farms surplus can be donated to food banks which, in turn, can help reduce money spent on food for low-income neighborhoods. In addition, when vacant land is utilized by urban agriculture activities, it can save maintenance costs for the city as illegal waste dumping and/or vandalism will decrease (Steele 2017).

It is important to note that there is a need for more research on economic outputs of urban farming, profitability of commercial food production in cities, long-term studies on employment opportunities and on neighborhoods' indicators, discussed above. Economic outcomes are the "least documented aspect of urban agriculture" (Hodgson et al. 2011).

Health Benefits

There are several aspects relevant to this topic that should be discussed, specifically, improved food security and physical and emotional well-being. Poorer neighborhoods often find themselves living in so-called "food deserts". USDA defines food deserts as "parts of the country vapid of fresh fruit, vegetables, and other healthful whole foods, usually found in impoverished areas" (*Nutrition Digest* 2019). To qualify as a food desert, "at least 500 people and/or at least 33 percent of the census tract's population must reside more than one mile from a supermarket or large grocery store (for rural census tracts, the distance is more than 10 miles)" (*Nutrition Digest* 2019). When analyzing the ability of urban agriculture to improve food security, one should look at the individual/household and municipal levels. Neither, though, should be understood as a substitute to traditional food retail (Santo et al. 2016). While not the whole answer to urban food deserts, farming in these locations provides participants with access to some fresh produce that otherwise is unavailable; thus, improving their food security and providing a better diet that includes more fresh fruits and vegetables (Zezza and Tasciotti 2008, Smith and Harrington 2014). Numerous studies report participants, their households, as well as their neighbors and friends, gaining access to a variety of seasonally appropriate foods while saving money for other needs (Corlett et al. 2003, Wakefield et al. 2007, Kortright and Wakefield 2011). However, more empirical evidence and longitudinal studies are necessary to support the claim that urban farming can significantly amend food security over time in poorer urban neighborhoods (Siegner et al. 2018). While there is accumulated evidence that supports increased produce consumption among the gardeners (Alaimo et al. 2008, Litt et al. 2011), experts suggest it is not an overall significant indicator of improved food security and/or quality of the diets in low-income households (Hallsworth and Wong 2013). Based on physical land availability in cities, along with other barriers, urban agriculture's capacity to ameliorate food security is not strongly supported; however, it can be a supplemental solution for the urban poor (Badami and Ramankutty 2015).

Improved access to fresh produce, fruits and vegetables, may lead to better nutrition, and thus can be considered a viable strategy to mitigate certain diet-related diseases, such as obesity and diabetes (Steele 2017). However, through many studies and research done, it appears that via urban farming activities most notable and well-documented gains are in improved mental health and physical health in participants. This is accomplished through acquiring a purpose, seeing the results of one's effort, and being physically active (Santo et al. 2016). In addition, simply being close to nature has been observed to be a stress-relieving activity (Armstrong 2000, Brown and Jameton 2000, Wolf and Robbins 2015). Increased green cover in cities associated with urban agriculture activities contributes to such ecosystem services as air filtration and temperature moderation, thus also benefiting the overall well-being of urban residents (Wolf and Robbins 2015).

Environmental Benefits

Besides social and human health benefits of urban agriculture, arguably, environmental advantages associated with this activity are the dominant ones. Among particular benefits to the environment are improved air quality and mitigation of the urban heat island effect. Extensive research studies support that replacing impervious surfaces with vegetation and trees, in particular, leads to cleaner air, as vegetation filters particulate matter, and to cooler temperatures during summer. For example, in the Phoenix metropolitan area, increased tree canopy cover by 25% leads to a total cooling of 7.9°F as compared to a bare neighborhood, and addition of greenery to residential backyards reduces average neighborhood temperatures by 0.4°F–0.5°F (Middel and Chhetri 2014). This phenomenon appears to be of particular importance for low-income neighborhoods as they tend to have fewer trees and grass cover and, thus, higher summertime temperatures (Jenerette et al. 2011).

Among other environmental advantages of urban agriculture are its ability to maintain and support biodiversity via creation of habitats; its contribution to recycling of nutrients when organic waste is turned into a resource – soil (see Figures 2 and 3); its ability to increase water infiltration, thus reducing stormwater runoff and decreasing the flooding potential (van Veenhuizen 2006, Santo et al. 2016).

FIGURE 2 Composting bin at aquaponics farm, Baltimore, MD, USA. Center for a Livable Future, JHU. (*Photo credit: N. Fath.*)

FIGURE 3   Composting bins at urban farm, Towson University, MD, USA. (*Photo credit: N. Fath.*)

Urban farming should also be looked at from the perspective of one of the most pressing challenges present day – scarcity of freshwater and agriculture being the largest user of freshwater. Some types of urban farming contribute to lessening this issue by collecting rainwater on site and using it for irrigation (see Figure 4). Others are capable of re-using wastewater, thus lessening the demand for freshwater supply (Haysom 2009). A particular type of urban agriculture, aquaponics, attempts to recycle water used for fish and produce production. While initially the system requires a significant input of water, the continued operation is based on this water being re-used and accompanied only by some expected losses due to evaporation, spillage, etc., which could be replenished by rainwater collection (Love et al. 2015). United Nations' Food and Agriculture Organization, FAO, is particularly invested in exploring the possibilities of integrating aquaponics in drier climate cities, such as in the Near East and North Africa (FAO 2019).

The overall reduced ecological footprint of the cities practicing urban agriculture leads to the discussion of its positive influence on one of the dominant topics discussed today, climate change. There are many challenges faced by growing cities and continued urbanization throughout the world; however, climate change is recognized as one of the most pressing challenges (IPCC 2018). The most obvious way how urban farming can lessen the contribution of food production to climate change is by reducing carbon footprint. Increase in urban green spaces, including fruit trees and crops grown, will lead to increased carbon sequestration (Thornbush 2015).

By practicing food production close to the consumers, the greenhouse gas emissions associated with the food transportation, storage, and cooling will be reduced (Kulak et al. 2013). Numerous case studies exist that support the above, and while this has not been quantified on a global scale, urban agriculture can and should be promoted as both climate change mitigation and adaptation measures.

As a concluding remark on the benefits of urban agriculture, it is important to note that the extent of these benefits directly depends on the policies in practice, the type of urban farming, and the local environment. Summary of the benefits is presented in Table 1.

FIGURE 4   Rain barrels at urban farm, Towson University, MD, USA. (*Photo credit: N. Fath.*)

TABLE 1   Benefits of Urban Agriculture

Categories	Benefits
Social	• Builds and strengthens communities • Reconnection with food production • Renders an educational venue • Promotes social, economic, and environmental activism • Promotes social inclusion
Environmental	• Increases urban green spaces • Improves air quality • Reduces Urban Heat Island effect • Promotes biodiversity • Recycles organic waste • Increases stormwater infiltration • Reduces runoff • Reduces water usage and reuses wastewater • Reduces carbon footprint
Health	• Improves food security in urban food deserts • Improves access to fresh food and, thus, overall nutrition • Encourages physical activity • Cleaner air → improves overall health • Increases urban green spaces → improves psychological well-being and mental health

Challenges of Urban Farming

Securing Land for Food Production in Cities

One of the major obvious challenges and constraints is the lack of space in world cities to be used for growing food. On the other hand, there are many cities where plenty of underutilized or abandoned land is available. Particularly, in the cities that have experienced post-industrial decline and out-migration, such as in the city of Detroit, USA. While thousands of acres of land have been given over to unemployed workers for food growing in American cities of Detroit and New York (van Veenhuizen 2006), it is imperative to enhance access to this available vacant land by creating appropriate policies. These policies will need to enable sustainable urban agriculture and remove unnecessary obstacles that exist currently in securing land for food production in cities as agriculture traditionally refers to the rural areas; thus, urban agriculture currently has no institutional home (RUAF Foundation 2019).

Additional problem is that worldwide urban land value typically rises, and therefore urban farmers have to look for free land to use. Municipal land-use regulations do not guarantee long-term use of these plots, and frequently only temporary farms or gardens can be established (Steele 2017). Given the notable time and initial money and resources necessary to be invested in developing a productive urban farm, the temporary status may not appear to be an attractive venue to pursue as they are vulnerable to redevelopment.

A relatively new phenomenon called "guerrilla gardening" has been growing around the world over the last two or three decades, and it arose in part as a response to the difficulty of securing land for food production in the cities. The name suggests that the gardening is done without the securing legal rights to the land that is being used, thus making them vulnerable.

Health Risks

While food production in urban areas promotes some substantial benefits to human health (they were discussed above), it also may present some health risks to the participants and the nearby community. Specifically, if the gardens are not organic, farmers and consumers may be exposed to doses of fertilizers and pesticides that are used. This risk, however, is not pertinent only to urban agriculture, but to rural agriculture as well.

One of the most common concerns associated with urban farming is the high probability that urban air and soil may be contaminated with particulate matter, heavy metals, and others. These may result from heavy traffic congestion that is typical for cities, from industrial activities that tend to be concentrated in urban areas, and possibly, from waste dumping areas in or around the cities. Accumulation of these pollutants in air or soil may negatively impact health of the farmers and consumers; however, those can be reduced if adequate measures and regulations are put in place (van Veenhuizen and Danso 2007).

Remediation Costs

As mentioned in the section above, it is plausible that some land plots in urban areas have higher levels of contamination, and thus it is imperative that soil samples are analyzed prior to farming activities. If soil contamination is confirmed, then land requires remediation. Current remediation methods include soil removal, washing, or capping. They are costly and therefore are limiting factors in growth of outdoor urban agricultural systems (Wortman and Lovell 2013).

Eliminating contact between food grown and the ground via construction of raised beds is a cost-effective way to establish food production on sites where soil has been contaminated (see Figure 5).

Energy Usage

Certain types of urban agriculture, particularly, indoor activities such as vertical farming, greenhouses, hydroponics, and aquaponics, do require substantial energy input for their proper operation, especially

FIGURE 5   Raised beds at urban farm, Towson University, MD, USA. (*Photo credit: N. Fath.*)

if these facilities are located in cold or water-scarce regions. In the world where 64.5% of electricity production still falls on fossil fuels, coal being the largest source, this is a concern (IEA 2019).

An aquaponics farm, which integrates hydroponics (soilless plant cultivation) with aquaculture (fish farming), where both components complement each other to generate two products at once, crops and fish, is a good practice to reduce agricultural water footprint as it recirculates water initially added to the system, but it has been proved that it requires a significant energy usage. While the food produced at the aquaponics farm in Baltimore, Maryland, USA did not contain antibiotics, synthetic pesticides, or chemical fertilizers, the energy use was remarkably high, particularly during the winter months. When input costs were compared to market prices on tilapia, it demonstrated a net economic loss (Love et al. 2015). High energy costs forced the shutdown of a hydroponics farm in Buffalo, New York and its relocation to southwest Texas (Santo et al. 2016).

Alternative types of energy should be considered where applicable, in particular, in respect to the construction of new greenhouse systems or retrofitting conventional greenhouses. Solar-powered greenhouse systems are more popular in Europe and can achieve up to 70% of energy saving (Taki et al. 2018).

Summary of the challenges and risks are presented in Table 2.

TABLE 2   Challenges of Urban Agriculture

Categories	Problems/Limitations
Social	• If air and soil contamination → high remediation costs • Unavailable land due to other uses or unsuitable • Expensive or inaccessible land • Lack of zoning regulations and policies, which include urban farming as part of future urban development • Increased fresh produce consumption does not notably lessen food insecurity • Increased fresh produce consumption does not notably improve diets overall
Environmental	• Some instances require high energy input • Likelihood of air and soil being contaminated • If in drier climates, then increased water demand (drinking water used for irrigation)
Health	• Potential health risks to farmers and consumers from soil contaminants and air pollutants if adequate preventative remediation measures not taken

Future Projections

Urban agriculture appears to be growing in popularity around the world, although the goals behind their establishment range from aesthetics to communal activity, to actual crop production in low-income "food deserts". Assessment of feasibility and sustainability or these food-producing practices should be further continued. It is implied that the land available for urban agriculture is limited given all the traditional land-uses in cities. Nevertheless, the productivity of that land can be high and it can achieve high yields, but currently material and labor resources are used inefficiently (McDougall et al. 2019).

It seems unlikely that urban agriculture can or will replace rural agriculture, as the majority of food calories in the form of cereal grains will continue to be produced outside the cities. However, urban agriculture offers an important supplement, particularly to food impoverished areas, it promotes a strong sense of community in terms of caring and investing in local place and educates people on the cycles of nature and where food comes from. With the growth of the industrial food production system in the mid-20th century in the United States, a huge disconnect arose between farming and most people. This gap can be breached by promoting farming practices in cities.

The scale of food production in cities can vary from small community gardens to large urban farms, such as the one that is currently under construction in Paris and is scheduled to open in 2020. It is projected to be the world's largest urban rooftop farm, which will also incorporate closed-water hydroponic vertical system spanning overall ~14,000 m² (150,695 ft²). They are expecting to cultivate more than 30 different plant species using entirely organic methods, with fruit and vegetable daily production amounting to ~1,000 kg in the high season (*The Guardian* 2019). This could become a globally acknowledged model for sustainable urban food production.

References

Alaimo, K., Packnett, E., Miles, R.A., & Kruger, D.J. 2008. Fruit and vegetable intake among urban community gardeners. *Journal of Nutrition Education and Behavior*, 40(2), 94–101.

Armstrong, D. 2000. A survey of community gardens in upstate New York: Implications for health promotion and community development. *Health and Place*, 6, 319–327.

Austin, D. 2019. Hazen S. Pingree Monument. HistoricDetroit.org. https://historicdetroit.org/buildings/hazen-s-pingree-monument

Badami, M.G. & Ramankutty, N. 2015. Urban agriculture and food security: A critique based on an assessment of urban land constraints. *Global Food Security*, 4, 8–15. doi:10.1016/j.gfs.2014.10.003

Berry, W. 2010. "The pleasures of eating." In *Food* edited by Brooke Rollins & Lee Bauknight, 21–28. Southlake: Fountainhead Press.

Brown, K.H. & Jameton, A.L. 2000. Public health implications of urban agriculture. *Journal of Public Health Policy*, 21(1), 20–39. doi:10.2307/3343472

Clinton, N. et al. 2018. A global geospatial ecosystem services estimate of urban agriculture. *Earth's Future*, 6, 40–60. doi:10.1002/2017EF000536

Corlett, J.L., Dean, E.A., & Grivetti, L.E. 2003. Hmong gardens: Botanical diversity in an urban setting. *Economic Botany*, 57(3), 365–379.

Deelstra, T. & Girardet, H. 2000. Urban Agriculture and Sustainable Cities. *Growing Cities, Growing Food: Urban Agriculture on the Policy Agenda*.

Drescher, A.W. 2001. The German Allotment Gardens – A Model for Poverty Alleviation and Food Security in Southern African Cities? Published in the *Proceedings of the Sub-Regional Expert Meeting on Urban Horticulture*, Stellenbosch, South Africa, Jan. 15–19. FAO/University of Stellenbosch. http://www.cityfarmer.org/germanAllot.html

FAO. 2019. Every Drop Counts. http://www.fao.org/fao-stories/article/en/c/1111580/

Guitart, D., Pickering, C., & Byrne, J. 2012. Past results and future directions in urban community gardens research. *Urban Forestry & Urban Greening*, 11(4), 364–373. doi:10.1016/j.ufug.2012.06.007

Hallsworth, A. & Wong, A. 2013. Urban gardening: A valuable activity, but.... *Journal of Agriculture, Food Systems, and Community Development*, 3(2), 11–14.

Haysom, G. 2009. Urban agriculture and food security. *The Sustainability Institute. Sustainable Neighbourhood Design Manual.* Chapter 8. https://pdfs.semanticscholar.org/9b05/a3ad15a9d1e-b57227e01e3387d0966888fec.pdf

Hodgson, K., Caton Campbell, M., & Bailkey, M. 2011. *Urban Agriculture: Growing Healthy, Sustainable Places.* Chicago, IL: American Planning Association Planning Advisory Service.

Holland, L. 2004. Diversity and connections in community gardens: A contribution to local sustainability. *Local Environment,* 9(3), 285–305. doi:10.1080/1354983042000219388

Holmer, R.J. & Drescher, A.W. 2005. In Christine Knie (ed.): Urban and Peri-Urban Developments – Structures, Processes and Solutions. Southeast Asian-German Summer School Program. 2005 in Cologne/Germany, Oct. 16–29: 149–155 Allotment Gardens of Cagayan de Oro: Their Contribution to Food Security and Urban Environmental Management.

Homem de Carvalho, J.L. 2001. Prove: Small Agricultural Production Virtualization Programme. *Urban Agriculture Magazine*, No 5. Appropriate Methods for Urban Agriculture. Leusden, RUAF.

IEA. 2019. Electricity Statistics. https://www.iea.org/statistics/electricity/

IPCC. 2018. Global Warming of 1.5°C. *Special Report.* https://www.ipcc.ch/sr15/

Jenerette, G.D., Harlan, S.L., Stefanov, W.L., & Martin, C.A. 2011. Ecosystem services and urban heat riskscape moderation: Water, green spaces, and social inequality in Phoenix, USA. *Ecological Applications,* 21(7):2637–2651.

Kortright, R. & Wakefield, S. 2011. Edible backyards: A qualitative study of household food growing and its contributions to food security. *Agriculture and Human Values,* 28(1), 39–53.

Kulak, M., Graves, A., & Chatterton, J. 2013. Reducing greenhouse gas emissions with urban agriculture: A life cycle assessment perspective. *Landscape Urban Plan,* 111, 68–78.

Litt, J.S., Soobader, M.-J., Turbin, M.S., Hale, J.W., Buchenau, M., & Marshall, J.A. 2011. The influence of social involvement, neighborhood aesthetics, and community garden participation on fruit and vegetable consumption. *American Journal of Public Health,* 101(8), 1466–1473.

Love, D.C., Uhl, M.S., & Genello, L. 2015. Energy and water use of a small-scale raft aquaponics system in Baltimore, Maryland, United States. *Aquacultural Engineering,* 68, 19–27. doi:10.1016/j.aquaeng.2015.07.003

McDougall, R., Kristiansen, P., & Rader, R. 2019. Small-scale urban agriculture results in high yields but requires judicious management of inputs to achieve sustainability. *PNAS,* 116(1), 129–134. doi:10.1073/pnas.1809707115

Middel, A. & Chhetri, N. 2014. City of Phoenix Cool Urban Spaces Project: Urban Forestry and Cool Roofs. Center for Integrated Solutions to Climate Challenges, Arizona State University. https://d3dqsm2futmewz.cloudfront.net/docs/dcdc/website/documents/NOAA_PHX_UrbanSpaces_Rep.pdf

Nutrition Digest. 2019. USDA Defines Food Deserts. American Nutrition Association. Vol. 28, No 2. http://americannutritionassociation.org/newsletter/usda-defines-food-deserts

RUAF Foundation. 2019. Urban Agriculture: What and Why? https://www.ruaf.org/urban-agriculture-what-and-why

Santo, R., Palmer, A., & Kim, B. 2016. Vacant Lots to Vibrant Plots: A Review of the Benefits and Limitations of Urban Agriculture. *CLF Report.* Johns Hopkins University. https://www.jhsph.edu/research/centers-and-institutes/johns-hopkins-center-for-a-livable-future/_pdf/research/clf_reports/urban-ag-literature-review.pdf

Schumm, L. 2018. America's Patriotic Victory Gardens: During Both World Wars, America's Agricultural Production Became a Powerful Military Tool. https://www.history.com/news/americas-patriotic-victory-gardens

Siegner, A., Sowerwine, J., & Acey, C. 2018. Does urban agriculture improve food security? Examining the nexus of food access and distribution of urban produced foods in the United States: A systematic review. *Sustainability*, 10(9), 2988. doi:10.3390/su10092988

Smith, V.M. & Harrington, J.A. 2014. Community food production as food security: Resource and market valuation in Madison, Wisconsin (USA). *Journal of Agriculture, Food Systems, and Community Development*, 4(2), 61–80.

Steele, K. 2017. Urban Farming: An Introduction to Urban Farming, from Types and Benefits to Strategies and Regulations. Vitalyst Health Foundation. www.vitalysthealth.org

Taki, M., Rohani, A., & Rahmati-Joneidabad, M. 2018. Solar thermal simulation and applications in greenhouse. *Information Processing in Agriculture*, 5(1), 83–113. doi:10.1016/j.inpa.2017.10.003

The Guardian. 2019. World's Largest Urban Farm to Open – on a Paris Rooftop. Aug. 13. https://www.theguardian.com/cities/2019/aug/13/worlds-largest-urban-farm-to-open-on-a-paris-rooftop

Thornbush, M. 2015. Urban agriculture in the transition to low carbon cities through urban greening. *AIMS Environmental Science*, 2(3), 852–867.

van Veenhuizen, R. & Danso, G. 2007. Profitability and Sustainability of Urban and Peri-Urban Agriculture. *Agricultural Management, Marketing and Finance Occasional Paper*. Food and Agriculture Organization of the United Nations. http://www.fao.org/3/a-a1471e.pdf

van Veenhuizen, R. 2006. Cities Farming for the Future, Urban Agriculture for Green and Productive Cities. International Institute of Rural Reconstruction and ETC Urban Agriculture. http://citeseerx.ist.psu.edu/viewdoc/download?doi=10.1.1.124.4555&rep=rep1&type=pdf

Voicu, I. & Been, V. 2008. The effect of community gardens on neighboring property values. *Real Estate Economics*, 36(2), 241–283.

Wakefield, S., Yeudall, F., Taron, C., Reynolds, J., & Skinner, A.L. 2007. Growing urban health: Community gardening in South-East Toronto. *Health Promotion International*, 22(2), 92–100.

Wolf, K.L. & Robbins, A.S. 2015. Metro nature, environmental health, and economic value. *Environmental Health Perspectives*, 123(5), 390–398.

Wortman, S.E. & Lovell, S.T. 2013. Environmental challenges threatening the growth of urban agriculture in the United States. *Journal of Environmental Quality*, 42(5), 1283–1294.

Zezza, A. & Tasciotti, L. 2008. *Does Urban Agriculture Enhance Dietary Diversity? Empirical Evidence from a Sample of Developing Countries*. Rome: FAO.

III

CSS: Case Studies of Environmental Management

III

Case
Studies of
International
Management

20

Cell Tower Procurement: Public School Placement

Joshua Steinfeld

Introduction

Personal wireless service providers, independently and through use of brokers, have installed cell towers at or close to public schools in the United States. The open space surrounding schools, typically baseball and football fields, allows for optimal transmission of radio frequency (RF) waves between towers. Furthermore, public school districts, in comparison with private enterprises, have been more easily won over in the cell tower proposal process. Surges in experimental research regarding health hazards to RF emission, especially the apparent increased susceptibility of children to RF radiation, have sparked controversy over the exact locations of cell towers.

The purpose of this manuscript is to present an argument opposing the placement of cell towers at public schools. First, scientific studies related to animals and humans are provided to show that RF waves may be harmful to humans, and there is concern over electromagnetism. Second, voluntary initiatives and cell tower proposal processes are discussed, highlighting the Rockville, Maryland, community's precedential victory in opposition to the placement of a cell tower at Wootton High School. Third, ethics and public policy considerations address the need for streamlining nationwide community efforts by amending the Telecommunications Act of 1996 to disallow the placement of cell towers near public schools. Fourth, areas for further research are presented, which include the compilation of data sets, tracking of exposed students at school, and new theory to address causality issues related to competing risks. Finally, the chapter concludes by providing commentary regarding invisible risks.

Impact and Study of RF Hazards

Recent Trends

Concerns regarding the safety of cell towers at school have risen across the country. In 2010, Vista del Monte Elementary School in Palm Springs, CA, was pegged as having a reputation for being a cancer school. In 2005, a cell tower was erected on campus. Since then, 12 people have been diagnosed with cancer, affecting those who worked closest to where the cell tower was installed, where the field-strength readings were highest.[1]

From 1975 to 2000, childhood cancer rates had increased dramatically by a rate of 32%. Some of the most severe and deadly cancers such as acute lymphocytic leukemia, brain, kidney, and bone cancer also increased considerably.[2] In 2004, there were 36 million prescriptions of sleeping medications. As of 2009, 56 million prescriptions were outstanding, a whopping 56% increase. In 2004, the number of residents using cell phones was 109 million. By 2009, the number was up to 271 million.[3] Public health officials and environmental experts alike have been searching for environmental stimuli that may be contributing to the increased childhood cancer rates and sleep deprivation. The placement of cell towers at schools has been an area of focus (Figures 1 and 2). This Florida community elementary school also has a narrow-band transmission device installed on the tower (Figure 3). Installation of narrow-band cell towers is generally more restrictive than wide-band cell towers because of the narrow band's compact, piercing wavelength. Electromagnetic stimuli emitted from cell devices have become an area of intense research interest.

Background Research

Just to get an idea of the strength of an electromagnetic force emitted from a cell tower in comparison with a more well-known object, cell phone technology operates on frequencies up to 3 gigahertz (GHz), and a microwave oven cooks food at 2,450 megahertz (MHz).[4] Three GHz is equal to 300,000 MHz! Coulomb's law states that the force of an electromagnetic field is proportional to the magnitude of the charge and inversely proportional to the square of the separation. From Coulomb's law, we can derive the Inverse law, which suggests that holding the magnitude of the charge constant, the electromagnetic force emitted on a subject increases exponentially as the subject moves closer to the source.[5] Using

FIGURE 1 Public elementary school.

FIGURE 2 Wide-band cell tower at elementary school.

FIGURE 3 Narrow-band cell tower at elementary school.

Ampere's law, which combines the magnetostatic equation for determining the magnitude of a magnetic field with Stokes' theorem dealing with surface area, it is possible to determine the strength of an electromagnetic force through a closed path that may be tangent or indirectly exposed to the source.[5,6] Coulomb's law and the resulting derivation of the Inverse law demonstrate key findings in the discussion of cell towers. The farther away the cell tower site is from the subject, the lower the field strength absorbed. Furthermore, due to the exponential nature of the Inverse law, being close means being really close. Ampere's law is especially helpful in determining indirect exposure to force strength in cities where waves regularly bounce off other buildings.

A given material is composed of atoms. Each atom consists of electrons orbiting a central positive nucleus. Electrons also spin around their own axis. The orbital array of activity occurring between magnetic forces of protons and electrons in atoms results in an organized disarray of unpredictability. This unpredictable orbital disarray is the normal, unaltered state of the atom. When an external RF field is applied, the bombardment of electrons (from the source) stimulates host atom movement changes (in the subject), and the atom has a magnetic moment.[7] The orbital path of the electrons is brought into a slight sense of organization, throwing normal behavior out of whack. It is the electron stimulus emitted from cell towers that has been of much focus in experimental studies.

Radio Frequency Research on Animals and Humans

Scientists and researchers have proceeded cautiously regarding the use of human subjects in testing the effects of RF exposure. As a result, numerous testing on animals has been done to learn about the effects of electromagnetic radiation on living organisms. A 1997 study on mice demonstrated the effects of radiation on prenatal development and resulted in a progressive decrease in the number of newborns per dam, ending in irreversible infertility. In a subsequent 1999 study, mice exposed for just 24–72 hours to weak electromagnetic waves increased the activity of natural killer cells by 130%. Meanwhile, exposure to microwave irradiation had no effect on the activity of natural killer cells.[8] Nonetheless, microwave stimulus interfered with cell immunity of mice, increasing T-cell proliferation in response to stimulus.[9]

Studies on other living organisms have also been conducted. In Germany, behavioral abnormalities were observed in a herd of dairy cows that grazed near a cell tower for over 2 years, leading to reduction in milk yield and increased health problems.[10] In Russia, the effects of electromagnetic radiation on sea urchin embryos were tested. Only sea urchins with preexisting weakened viability were impacted by the electromagnetic radiation, in which case the electromagnetic radiation stimulated the onset of early development of embryos.[11]

While much more rare than electromagnetic testing on animals, some testing on humans has taken place. It was discovered that electromagnetic fields affect the central nervous system in humans because visual reaction time was prolonged and scores on short-term memory tests were lower in high-intensity exposure test groups.[12] Also, in a controlled study aimed to investigate the impact of low-force electromagnetic fields on healthy humans, human subjects were exposed to a 900-MHz electromagnetic field and intermittently pulsed with 217 MHz. It was determined that low-force fields have no effect on nocturnal hormone secretion under polysomnographic controls. However, cortisol production increased, which is transient by classification, indicating the organism (human subject in this case) adapted to the stimulus.[13] It was unclear if any mental impediments or genetic responses may have taken place in addition to the increase in cortisol production, but it was certain that the human subjects endured a cellular response. Nonetheless, current research indicates that genes are at risk, even at low-force electromagnetic fields.

Genes that ward off cancer and other illness may be inhibited when a cell receives stimuli from the environment. The National Institute of Environmental Health Sciences has been using genomic techniques to determine the behavior of promoter-proximal paused polymerase (Pol II) with and without environmental stimuli.[14] Pol II is known to have a role in fighting disease. Transcriptional responses to environmental stimuli can cause alterations in Pol II distribution, gene expression, and epigenetic chromatin signatures, leading to transcription dysregulation that can cause etiology of cancers.[15] Additionally, recent work has revealed that signal-response pathways are loaded with Pol II prior to final gene activation, further enhancing the opportunity for cellular changes to take place as a result of harmful environmental stimuli.[16,17] Molecules transported from environmental stimuli can inhibit the signal-response pathways' ability to pause release of Pol II.[18] Pol II pausing is necessary in providing an accessible chromatin architecture for gene promoters that inevitably fight disease.[19,20] It has been determined that Pol II pausing facilities' precise control and coordination of genes is a crucial regulatory step in rate-limiting the expression of DNA damage responsive genes.[21]

The extent to which scientists and doctors alike understand the impact of RF waves on humans is unquestionable in its fineness. For example, RF waves are routinely manipulated in clinical medicine to achieve exact thermal dosimetry and thermal pattern poisoning of tumors. The important nuance to remember is that the beneficial uses of RF waves in medicine are based on the destructive qualities of the high-energy RF waves.[22]

Voluntary Initiative

Students Against Cell Towers

In 1963, a group of citizens became activists in opposition to the proposal of a Con Edison power plant on Storm King Mountain in New York.[23] If the Storm King success story gave life to an entire environmental movement, then the success of Students Against Cell Towers (SACT) in opposing a cell tower at Wootton High School in Rockville, Maryland, solidified the environmental movement's existence by the hundreds of cell tower opposition advocacy groups that spawned across the nation since the SACT community voluntary initiative spanned from 2003 to 2005. Although SACT is no longer an active organization, its over 100 former members carry with them knowledge of the vital considerations regarding placement of cell towers at schools.

In 2000, Cingular Wireless began an aggressive campaign to install cell towers at public schools in Montgomery County, Maryland. The high schools were targeted first, perhaps because younger children are known to be especially susceptible to the radiation emitted from cell towers, although we do not know for sure what Cingular's strategy was. Additionally, Cingular Wireless first targeted public high schools located in communities with relative economic disadvantages and therefore less likelihood of organizing against a cell tower proposal. Cingular installed cell towers at Wheaton High School, Sherwood High School, and Kennedy High School. The students in these school districts come from families with median household incomes of $55,562, $57,260, and $60,296, respectively. It was not until 2003 that Cingular approached the wealthier Wootton and Walter Johnson High Schools, with median household incomes of $74,655 and $77,568, respectively.[24] Cingular never approached Whitman High School and Churchill High School with cell tower proposals; families of students in these schools have median household incomes of $113,788 and $140,222, respectively. Cingular Wireless' schematic timeline for public school cell tower proposals aimed to first test the will of communities who were economically and educationally disadvantaged before targeting wealthier, resourceful communities. Eventually, the residents of the Wootton and Walter Johnson school districts successfully rejected the cell towers due to the organization of a community voluntary initiative.

Cingular Wireless hired an experienced attorney to implement objectives related to the Wootton cell tower proposal, the same attorney who had previously handled the installment of cell towers at ten other public schools in Montgomery County. With little to no resistance coming from the Montgomery County public schools where cell tower installation was already in place, there was no need for Cingular to hire more than one person, an attorney rightfully so, to execute public school cell tower proposals in Montgomery County. An emphasis is placed on Cingular's need for just a one-man proposal team to demonstrate two things. First, it may be surprising how little publicity and resistance cell tower proposals typically receive, considering how big and visible they are. Second, Cingular's one-man show enabled Cingular to circumvent the proposal process. According to the Telecommunications Act of 1996, personal wireless service providers are required to notify in writing students, employees, and local residents of any proposal to erect a cell tower on public school property. Cingular's attorney had the ability to make sole judgment in his decision to refuse to notify the community regarding the plan to install a cell tower at Wootton High School.

In attempting to reject Cingular's proposal to erect a cell tower at Wootton High School, local residents formed a coalition called SACT. Despite any protections that may be offered by the Telecommunications Act of 1996, it clearly states that cell towers may not be rejected because of health concerns of nearby subjects. However, since Wootton High School is situated in a valley with Frost

Middle School perched up on the adjacent valley ridge, the RF waves emitted from the cell tower at Wootton would be passing through Frost Middle School, creating the allowance for a new interpretation of the act and a platform to remain steadfast. Next, it was discovered by SACT that the principal of Wootton High School had the exclusive authority to decide on whether to allow a 150-ft cell tower to be erected next to the football field at Wootton. SACT contacted local neighborhood associations, cluster school principals and administrators of Frost Middle School and Fallsmead Elementary School, Rockville City management, Montgomery County executive offices, parent–teacher association groups, and other perceived interest groups to spread word of the issue and provide scientific research related to health risks of RF exposure. Through discussion with local government and administration, SACT realized that the community had the right to a town hall meeting prior to the consideration of a cell tower proposal. Over 100 advocates holding greater than 1,400 petitions standing against the placement of a cell tower at Wootton High School arrived at the town hall meeting held in the Wootton library to greet Cingular's attorney and his science expert, with state representatives and media in attendance anxiously awaiting the confrontation. When Cingular's scientific expert utilized research no more recent than the 1960s, it became clear even to Cingular's attorney that SACT's research was plausible and that cell tower radiation may indeed be harmful to humans. For the benefit of Cingular's attorney, it was likely that he did not believe in his own company's stance; instead, his involvement was probably based on solidaristic group loyalties.[25]

An insight deserving mention was a key leadership tactic used by SACT to retrieve petition signatures. One of the exemplary practices of leadership is to inspire a shared vision.[26] Considering the perceived lack of concrete scientific data regarding human exposure to cell towers, one of the most effective ways to gain support for the petitions was to focus on the negative aesthetics of a 150-ft tower that would be visible from a Rockville resident's nearby home. Realizing the eyesore created by cell towers and the fact that opposition due to aesthetic concerns frequently arise, personal wireless service providers have begun camouflaging the cell towers (Figure 4). Figure 4 shows a cell tower with elaborate camouflage meant to make the cell tower look like a tree. After further survey of the surrounding area, it was discovered the camouflage was an attempt to help the cell tower blend in with a tree line overlooked by a nearby middle school (Figure 5). The canvas makes the cell tower more difficult to notice and provides personal wireless service providers with a prompted solution to aesthetic concerns.

A key ingredient behind this chapter's explicit focus on the placement of cell towers at public schools as opposed to all schools, in general, was illustrated by the Wootton principal's authority to decide for or against installation of the cell tower. If Wootton were a private school, the principal may not have been dictated the authority of decision maker or, in the event that authority was dictated, may not have had the wherewithal to acknowledge responsibility to local residents not affiliated with the school. SACT was a community effort predominately fueled by advocates not directly associated with Wootton High School. The idea that public school principals and all public school administrators, in general, are public servants lends additional support for the case against cell towers at public schools in particular. However, the stance against cell towers at public schools applies to all schools because of the RF-related health hazards.

Discussion of Cell Tower Placement Process

The Telecommunications Act of 1934 established the Federal Communications Commission (FCC) as the regulatory authority over communications activities in the United States. Because digital cell phones were not available to the public until 1988, controversy over the placement of cell towers is a new phenomenon. It was not until the Energy Policy Act of 1992 and the Telecommunications Act of 1996 when the federal government realized the issues of RF signal strength and tower placement, respectively. In 2003, SACT became one of the first nongovernmental organizations to address hazards related to cell

FIGURE 4　RF tower with elaborate camouflage.

FIGURE 5　The tallest tree in this image is the cell tower.

tower placement near public institutions. However, a number of transformations in the telecommunications industry have taken place since the Telecommunications Act of 1934, which have shaped the current regulatory environment. The years 1945 and 1952 marked the first major oversight by the FCC on over-the-air television, regulating the spectrum allocations and color standards. In 1968, telecommunications service providers were authorized by the FCC to attach equipment to preexisting above-the-ground electrical lines. And, in 1992, the FCC ruled to let the market decide the appropriate standards for digital cell phones and related equipment.[27]

The political actors typically involved in the process of determining the placement of cell towers complicate the ability of community voluntary initiatives to succeed in opposing the placement of cell towers. Achieved by a 1999 amendment to the Telecommunications Act of 1934, the local government has authority over state and federal governments on the issue of tower placement.[28] However, the amendment to the Telecommunications Act of 1934 is pursuant to the Telecommunications Act of 1996 specified requirement of the federal government to assist licensees' pursuit of preferred sites.[29] Federal involvement in the tower placement process results in streamlining of policy action. Streamlining results in the shrinking of the policy window for community advocates, which reduces community advocates' opportunity to introduce their own policies.[30] Additionally, personal wireless service providers sometimes hire independent facilities siting companies who offer comprehensive tower placement services, from lobbying of local government and communities to addressing zoning regulatory concerns. Leasing of sites and fulfillment of regulatory and registration requirements may also be taken care of by facilities siting companies (Figures 6–8). "The local zoning authorities should therefore be aware that a facilities siting company may not be seeking the sites that are of most interest to particular Commission licensees [personal wireless service providers], but rather seek general sites on highly elevated locations in the hopes of leasing the sites, in turn, to Commission licensees."[31] It would be intuitive to reason that the existence of broker special interests in the placement of cell towers would be an additional obstacle to voluntary initiatives striving to oppose cell towers. However, the contrary is sometimes true. Policy brokers, such as independent facilities siting companies, are interested in maintaining a sustainable level of conflict in order for services being offered to remain in high demand.[32] Lingering, yet not overpowering, community opposition is welcomed by facilities siting companies. To combat the influence of the federal government and policy brokers in the tower placement process, community organizations need to adopt a policy of political efficacy, arising from political participation as a means to exert influence.[33] Political efficacy is especially important when it comes to the upholding of FCC guidelines on tower height and field strength. According to the Energy Policy Act of 1992, electric and magnetic field strength must be made public.[34] Yet, according to the Code of Federal Regulations pertaining to personal communications services, height-above average-terrain (HAAT) and field-strength guidelines may be waived if all parties involved agree.[35] If the community does not involve local government, the community will have no voice on the issue, and the already-flaccid federal guidelines will leave school children and staff insurmountably exposed.

FIGURE 6   Brokers lease to wireless service providers.

FIGURE 7 Warning of RF emission.

FIGURE 8 Record keeping by the FCC.

Procurement and Contracting

The cell tower placement process as it involves local, state, and federal governance can be a separate, yet parallel set of activities to procurement and contracting processes. While law and statutes specific to cell tower placement may be directed at issues of allowability, the procurement and contracting functions deal specifically with the actual purchase in terms of cell tower specifications, features, pricing, contract terms and conditions between the public and private entity, solicitation for prospective vendors, competition for contract award, and contract administration, so that the private entities fulfill its obligations for the duration of the contract, among other procurement activities. In order to execute these functions, local government procurement units, which are typically housed in a department of procurement, finance, or budgeting, engage a contracting process involving pre-solicitation, solicitation, proposal evaluation, contract award, contract administration, and contract close-out.[36]

The pre-solicitation process involves procurement planning and communication with potential vendors who aim to install cell towers and provide cellular service to the area. Here, the government engages a process of discovery in which it learns about products and services and may hold formal or informal discussions with vendors. These discussions may involve policy matters, issues of feasibility for placement or service provision, the technology, and initial concerns regarding contract terms and conditions that may eventually follow. Generalities regarding the process for government's selection of the vendor for actual cell tower placement may also be discussed. These discussions may take place between a procurement agent and company representative over the phone, an in-person meeting, in writing, or at a trade show or conference. The pre-solicitation phase also enables government to prepare its formal solicitation document according to vendor concerns and considerations in conjunction with its own needs and priorities.

Oftentimes, wireless service providers may initially contact government to place cell towers on public properties. However, once government decides that it will entertain the possibility of placing a cell tower on public property, then it typically must engage a competitive process that allows the various companies an opportunity to compete for the contract. At the next phase, solicitation, the government may issue a request for proposals (RFP) or an invitation to bid (ITB). The RFP seeks proposals from vendors and provides them an opportunity to win the contract by setting forth a product and service to government, including various factors such as price, specifications, and commitments. Meanwhile, an ITB is a more rigid form of solicitation, in which the government sets forth its required criteria and the vendors simply respond regarding how they would perform to contract.

The justification behind having vendors compete for the contract award is so that government can be in a position to choose from the best product and service, i.e. the vendor who will commit the greatest amount of revenues at the most favorable contract terms and conditions for government and its community stakeholders. When vendors are required to compete for the contract award, then they will aim to increase the rent that they will commit to pay to government, or the public entity, for use of the property, as well as bolster their services or offerings in order to be the more attractive vendor. Furthermore, the vendors may make special accommodations in attempts to ameliorate public concerns that may have been vocalized during the pre-solicitation process such as issues of tower unsightliness, proximity, equipment dimensions, transmission power, servicing, and maintenance.

Once the RFPs or responses to the ITB have been submitted by the vendors to the procuring entity, then the procuring entity must evaluate these proposals to decide which vendor will be awarded with the contract. In this proposal evaluation phase, a team is formed to evaluate the documents submitted by the vendors. It is important to note that the proposal evaluation team members, per public procurement mandate, are usually required to consist of public personnel that were not involved in the pre-solicitation or solicitation phases in order to avoid conflict of interest. It may be the case that various personalized relationships were formed between the procurement agent and company representative during these early discussions, and it is important that the proposals are evaluated objectively according to the actual product and service to be delivered without consideration of prior personal relationships. Once the proposal evaluation team is in place, the evaluations are made based on factors of responsibility and responsiveness.

Responsibility refers to matters of vendor track record including their past performance on contracts with other public entities and the vendor's financial capitalization. The notions are that a vendor should be well capitalized in order to ensure that it has the resources to fulfill contractual obligations for the duration of the contract term and that past performance on contracts is an indicator of a company's ability to deliver its products and services as promised. Responsiveness deals with the extent to which a vendor's proposal reflects the needs as set forth by the public entity in the language of the RFP or ITB. A scoring methodology is then typically developed by the proposal evaluation team such as a weighted or other numerical scoring schemata, which assigns scores to various facets of the vendor proposals. The proposal that receives the highest score by the proposal evaluation team will be awarded with the contract.

Next, the contract award phase involves a debriefing process so that the winning and losing bidders are notified of the proposal evaluation team's selection. In this phase, any deviations between the vendor's proposal submitted and the RFP or ITB are addressed and a negotiation process may ensue between the selected vendor and procuring entity regarding actual contract terms and conditions. For example, some of the requirements set forth by the public entity in the RFP or ITB for the cell tower may be questioned by the vendor. Since the vendor has expertise regarding its own products and services, including the technology and equipment, it may be that the public entity is requiring features or specifications of the cell tower that may not be reasonable or feasible according to the vendor. Oppositely, the vendor's proposal as evaluated may lack some aspects of responsibility or responsiveness that need to be addressed, which may include community concerns or interests regarding cell towers, and it may be that the public entity has the greater knowledge and awareness in this particular political area. Of course, any deviations between the vendor's proposal and the RFP or ITB may reflect modifications in price, i.e., the rent that the public entity will be charging. If there are aspects of the RFP or ITB that the vendor cannot guarantee or conform to, then the public entity may offset those deficiencies by charging higher rent or setting forth other collateral commitments.

The first three phases of contracting involve pre-award processes, while contract administration and close-out are two phases that deal with post-award processes. Contract administration involves the actual delivery of products and services according to the contractual agreement known as performance to contract. This involves planning, preparation, and installation of the cell tower, and the administrative functions and procedures that will be involved such as oversight of construction, facilities monitoring, inspection, payment of rent, and other facets of interaction, communication, and coordination between the vendor and procuring entity. If the products and services delivered by the vendor are not consistent with the contract terms and conditions, which may include the manner that tower construction is taking place, the actual tower, or its servicing, then various modifications to the contract may be stipulated by the procuring entity. Meanwhile, if the procuring entity does not deliver its property site in the condition as promised, then the procuring entity may need to institute various change orders that could force it to reduce or avoid its rent payment obligations from the vendor for a specified time period or allow the vendor to make changes to its products or services. Thus, various change clauses will be written into the contract during negotiations in the event that unintended scenarios or problems arise during contract administration. Of course, it is difficult if not impossible to imagine and surmise all the potential scenarios and obstacles that may surface for a given procurement. If the vendor fails to adhere to the contract terms and conditions or does not adequately respond to the procuring entity's inquiry into contract violations, then the contract may be terminated by the procuring entity.

The final procurement phase is contract close-out. This phase involves quality assurance that the vendor has delivered as promised by installing a cell tower according to the requirements set forth by the procuring entity in the contract, and that servicing and maintenance are being executed accordingly. Any accounts receivable or rent payments owed to the public entity are accounted for and contract evaluation takes place to ensure that all contractual obligations are satisfied. Contract close-out is also important because of the lessons learned. The procuring entity, as the procurement unit within the government's department of procurement, finance, or budgeting, likely has other potential property sites in the municipality that could be attractive to wireless service providers. Hence, the lessons learned from the procurement and contracting process on the current contract may be applied and incorporated by the procuring entity into future cell tower procurements.

Meanwhile, the contract close-out phase also entails evaluation of the vendor on their performance to the contract, which could subsequently serve as performance evaluation data for the same or other procuring entities as they evaluate any future proposals by the vendor for aspects of responsibility. Whereas the early years of cell tower placement on public school and other public properties involved much more devolved processes that were more inconsistently applied on a case by case basis, even within the same municipality, the prevalence and debate surrounding cell tower placement has partially shifted the public authority on cell tower placement from public schools and school boards, and those public units or divisions having operations on the actual property, to procurement departments acting as agents on behalf of public schools or other public units. As a result, lessons learned from cell tower procurements can be more aptly applied to future contracts in order to address previous contract failures that may involve public concern and outcry. Essentially, the cell tower procurement is managed by the same procuring entity within a given municipality. Therefore, it has the opportunity to develop policy, politics, and process expertise from experience similar to the private entities who repeatedly write cell tower proposals to government.

The procuring entity does face a myriad of challenges though. The procuring entity must comply with local, state, and federal procurement law and policy, in addition to any local, state, and federal law and policy governing cell tower placement specifically. Additionally, there must be due consideration given to voluntary initiatives and advocacy groups. At times, these laws and policies may be conflicting or difficult to streamline into procurement and contracting methods.

Ethics and Public Administration

The three elements of corporate social responsibility are market actions, externally mandated actions, and voluntary actions.[36] Personal wireless service providers fail to address all three elements of social responsibility when dealing with the placement of cell towers. First, personal wireless service providers have poorly responded to market actions in their use of policy brokers to ensnare the tower placement process. Second, the mandated actions of the FCC related to HAAT and field-strength guidelines allow regulatory thresholds to be exceeded if no opposing voluntary organization is present at scheduled hearings. Third, voluntary actions that aim to avoid students' exposure to cell towers are not taken. In fact, the current trend is just the opposite, in which schools are targeted because of the surrounding open space that allows for enhanced RF wave transmission (Figures 9 and 10).

The ability of voluntary initiatives to oppose the placement of cell towers at public schools relies upon the formation of a nucleus of zealous participants; charismatic leadership alone does not result in the

FIGURE 9   RF tower overlooking playing fields.

FIGURE 10　RF tower installed adjacent to football field.

type of rapid expansion of the coalition that is necessary to fight seasoned corporate interests such as personal wireless service providers.[37] In addition, the media may be needed to intervene in facilitating changes in public perceptions. The media can help provide a transition and a way for the public to digest new policy initiatives.[38] In some cases, even with high visibility, the strongest coalitions are unable to defeat polyarchal interest groups on a particular issue, be it the placement of a cell tower at yet another school. Success depends on forging relationships with government officials as much as administrative competence.[39] Because of the variations in local government across the nation, including inconsistencies in the law and process governing zoning and other enforcement departments, there is a need to adopt a customizable approach when attempting to oppose a cell tower.

The SACT's ability to reject a cell tower proposal at Wootton High School but failure to impact policy change nationally brings into question the federalist debate. Hamilton, Madison, and Jay desired a strong central authority in their staunch support of federal government and ratification of the U.S. Constitution.[40] Opposing the federalists, Patrick Henry led the antifederalist approach arguing for decentralization and states' rights.[41] The federal government's determination to simultaneously support personal wireless service providers' cell tower installation campaign at public schools while allowing local communities to decide for themselves is rooted in both federalist and antifederalist modes. Nonetheless, the federal government's dual role is authoritarian.[42]

The cry for federal regulation disallowing placement of cell towers at or near public schools involves regulatory policy making, which inevitably will indulge or deprive one specific interest or another.[43] The regulatory approach offered does not necessarily favor some sort of Weberian chain-of-command hierarchy originating from the top down[44] but rather prefers reactionary cultural movement that reflects Thelen and Mahoney's,[45] Hacker's,[46] and Sheingate's[47] ideas on institutional change according to evolving assumptions of administrative and technological environments. If the dominant approach were favored, the solution would be to charge personal wireless service providers higher rental fees to compensate for any associated health care costs that may result.[48,49] Other rational socioeconomic approaches would seek to place value on the cost of the loss of human life and monetary damages due to the terminal pain and suffering induced by cell towers. Unfortunately, the most commonly observed failure in public management and the one associated with the egalitarian approach is a lack of ability to exert authority.[50]

Areas for Further Research

A logical area for further research is to track and survey humans. Tracking students who attend schools with cell towers on premises is the best way to determine the danger of cell towers to students. Tracking students while in school and in years beyond may help determine whether cell towers are indeed harmful to children or humans in general. An effective data set for a future study would need to take into account differentiation between student exposure at elementary, middle, and high schools, considering that young children are known to be more susceptible to RF exposure. Additionally, the data would need to differentiate between students who had many instances of intense exposure, such as student-athletes on a sports field containing a cell tower, and students who had fewer instances of intense exposure. Previously, this type of study was not possible. It was only since 2002 that the personal wireless service providers started targeting schools. Today, cell towers are erected on school properties across the country. However, obtaining access to student records for data collection of this type of study invokes the support of government. The resources and sheer number of people who would need to be mobilized by such a study requires congressional backing. Furthermore, the scientific community has questioned the methods that would underwrite such a study. Areas of inquiry that have been considered to be obstacles to a human study of this type are the competing risks when experimenting for causality. Oftentimes, it is difficult to determine which risk is the source of the illness. For example, did cancer clusters in the area form as a result of contaminated drinking water or exposure to a cell tower? New research indicates that margins of error in causality can be reduced. Building on time-dependent predictive accuracy measures,[51,52] the coefficient of the distribution of false positives among event-free subjects can be adjusted to reflect nearest neighbor (in our case, the highly exposed student-athletes on the sports fields containing cell towers) estimation of the distribution of input variables representing true positive incidents and length of exposure of competing risks in order to help determine causality.[53] The false-positive value is manipulated by a coefficient that is calculated from the estimate of true positive incidents in order to offset causality miscalculations stemming from overlapping incident rates of competing risks. Unconcerned citizens and proponents of cell tower placement on school property argue that cancer clusters related to cell towers have not emerged. However, it is impossible to predict what the health impacts may be over time. An evaluation technique is needed for cumulating, comparing, and contrasting varied results in order to establish an applied theory and framework.[54]

Conclusion

Just 2 years after the victory at Storm King Mountain paved the way for environmentalism, Olson's *The Logic of Collective Action* (1965) opened our eyes to the intuitiveness of equality and why the events of the 1960s were unlikely to ever recur.[55] Much different from the equal rights movement, the successful voluntary initiative at Wootton High School has not led to nationwide legislation and is embedded in an issue that is invisible. The inability to see the RF waves somehow precludes from our psyche the notion of harm. Invisible risks skew the indifference curve that guides our behavior in responding to risk. An indifference map is our collection of indifference curves and helps to shape our order of preferences.[56] The typical reaction to an invisible risk is a delay in response. After a 5-year pause, personal wireless service providers are once again proposing cell towers at public schools in Montgomery County, Maryland, this time at Whitman High School in Bethesda, Maryland.[57] Bethesda community advocates are citing Wootton's 2005 rejection of a cell tower as precedent in Montgomery County.[58] Currently, only the ten cell towers that were installed prior to the Wootton campaign exist in Montgomery County. Also, in the sense of a health-wise decision at Wootton, why have the ten previously erected towers not been taken down?

Cell towers should not be placed at schools. Scientific experimental research on animals and humans is conclusive that RF waves increase animal and human risk to cancer and other illnesses. Also, in the case of public schools, students have little to no choice in deciding whether or not to attend a particular school. To address personal wireless service providers' desire to bolster cell phone connectivity in communities, state and local governmental zoning boards should work together to designate specific areas where the placement of cell towers will be permitted. Assuming no cell tower is placed at or near a public school, the designated zones will enable the public to make their own decision regarding living or spending time near cell tower sites.

Community voluntary initiatives are the answer if the goal is to reject a cell tower proposal. Federal governments' involvement, not from the standpoint of assisting personal wireless service providers meet their objective of cell tower placement in a given community but from the standpoint of bringing together the nationwide advocacy groups and coalitions that have both successfully and unsuccessfully defended their schools from RF wave penetration, is the long-term solution to keeping schools safe from cell towers. Also, although the prospect of fair gamesmanship in the process of appeal against cell tower placement at public schools was not presented to be optimistic, consideration of bureaucratic red tape, or in this case, purposeful lack thereof, is essential to successfully implementing a strategy that leads to the rejection of a personal wireless service provider's cell tower proposal. In any case, policies that engage citizens in their own communities and ask them to do their own policy analysis are generally more preferable to those policies that do not.[59]

References

1. Milham, S. *Dirty Electricity, Electrification and the Diseases of Civilization*; Universe: New York, 2010.
2. The Stop Cancer before It Starts Campaign, Press Report, May 8, 2003, Cancer Prevention Coalition, available at http://preventcancer.com/publications/pdf/M803.htm (accessed September 2011).
3. LeBeau, C. *Insomnia, Fatigue, and Cell-Phone Towers*; Vital Health Publications: West Allis, WI, 2010.
4. Lai, H. Biological effects of radiofrequency radiation from wireless transmission towers. In *Cell Towers, Wireless Convenience? Or Environmental Hazard?* Levitt, B., Ed.; New Century Publishing: Markham, Canada, 2000; 65.
5. Schwartz, M. *Principles of Electrodynamics*; McGraw Hill: New York, 1972; 2–0, 14–49.
6. Edminister, J.; Nahri-Dekhordi, M. *Electromagnetics*, 3rd Ed.; McGraw Hill: New York, 2011; 174.
7. Sadiku, M. *Elements of Electromagnetics*; Oxford University Press: New York, 2010; 350.
8. Fesenko, E.; Novoselova, E.; Semiletova, N.; Aganova, T.; Sadovnikov, V. Stimulation of murine natural killer cells by weak electromagnetic waves in the centimeter range. *Biofizika* 1999, 44 (4), 73–41.
9. Fesenko, E.; Makar, V.; Novoselova, E.; Sadovnikov, V. Microwaves and cellular immunity. Effect of whole body microwave irradiation on tumor necrosis factor production in mouse cells. *Bioelectrochem. Bioenerg.* 1999, 49 (1), 29–35.
10. Loscher, W. Extraordinary behavior disorders in cows in proximity to transmission stations. *Der Praktische Tierarz* 1998, 79 (1), 437–444.
11. Galat, V.; Mezhevikina, L.; Zubin, M.; Lepikhov, K.; Khramov, R.; Chailakhian, L. Effect of millimeter waves on the early development of the mouse and sea urchin embryo. *Biofizika* 1999, 44 (1), 137–140.
12. Chiang, H.; Yao, G.; Fang, Q.; Wang, K.; Lu, D.; Zhov, Y. Health effects of environmental electromagnetic fields. *J. Bioelectr.* 1989, 8 (1), 127–131.
13. Mann, K.; Wagner, P.; Brunn, G.; Hassan, F.; Hiemke, C.; Roschke, J. Effects of pulsed high-frequency electromagnetic fields on the neuroendocrine system. *Neuroendocrinology* 1998, 67 (2), 139–144.
14. Nechaev, S.; Fargo, D.; dos Santos, G.; Liv, L.; Gao, Y.; Adelman, K. Global analysis of short RNA's reveals widespread promoter-proximal stalling and arrest of Pol II in Drosophila. *Science* 2010, 327 (5963), 335–338.
15. Adelman, K. Transcriptional Responses to the Environment Group, Chromatic Signatures and Gene Expression. Principal Investigator. National Institute of Environmental Health, available at http://niehs.nih.gov/research/atniehs/labs/lmc/tre/index.cfm (accessed September 2011).
16. Muse, G.; Gilchrist, D.; Nechaev, S.; Shah, R.; Parker, J.; Grissom, S.; Zeitlinger, J.; Adelman, K. RNA polymerase is poised for activation across genome. *Nat. Genet.* 2007, 39 (12), 1507–1511.

17. Zeitlinger, J.; Stark, A.; Kellis, M.; Hong, J.; Nechaev, S.; Adelman, K.; Levin, M.; Young, R. RNA polymerases stalling at developmental control genes in the Drosophila embryo. *Nat. Genet.* 2007, 39 (12), 1512–1516.

18. Boettiger, A.; Ralph, P.; Evans, S. Transcriptional regulation: Effects of promoter proximal pausing on speed, synchrony, and reliability. *Comput. Biol.* 2011, 7 (5), 1–14.

19. Gilchrist, D.; Nechaev, S.; Lee, C.; Gosh, S.; Collins, J.; Li, L.; Gilmour, D.; Adelman, K. NELF-mediated stalling of Pol II can enhance gene expression by blocking promoter-proximal nucleosome assembly. *Genes Dev.* 2008, 22 (14), 1921–1933.

20. Gilchrist, D.; dos Santos, G.; Fargo, D.; Xie, B.; Gao, Y.; Li, L.; Adelman, K. Pausing of RNA polymerase II disrupts DNA-specified nucleosome organization to enable precise gene regulation. *Cell* 2010, 143 (4), 540–541.

21. Adelman, K.; Kennedy, M.; Nechaev, S.; Gilchrist, D.; Muse, G.; Chineov, Y.; Rogatsky, I. Immediate mediators of the inflammatory response are poised for gene activation through RNA polymerase stalling. *Proc. Nat. Acad. Sci.* 2009, 106 (43), 18207–18212.

22. Kasevich, R. Brief overview of the effects of electromagnetic fields on the environment. In *Cell Towers, Wireless Convenience? Or Environmental Hazard?* Levitt, B., Ed.; New Century Publishing: Markham, Canada, 2000; 170.

23. Anzevino, J. Preserving scenic and historic sites: The dilemma of siting cell towers and antennas in sensitive areas. In *Cell Towers, Wireless Convenience? Or Environmental Hazard?* Levitt, B., Ed.; New Century Publishing: Markham, Canada, 2000; 169.

24. Public School Review. Public Elementary, Middle, and High Schools, Data, 2011, available at http://www.publicschoolreview.com (accessed September 2011).

25. Dunleavy, P. *Democracy, Bureaucracy, and Public Choice*; Prentice Hall: London, 1991; 28.

26. Kouzes, J.; Posner, B. *The Leadership Challenge*, 4th Ed.; Jossey-Bass: San Francisco, 2007; 16–18, 142–143.

27. Ismail, S. Transformative Choices: A Review of 70 Years of FCC Decisions, FCC Staff Working Paper 1; Federal Communications Commission: Washington, DC, 2010; 1, 18, 19.

28. A Bill to Amend the Communications Act of 1934, 106th Congress, 1st Session; U.S. Senate: Washington, DC, 1999; S.1538.

29. New Wireless Tower Siting Policies, Fact Sheet #1; Wireless Telecommunications Bureau, Federal Communications Commission: Washington, DC, 1996; 1–2.

30. Kingdon, J. *Agendas, Alternatives, and Public Policies*, 2nd Ed., Longman: New York, 2003; 165.

31. National Wireless Tower Siting Policies, Fact Sheet #2; Wireless Telecommunications Bureau, Federal Communications Commission: Washington, DC, 1996; 8.

32. Sabatier, P. An advocacy coalition framework of policy change and the role of policy-oriented learning. *Policy Sci.* 1988, 21 (1), 141.

33. Sabatier, P. Political science and public policy. *Political Sci. Polit.* 1991, 24 (2), 145.

34. Energy Policy Act of 1992, Section 2118, Electric and Magnetic Fields Research and Public Information Dissemination (RAPID), Public Law 102–486; 42 U.S.C. 13478.

35. Code of Federal Regulations, Title 47, Calculation of Height above Average Terrain and Field Strength Limits; Personal Communications Services, 2010; 24.53, 24.236.

36. Thai, K.V. *Introduction to Public Procurement.* Herndon: Virginia, 2013.

37. Steiner, J.; Steiner, G. *Business, Government, and Society*; McGraw Hill: New York, 2012; 131.

38. Downs, A. *Inside Bureaucracy*; Little, Brown, and Company: Boston, 1967; 7, 9.

39. Sabatier, P.; Mazmanian, D. The implementation of public policy: A framework of analysis. *Policy Stud.* 1980, 8 (4), Special no. 2, 550.

40. Nalbandian, J. Reflections of a "pracademic" on the logic of politics and administrations. *Pub. Admin. Rev.* 1994, 54 (6), 531.

41. Kesler, C.; Rossiter, C. *The Federalist Papers*; New America Library: New York, 1999; vii–xii.

42. Henry, P. Legitimate government. In *The Anti-Federalist Papers*. Borden, M., Ed.; Michigan State University Press: East Lansing, MI; Par. 5.

43. Steinfeld, J. American authoritarian democracy: Vietnam War, Kosovo War, and Overseas Contingency Operation. In *Proceedings of the 32nd Annual Southeastern Conference for Public Administration*, New Orleans, LA, Sept. 21–24, 2011.

44. Miller, H. Weber's action theory and Lowi's policy types in formulation, enactment, and implementation. *Policy Stud.* 1990, 18 (4), 895.

45. Weber, M. *The Theory of Social and Economic Organization*; The Free Press: New York, 1947.

46. Thelen, K.; Mahoney, J. *Explaining Institutional Change: Ambiguity, Agency, and Power*; Cambridge University Press: Cambridge, England, 2010.

47. Hacker, J. Policy drift: The hidden politics of welfare state retrenchment. In *Beyond Continuity: Institutional Change in Advanced Political Economies*; Streeck, W., Thelen, K., Eds.; Oxford University Press: Oxford, 2005; 17.

48. Sheingate, A. Rethinking rules: Creativity in the House of Representatives. In *Explaining Institutional Change: Ambiguity, Agency, and Power*; Thelen, K., Mahoney, K., Eds.; Cambridge University Press: Cambridge, England, 2010.

49. Weber, M. *Economy and Society, an Outline of Interpretive Sociology*; Roth, G., Wittich, C., Eds.; University of California Press: Berkeley, CA, 1978; Vol. 1.

50. Weber, M. *Economy and Society, an Outline of Interpretive Sociology*; Roth, G., Wittich, C., Eds.; University of California Press: Berkeley, CA, 1978; Vol. 2.

51. Hood, C. *The Art of the State: Culture, Rhetoric, and Public Management*; Oxford University Press: Oxford, 1998; 40.

52. Heagerty, P.; Lumley, T.; Pepe, M. Time-dependent ROC curves for censored survival data and diagnostic marker. *Biometrics* 2000, 56 (1), 337–344.

53. Heagerty P.; Zheng, Y. Survival mode predictive accuracy and ROC curves. *Biometrics* 2005, 61 (1), 921.

54. Saha, P.; Heagerty, P. Time-dependent predictive accuracy in the presence of competing risks. *Biometrics* 2010, 66 (4), 999–1011.

55. Lowi, T. American business, public policy, case-studies, and political theory. *World Polit.* 1964, 16 (4), 688.

56. Hirschman, A. *Shifting Involvements: Private Interests and Public Affairs*; Blackwell: Oxford, 1985; 79.

57. Anderton, C.; Carter, J. *Principles of Conflict Economics*; Cambridge University Press: New York, 2009; 29.

58. Barnes, A. Cell Tower Proposed for Walt Whitman High School; USA 9 News: Bethesda, MD, 2010, available at http://wusa9.com/news/local/stay.aspx?stayid=95703&catid=158 (accessed September 2011).

59. Cropper, M. No cell tower at Whitman. Rockville Gazette, 2011, available at http://ww2.gazette.net/stories/03102010/montlet175220_32598.php (accessed September 2011).

60. Clarke, J.; Ingram, H. A founder: Aaron Wildavsky and the study of public policy. *Policy Stud.* 2010, 38 (3), 574.

21

Community-Based Monitoring: Ngarenanyuki, Tanzania

Aiwerasia V.F.
Ngowi, Larama
M.B. Rongo, and
Thomas J. Mbise

Introduction

Community-based monitoring was initiated in Ngarenanyuki, Tanzania, to study the impacts of pesticides on health and the environment. This entry is organized and divided into the following main sections: "Introduction," "Methodology," "Results and Discussion," and "Conclusion." Illustrations are included in the "Introduction," "Methodology," and "Results and Discussion" sections. The main goal of the study was to reduce exposure to pesticides among the farmers in Ngarenanyuki by training the farmers on health impacts related to exposure to pesticides, how to monitor such impacts, and how to reduce the risks.

Incidences of poisoning from pesticides are estimated to be highest in developing countries, despite the higher use of pesticides in developed countries.[1,2] The monitoring of pesticides and their health impacts on farmers and the public in general, which is normally performed by qualified researchers, is not sufficiently practiced in many developing countries, owing to financial constraints and to competing research interests. It therefore makes more sense to empower communities themselves to monitor the impact of pesticides and to take decisions that might reduce the risks to themselves and to their environment. Community-based monitoring of the impacts from pesticides enable those communities to determine whether or not the chemicals they are already exposed to, or might be exposed to, present any sort of hazard to their health and a potential threat to their environment.

Community pesticide-surveillance methods have been successfully used in the Asia Pacific[3] and could therefore be considered appropriate in Tanzania and other Southern African countries, both for establishing better data on the extent of pesticide poisoning and to raise awareness among farmers themselves. Using their own system of observation and evaluation of risks, for example, Malaysian plantation workers have developed the Community Pesticides Action Monitoring (CPAM) approach, in which they succeeded in documenting the health effects of airborne pesticides, identifying paraquat in particular as a major problem.[4]

They then proceeded to take action to prevent further exposure of plantation workers to paraquat. Communities in Kasargod District, Kerala (India), after investigation, monitoring, and documentation, using a CPAM approach, identified endosulfan as the major pesticide causing health and environmental problems and subsequently called for the ban on endosulfan to prevent further exposure and damage to the communities.[5]

Incorrect pesticide handling and management is thus known to be unsafe to both human and environmental health and jeopardizes biodiversity.[6] It is also further evident that most rural communities in Tanzania depend on farming and agribusinesses to earn their living. However, traditional agricultural production in the country has come under continuous pressure from globalization and other market forces, with the result that high-input agriculture has come to play an increasingly major role in the economies of rural communities. Although the use of pesticides in combating pests and diseases is widely encouraged among the farmers in these communities to promote production, less emphasis has been placed on safety practices and the proper handling and management of materials.

Ngarenanyuki Ward gives an example of a community in Tanzania where the majority of vegetable farmers believed that, without pesticide use, crop production would have been impossible. In a previous study carried out by the Work and Health in Southern Africa (WAHSA) team (unpublished), it was found that mixing three to five different types of pesticides in a single spray mix was a common practice in Ngarenanyuki, and that farmers did not understand what was written on the label or the meaning of the colors on the containers. They simply applied pesticides because a neighbor had applied them, and not because they had identified a particular pest problem. The farmers were also found to have mixed pesticides without following the doses recommended on the label, sometimes doubling or trebling the dosage regardless.

Retail outlets for pesticides in Ngarenanyuki were also found not to have been registered with the regulatory authority and therefore appeared to be selling pesticides illegally. It was further noted that, because the shop owners tended to repackage or dispense pesticides in other containers, they were observed on occasion to be left with empty pesticide containers, which they apparently destroyed by burning them at the marketplace. Farmers were also observed in some instances to have stocked substantial amounts of pesticides to cater for the whole year, owing to perceived shortages in the local village and the reported distances they had to travel in order to purchase pesticides in towns, frequently mentioned as Arusha or Moshi. Some of these stored pesticides were also observed to have become obsolete, which were then likely to create fresh problems of disposal as shown in Figure 1.

The unintended outcomes of pesticide exposures are difficult to reverse once they have been established and are in themselves expensive. Although advances in acute pesticide-poisoning surveillance and treatment in developed countries have led to some achievements in control, pesticide poisoning remains a public health problem globally, particularly in developing countries.[7] Those applying pesticides need to understand the effects of these chemicals to the environment and to their own health and the resulting costs. Alternative pest management strategies that are cheaper and friendlier to end users and the environment need to be promoted.

A workshop was organized for Southern Africa Development Community (SADC) registrars of pesticides in Arusha on October 13–14, 2006. Participants suggested that WAHSA-TPRI (Table 1) should pilot the tool used in the Asia Pacific to establish a systematic mechanism for pesticide monitoring and data collection, with a view to determining the extent of pesticide exposures, injuries, and diseases at the community level. WAHSA-TPRI then selected Ngarenanyuki as the study area, based on their working experience in Northern Tanzania in health hazards posed by pesticides and on the knowledge that farmers in the area were especially at risk with regard to pesticide poisoning.

FIGURE 1   Hazardous practices observed in Ngarenanyuki, 2006–2007. (left) Haphazard disposal of empty containers. (right) Dispensing/repackaging of pesticides in retail shops.

TABLE 1   The WAHSA Program Was Established in October 2004 as a Regional Initiative in Southern Africa to Build Capacity in the Region in Occupational Health

One of its key programs was its project *Action on Health Impacts of Pesticides,* which aimed to:
• Improve pesticide-safety materials for the SADC region
• Intervene to reduce pesticide usage
• Improve on agricultural policies and pesticide registration
• Enhance knowledge and improve surveillance about pesticide exposures and health impacts in the region
• Foster a strong regional network for information exchange and consultation

This entry reports on the process involved in the establishment of a community-monitoring team in Ngarenanyuki and on the preliminary results of the monitoring exercise. The authors hope that the findings will facilitate a process to identify those resources required to reduce pesticide use, the development of an action plan to access and mobilize these resources, and the further establishment of an effective system of communication among members of the community on pesticide use and access to any other information with regard to pesticide poisoning.

Methodology

This initiative was intended to pilot the CPAM approach that has already been used successfully in Asian countries. The WAHSA-TPRI Team was trained on the subject through their link with a nongovernmental organization, AGENDA for Environment and Responsible Development, and was then employed to mobilize the community in Ngarenanyuki.

The initiative adopted a participatory research methodology by involving farmers in the collection and analysis of pesticide-related data. Data collection tools were developed by making use of Community Pesticide Action Kits (CPAKs), and training materials were developed in the regional language of Swahili. As a result of these activities, the community was sensitized and a subsequent rise in awareness was noted. CPAKs were produced by an ASEAN team of citizens' groups and farmer schools as a tool for action and advocacy, encouraging community education/empowerment. It contains modules that address various aspects of concern such as Warning! Pesticides are a Danger to Your Health; Breaking the Silence: Pesticides in Plantations; Profiting from Poisons: The Pesticides Industry; Drop Pesticides! Build a Sustainable World; Pesticides Destroy our World; Women and Pesticides; Keeping

Watch: Pesticides Laws; How to say NO! to Pesticides: Community Organizing; and Seeking out the Poisons: A Guide to Community Monitoring. The modules are not complete in themselves but need additional materials in local languages.

After securing the community's consent and the involvement of farmers' representatives and communities in capacity building, the program of community-based data collection and analysis was started and the monitoring exercise was implemented. Selected farmers worked in collaboration with the WAHSA experts to monitor and record issues related to pesticide use and exposures in the Ngarenanyuki villages.

Through a series of village meetings, the ward government in collaboration with the ward extension officer invited the farmers to participate in the training. Thirty farmers were selected by the farmers themselves from two villages (Uwiro and Olkung'wado) to represent each subvillage. No farmers were selected from Ngabobo, Kisimiri Juu, and Kisimiri Chini as communication became difficult. Ngarenanyuki as a whole is situated between Mt. Kilimanjaro and Mt. Meru, the first and third tallest mountains in Africa, respectively. The terrain thus consists of rocks, hills, valleys, rivers, and streams, which make some areas impassable during the wet season. One of the villages left out did not cooperate well with the others as they are believed to grow cannabis, a plant that is illegal in Tanzania.

A 6-day training of 25 representative farmers from the Ngarenanyuki was conducted by WAHSA-TPRI scientists, who had expertise in agricultural extension, agronomy, toxicology, entomology, plant pathology, and environmental science, and covered the following topics:

1. Pesticide use and their impacts on human health and environment, where farmers learned about pesticide use around the world, including examples of the negative impacts of pesticides on human health and the environment.
2. Pesticide identification and classification according to their acute toxicity, where farmers learned to identify the types of pesticides used and how they are classified according to their acute toxicity by the World Health Organization (WHO). One such classification the farmers learned was according to the different chemical families, such as organochlorines, organo-phosphates, carbamates, pyrethroids, and so on.
3. Pesticide label identification and interpretation, where farmers learned to read and understand pesticide labels, the various pictograms, and colored warning signs on containers, including the interpretation of the various toxicity symbols.
4. Pesticide handling and management, where farmers learned how to handle pesticides properly to safeguard themselves, their families, their neighbors, and their surroundings. They learned to observe which pesticides are used in their area, how they are used, and to observe the protective measures that are taken during mixing and application. They also learned how they might get contaminated during the handling of pesticides.
5. Pesticide storage and disposal of empty pesticide containers, where farmers learned about proper storage and disposal of surplus pesticides and their empty containers. They also learned how the improper storage of pesticides and the careless disposal of empty containers could form a risk for children, foodstuff, freshwater supplies, farm animals, and so on.
6. Recognizing the signs and symptoms of pesticide poisoning, where farmers learned the different signs and symptoms of pesticide poisoning and how to recognize them. They also learned how to distinguish these from other signs and symptoms that are simply due to poor health.
7. Pest identification and management, where farmers learned how to identify different insects, distinguishing genuine pests from more beneficial insects and symptoms of common vegetable diseases. Farmers were introduced to the basic principles of pest control methods. They were thus equipped with a practical knowledge of insects as an important component in pest management and on how to protect their crops from insect attack with a view to reducing the insecticide load on the environment.

8. Introduction to Integrated Pest Management (IPM), where farmers learned the principles of IPM as a sustainable approach to managing pests by combining biological, cultural, physical, and chemical tools in a way that minimized financial, health, and environmental risks. Farmers were further informed that one of the primary missions of IPM was to assist them in producing profitable crops, using environmentally and economically sound approaches.

9. Spraying equipment and techniques, where farmers learned about spray equipment [such as the knapsack sprayer, the motorized ultra low volume (ULV) sprayer, and so on]. They also learned about the handling, maintenance, and spraying techniques with regard to this equipment.

10. Reducing pesticide costs, where farmers learned to assess actual costs of pesticide use to include direct and indirect costs.

11. Participatory data collection methodology, where farmers learned about methods of data collection and analysis. They were introduced to the kind of data needed and data collection procedures, using different techniques with different data collection tools. Demonstration and practical sessions on how to handle and record data were also held with the farmers.

Establishment of Community Pesticide Monitoring Team

The 25 trained representative farmers were divided into teams of at least three people each, who then became the focal point for monitoring and recording of all pesticide incidences in Ngarenanyuki, and who also worked closely with the WAHSA-TPRI Team, including those who had been working with communities in the Arumeru district in research and training in their respective fields.

Data Collection

Consent forms were developed to be completed by those individuals who agreed to participate in the Community Monitoring Project. Three data collection tools were developed: a questionnaire, a checklist, and a self-surveillance form. They were designed to cover all areas of interest in community pesticide monitoring through interviews, observation, and self-examination of pesticide exposure.

The tools were pretested for validity and consistence in Mlangarini ward with a sample of 30 farmers. Adjustments and other improvements were made to the tools prior to final data collection in Ngarenanyuki. A self-surveillance form without pictograms was preferred, owing to some confusion arising from the meaning of the pictograms.

Farmers were organized into teams to conduct crop surveys and recognize damages, assess losses, and collect insect pests for identification. The farmers went out into their respective villages to collect information on pesticides used, perceptions on pesticide hazards, poisoning, and symptoms using the questionnaire. They also used the checklist to observe and record pesticides available in the area, means of storage, and use of protective equipment. Each farmer contacted by the team members was asked to do self-surveillance and record pesticide use conditions and practices, as well as poisoning signs and symptoms experienced.

The Community Pesticide Monitoring Teams needed technical support and close follow-up to ensure consistency in data collection and in transferring the knowledge gained to the entire Ngarenanyuki community. However, the teams were fully prepared in getting the message across to the community and to involve them in providing relevant information regarding pesticide issues. Moreover, the village leaders were made responsible for making a close follow-up of the team and the villagers involved.

Data Analysis

Analysis of information collected was performed using two different approaches. Structured interviews were conducted by the farmers using questionnaires, and the information was tallied, before it was tidied up and analyzed with the aid of the SPSS (Statistical Package for the Social Sciences) computer software to obtain frequencies. The data that were collected through observation on the basis of check-lists were manually analyzed by the farmers themselves using flip charts and colored pens.

Results and Discussion

The training of the farmers was meant to prepare the community in taking responsibility themselves for monitoring the negative impacts of pesticides in their area. Subject matter specialists conducted the training with the aim of building the capacity of participants to make the right judgement and decisions when dealing with pest and pesticide issues. During the training sessions, farmers expressed keen interest in learning how to recognize insects (beneficial and harmful), to recognize the signs and symptoms of pesticide exposure, to practice safer pesticide handling and management, to understand proper spraying techniques and the maintenance of knapsacks, and to understand the benefits of participatory data collection techniques.

General Information

The data from the farmers in Ngarenanyuki Ward were collected between February and April 2007 by the farmers trained on pesticide monitoring and analyzed using the SPSS computer software. While the majority of the 120 farmers were males (90%), the average highest education level recorded was that of primary education (76%); hence, functional literacy was not a problem in this community. Agriculture (98.3%) was the major income-generating activity, although some farmers also kept livestock.

Pesticides Used in Ngarenanyuki

Thirty different types of pesticides commonly used in Ngarenanyuki were identified by the farmers. The major groups of pesticides used included insecticides, fungicides, and, to a lesser extent, herbicides. The most widely used insecticide and fungicide were chlorpyrifos (72.5%) and mancozeb (69.2%), respectively. Only 36.6% of the pesticides used in Ngarenanyuki had full registration, while some had provisional or experimental registration. It is mandatory for pesticides intended for use in Tanzania to go through a registration process, which involves efficacy and quality tests before they are approved for general, restricted, or experimental use. Pesticides under experimental use are not expected to be sold in retail shops. There was also the presence of one class 1b pesticide (chlorfenvinphos) and banned/restricted pesticides such as DDT.

Pesticide Availability, Affordability, and Application

The majority of pesticides (86.7%) were locally available in Ngarenanyuki, and a considerable proportion of farmers (65.8%) could afford to buy pesticides. Those unable to buy mostly obtained their pesticides on credit, and paid after harvesting. Most farmers (68.3%) claimed to have a pesticide application timetable, the most prominent approach being that of applying pesticides whenever insects or disease symptoms appeared.

Pesticide Mixing

The majority of farmers (90%) mixed more than one pesticide in a single application. The main reason given for mixing was to kill all pests and diseases at a go and to improve the quality of leaves and fruits (54%) in the field. A few (25%) said they preferred mixing to ease the workload and in order to cover larger areas with one treatment, while some (55%) said they had simply followed the pesticide retailer's advice. The mixing exercise was widely done in respective farms (89%). The common mixtures normally contained more than one fungicide and one insecticide, although some mixtures were found with around three fungicides and two insecticides.

Frequency of Pesticide Application, Number of Risk Days per Year, Spraying Equipment, and Pesticide Storage

The scale of environmental pollution was fairly evident as pesticides could be smelled all over the farms and in nearby residential areas, causing health problems (such as cough, sneezing, excessive difficulty breathing, and chest pains) to both sprayers and those who found themselves in the path of the sprays. The farmers worked out 52 risk days per year, as the majority (73%) of 120 farmers applied pesticides once a week and fewer (18%) applied the pesticides twice a week. The most common spraying equipment was the knapsack (76%), while in some cases (21%), buckets were also used.

Most respondents (57%) stored pesticides in a pesticide store, and in some cases, storage took place in sitting or living rooms (12%), in general stores (13%), and in bedrooms (7%). Pesticides were also found to be stored in toilets (1%). The choice of storage areas was often determined by their offering protection against thieves.

Adherence to Pesticide Label Instruction

The study by the farmers revealed that many of them did read the instructions on the pesticide containers, but only few actually followed the instructions as shown in Table 2. An example is the mixing of ULV formulations in water sprays while instructions given on the label are for direct application without dilution. The following were the arguments put forward for not following instructions: that some labels were only written in English, that the farmers were not familiar with conventional signs and symbols, and that some containers had no labels at all, having been dispensed from another container. The repackaging and dispensing of smaller quantities was found to be a common, albeit illegal, practice and it was felt that this needed greater attention, since this practice has negative implications for efforts to reduce the worst effects of pesticide poisoning by the implementation of proper labeling and instructions.

Disposal of Pesticide Containers

The major mode of disposal of empty pesticide containers by most of the farmers was by simply throwing the containers away in the farm surroundings and by burning. It was observed that some empty containers from the pesticide retail shops were also thrown or burnt at the marketplace. It was also revealed that some farmers did reuse empty pesticide containers for domestic purposes such as buying cooking oil and kerosene and for local brewing (Table 3).

TABLE 2   Adherence to Pesticide Label Instruction by Farmers in Ngarenanyuki during a Pevious Farming Season (December 2006 to March 2007)

Response toward Pesticide Label Instructions	Number of Farmers (N = 120)	% of Farmers
Always read instructions	72	60
Follow instructions	45	38
Sometimes follow instructions	34	28
Sometimes read instructions	28	23
Trained on pesticide issues	16	13
Get information on pesticides	7	6

TABLE 3   Modes of Disposal of Empty Pesticide Containers in Ngarenanyuki (December 2006 to March 2007)

Mode of Disposal	Number of Farmers (N = 120)	% of Farmers
Burn	41	34
Throw away on the farm	35	29
Bury in the farm surroundings	18	15
Sell back to pesticide vendors	8	7
Throw in the toilet	7	6
Use for other domestic uses	2	2

Pesticide Poisoning

The self-surveillance form was used without pictograms, to record the signs and symptoms of pesticide poisoning, owing to the confusion arising from the use of pictograms. During the pretest, farmers did not understand what the pictograms meant, and since the majority were able to read and write, it was agreed that there would be no need to include picto-grams in the surveillance form and that the list of signs and symptoms provided in the form was sufficient until proper research had been undertaken to determine what visual aid would be considered appropriate for the target audience.

The majority of the farmers (69.2%) had experienced pesticide poisoning in the previous farming season, owing to exposure, much of which had occurred more than 3 times to a single farmer. Pesticide poisoning was characterized by signs and symptoms known from previous studies to be related to pesticide exposures. Cypermethrin–profenofos mixture and profenofos were mostly associated with poisoning, and the action taken by many of those exposed (43.3%) was to drink milk, while a few respondents had attended hospitals for a proper medical examination. A considerable high proportion (57.5%) had been admitted more than 3 times, owing to pesticide poisoning (Table 4). Validation of poisoning through biological monitoring was not possible during this pilot stage but has been planned in future surveillances.

The action taken by the 25 farmers who fully participated in the pilot study was to intensify the training by initiating capacity-building sessions in all villages in Ngarenanyuki. They held community pesticide monitoring training in every village meeting, gave feedback to the WAHSA-TPRI Team on the farmers' reaction, and suggested what further input they needed from the team.

Conclusion

This pilot enabled the building of Ngarenanyuki farmers' capacities to assess their own health and environment as far as pesticides were concerned, analyze the situation, develop a plan of action, and work toward improving their condition. It facilitated the farmers' capacity so that they could take control and work with pesticides more safely and so become healthier. This program therefore works to benefit not only the farming community in the long run but also those consumers who would otherwise be forced to eat contaminated crops, and it contributes to the health of the environment as a whole.

Relevant data relating to pesticides, such as their availability, their usage, the farmers' handling practices, risk perception, and behavior, all gathered during the pilot project, enabled farmers in Ngarenanyuki and the WAHSA-TPRI Team to properly document the incidents and adverse events resulting from pesticide use. The initial evaluation of the association of the observed adverse event and pesticide exposure revealed that different pesticide-related tasks gave rise to signs and symptoms of pesticide poisoning and that skin and eye problems, for example, needed more attention during interventions.

The impact of the sensitization and awareness-raising seminars has been dramatic. The disposal site at the marketplaces vanished and the mistake of mixing ULV formulations with water has also been abandoned by the trained farmers and their associates. The farmers realized that the formulation was suspended in water and they were spraying water in some areas instead of pesticide. The formation

TABLE 4 Pesticide Poisoning, Circumstances, and Action Taken in Ngarenanyuki (December 2006 to March 2007)

Event		Number of Farmers (N = 120)	% of Farmers
Pesticide poisoning	Affected by pesticides in the last farming season	83	69
	Not sure	18	15
Occurrence of effects in the last farming season	Once	11	9
	Twice	12	10
	Thrice	9	8
	More than three times	26	22
Pesticides used	Profenofos + cypermethrin	32	27
	Profenofos	25	21
	Mancozeb (Dithane)	14	12
	Endosulfan	14	12
	Triadimenol	12	10
	Chlorothalonil	12	10
	Lambda-cyhalothrin	8	7
	Mancozeb (Ivory)	8	7
	Deltamethrin	1	1
	Copper sulfate	1	1
Action taken after pesticide Poisoning	Drank milk	52	43
	Went to the hospital	34	28
	No action taken	4	3
	Washed with water	2	2
Number of times admitted due to pesticide poisoning	Once	20	17
	Twice	23	19
	Thrice	8	7
	More than three times	69	58

of the community monitoring teams enhanced the whole process of data collection and action being taken. This has also provided a base for the sustainability of the project as the team continues to be in close contact and collaboration with the WAHSA-TPRI Team through the training of other farmers and in responding to their queries on pests, pesticides, monitoring pesticide use, their application, and the disposal of obsolete pesticides and empty pesticide containers.

The project was well received by the Ngarenanyuki community and has shown that if it is applied elsewhere, it will help in changing risk behaviors and in reducing the negative impact of pesticide exposures in communities. It is therefore recommended that the program be implemented systematically in Ngarenanyuki and be extended to other communities in Tanzania such as Mang'ola, in Karatu District, where the current use of pesticides appears to be indiscriminate.

Acknowledgments

We are indebted to the small-scale vegetable farmers in Ngarenanyuki for their cooperation. This project could not have been implemented without support from AGENDA for Environment and Responsible Development for which we are grateful. We appreciate the work done by the WAHSA-TPRI Team and the contribution of experts from Selian Agricultural Research Institute (Dr. Hussein Man-soor) and Mikocheni Research Institute (Dr. Ruth Minja) during the farmers' training.

References

1. WHO. *Public Health Impact of Pesticide Used in Agriculture*; Geneva, 1990.
2. ILO Chemicals in the Working Environment. In *World Labour Report*; International Labour Office: Geneva, Switzerland, 1994.
3. Murphy, H.H.; Hoan, N.P.; Matteson, P.; Abubakar, A.L. Farmers' self-surveillance of pesticide poisoning: A 12-month pilot in northern Vietnam. Int. J. Occup. Environ. Health **2002**, *8* (3), 201–211.
4. Rengam, S. Breaking the silence: Women struggle for pesticide elimination. In *Silent Invaders*; Jacobs, M., Dinham, B., Eds.; Zed Books Ltd.: London, 2003.
5. Quijano, R.F. *Endosulfan Poisoning in Kasargod, Kerala, India: Report on a Fact-Finding Mission*; Pesticide Action Network Asia and the Pacific: Penang, 2002.
6. Ngowi, A.V.F.; Mbise, T.J.; Ijani, A.S.M.; London, L.; Ajayi, O.C. Smallholder vegetable farmers in Northern Tanzania: Pesticides use practices, perceptions, cost and health effects. Crop Prot. **2007**, *26*, 1617–1624.
7. Ngowi, A.V.F. *Health Impact of Exposure to Pesticides in Agriculture in Tanzania*; PhD Thesis, Acta Universitatis Tamperensis 890; University of Tampere, 2002.

22

Developing Countries: Pesticide Health Impacts

Aiwerasia V.F.
Ngowi, Catharina
Wesseling, and
Leslie London

Introduction

The use of potentially hazardous chemicals is increasing in developing countries whose populations have the least capacity to protect themselves. Hundreds of thousands of people die annually from the effects of use, misuse, or accidental exposures to pesticides.[1,2] Developing nations in Africa, Asia, and Latin America comprise more than 75% of the total world population, use 25% of the world's pesticides, yet account for 99% of deaths caused by these toxic agents.[3]

Health Impacts

Acute Pesticide Poisoning

Two decades ago, the World Health Organization estimated that three million cases of acute pesticide poisoning resulting in 220,000 deaths occur worldwide each year, the majority in developing countries.[3] However, it is well recognized that these figures are an underestimate because of underdiagnosis and/or underreporting. Diagnostic difficulties are prominent in developing countries,[4-6] owing to insufficient medical training and high background levels of ill health.

Organophosphorus insecticides are the most common agents involved in acute pesticide poisonings, accounting for between 50% and 80% of all poisonings in Asia[7] and are a major public health concern in most African countries, where approximately 80% of the workforce is involved in agricultural work. In Central America, pesticides identified as causing most poisonings between 1992 and 2000 were paraquat; aluminum phosphide; the organophosphates methyl parathion, methamidophos, monocrotophos, chlorpyrifos, terbufos, and ethoprophos; the carbamates carbofuran, methomyl and aldicarb; and endosulfan.[8]

Part of the reason for this picture is the continued use in developing countries of pesticides no longer registered for use in the developed world, because of their high toxicity, and the substitution of persistent organochlorines with organophosphate insecticides.

Fatality rates and lifelong disability resulting from pesticide poisoning in developing countries are exacerbated by poor diagnosis and delayed treatment, resulting in both human suffering and economic losses.

High rates of unintentional poisoning, mostly occupational, have been reported in rural agricultural working and urban populations worldwide.[9,10] Nearly 66,000 cases of acute pesticide poisonings occur annually in Nicaragua.[11] Mass poisonings by pesticides in developing countries have typically resulted in high numbers of fatalities.

In the remote Andean village of Tauccamarca in October 1999, 42 children were poisoned after eating a school breakfast contaminated with the organophosphate pesticide methyl parathion, resulting in 24 deaths before the children could reach medical treatment.[12]

However, it is only a limited number of the most extreme cases in developing countries, which appear to be documented. Less high-profile cases are common but unrecorded. For example, a methomyl-poisoning incident involving 11 female flower farm workers in Arusha, Tanzania, in March 2004 was reported in the press, but absence of adequate local investigation mechanisms prevented its documentation in the peer-reviewed literature.

Deliberate self-harm is a major problem in the developing world. Pesticides are commonly used as suicidal agents throughout developing nations and are associated with high mortality rates, causing an estimated 300,000 deaths annually in the Asia Pacific.[1,2,8,9,13] In India, suicide using aluminum phosphide was reported as so common that postmortem examinations on deceased bodies were said to be routinely conducted by staff wearing respirators for personal protection from released fumes.[14]

Underlying factors that make individuals at risk for self-harm are both social (including domestic problems, poverty, social isolation, and financial hardship) and medical.[1] Farmer indebtedness, widespread in many developing countries characterized by unequal economic systems, is an important factor driving high rates of suicide. More recent findings suggest that pesticides, particularly organophosphates, may be more than agents in suicidal attempts; they are also part of the causal pathway because of their neurotoxicity and the possible links between organophosphate exposure, depression, and impulsivity, mediated through effects on neurotransmitters such as serotonin.[9] In a context where the above social risk factors for depression are common in developing countries, further exposure to neurotoxic pesticides may substantially increase the risks of suicide.

Chronic Health Impacts Unknown

Although long-term consequences of pesticide poisoning are well recognized in the literature, relatively few studies of long-term health effects of pesticide exposure have been conducted among working populations in developing countries. Underdiagnosis is accentuated for long-term health consequences that require greater diagnostic capacity. Dermal exposure routes for developing country workers are also common but are an underdocumented yet critical pathway for systemic poisonings, both acute and chronic. Consequently, the extent of chronic health impacts of pesticides in developing country workers is poorly characterized.

However, there is little reason to believe that their impact would be any less than that in developed countries. Indeed, high levels of background morbidity and poor social conditions are likely to aggravate pesticide toxicity. For example, research among South African farm workers highlighted the link between chronic lifetime undernutrition, organophosphate exposure, and impaired neurological performance on tests of vibration threshold.[15]

Azoospermia (absent sperm), oligospermia (low sperm count in semen), and low fertility have been documented in more than 26,000 workers previously exposed to 1,2-dibromo-3-chloropropane (DBCP) on banana and pineapple plantations in more than 12 countries.[16]

Weak Surveillance for Hazards and Impact

Although a critical public health tool for the control of pesticide poisoning, surveillance in developing countries is bedeviled by multiple problems such as lack of access to health care for poisoning survivors, lack of human resources, diagnostic skills and equipment to identify cases, and weak information systems. Acute poisoning rates are consequently underestimated and may selectively undercount certain types of poisoning (occupational circumstances) and certain risk groups (women and migrant workers). Lack of professional competence and conflict of interest arising from compensation system levies may also lead occupational poisonings to be misreported as suicide.[10] As a result, inferences from review of flawed data may lead to mistaken policy decisions.[17]

To improve information on the extent of pesticide poisoning in developing countries, surveillance systems for acute health effects from pesticides are being established in developing nations. In 1998, almost 6000 pesticide poisonings were reported in five of the seven Central American countries generating an estimate (corrected for underreporting) of 30,000 pesticide poisonings annually in the region.[11] Poisoning rates reported in an intensified surveillance intervention in South Africa increased 10-fold in the study area compared to a control area.[17] Recently, WHO in collaboration with partners initiated a community intervention, the Global Public Health Initiative, to prevent self-harm by pesticide poisoning.[18]

Weak Regulation and Enforcement

Vulnerable economies and weak infrastructure in developing nations hinder their ability to regulate the use of pesticides, particularly when macroeconomic pressures promote deregulation and restrict public spending required to implement regulatory controls. As a result, marketing and advertising of pesticides are often uncontrolled. Incorrectly labeled or unlabeled formulations, including readymade solutions in soft drink bottles and other containers, are commonly sold at open stands. In South Africa, the repackaging of aldicarb granules into small-volume packets sold by street vendors[19] for domestic pest control has been linked to increasing numbers of poisonings in urban areas. Low retail prices, sometimes associated with subsidy policies, promote risky pesticide use. Weaknesses in sustainable international and national agricultural and chemical management policies manifest in a reliance on "safe-use" strategies. Yet, evidence has shown that the so-called "good agricultural practices" and "safe use" are ineffective in controlling risks in developing countries, principally because many measures assumed to enable safe use are not feasible in developing countries, particularly under tropical or adverse climatic conditions.[20]

Low Levels of Worker and Community Awareness

Farmers and farm workers rarely have access to adequate training in pesticide safety or advice on the complicated management of pesticides. Hot climates are a disincentive to use of protective clothing, and many workers and farmers lack access to water for washing hands or exposed skin, increasing the risks of contamination. Recognition of pests and their predators is generally low, leading to overreliance on routine pesticide applications to control pests; knowledge of product selection, application rates, and timing is poor; different products are often combined in the belief that the effect will be greater; re-entry periods after spraying are not known; and without knowledge of alternatives, farmers often assume that the only solution to pest problems is to spray more frequently.[21]

Pesticides are often stored improperly in or around farmers' homes, increasing family member's access.[21,22] In some instances, empty pesticide containers are reused to store water and food, resulting in serious poisonings.

Import/Export of Banned and Restricted Compounds

Pesticides banned or restricted in developed countries are often easily available in developing countries. These include pesticides causing significant acute and chronic morbidity (such as class I and II organophosphates and paraquat) and organochlorines earmarked for eradication under the Stockholm Convention on persistent organic pollutants (POPs) (particularly dieldrin, lindane, and chlordane). Endosulfan, a candidate pesticide for inclusion under the POPs treaty, has been responsible for a series of poisonings in Benin[21] and developmental impacts on children in Kerala, a state in India.[21,23]

The use of *p-p'*-dichlorodiphenyltrichloroethane (DDT) continues to be permitted for malaria control in developing countries, where malaria remains endemic, despite its known hazards for wildlife and controversial adverse effects on human health.[24,25] Ironically, DDT use in Africa has increased since the Stockholm Convention came into effect.[26] As a result, it is still produced for export in at least three countries. Because of its ongoing usage for public health vector control, unauthorized use for agricultural purposes remains a concern in developing countries, particularly where regulatory controls are weak. The presence and persistence of DDT and its metabolites worldwide are still problems of great global relevance to public health.

Although the Prior Informed Consent (PIC) procedure, upgraded in status from a voluntary agreement to an international convention known as the Rotterdam Convention, seeks to protect developing countries from harms arising from import of chemicals banned or restricted in exporting countries, the effectiveness of the Convention has been questioned. For example, the data requirement for a Severely Hazardous Pesticide Incident report, used under the Convention to add a pesticide onto the controlled list, lacks mechanisms suited to developing country conditions. This is usually because developing countries lack the infrastructure to collect the required data. Similarly, the process of adding a pesticide onto the POPs list (e.g., endosulfan) is often met in practice with strong resistance. Even when pesticides are restricted by the POPs and Rotterdam conventions, compliance with the obligations contained in the conventions may often be poor, despite developed countries ratifying the conventions. The PIC and POP secretariat could assist local contacts such as the Designated National Authorities (DNAs) in developing simple tools that will be used to collect relevant data and help in the establishment of suitable mechanisms for the flow of data and information in member countries.

Over the past few years, pressure from non-governmental organizations and discussions within the Food and Agriculture Organization and other intergovernmental bodies has recognized that greater effort must be put into restricting the availability of pesticides based not only on their acute toxicity, traditionally measured through the WHO classification system, but also on the capacity of particular pesticides to cause long-term toxic effects with chronic exposure. These initiatives, particularly the call for a progressive ban on highly hazardous pesticides, offer some hope for better protections for developing country populations, but are still in development.

Lack of Technical and Laboratory Capacity

Many developing countries suffer from a lack of human and technical resources, aggravated by the global brain drain and weak economies. As a result, few developing countries are able to monitor pesticide residues. Most developing countries do not have laboratories capable of conducting analyses for pesticides and their residues, particularly at standards that meet good laboratory practice. Where laboratory capacity is available, it is usually to service residue testing of agricultural exports destined for consumers in developed countries. Produce grown for domestic consumption is rarely monitored.

Environmental media such as water and soil are rarely tested, and, even then, usually only on a research basis. Isolated studies of lactating women in Southern Africa have confirmed the presence of high levels of DDT metabolites in breast milk in populations living in malaria-endemic areas subject to DDT applications. Yet, despite provisions arising from the POPs treaty to undertake routine testing to monitor the impact of DDT use, there is no system for biological monitoring for DDT metabolites in place in Southern Africa. As a result, many infants in the region are substantially exposed through cross-placental transfer and breastfeeding, with potential adverse impacts on childhood neurodevelopment.

Research capacity to identify problems and develop prevention strategies is also constrained by limited investments in capacity building in relevant scientific fields in developing countries. As a result, there is neither proactive monitoring nor information systems usage to effect adequate responses to pesticide problems identified. It is critical to foster South–South learning to promote best practice because what applies in the north may be different to what happens in the south. There is a need to build southern capacity because reliance on the north at times perpetuates many of the problems. Indeed, there is much expertise in the south from which both north and south can learn.

Pest Control Policies

Unlike many developed countries, agricultural policies in many developing countries have emphasized short-term economic gains at the expense of environmental sustainability or human health. Few developing countries have adopted integrated pest management or pest reduction strategies. The dominant "pesticide culture" assumes that the use of pesticides to control pest as the first option is the norm, is reinforced by advertising and marketing practices, and is often encouraged by agricultural credit policies and development aid. Much needs to be done to enhance research and development to support pesticide reduction for agriculture and public health, and to strengthen the capacity in developing countries to develop monitoring systems and research capacity to deal with the problems of pesticides in developing nations. Reducing deaths from pesticide poisoning through restrictions on the availability of pesticides can be accomplished based on a prior evaluation of national agricultural needs and the development of a plan to encourage substitution with less toxic pesticides without loss of agricultural output,[27] bearing in mind that policies aiming towards sustainable chemical-free agriculture would be the ideal long-term solution.

Conclusions

Underestimations of acute and long-term effects of pesticide in developing countries occur due to under-diagnosis and/or underreporting. The impact of pesticide poisoning is also unknown because of weak surveillance for hazards and impact, import/export of banned or restricted compounds, lack of technical and laboratory capacity, weak regulations and enforcement, low level of worker and community awareness, and inappropriate pest control policies. Enhancing research and development to support pesticide reduction for agriculture and public health and strengthening capacity to develop monitoring systems are critically important for developing countries to deal with problems concerning pesticides.

References

1. Konradsen, F.; van der Hoek, W.; Cole, D.C.; Hutchinson, G.; Daisley, H.; Singh, S.; Eddleston, M. Reducing acute poisoning in developing countries—Options for restricting the availability of pesticides. Toxicology **2003**, *192* (2–3), 249–261.
2. Konradsen, F.; Dawson, A.H.; Eddleston, M.; Gunnell, D. Pesticide self-poisoning: Thinking outside the box. Lancet **2007** *369* (9557), 169–170.
3. WHO. *Public Health Impact of Pesticide Used in Agriculture;* World Health Organization and United Nations Environment Programme: Geneva, 1990.
4. Corriols, M.; Marin, J.; Berroteran, J.; Lozano, L.M.; Lundberg, I.; Thörn, A. The Nicaraguan Pesticide Poisoning Register: Constant underreporting. Int. J. Health Serv.**2008** *38* (4), 773–87.
5. Mbakaya, C.F.L.; Ohayo-Mitoko, G.J.A.; Ngowi, A.V.F.; Mbabazi, R.; Simwa, J.M.; Maeda, D.N.; Stephens, J.; Hakuza, H. The status of pesticide usage in East Africa. Afr. J. Health Sci. **1994**, *1*, 37–41.

6. London, L.; Myers, J.E. Critical issues for agrichemical safety in South Africa. Am. J. Ind. Med. **1995**, *27*, 1–14.

7. He, F.; Xu, H.; Quin, F. Intermediate myasthenia syndrome following acute organophosphates poisoning—An analysis of 21 cases. Hum. Exp. Toxicol. **1998**, *17*, 40–45.

8. Henao, S.; Arbelaez, M.P. Epidemiologic situation of acute pesticide poisoning in Central America, 1992–2000. Epidemiol. Bull. **2002**, *23* (3), 5–9.

9. London, L.; Flisher, A.J.; Wesseling, C.; Mergler, D.; Kromhout, H. Suicide and exposure to organophosphate insecticides: Cause or effect. Am. J. Ind. Med. **2005**, *47*, 308–321.

10. Wesseling, C. Multiple health problems in Latin America. In *Silent Invaders: Pesticides, Livelihoods and Women's Health*; Jacobs, M., Dinham, B., Eds.; ZED Books: London, 2003.

11. Corriols, M.; Marin, J.; Berroteran, J.; Lozano, L.M.; Lundberg, I. Incidence of acute pesticide poisonings in Nicaragua: A public health concern. Occup. Environ. Med. **2009**, *66*, 205–210.

12. Rosenthal, E. The tragedy of Tauccamarca: A human rights perspective on the pesticide poisoning deaths of 24 children in the Peruvian Andes. Int. J. Occup. Environ. Health **2003**, *9*, 53–58.

13. Chowdhury, A.N.; Banerjee, S.; Brahma, A.; Biswas, M.K. Pesticide poisoning in non-fatal deliberate self-harm: A public health issue. Study from Sundarban delta, India. J. Psychiatry **2007**, *49*, 262–266.

14. Levine, R.S.; Doull, J. Global estimates of acute pesticide morbidity and mortality. Rev. Environ. Contam. Toxicol. **1992**, *129*, 29–50.

15. London, L. Occupational epidemiology in agriculture: A case study in the Southern African context. Int. J. Environ. Occup. Health **1998**, 4, 245–256.

16. Slutsky, M.; Levin, J.L.; Levy, B.S. Azoospermia and oligospermia among a large cohort of DBCP applicators in 12 countries. Int. J. Occup. Environ. Health **1999**, *5* (2), 116–122.

17. London, L.; Bailie, R. Challenges for improving surveillance for pesticide poisoning: Policy implications for developing countries. Int. J. Epidemiol. **2001**, *30* (3), 564–570.

18. WHO. *Safer Access to Pesticides: Community Interventions*; World Health Organization and International Association for Suicide Prevention: Geneva, 2006.

19. Rother, A. Falling through the regulatory cracks: Street selling of pesticides and urban youth in South Africa. Int. J. Occup. Environ. Health **2010**, *16*, 202–213.

20. Wesseling, C.; Ruepert, C.; Chavarri, F. Safe use of pesticides: A developing country's point of view. In *Encyclopedia of Pest Management;* Marcel Dekker, Inc.: New York, 2003.

21. Ngowi, A.V.F.; Maeda, D.N.; Wesseling, C.; Partanen, T.J.; Sanga, M.P.; Mbise, G. Pesticide handling practices in agriculture in Tanzania: Observational data on 27 coffee and cotton farms. Int. J. Occup. Environ. Health **2001**, *7*, 326–332.

22. Dinham, B.; Malik, S. Pesticides and human rights. Int. J. Occup. Environ. Health **2003**, *9*, 40–52.

23. Saiyed, H.; Dewan, A.; Bhatnagar, V.; Shenoy, U.; Shenoy, R.; Rajmohan, H.; Patel, K.; Kashyap, R.; KuLkarni, P.; Rajan, B.; Lakkad, B. Effect of endosulfan on male reproductive development. Environ. Health Perspect. **2003**, *111*, 1958–1962.

24. Bouwman, H.; Becker, P.J.; Schutte, C.H.J. Malaria control and longitudinal changes in levels of DDT and its metabolites in human serum from KwaZulu. Bull. W. H. O. **1994**, *72* (6), 921–930.

25. Eskenazi, B.; Chevrier, J.; Rosas, L.G.; Anderson, H.A.; Bornman, M.S.; Bouwman, H.; Chen, A.; Cohn, B.A.; de Jager, C.; Henshel, D.S.; Leipzig, F.; Leipzig, J.S.; Lorenz, E.C.; Snedeker, S.M.; Stapleton, D. The Pine River statement: Human health consequences of DDT use. Environ. Health Perspect. **2009**, *117*, 1359–1367.

26. van den Berg, H. Global status of DDT and its alternatives for use in vector control to prevent disease. Environ. Health Perspect. **2009**, *117*, 1656–1663.

27. Manuweera, G.; Eddleston, M.; Egodage, S.; Buckley, N.A. Do targeted bans of insecticides to prevent deaths from selfpoisoning result in reduced agricultural output? Environ. Health Perspect. **2008**, *116* (4), 492–495.

Insulation: Facilities

Wendell A. Porter

Introduction

Insulation is rated in terms of thermal resistance, called the *R*-value, which indicates the resistance to heat flow. Although insulation can slow all types of heat flow—conduction, convection, and radiation—its greatest impact is on conduction.

The higher the *R*-value is, the greater the insulation effectiveness is.[1,2] The *R*-value of thermal insulation depends on the type of material, the thickness, and the density. When calculating the *R*-value of a multilayered installation, the *R*-values of the individual layers are added.

The effectiveness of an insulated wall or ceiling also depends on how and where the insulation is installed. For example, compressed insulation will not give its full rated *R*-value. The overall *R*-value of a wall or ceiling will also be somewhat different from the *R*-value of the insulation itself because some heat flows around the insulation through the studs and joists thermal bridging. With careful design, this short-circuiting can be reduced.

The key to an effective insulation system is proper installation of quality insulation products. A building should have a continuous layer of insulation around the entire building envelope (Figure 1). Studies show that improper installation can cut performance by 30% or more.

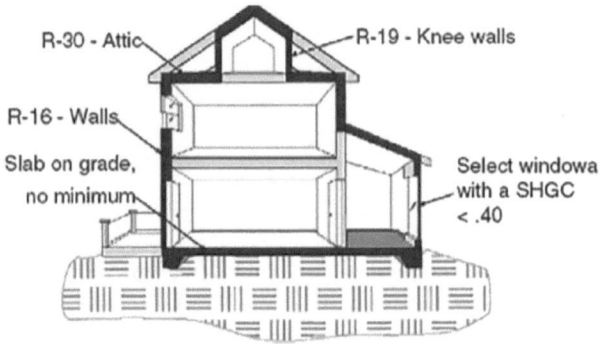

R-30 - Attic
R-19 - Knee walls
R-16 - Walls
Slab on grade,
no minimum
Select windowa
with a SHGC
< .40

FIGURE 1   Building envelope insulation.

Insulation Materials

The wide variety of insulation materials makes it difficult to determine which products and techniques are the most cost effective (Table 1). Whatever product is chosen, install it per the manufacturer's specifications.

Here are short descriptions of a few of the insulation products available today:

- Fiberglass insulation products come in batt, roll, and loose-fill form, as well as a semirigid board material. Many manufacturers use recycled glass in the production process of fiberglass building insulation, with most using between 20% and 30% recycled glass in their product. Fiberglass is used for insulating virtually every building component—from walls to attics to ductwork.
- The term mineral wool refers to both slag wool and rock wool. Slag wool is manufactured from industrial waste product. It is primarily (~75%) produced from iron ore blast furnace slag, a by-product of smelting. Rock wool is fireproof and produced from natural rocks—basalt primarily—under high heat. Mineral wool insulation is available as a loose-fill product, batts, semirigid, or

TABLE 1   Comparison of Insulating Materials

Material	Typical R-Value (per inch)
Batts, blankets, and loose-fill insulation	
Mineral wool and fiberglass	2.2–4.0
Cellulose (loose-fill)	3.0–3.7
Cotton (batts)	3.0–3.7
Perlite (loose-fill)	2.5–3.3
Foam insulation and sheathing	
Polyisocyanurate	6.0–6.5
Closed-cell, spray polyurethane	5.8–6.8
Open-cell, low-density polyurethane	3.6–3.8
Extruded polystyrene	5.0
Molded expanded polystyrene (beadboard)	4.0
Fiberboard sheathing (blackboard)	1.3
Air-krete	3.9
OSB sheathing (3/8 in.)	0.5
Foil-faced OSB	Depends on installation
Polyicynene	3.6

rigid board. Usage of this product has decreased as more and more building codes require active sprinklering of buildings.

- Cellulose insulation, primarily made from post-consumer recycled newsprint with up to 20% ammonium sulfate and/or borate flame retardants, is installed in loose-fill, wall-spray (damp), dense-pack, and stabilized forms. Because of its high density, cellulose can help reduce air leaks in wall cavities, but air sealing other areas of air infiltration, such as under wall plates and band joists, must be performed to obtain an effective air barrier. However, given certain conditions and applications, cellulose may hold moisture.
- Molded expanded polystyrene (MEPS), often known as beadboard, is a foam product made from molded beads of plastic. MEPS is used in several alternative building products discussed in this entry, including insulated concrete forms and structural insulated panels (SIPs).
- Extruded polystyrene (XPS), also a foam product in rigid board form, is a homogenous polystyrene produced primarily by three manufacturers with characteristic colors of blue, pink, and green.
- Polyisocyanurate, foil-faced rigid board, is insulating foam with one of the highest available *R*-values per inch.
- Closed-cell, high-density spray polyurethane is used both for cavity insulation and as insulating roofing materials [often referred to as spray polyurethane foam (SPF)]. It has structural properties, good adhesive properties, and good compressive strength.
- Open-cell, low-density polyurethane foam is used primarily to seal air leaks and provide an insulating layer. Produced primarily from petrochemicals, some of these products are now manufactured in part from soybeans.
- Aerated concrete, including lightweight, autoclaved (processed at high temperature) concrete, can provide a combination of moderate R-values and thermal mass for floors, walls, and ceilings, in addition to structural framing.
- Reflective insulation is often used between furring strips on concrete block walls to reflect the heat. Note that reflective insulation products differ from radiant barriers in that they include a trapped air space as part of the product. These trapped air spaces may be a result of the way the reflective insulation is manufactured or installed.

Determine actual R-values and costs from manufacturers or local suppliers.

Note that many new types of insulation are rapidly becoming incorporated into conventional construction. However, always research a material's characteristics and suitability to a particular situation before buying any new product. For instance, many new insulation products require covering for fire rating.

Insulation and the Environment

There has been considerable study and debate about the potential negative environmental and health impacts of insulation products.[3] These concerns range from detrimental health effects for the installer to depletion of the earth's ozone layer.

- Fiberglass and mineral wool—questions about effects on health from breathing in fibers. In 2001, the International Agency for Research on Cancer changed its classification for fiberglass and mineral wool from "possible human carcinogen" to "not a known human carcinogen."
- Cellulose—concerns about dust inhalation during installation to VOC emissions from printing inks (these are now almost entirely vegetable-based) and limited evidence of toxicity of boric acid flame retardants. Long-term fire retardancy is unknown. Limited health and safety research has been performed on these products.
- Foam products and chlorofluorocarbons—for years, many foam products contained chlorofluorocarbons (CFCs), which are quite detrimental to the earth's ozone layer. The CFCs were the blowing agent that helped create the lightweight foams. Current blowing agents are

- Expanded polystyrene—pentane, which has no impact on ozone layer, but may increase the potential for smog formation.
- Extruded polystyrene, polyisocyanurate, and poly-urethane—use primarily hydrochloro-fluorocarbons (HCFCs), which are 90% less harmful to the ozone layer than CFCs. Some companies are moving to non-HCFC blowing agents.
- Open-cell polyurethane, including the products made by Icynene, Inc. and Demilec, Inc., as well as the newer soy-based foams—use water, which is much less detrimental than other blowing agents (Table 2).

Insulation Strategies

In general, commonly used insulation products are the most economical. Prices can vary according to in staller and location. Review all of the choices, as they offer different *R*-values, suggested uses, and environmental and health considerations.

TABLE 2 Comparison of Insulation Materials (Environmental Characteristics and Other Information)

Type of Insulation	Installation Method(s)	R-Value per Inch (RSI/m)[a]	Raw Materials	Pollution from Manufacture	Indoor Air Quality Impacts	Comments
Fiber insulation Cellulose	Loose fill; wall-spray (damp); dense-pack; stabilized	3.0–3.7 (21–26)	Old newspaper, borates, ammonium sulfate	Vehicle energy use and pollution from newspaper recycling	Fibers and chemicals can be irritants. Should be isolated from interior space	High recycled content; very low embodied energy
Fiberglass	Batts; loose fill; semi-rigid board	2.2–1.0 (15–28)	Silica sand; limestone; boron; recycled glass, phenol formaldehyde resin or acrylic resin	Formaldehyde emissions and energy use during manufacture; some manufactured without formaldehyde	Fibers can be irritants, and should be isolated from interior spaces. Formaldehyde is a carcinogen. Less concern about cancer from respirable fibers	
Mineral wool	Loose fill; batts; semi-rigid or rigid board	2.8–3.7 (19–26)	Iron-ore blast furnace slag; natural rock; phenol formaldehyde binder	Formaldehyde emissions and energy use during manufacture	Fibers can be irritants, and should be isolated from interior spaces. Formaldehyde is a carcinogen. Less concern about cancer from respirable fibers	Rigid board (e.g., Roxul) can be an excellent foundation drainage and insulation material
Cotton	Batts	3.0–3.7 (21–26)	Cotton and polyester mill scraps (especially denim)	Negligible	Considered very safe	Two producers; also used for flexible duct insulation
Perlite	Loose fill	2.5–3.3 (17–23)	Volcanic rock	Negligible	Some nuisance dust	

(*Continued*)

TABLE 2 (*Continued*) Comparison of Insulation Materials (Environmental Characteristics and Other Information)

	Conversion Factors		
R-Value Conversions	To Get	Multiply	By
Thermal resistance (R)	RSI (m2 C/w)	R (ft2 hF/Btu)	0.1761
Insulation R/unit thickness	RSI/mm	R/in.	0.00693

In the chart the heading is R-value per inch (RSI/m); to obtain this number, the RSI/mm is divided by 1000.

Foam insulation

Polyisocyanurate	Foil-faced rigid boards; nail-base with OSB sheathing	6.0–6.5 (42–45)	Fossil fuels; some recycled PET; pentane blowing agent; TCPP flame retardant; aluminum facing	Energy use during manufacture	Potential health concerns during manufacture. Negligible emissions after installation	Phaseout of HCFC ozone-depleting blowing agents completed
Extruded polystyrene (XPS)	Rigid board	5.0 (35)	Fossil fuels; HCFC-142b blowing agent; HBCD flame retardant	Energy use during manufacture. Ozone depletion	Potential release of residual styrene monomer (a carcinogen) and HBCD flame retardant	Last remaining insulation material with ozone-depleting blowing agents
Expanded polystyrene (EPS)	Rigid board	3.6–1.4 (25–31)	Fossil fuels; pentane blowing agent; HBCD flame retardant	Energy use during manufacture	Potential release of residual styrene monomer (a carcinogen) and HBCD flame retardant	
Closed-cell spray polyurethane	Spray-in cavity-fill or spray-on roofing	5.8–6.8 (40–4.7)	Fossil fuels, HCFC-141b (through early 2005) or HFC-245fa blowing agent; nonbrominated flame retardant	Energy use during manufacture, global-warming potential from HFC blowing agent	Quite toxic during installation (respirators or supplied air required). Allow several days of ailing out prior to occupancy	
Open-cell, low-density polyurethane	Spray-in cavity-fill	3.6–3.8 (25–27)	Fossil fuels and soybeans; water as a blowing agent; nonbrominated flame retardant	Energy use during manufacture	Quite toxic during installation (respirators or supplied air required). Allow several days of ailing out prior to occupancy	
Air-Krete	Spray-in cavity-fill	3.9 (27)	Magnesium oxide from seawater; ceramic talc	Negligible	Considered very safe	Highly fire-resistant; inert; remains friable

(Continued)

TABLE 2 (*Continued*) Comparison of Insulation Materials (Environmental Characteristics and Other Information)

		Conversion Factors				
R-Value Conversions		To Get		Multiply		By
Radiant barrier						
Bubble back	Stapled to framing	Depends on installation	Aluminum; fossil fuels	Energy use during manufacture	Minimal offgassing from plastic	Exaggerated R-value claims have been common
Foil-faced	Stapled to framing;	Depends on installation	Aluminum; fossil	Energy use during	Minimal offgassing from	Exaggerated Æ-value claims
polyethylene foam	requires air' space for radiant benefit		fuels; recycled polyethylene	manufacture	polyethylene	have been common. Recycled content in some
Foil-faced paperboard sheathing	Stapled to framing; requires air' space for radiant benefit	Depends on installation	Aluminum; fossil fuels; recycled paper	Energy use during manufacture	Considered very safe	High recycled content. Structural sheathing available (e.g., Thermo-Ply'))
Foil-faced OSB	Most common as attic sheathing	Depends on installation	Wood fiber; formaldehyde binder in OSB; aluminum	Energy use and VOC emissions during manufacture	Formaldehyde emissions	Primary benefit is reduced heat gain

[a] RSI/m: The standard unit of measurement in the United States has been the Imperial unit. The country is converting to the International System (SI) unit—or metric standard—which predominates internationally. To differentiate like terms, you may find "SI" added to the term symbol. For example, RSI refers to the Æ-value in International System (metric) units.

Critical Guidelines

When installing any insulating material, the following guidelines are critical for optimum performance[4]:

- Seal all air leaks between conditioned and unconditioned areas
- Obtain complete coverage of the insulation, especially around doors and windows
- Minimize air leakage through the material with air sealing measures
- Avoid compressing insulation
- Avoid lofting (installing too much air) in loose-fill products

Foam Insulation Strategies

Foam products are primarily economical when they can be applied in thin layers as part of a structural system or to help seal air leaks. Examples include:

- Exterior sheathing over wall framing
- Forms in which concrete can be poured
- As part of a structural insulated panel for building walls
- Spray-applied foam insulation

Floor Insulation

Slab-on-Grade

Slab-on-grade floors consist of a concrete slab poured over at least four inches of compacted gravel or sand and a layer of 10-mil polyethylene used as a vapor barrier. In hot, humid climates, most buildings are built with concrete slab-on-grade.

For colder climates, slabs lose energy as a result of heat conducted outward toward the perimeter of the slab. Insulating the exterior edge of the slab with R-10 rigid insulation can reduce winter heating bills by 10%–20%.

Raised Floor

Raised floor systems (wood and concrete) have specific requirements depending on climate zones. Consult the local building code for specific details.

Wall Insulation

Walls are the most complex component of the building envelope to provide adequate thermal insulation, air sealing, and moisture control.

Concrete Wall Insulation

Foundation walls and other masonry walls are usually built of concrete block or poured concrete. Insulating concrete block walls is more difficult than insulating framed walls.

Insulating Concrete Block Cores

Builders can insulate the interior cores of concrete block walls with insulation such as:

- Vermiculite R-2.1 per inch (See Figure 2)
- Polystyrene inserts or beads R-4.0–5.0 per inch
- Polyurethane foam R-5.8-6.8 per inch

Unfortunately, as shown in Figure 2, the substantial thermal bridging in the concrete connections between cores continues to depreciate the overall *R*-value. This approach is only a partial solution to providing a quality, well-insulating wall. Other techniques, as explained in the next few pages, provide more cost-effective solutions.

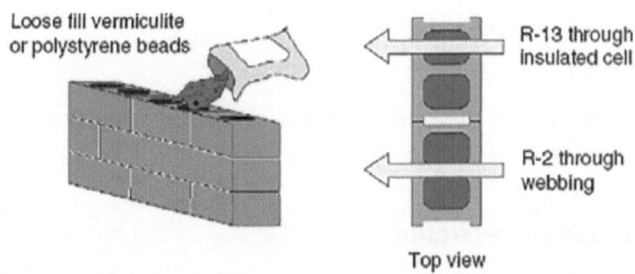

FIGURE 2 Insulating concrete block cores (R-4–R-6 overall).

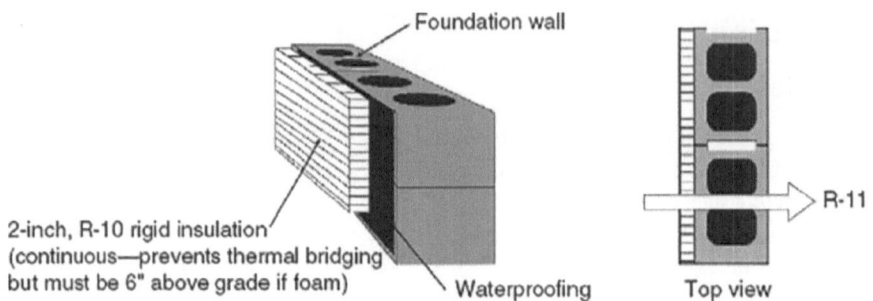

FIGURE 3 Exterior foam insulation (R-11–R-12 overall).

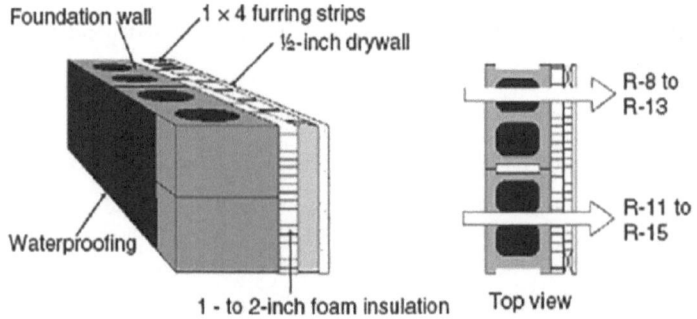

FIGURE 4 Interior foam insulation (R-10–R-14 overall).

Exterior Rigid Fiber Glass or Foam Insulation

Rigid insulation is generally more expensive per *R*-value than mineral wool or cellulose, but its rigidity is a major advantage (Figure 3). However, it is difficult and expensive to obtain R-values as high as in framed walls.

Interior Foam Wall Insulation

Foam insulation can be installed on the interior of concrete block walls (Figure 4); however, it must be covered with a material that resists damage and meets local fire code requirements. Half-inch drywall will typically comply, but furring strips will need to be installed as nailing surfaces. Furring strips are usually installed between sheets of foam insulation; however, to avoid the direct, uninsulated thermal bridge between the concrete wall and the furring strips, a continuous layer of foam should be installed underneath or on top of the nailing strips.

Interior Framed Wall

In some cases, designers will specify a framed wall on the interior of a masonry wall (Figure 5). Standard framed wall insulation and air-sealing practice can then be applied.

Insulated Concrete Form Systems

Insulated concrete forms (ICFs) are permanent rigid plastic foam forms that are filled with reinforced concrete to create structural walls with significant thermal insulation (Figure 6). The foam is typically

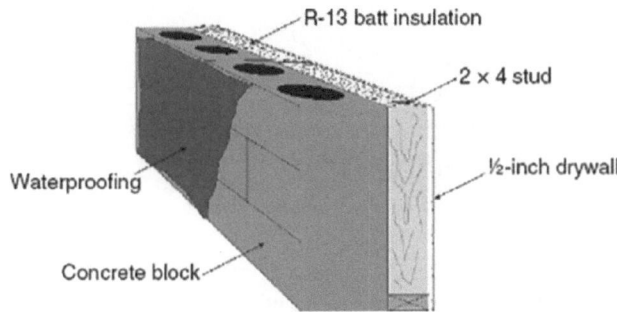

FIGURE 5 Interior framed wall insulation (R-11–R-13 overall).

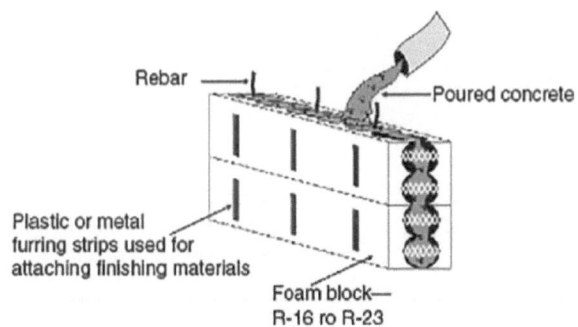

FIGURE 6 Insulated concrete foam system (R-17–R-24 overall).

either expanded polystyrene (EPS) or extruded polystyrene (XPS) and occasionally polyurethane, but it may also be made from a composite of cement and foam insulation or a composite of cement and processed wood.[5]

The concrete will be one of several shapes: flat, waffle-or screen-grid, or post-and-beam, depending on the specific form design. The Portland Cement Association (PCA) reports that in 1994, 0.1% of all new homes used ICFs in above-grade walls (about 1100 new homes). That number rose to 1.2% in 1999, 2.7% in 2001, and increased to 3.8% in 2002, which would be 50,639 homes, according to U.S. Census Data.

Above-grade ICF walls cost more to build than typical wood-framed walls. As wood-framed walls approach the thermal insulation value of ICFs, the cost differential will decrease. In most cases, materials' costs (concrete and forms) are primarily responsible for increased costs, while labor costs are often similar to wood framing. Cost premium depends on relative material prices, labor efficiency for each system, engineering necessity, and its effect on other practices or trades, among other factors.

The cost premium for ICF houses is smaller in areas such as high-wind regions, which require additional labor, time, and materials for special construction of woodframed houses. According to an NAHB Research Center study, costs are estimated to increase by 1%–8% of total house cost over a wood-framed house.

Lightweight Concrete Products

Lightweight, air-entrained concrete is an alternative wall system (Figure 7). Autoclaved aerated concrete (AAC), sometimes referred to as precast autoclaved aerated concrete (PAAC), which can be shipped either as blocks or panels, combines elevated *R*-values (compared to standard concrete) with thermal mass.

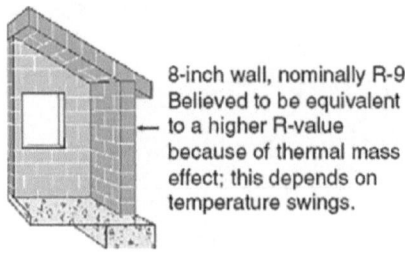

8-inch wall, nominally R-9.
Believed to be equivalent
to a higher R-value
because of thermal mass
effect; this depends on
temperature swings.

FIGURE 7   Lightweight concrete products (R-1.1 per inch plus).

TABLE 3   2 × 4 Framed Wall Problems and Solutions

Problem	Solution
Small space available for insulation	Install continuous exterior foam sheathing and medium (R-13) to high (R-15) density cavity insulation
Enclosed cavities are more prone to cause condensation, particularly when sheathing materials with low R-values are used	Install a continuous air barrier system. Use continuous foam sheathing
Presence of wiring, plumbing, ductwork, and framing members lessens potential R-value and provides pathways for air leakage	Locate mechanical systems in interior walls; avoid horizontal wiring runs through exterior walls; use air sealing insulation system

2 × 4 Wall Insulation

Throughout the United States, debates on optimal wall construction continue.[6] Table 3 summarizes typical problems and solutions in walls framed with 2 × 4 studs. In addition to standard framing lumber and fasteners, the following materials will also be required during construction:

- Foam sheathing for insulating headers.
- 1 × 4 or metal T-bracing for corner bracing (Figure 8).

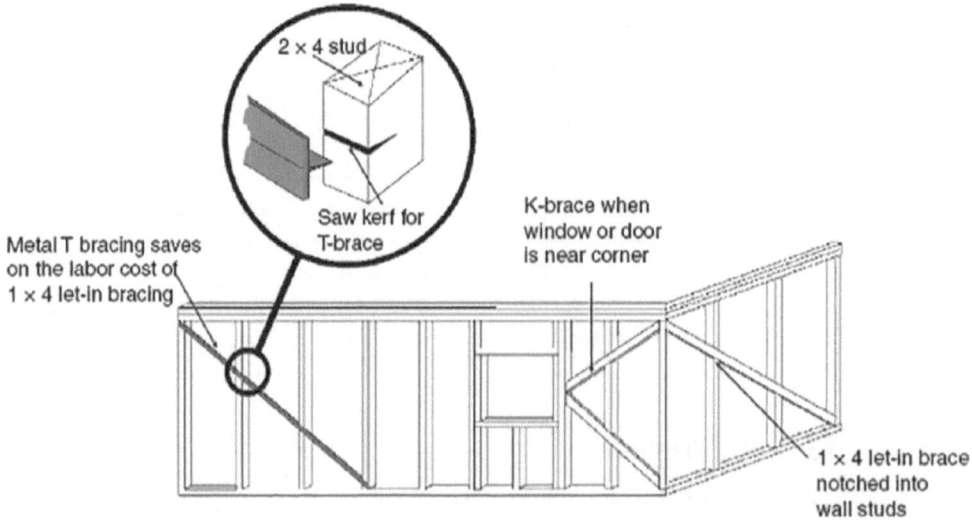

2 × 4 stud

Saw kerf for
T-brace

K-brace when
window or door
is near corner

Metal T bracing saves
on the labor cost of
1 × 4 let-in bracing

1 × 4 let-in brace
notched into
wall studs

FIGURE 8   Let-in bracing.

- R-13 batts for insulating areas during framing behind shower/tub enclosures and other hidden areas.
- From the Florida Building Code, Residential (FBC-R): "R307.2 Bathtub and shower spaces. Bathtub and shower floors and walls above bathtubs with installed shower heads and in shower compartments shall be finished with a nonabsorbent surface. Such wall surfaces shall extend to a height of not less than 6 ft above the floor."
- Caulking or foam sealant for sealing areas that may be more difficult to seal later.

Avoid Side Stapling

Walls are usually insulated with batts that have attached vapor retarder facing. Many builders question whether it is best to side staple or face staple batt insulation. The common arguments are that face stapling results in less compression, while side stapling interferes less with drywall installation.

The ideal solution should focus on where the kraft paper (vapor retarder) is rather than on how it is installed.[7]

The face stapling question is an appropriate question in northern or "heating-dominated" climates. In northern areas, vapor retarders should be installed on the "warm" side of the wall cavity. In southern or "cooling-dominated" climates, the vapor retarder should be on the outside surface of the wall cavity. Because of this, the use of unfaced batts is recommended in hot-humid climates (Figure 9).

Unfaced batts are slightly larger than the standard 16-or 24-inch stud spacing and rely on a friction-fit for support. Because unfaced batts are not stapled, they can often be installed in less time. In addition, it is easier to cut unfaced batts to fit around wiring, plumbing, and other obstructions in the walls.

Blown Loose-Fill Insulation

Loose-fill cellulose, fiberglass, and rock wool insulation can also be used to insulate walls. This insulation is installed with a blowing machine and held in place with a glue binder or netting (Figure 10). This technique can provide good insulation coverage in the stud cavities; however, it is very important that excess moisture in the binder be allowed to evaporate before the wall cavities are enclosed by an interior finish. Keep in mind that insulation products are not replacements for proper air sealing technique.

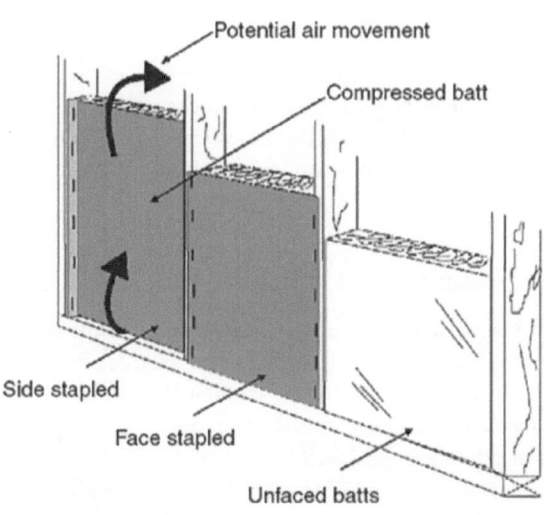

FIGURE 9   Insulating walls with batts.

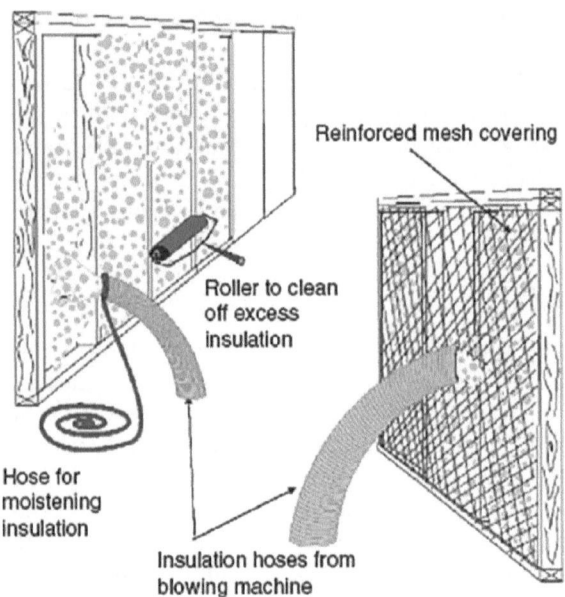

FIGURE 10   Blown sidewall insulation options.

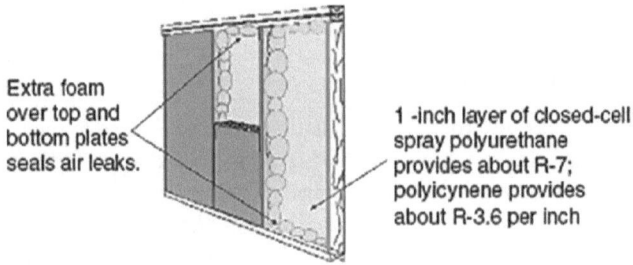

FIGURE 11   Blown foam insulation.

Blown Foam Insulation

Some insulation contractors are now blowing polyurethane or polyicynene insulation into the walls and ceilings of new buildings (Figure 11). This technique provides high R-values in relatively thin spaces and seals air leaks effectively. The economics of foam insulation should be examined carefully prior to their application.

Structural Insulated Panels

Another approach to wall construction is the use of structural insulated panels (SIP), also known as stress-skin panels (Figure 12).[8] They consist of 4-inch or 6-inch thick foam panels onto which structural sheathing, such as oriented strand board (OSB), cement fiber board, or various types of metal have been attached. They reduce labor costs, and because of the reduced framing in the wall, they have higher R-values and less air leakage than standard walls.

SIPs are generally 4 feet wide and 8 to 12 feet long. There are a wide variety of manufacturers, each with its own method of attaching panels together. Procedures for installing windows, doors, wiring, and plumbing have been worked out by each manufacturer. Some SIPs come from the factory with

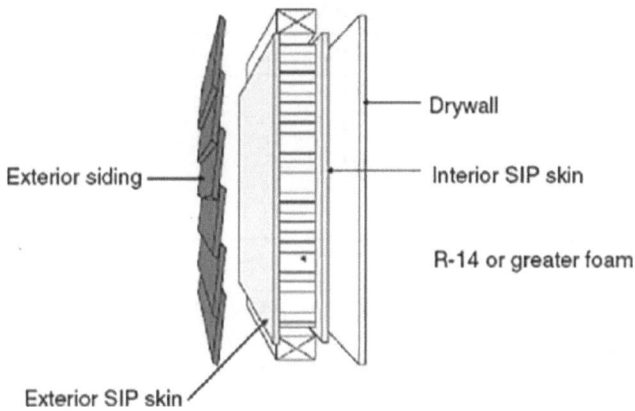

Drywall

Exterior siding

Interior SIP skin

R-14 or greater foam

Exterior SIP skin

FIGURE 12   Structural insulated panels (SIP)

preinstalled windows. In addition to their use as wall framing, SIPs can also be used in ceilings, floors, and roofs.

While buildings constructed with SIPs may be more expensive than those with standard framed and insulated walls, research studies have shown that SIP-built buildings have higher average insulating values per inch than most commonly used insulation materials. Due to their typical modular style of construction, infiltration losses are also reduced. Thus, they can provide substantial energy savings. Be sure to follow local building codes with regard to termites, including leaving a 6" inspection zone above finished grade.

The performance of any SIP wall depends on its component materials and installation processes. There are a few important variables to take into consideration when building with SIP systems:

1. Panel fabrication (proper panel gluing, pressing, and curing) is critical to prevent delamination.
2. Panels must be flat, plumb, and have well-designed connections to ensure tightness of construction.
3. Though SIPs offer ease of construction, installers may need training in installing the system being used.
4. Fire rating of SIP materials and air-tightness of SIP installation affect the system's fire safety.
5. There may be potential insect and rodent mitigation issues, depending on SIP materials and construction.
6. Proper HVAC design and installation must take the SIP system being used into account.

Metal Framing

Builders and designers are well aware of the increasing cost and decreasing quality of framing lumber. As a consequence, interest in alternative framing materials, such as metal framing, has grown. While metal framing offers advantages over wood, such as consistency of dimensions, lack of warping, and resistance to moisture and insect problems, it has distinct disadvantages from an energy perspective.

Metal framing is an excellent conductor of heat. Buildings framed with metal studs and plates usually have metal ceiling joists and rafters, as well. Thus, the entire structure serves as a highly conductive thermal grid. Insulation placed between metal studs and joists is much less effective due to the extreme thermal bridging that occurs across the framing members.

The American Iron and Steel Institute is well aware of the challenges involved in building an energy efficient steel structure. In their publication *Thermal Design Guide for Exterior Walls* (Publication RG-9405), the Institute provides information on the thermal performance of steel-framed buildings. Table 4 summarizes some of their findings.

TABLE 4   Effective Steel Wall *R*-Values

Cavity Insulation	Sheathing	Effective Overall R-Value
11	2.5	9.5
11	5	13
11	10	18
13	2.5	10
13	5	14

Moisture-related problems have been reported in metal frame buildings that do not use sufficient insulated sheathing on exterior walls. Metal studs cooled by the air conditioning system can cause outdoor air to condense, leading to mildew streaks (or ghosting), where one can see the framing members on the inside and outside of a home. In winter, studs covered by cold outside air can also cause streaking. Attention to proper insulation techniques can alleviate this problem.

2 × 6 Wall Construction

There has been interest in hot, humid climates in the use of 2 × 6s for construction. The advantages of using wider wall framing are:

- More space provides room for R-19 or R-21 wall insulation.
- Thermal bridging across the studs is less of a penalty due to the higher *R*-value of 2 × 6s.
- Less framing reduces labor and material costs.
- There is more space for insulating around piping, wiring, and ductwork.

Disadvantages of 2 × 6 framing include:

- Wider spacing may cause the interior finish or exterior siding to bow slightly between studs.
- Window and door jambs must be deeper, resulting in additional costs.
- Walls with substantial window and door area may require almost as much framing as 2 × 4 walls and leave relatively little area for actual insulation.

The economics of 2 × 6 wall insulation are affected by the number of windows in the wall because each window opening adds extra studs and may require the purchase of a jamb extender. Walls built with 2 × 6s having few windows provide positive economic payback. However, for walls in which windows make up over 10% of the total area, the economics become questionable.

Ceilings and Roofs

Attics over flat ceilings are usually the easiest part of a building's exterior envelope to insulate. They are accessible and have ample room for insulation. However, many homes use cathedral ceilings that provide little space for insulation.[9] It is important to insulate both types of ceilings properly.

Attic Ventilation

In the summer, properly designed ventilation reduces roof and ceiling temperatures, thus potentially saving on cooling costs and lengthening the life of the roof. In winter, roof vents expel moisture which could otherwise accumulate and deteriorate insulation or other building materials.

At present, several research studies are investigating whether attic ventilation is beneficial. For years, researchers have believed the cooling benefits of ventilating a well-insulated attic to be negligible. However, some experts are now questioning whether ventilation is even effective at moisture

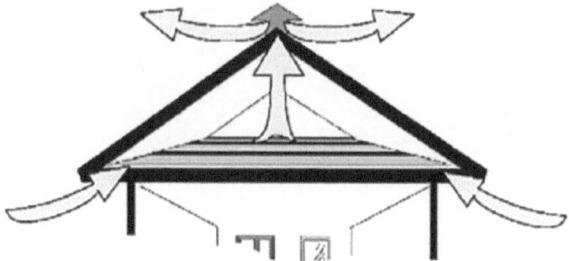

FIGURE 13   Attic ventilation through soffit and ridge vents.

removal. The Florida Building Code, Residential now provides provisions for "conditioned attic assemblies" (Section R806.4) as long as certain conditions are met. When attic ventilation is provided, ventilation openings shall be provided with corrosion-resistant wire mesh with 1/8 in. minimum to 1/4 in. maximum openings. Total net free ventilation area shall not be less than 1–150 of the area of the space ventilated. An exception for 1–300 is provided in the code.

Vent Selection

The amount of attic ventilation needed is based on state building code requirements. If ventilating the roof, locate vents high along the roof ridge and low along the eave or soffit. Vents should provide air movement across the entire roof area (Figure 13). There are a wide variety of products available, including ridge, gable, soffit, and mushroom vents.

To allow for proper airflow in attic spaces, it is common practice to install a rafter baffle at the soffit. This will prevent insulation from sealing off the airflow from the soffit vent to the attic space.

The combination of continuous ridge vents along the peak of the roof and continuous soffit vents at the eave provides the most effective ventilation. Ridge vents come in a variety of colors to match any roof. Some brands are made of plastic and covered by cap shingles to hide the vent from view.

Manufacturer or product testing is being performed by a variety of organizations to verify leak-free operation of continuous ridge vents in high wind situations. Care should be taken to ensure that the vents chosen are appropriate for hurricane-prone areas.

Powered Attic Ventilator

Electrically powered roof ventilators can consume more electricity to operate than they save on air conditioning costs and are not recommended for most designs. Power vents can create negative pressures in the home, which may have detrimental effects, such as (Figure 14):

- Drawing outside air into the home
- Removing conditioned air from the home through ceiling leaks and bypasses
- Pulling pollutants, such as radon and sewer gases, into the home
- Backdrafting fireplaces and fuel-burning appliances

Attic Floor Insulation Techniques

Either loose-fill or batt insulation can be installed on an attic floor. Generally, blowing loose-fill attic insulation is usually less expensive than installing batts or rolls. Most attics have either blown fiberglass, rock wool, or cellulose. Ceilings with a rise greater than 5 and a run of 12 (5 over 12) should not be insulated with blown-in insulation.

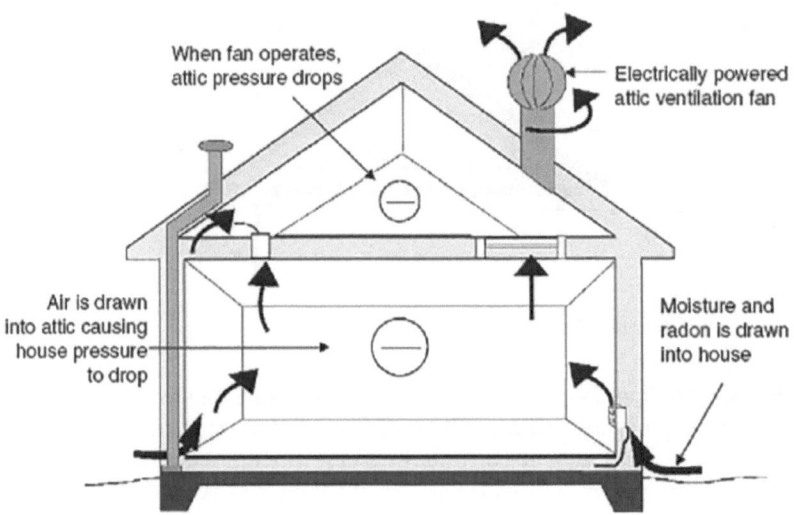

FIGURE 14   Attic ventilation through powered ventilation.

Loose-Fill Attic Insulation

Steps for installing loose-fill attic insulation[10]:

7. Seal attic air leaks, as prescribed by fire and energy codes.
8. Follow manufacturer's and state building code clearance requirements for heat-producing equipment found in an attic, such as flues or exhaust fans. One example of attic blocking is shown later in this entry.
9. Use baffles to preserve ventilation space at eave of roof for soffit vents.
10. Insulate the attic hatch or attic stair. Foam boxes are available for providing a degree of insulation over a pull-down attic stairway.
11. Determine the attic insulation area; based on the spacing and size of the joists, use the chart on the insulation bag to determine the number of bags to install. Table 5 shows a sample chart for cellulose insulation. Cellulose is heavier than fiberglass for the same *R*-value. Closer spacing of roof joists and thicker dry-wall is required for larger *R*-values. Check this detail with the insulation contractor. Weight limits and other factors at R-38 insulation levels are shown in Table 5 for the three primary types of loose fills.
12. Avoid fluffing the insulation (blowing with too much air) by using the proper air-to-insulation mixture in the blowing machine. A few insulation contractors have "fluffed" (added extra air to) loose-fill insulation to give the impression of a high *R*-value. The insulation may be the proper depth, but if too few bags are installed, the *R*-values will be less than claimed.

TABLE 5   Blown Cellulose in Attics

R-Value at 75°F	Minimum Thickness (in.)	Minimum Weight (lb/ft2)	2 × 6 Joists Spaced 24 in. on Center		2 × 6 Joists Spaced 16 in. on Center	
			Coverage per 25-lb Bag (ft2)	Bags per 1000 ft2	Coverage per 25-lb Bag (ft2)	Bags per 1000 ft2
R-40	10.8	2.10	12	83	13	77
R-32	8.6	1.60	16	63	18	56
R-24	6.5	0.98	21	48	23	43
R-19	5.1	0.67	37	27	41	24

13. Obtain complete coverage of the blown insulation at relatively even insulation depths. Use attic rulers (obtainable from insulation contractors) to ensure uniform depth of insulation.

Batt Attic Insulation

Steps for installing batt insulation:

14. Seal attic air leaks, as prescribed by fire and energy codes.
15. Block around heat-producing devices, as described in Step 2 for loose-fill insulation.
16. Insulate the attic hatch or attic stair as described in Step 4 for loose-fill insulation.
17. Determine the attic insulation area based on the spacing and size of the joists, order sufficient R-30 insulation for the flat attic floor. Choose batts that are tapered—cut wider on top—so that they cover the top of the ceiling joists. (See Figure 15).
18. When installing the batts, make certain they completely fill the joist cavities. Shake batts to ensure proper loft. If the joist spacing is uneven, patch gaps in the insulation with scrap pieces. Try not to compress the insulation with wiring, plumbing, or ductwork. In general, obtain complete coverage of full-thickness, noncompressed insulation.
19. Attic storage areas can pose a problem. If the ceiling joists are shallower than the depth of the insulation (generally less than 2 ×10s), raise the finished floor using 2 × 4s or other spacing lumber. Install the batts before nailing the storage floor in place (see Figure 16).

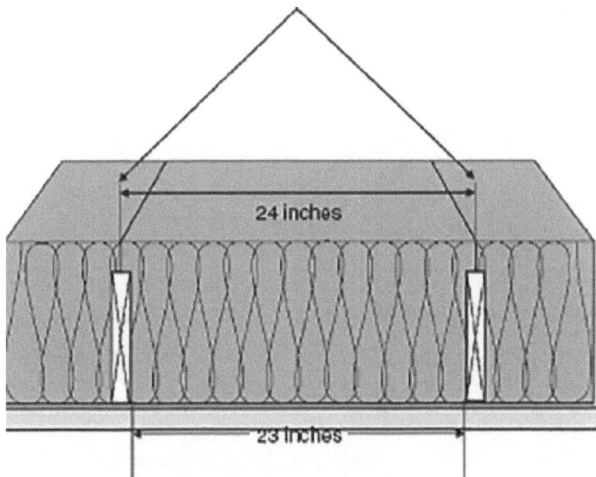

FIGURE 15   Full-width ceiling batt insulation.

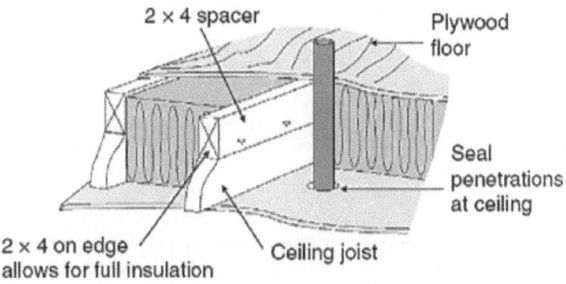

FIGURE 16   Ceiling insulation under attic storage floor.

Note that often attic framing is not designed for storage. Check engineered loads of framing before increasing loads and piggy-backing ceiling joists.

Preventing Air Flow Restrictions at the Eave

One problem area in many standard roof designs is at the eave, where there is not enough room for full R-30 insulation without preventing air flow from the soffit vent or compressing the insulation, which reduces its *R*-value. Figures. 17 and 18 show several solutions to this problem. If using a truss roof, purchase raised heel trusses that form horizontal overhangs. They should provide adequate clearance for both ventilation and insulation.

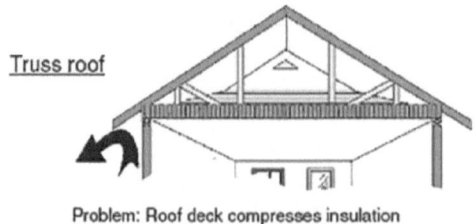

Problem: Roof deck compresses insulation
and blocks air flow from soffit vent

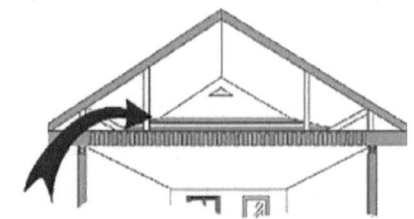

Solution: Raised heel trusses—insulation not
compressed; air flow path is open

FIGURE 17   Soffit air ventilation—raised heel trusses.

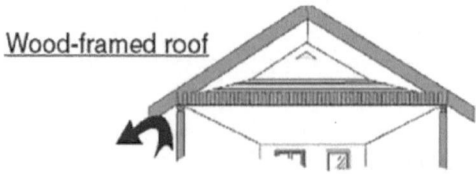

Problem: Roof compresses insulation at eave
and blocks air flow from soffit vent

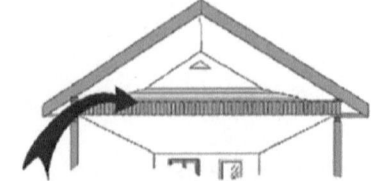

Solution: Raised top plate—insulation not
compressed; air flow path is open

FIGURE 18   Soffit air ventilation—raised top plate.

In stick-built roofs, where rafters and ceiling joists are cut and installed on the construction site, an additional top plate that lies across the top of the ceiling joists at the eave will prevent compression of the attic insulation. Note: This needs to be a double plate for bearing unless rafters sit directly above joists. The rafters sitting on this raised top plate allow for both insulation and ventilation.

Cathedral Ceiling Insulation Techniques

Cathedral ceilings are a special case because of the limited space for insulation and ventilation within the depth of the rafter. Fitting in a 10-in. batt (R-30) and still providing ventilation is impossible with a 2 × 6 or 2 × 8 rafter (R-19 or R-25, respectively).

Building R-30 Cathedral Ceilings

Cathedral ceilings built with 2 × 12 rafters can be insulated with standard R-30 batts and still have plenty of space for ventilation. Some builders use a vent baffle between the insulation and roof decking to ensure that the ventilation channel is maintained.

If 2 × 12s are not required structurally, most builders find it cheaper to construct cathedral ceilings with 2 × 10 rafters and high-density R-30 batts, which are $8^{1/4}$ in. thick (Table 6).

Some contractors wish to avoid the higher cost of 2 × 10 lumber and use 2 × 8 rafters. These roofs are usually insulated with R-19 batts.

In framing with 2 × 6 and 2 × 8 rafters, higher insulating values can be obtained by installing rigid foam insulation under the rafters. Note that the rigid foam insulation must be covered with a fire-rated material when used on the interior of the building. Drywall usually meets this requirement.

Scissor Trusses

Scissor trusses are another cathedral ceiling framing option. Make certain they provide adequate room for both R-30 insulation and ventilation, especially at their ends, which form the eave section of the roof.

Any sized rafter; blown-in cellulose, fiberglass, or rock wool held in place; provide 1 in. ventilation space above.

Difficulties with Exposed Rafters

A cathedral ceiling with exposed rafters or roof decking is difficult and expensive to insulate well. Often, foam insulation panels are used over the attic deck, as shown in Figure 19. However, to achieve R-30, four to seven inches of foam insulation are needed. Ventilation is also a problem.

In homes where exposed rafters are desired, it may be more economical to build a standard, energy efficient cathedral ceiling, and then add exposed decorative beams underneath. Note that homes having tongue-and-groove ceilings can experience substantially more air leakage than solid, drywall ceilings. Install a continuous air barrier, sealed to the walls above the tongue-and-groove roof deck and held in place; provide 1 in. ventilation space above.

TABLE 6   Cathedral Ceiling Insulation Options

Rafter	Batt
2 × 8	R-19
2 × 10	R-25
2 ×10	Moderate density R-30
2 × 12	Standard density R-30

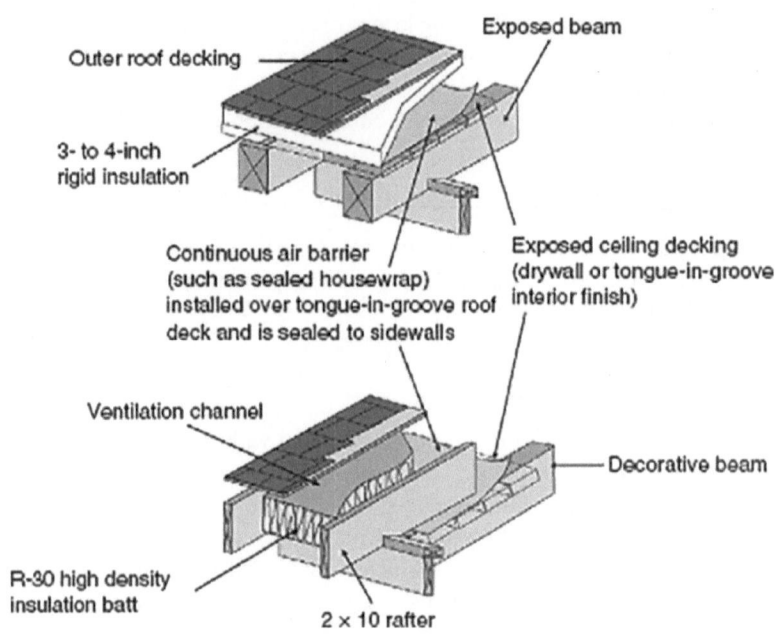

FIGURE 19   Insulating exposed rafters.

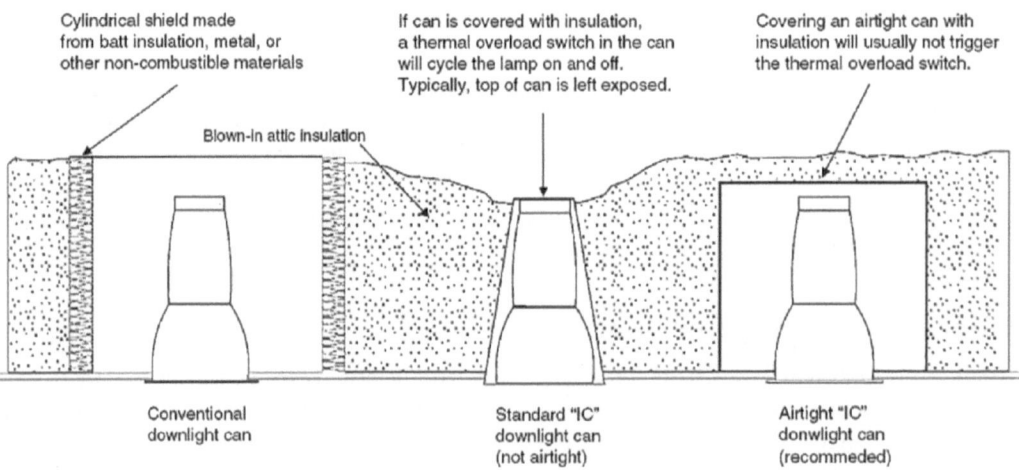

FIGURE 20   Recessed lighting insulation.

Recessed Lights

Standard recessed fixtures require a clearance of several inches between the sides of the lamp's housing and the attic insulation. In addition, insulation cannot be placed over the fixture. Even worse, recessed lights leak considerable air between attics and the home.

Insulated ceiling (IC) rated fixtures have a heat sensor switch that allows the fixture to be covered—except for the top—with insulation (see Figure 20 for the proper insulation methods for these fixtures). However, these units also leak air. If you have to use recessed lights, install airtight IC-rated fixtures. There are alternatives to recessed lights, including surface-mounted ceiling fixtures and track lighting, which typically contribute less air leakage to the home.

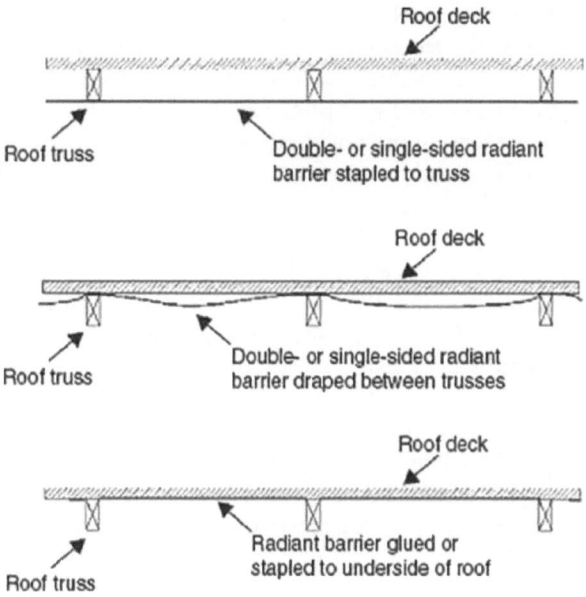

FIGURE 21   Radiant barrier configuration

Radiant Heat Barriers

Radiant heat barriers (RHB) are reflective materials that can reduce summer heat gain via the insulation and building materials in attics and walls. RHBs work two ways: first, they reflect thermal radiation well; and, second, they emit (give off) very little heat. RHBs should always face a vented airspace and be installed to prevent dust build-up. They are usually attached to the underside of the rafter or truss top chord or to the underside of the roof decking. Acceptable attic radiant barrier configurations can be found in Figure 21.

How Radiant Barrier Systems Work

A radiant barrier reduces heat transfer. Thermal radiation, or radiant heat, travels in a straight line away from a hot surface and heats any object in its path.

When sunshine heats a roof, most of the heat conducts through the exterior roofing materials to the inside surface of the roof sheathing. Heat then transfers by radiation across the attic space to the next material—either the top of the attic insulation or the attic floor. A radiant barrier, properly installed in one of many locations between the roof surface and the attic floor, will reduce radiant heat flow. Thermal insulation on the attic floor resists the flow of heat through the ceiling into the living space below. The rate at which insulation resists this flow determines the insulation's R-value. The amount of thermal insulation affects the potential radiant barrier energy savings. For example, installing a radiant barrier in an attic that already has high levels of insulation (R-30 or above) would result in much lower energy savings than an attic insulated at a low level (R-11 or less).

All radiant barriers use reflective foil that blocks radiant heat transfer. In an attic, a radiant barrier that faces an air space can block up to 95% of the heat radiating down from a hot roof. Only a single, thin, reflective surface is necessary to produce this reduction in radiant heat transfer. Additional layers of foil do little more to reduce the remaining radiant heat flow.

Conventional types of insulation consist of fibers or cells that trap air or contain a gas to retard heat conduction. These types of insulation reduce conductive and radiant heat transfer at a rate determined by their

R-value, while radiant barriers reduce only radiant heat transfer. There is no current method for assigning an *R*-value to radiant barriers. The reduction in heat flow achieved by the installation of a radiant barrier depends on a number of factors, such as ventilation rates, roof reflectivity, ambient air temperatures, geographical location, the amount of roof solar gains, and the amount of conventional insulation present.

Several factors affect the cost effectiveness of installing a radiant barrier. You should examine the performance and cost savings of at least three potential insulation options: adding additional conventional insulation, installing a radiant barrier, and adding both conventional insulation and a radiant barrier.

In 1991 (revised June, 2001), the U.S. Department of Energy (DOE) published the Radiant Barrier Attic Fact Sheet, which shows how to calculate the economics of radiant barriers and added ceiling insulation. It includes an Energy Savings Worksheet with an example. The worksheet is part of the fact sheet and can be found at http://www.ornl.gov/sci/roofs+walls/radiant/rb_05.html.[11]

Because radiant barriers redirect radiant heat back through the roofing materials, shingle temperatures may increase between 1 and 10 F (0.6 C-5.6 C). This increase does not appear to exceed the roof shingle design criteria. The overall effect on roof life, if any, is not known.

Remember, radiant barriers are most effective in blocking summer radiant heat gain and saving air-conditioning costs. Although the radiant barrier may be somewhat effective in retaining heat within a cold-climate home, it may also block any radiant winter solar heat gain in the attic.

Conclusion

Buildings are insulated to help moderate the environment that we live and work in. As utility costs rise, this aspect of our constructed environment becomes more and more important. Many new materials have entered the market in the past few years, each having their own particular advantages and disadvantages. This entry has described in detail the importance of installation practices and their effect on overall performance. Emphasis was placed on properly defining the building envelope and different methods and materials that can be used to effectively insulate and, therefore, thermally isolate the constructed environment from the daily extremes produced by local weather.

New techniques and materials will evolve; however, some guidelines will remain the same:

- Choose an insulation material with characteristics suitable for the climate region.
- Use the proper R-value for the climate region where each particular structure is being constructed.
- Apply the recommended thickness, or R-value, to each building component, such as ceiling, walls, and floors.
- Install each particular insulation product in a proper fashion, following the manufacturer's recommendations and instructions, leaving no voids, and filling each building cavity to the level required for the building site.

Insulation materials and techniques play a large part in minimizing energy consumption and maximizing human comfort. Recent advances in both materials and techniques will help further this trend.

Acknowledgments

Many thanks to Kyle Allen, an engineering student at the University of Florida, because without his persistent efforts, the publication of this entry would not have been possible.

References

1. *Energy Efficient Building Construction in Florida,* Florida Energy Extension Service, University of Florida, 2005. Adapted with permission, Gainesville, FL.
2. Jeffrey, T.; Dennis, C. *A Builder's Guide to Energy Efficient Homes in Georgia*; Georgia Environmental Facilities Authority; Division of Energy Resources: Atlanta, GA, 1999.

3. Environmental Building News. 2005. *Insulation Materials— Summary of Environmental and Health Considerations.* 122 Birge Street, Suite 30, Brattleboro, VT 05301. Phone: (800) 861–0954. Web site: http://www.BuildingGreen.com.

4. Federal Trade Commission. *Labeling and Advertising of Home Insulation (R-Value Rule) 16CFR460.* Available online at: http://www.ftc.gov/bcp/rulemaking/rvalue/16cfr460.htm Federal Register Notice (amendments) http://www3.ftc.gov/os/2005/05/05031homeinsulationfrn.pdf.

5. ToolBase Services. *Insulating Concrete Forms.* http://www.toolbase.org/techinv/techDetails. aspx?technologyIDZ97.

6. Oak Ridge National Laboratory, *Wall Insulation.* October 2000. Available online at: http://www. eere.energy.gov/buildings/info/documents/pdfs/26451.pdf.

7. Southface Energy Institute. March 2002. *Wall Insulation.* Available online at: http://www. southface.org/web/resources&services/publications/factsheets/26_insulatewalls4PDF.pdf.

8. Knowles, H. 2005. *Performance Under Pressure: Structural Insulated Panel (SIP) Walls.* http:// www.energy.ufl.edu/factsheets.htm.

9. Desjarlais, A.O.; Petrie, T.W.; Stovall, T. In *Comparison of Cathedralized Attics to Conventional Attics: Where and* When do Cathedralized Attics Save Energy and Operating Costs? *Performance of Exterior Envelopes of Whole Buildings, IX International Conference ASHRAE: Clearwater, FL,* 2004.

10. Oak Ridge National Laboratory, *Ceilings and Attics.* February 2000. Available online at: http:// eber.ed.ornl.gov/DOE-ceilingsattics_771.pdf.

11. U.S. Department of Energy. *Radiant Barrier Attic Fact Sheet.* Available online at: http://www.ornl. gov/sci/roofsC-walls/radiant/rb_05.html.

Bibliography

1. Air Conditioning Contractors of America. Appendix A-5, detailed infiltration estimate. In *Manual J, Load Calculation for Residential Winter and Summer Air Conditioning,* 8th Ed.; May be ordered online at: http://www.accaconference.com/Merchant2/merchant.mv?ScreenZPROD&Store_Code ZACCOA&Product_CodeZ33-8&Category_CodeZ.

2. Jeffrey, T.; Dennis, C. *Builder's Guide to Energy Efficient Homes in Louisiana*; Louisiana Department of Natural Resources: Louisiana, 2002.

3. Insulation Contractors Association of America. *Technical Bulletin No. 18, Reflective Insulations and Radiant Barriers.* Alexandria, VA, July 1997.

4. Oak Ridge National Laboratory, *Attic Access.* February 2000. Available online at: http://eber. ed.ornl.gov/Atticaccess_DOE_GO10099–768.pdf.

5. Office of Energy Efficiency and Renewable Energy (EERE) 1–877–337–3463 (EERE-INF) http:// www.eere.energy.gov and http://www.eere.energy.gov/consumer/.

6. U.S. Department of Energy. *Insulation and Weatherization.* Available online at: http://www. pueblo.gsa.gov/cic_text/housing/energy-savers/insulation.html.

7. U.S. Department of Energy. *Insulation Fact Sheet.* Available online at: http://www.ornl.gov/ roofsCwalls/insulation.

8. U.S. Department of Housing and Urban Development, *Volume 2—The Rehab Guide: Exterior Walls.* Washington, DC, August 1999. HUD-DU100C000005956 Available online at: http://www. huduser.org/publications/destech/walls.html.

9. U.S. Department of Housing and Urban Development, *Volume 3—The Rehab Guide: Roofs.* Washington, DC, March 1999. HUD-DU100C000005956, Available online at: http://www.huduser. org/publications/destech/roofs.html.

DIA: Diagnostic Tools: Monitoring, Ecological Modeling, Ecological Indicators, and Ecological Services

24

Environmental Accounting: A Tool for Supporting Environmental Management and Nature Conservation

Pier Paolo Franzese, Elvira Buonocore, and Giovanni F. Russo

Environmental Management and Nature Conservation

Healthy ecosystems are capable of maintaining their structures and functions, ensuring the generation and maintenance of natural capital stocks and ecosystem services flows (Buonocore et al., 2018; Pauna et al., 2018).

There is a growing evidence that biodiversity increases the stability of ecosystem functioning, representing the basis for the generation of ecosystem services vital for human well-being (Teixeira et al., 2019; Vihervaara et al., 2019). Therefore, biodiversity loss represents one of the major threats to humanity with potential impacts on both nature and human economy (Cardinale et al., 2012; TEEB, 2010).

In line with the objectives of the Convention on Biological Diversity (CBD) striving to address the issue of biodiversity loss, ensure the sustainable use of natural resources, and allow an equitable share of benefits, there has been significant progress toward conservation targets (UNEP-WCMC, IUCN, and NGS, 2018). Nonetheless, efforts are still needed to increase the effectiveness of nature conservation actions (Hoffmann et al., 2018).

Many international policy processes, including the CBD and the UN 2030 Agenda for Sustainable Development, recognize protected areas as successful management tools for protecting biodiversity while ensuring the sustainable exploitation of natural resources. When effective management measures are in place, protected areas are capable of meeting the multitude of objectives they are designed for, supporting the achievement of both local and large-scale sustainability goals.

Over the past decade, there have been increasing research efforts to assess the value of natural resources, also exploring how these values can be embedded into decision making (Costanza et al., 2014; European Union, 2014; Franzese et al., 2014, 2019).

Assessing the biophysical and economic value of natural capital and ecosystem services provides a deeper understanding of ecosystems functioning while supporting managers and policy makers in charge for implementing management plans and policies rooted in the principle of sustainable development (Maes et al., 2016; TEEB, 2010).

These assessments can be particularly useful in the case of protected areas to establish a baseline to monitor changes over time, assess the consequences of management decisions and changes, and demonstrate their importance in achieving international conservation targets (Neugarten et al., 2018).

Environmental Accounting

Environmental accounting is a tool useful for exploring three main dimensions related to the exploitation of natural resources both in the context of nature and human economy: environmental costs, impacts, and benefits (Häyhä and Franzese, 2014).

In particular, when applied in the context of nature conservation, environmental accounting is useful to address multiple aspects dealing with the exploitation of natural resources, among which (1) the assessment of environmental costs sustained by ecosystems for the generation and maintenance of natural capital stocks, (2) the assessment of goods and services that humans receive from ecosystems (i.e., the ecosystem services), and (3) the assessment of the impacts generated by human activities for the exploitation of natural resources (Figure 1).

Environmental accounting takes into account both direct and indirect environmental costs as well as local and global impacts. In general, environmental accounting methods can be assigned to two broad categories: (1) upstream methods, focusing on the cumulative amount of environmental resources supporting the investigated system, and (2) downstream methods, more concerned with the potential environmental impacts due to the system's emissions.

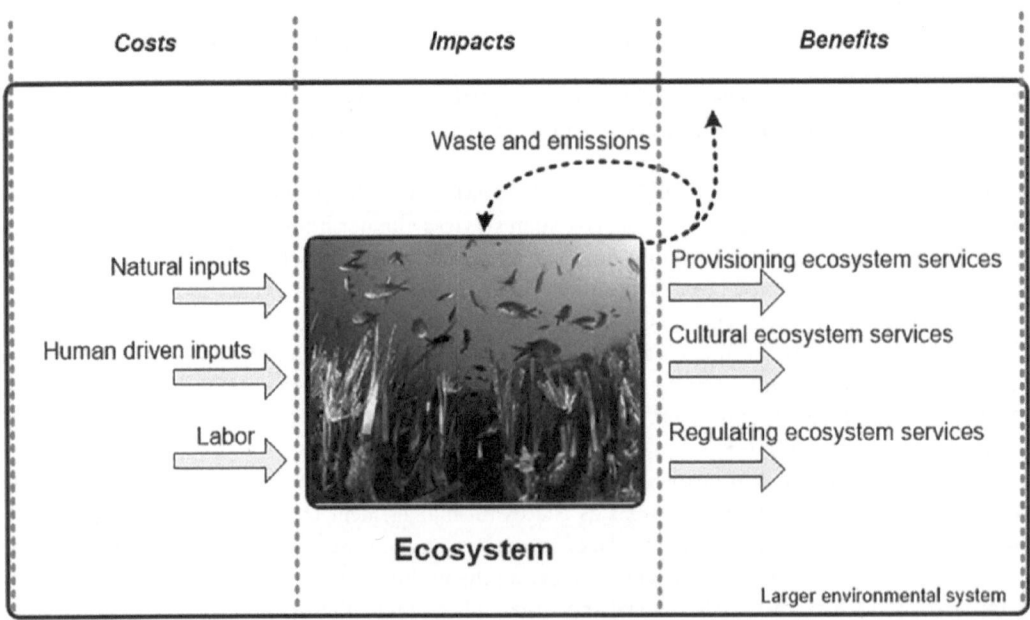

FIGURE 1 Three main dimensions of environmental accounting: assessment of environmental costs, impacts, and benefits.

The Emergy Accounting Method

The emergy accounting method (Odum, 1988, 1996) aims at calculating indicators of environmental performance and sustainability accounting for both natural and economic resources used up within ecosystems and human-dominated process.

Emergy is a "donor-side" approach assessing the value of goods and services in terms of work of biosphere invested for their generation. According to this method, the greater the cumulative environmental support to generate natural resources, the higher is their emergy value.

The emergy method takes into account: (1) natural input flows (e.g., solar radiation, wind, rain, and geothermal flow), (2) human-driven material and energy flows (e.g., machineries, fuels, chemicals), and (3) human labor and services converging to a process (Brown and Ulgiati, 2004).

According to this method, different forms of energy, materials, human labor, and economic services are accounted for in terms of their solar emergy, defined as the total amount of solar available energy (exergy) directly or indirectly required to make a given product or support a given flow, and measured in sej (solar emergy joules). The solar emergy required to generate one unit of product or service is referred to as Unit Emergy Value (UEV), expressed in sej/J in the case of energy flows or in sej/g in the case of mass flows.

The emergy method is deeply rooted in the concept of resource quality, i.e. the awareness that different energy forms have a different ability to do useful work even when their heat content is the same. Instead of only looking at what can be extracted out of a resource (exergy), the emergy accounting method focuses on what it takes for the biosphere to make and for human economy to process a given resource (Franzese et al., 2009).

The additional work provided by human activities to refine a raw resource adds up to its quality by making it more suitable to the final user. According to the emergy donor-side perspective, what makes a resource valuable is therefore both the environmental and human work investment for its production (Franzese et al., 2009).

An emergy assessment is typically implemented through the following main steps:

1. Identification of the spatial and temporal boundaries of the investigated system.
2. Modeling of the investigated system through an energy diagram drawn according to a standardized energy systems language (Odum, 1996).
3. Inventory of mass and energy flows supporting the investigated system.
4. Conversion of input flows into emergy units by using appropriate UEVs factors.
5. Calculation of the total emergy supporting the system.
6. Calculation of emergy-based indicators.

To calculate emergy-based indicators, all the input flows to a system are clustered into three main categories: the local Renewable flows (R), the local Non-renewable flows (N), and the flows Imported from outside the system (I). R, N, and I can be summed to account for the total emergy input (U) to a system.

In addition to the total emergy input (U) and the UEV of the system's output, the main emergy-based indicators used to describe the environmental performance and sustainability of the investigated system are: the Environmental Loading Ratio (ELR), the Emergy Yield Ratio (EYR), and the Emergy Sustainability Index (ESI).

The ELR compares the amount of non-renewable (N) and imported emergy (I) to the amount of locally available renewable emergy (R). In the case of a natural ecosystem, there would be no investments from outside and the system would be driven only by local renewable resources, thus having an ELR=0.

The EYR is a measure of the ability of a process to exploit and make available local resources by investing outside resources. The lowest possible value of the EYR is 1, situation in which the emergy converging to generate the yield does not differ significantly from the emergy invested from outside the system.

Finally, the ESI is an aggregated index calculated by the ratio of the EYR (sensitive to the outside-versus-local emergy use) and the ELR (sensitive to the non-renewable-versus-renewable emergy use) (Brown and Ulgiati, 2004).

Emergy flows can be converted into non-market "currency equivalents" by using the Emergy-to-Money Ratio (EMR) defined as the ratio between the total emergy (U) supporting a nation and its gross domestic product (GDP) in the same year (Lou and Ulgiati, 2013). This ratio represents the average amount of emergy needed to generate one unit of money in the national economy.

It is important to remark that, while the conversion of emergy units into money units can facilitate the communication of the results of emergy studies in political and socio-economic contexts, these figures expressed in money units should never be interpreted by applying a user-side perspective as they are non-market economic values.

Emergy Accounting and Marine Protected Areas

Emergy accounting has been widely applied to investigate natural, human-dominated, and man-made ecosystems. Many studies focused on the emergy assessment of natural capital and ecosystem services at local, regional, and national scales (Berrios et al., 2017; Campbell and Brown, 2012; Coscieme et al., 2014; Franzese et al., 2008, 2015; Huang et al., 2011; Lu et al., 2007, Ulgiati et al., 2011). Among them, recent studies focused on the assessment of matter and energy flows supporting the generation of natural capital stocks in marine protected areas. In particular, Vassallo et al. (2017) developed a biophysical and trophodynamic environmental accounting model based on emergy accounting and aimed at assessing the value of natural capital in marine protected areas. This model was articulated in three main steps: (1) trophodynamic analysis, providing an estimate of the primary productivity used to support the benthic trophic web within the study area; (2) biophysical accounting, providing an estimate of the biophysical value of natural capital; and (3) monetary conversion, expressing the biophysical value of natural capital into monetary units. This last step does not change the biophysical feature of the assessment, but instead it has the merit of allowing an easier understanding and effective communication of the ecological value of natural capital in socio-economic contexts.

Franzese et al. (2017) assessed the biophysical and economic value of natural capital in a Mediterranean marine protected area by using the emergy accounting method. In this study, the value of natural capital stocks in the main habitats included in the marine ecosystem was estimated in terms of work of biosphere invested for their generation and maintenance. In addition, using a GIS tool, the distribution of natural capital value in the protected area was made spatially explicit and integrated with the bionomic map and the zonation showing different protection levels.

The overlap of these different information layers allowed the identification of sites where natural capital value is more concentrated, supporting local managers and policy makers in charge for developing sustainable management strategies.

Picone et al. (2017) assessed the value of both autotrophic and heterotrophic natural capital stocks in a Mediterranean marine protected through emergy accounting. In addition, by using Marxan software, the results of the environmental accounting were integrated with spatial data on main human uses. This approach highlighted the importance of developing marine spatial planning taking into account both conservation measures and human activities devoted to resources exploitation.

Buonocore et al. (2019) jointly applied the emergy and eco-exergy methods to account for natural capital value in two marine protected areas located in Southern Italy. The emergy method allowed the assessment of natural capital in terms of direct and indirect solar energy flows invested for its generation, while the eco-exergy method accounted for the chemical energy stored in organic matter and the genetic information embodied in biomass stocks. This study showed that the sciaphilic hard bottom is characterized by a high emergy density value, confirming the high effort of nature in the formation of this habitat, suggesting the need for its proper consideration in the development of conservation strategies. Instead, *Posidonia oceanica* seagrass bed showed the highest value of eco-exergy density due to two main factors: (1) the large biomass density of *Posidonia oceanica* and (2) the high ß value of *Posidonia oceanica* reflecting its evolutionary history that involved the acquisition of key adaptations for the successful colonization of marine environments.

The integration between emergy and eco-exergy methods resulted in a useful approach to shed light on different biophysical measures of value based on different features of natural resources.

In all these studies, emergy accounting provided an alternative measure of value of natural resources complementary to more conventional economic studies. In addition, the integration of emergy accounting with other environmental accounting methods and tools allowed the generation of new scientific information useful for supporting the management of marine protected areas conceived to strike a balance between multiple human uses of the marine environment and nature conservation goals.

Concluding Remarks

Environmental accounting is a useful tool capable of addressing multiple aspects dealing with natural resources and their exploitation by human activities.

Among different environmental accounting methods, emergy accounting allows the assessment of nature's value applying a biophysical and geocentric perspective focusing on the work of biosphere supporting the generation of raw natural resources and refined products. The outcomes of emergy studies can complement those of more conventional economic analysis based on an anthropocentric perspective and on market laws.

A multicriteria approach to environmental accounting can address the plurality of nature values, supporting managers and policy makers in charge for sustaining the socio-economic development while ensuring the sustainable (long-term) management of natural resources.

References

Berrios F., Campbell D.E., Ortiz M., 2017. Emergy evaluation of benthic ecosystems influenced by upwelling in northern Chile: Contributions of the ecosystems to the regional economy. *Ecological Modelling* 359: 146–164.

Brown M.T., Ulgiati S., 2004. Emergy analysis and environmental accounting. In: *Encyclopedia of Energy*, C. Cleveland Editor, Academic Press: Elsevier, Oxford, UK, pp. 329–354.

Buonocore E., Picone F., Russo G.F., Franzese P.P., 2018. The scientific research on natural capital: A bibliometric network analysis. *Journal of Environmental Accounting and Management* 6(4): 374–384.

Buonocore, E., Picone, F., Donnarumma, L., Russo, G.F., Franzese, P.P., 2019. Modeling matter and energy flows in marine ecosystems using emergy and eco-exergy methods to account for natural capital value. *Ecological Modelling* 392: 137–146.

Campbell, E.T., Brown, M.T., 2012. Environmental accounting of natural capital and ecosystem services for the US National Forest System. *Environment, Development and Sustainability* 14: 691–724.

Cardinale, B. J., Duffy, J. E., Gonzalez, A., Hooper, D. U., Perrings, C., Venail, P., Narwani, A., Mace, G.M., Tilman, D., Wardle, D.A., Kinzig, A.P., Daily G.C., Loreau, M., Grace, J.B., Larigauderie, A., Srivastava, D.S., Naeem, S., 2012. Biodiversity loss and its impact on humanity. *Nature* 486: 59–67.

Coscieme, L., Pulselli, F.M., Marchettini, N., Sutton, P.C., Anderson, S., Sweeney, S., 2014. Emergy and ecosystem services: A national biogeographical assessment. *Ecosystem Services* 7: 152–159.

Costanza, R., de Groot, R., Sutton, P., van der Ploeg, S., Anderson, S.J., Kubiszewski, I., Farber, S., Turner, R.K., 2014. Changes in the global value of ecosystem services. *Global Environment Change* 26: 152–158.

EU, 2014. *Mapping and Assessment of Ecosystems and Their Services. Indicators for Ecosystem Assessment under Action 5 of the EU Biodiversity Strategy to 2020*. European Union. ISBN 978-92-79-36161-6.

Franzese, P.P., Brown, M.T., Ulgiati, S., 2014. Environmental accounting: Emergy, systems ecology, and ecological modelling. *Ecological Modelling* 271: 1–3.

Franzese, P.P., Buonocore, E., Donnarumma, L., Russo, G.F., 2017. Natural capital accounting in marine protected areas: The case of the Islands of Ventotene and S. Stefano (Central Italy). *Ecological Modelling* 360: 290–299.

Franzese, P.P., Buonocore, E., Paoli, C., Massa, F., Stefano, D., Fanciulli, G., Miccio, A., Mollica, E., Navone, A., Russo, G.F., Povero, P., Vassallo, P., 2015. Environmental accounting in marine protected areas: The EAMPA project. *Journal of Environmental Accounting and Management* 3(4): 324–332.

Franzese, P.P., Liu, G., Aricò, S., 2019. Environmental accounting models and nature conservation strategies. *Ecological Modelling* 397: 36–38.

Franzese, P.P., Russo, G.F., Ulgiati S., 2008. Modelling the interplay of environment, economy and resources in marine protected areas. A case study in Southern Italy. *Ecological Questions* 10: 91–97.

Franzese, P.P., Rydberg, T., Russo, G.F., Ulgiati, S., 2009. Sustainable biomass production: A comparison between gross energy requirement and emergy synthesis methods. *Ecological Indicators* 9(5): 959–970.

Häyhä, T., Franzese, P.P., 2014. Ecosystem services assessment: A review under an ecological-economic and systems perspective. *Ecological Modelling* 289: 124–132.

Huang, S., Chen, Y., Kuo, F., Wang, S., 2011. Emergy-based evaluation of peri-urban ecosystem services. *Ecological Complexity* 8: 38–50.

Hoffmann, S., Beierkuhnlein, C., Field, R., Provenzale, A., Chiarucci, A., 2018. Uniqueness of protected areas for conservation strategies in the European Union. *Scientific Reports* 8: 6445.

Lou, B., Ulgiati, S., 2013. Identifying the environmental support and constraints to the Chinese economic growth – An application of the emergy accounting method. *Energy Policy* 55: 217–233.

Lu, H., Campbell, D., Chen, J., Qin, P., Ren, H., 2007. Conservation and economic viability of nature reserves: An emergy evaluation of the Yancheng Biosphere Reserve. *Biological Conservation* 139: 415–438.

Maes, J., Liquete, C., Teller, A., Erhard, M., Luisa, M., Barredo, J. I., Grizzetti, B., Cardoso, A., Somma, F., Petersen, J., Meiner, A., Gelabert, E.R., Zal, N., Kristensen, P., Bastrup-Birk, A., Biala,K., Piroddi, C., Egoh, B., Degeorges, P., Fiorina, C., Santos-Martín, F., Naruševičius, V., Verboven, J., Pereira, H. M., Bengtsson, J., Gocheva,K., Marta-Pedroso, C., Snäll, T., Estreguil, C., San-Miguel-Ayanz, J., Pérez-Soba, M., Grêt-Regamey, A., Lillebø, A.I., Malak, D.A., Condé, S., Moen, J., Czúcz, B., Drakou, E.G., Zulian, G., Lavalle, C., 2016. An indicator framework for assessing ecosystem services in support of the EU biodiversity strategy to 2020. *Ecosystem Services* 17: 14–23.

Neugarten, R.A., Langhammer, P.F., Osipova, E., Bagstad, K.J., Bhagabati, N., Butchart, S.H.M., Dudley, N., Elliott, V., Gerber, L.R., Gutierrez Arrellano, C., Ivanić, K.-Z., Kettunen, M., Mandle, L., Merriman, J.C., Mulligan, M., Peh, K.S.-H., Raudsepp-Hearne, C., Semmens, D.J., Stolton, S., Willcock, S., 2018. *Tools for Measuring, Modelling, and Valuing Ecosystem Services: Guidance for Key Biodiversity Areas, Natural World Heritage Sites, and Protected Areas.* IUCN: Gland, Switzerland, 70 pp.

Odum, H.T., 1988. Self organization, transformity and information. *Science* 242: 1132–1139.

Odum, H.T., 1996. *Environmental Accounting: Emergy and Environmental Decision Making.* John Wiley & Sons: New York, 369 pp.

Pauna, V.H., Picone, F., Le Guyader, G., Buonocore, E., Franzese, P.P., 2018. The scientific research on ecosystem services: A bibliometric analysis. *Ecological Questions* 29 (3): 53–62.

Picone, F., Buonocore, E., D'Agostaro, R., Donati, S., Chemello, R., Franzese, P.P., 2017. Integrating natural capital assessment and marine spatial planning: A case study in the Mediterranean sea. *Ecological Modelling* 361: 1–13.

TEEB, 2010. *The Economics of Ecosystems and Biodiversity for Local and Regional Policy Makers.* TEEB: Bonn, Germany.

Teixeira, H., Lillebø, A.I., Culhane, F., Robinson, L., Trauner, D., Borgwardt, F., Kuemmerlen, M., Barbosa, A., McDonald, H., Funk, A., O'Higgins,T., Tjalling, J., der Wal, V., Piet, G., Hein, T., Arévalo-Torres, J., Iglesias-Campos, A., Barbière,J., Nogueira, A.J.A., 2019. Linking biodiversity to ecosystem services supply: Patterns across aquatic ecosystems. *Science of the Total Environment* 657: 517–534.

Ulgiati, S., Zucaro, A., Franzese, P.P., 2011. Shared wealth or nobody's land? The worth of natural capital and ecosystem services. *Ecological Economics* 70 (4): 778–787.

UNEP-WCMC, IUCN and NGS, 2018. *Protected Planet Report 2018*. UNEP-WCMC, IUCN and NGS: Cambridge UK; Gland, Switzerland; and Washington, DC.

Vassallo, P., Paoli, C., Buonocore, E., Franzese, P.P., Russo, G.F., Povero, P., 2017. Assessing the value of natural capital in marine protected areas: A biophysical and trophodynamic environmental accounting model. *Ecological Modelling* 355: 12–17.

Vihervaara, P., Franzese, P.P., Buonocore, E., 2019. Information, energy, and eco-exergy as indicators of ecosystem complexity. *Ecological Modelling* 395: 23–27.

25

Remote Sensing: Pollution

Massimo
Antoninetti

Introduction

The earth's environment is placed under constant stress by various human activities and natural processes.

Resource management activities for conservation or sustainable use rely on several critical components: accurate and up-to-date information, an understanding of "how" the environment and its components function, and an ability to continuously and constantly monitor the environment's components and processes. Consequently, there is a need for spatial information in the form of baseline mapping and inventory, monitoring programs, and predictive models. Satellite remote sensing (RS) is going to play a major role in fields related to place and space. Thanks to this innovative technology, it is possible to monitor and map rapidly changing phenomena on the surface of our planet due to its broad spectral range, affordable cost, and rapid coverage of large areas. Synergistic use of RS and ancillary data can create a geographic information system (GIS) database, which can be used to store, process, and retrieve environmental data.

This is not a comprehensive review, as many entries have been published on the topics covered. We have attempted to emphasize a few important applications, recent reviews, and recent papers that provide the reader with an overview of the main literature on each topic.

Remote Sensing

Remote sensing is generally defined as the set of techniques, instruments, and interpretative tools able to expand and improve the perceptive ability of the human eye, collecting information about an object without being in physical contact with the object itself.[1] Consequently, RS allows the obtaining of quantitative and qualitative measures from a distance of the surface of objects, and, under some specific conditions, also of the subsurface. The most frequently measured quantity is the electromagnetic energy in one or more regions of the electromagnetic spectrum reflected or emitted from the earth's surface.[2] The primary law governing radiation is the Planck radiation law, which governs the intensity of radiation emitted by unit surface area into a fixed direction (solid angle) from a blackbody as a function of wavelength for a fixed temperature. Conceptually, a blackbody is an ideal radiator of thermal radiant energy, and consequently, also an ideal absorber. The Planck law can be expressed through the following equation:

$$M_\lambda = \frac{\varepsilon C_1}{\lambda^5 \left[\exp\left(\dfrac{C_2}{\lambda T} \right) - 1 \right]}$$

(1)

where M_λ is the spectral radiant exitance (W cm^{-2}µm^{-1}); ε = emittance (emissivity), dimensionless; C_1 is the first radiation constant, 3.7413×10^8 [W (µm)^4]/m^2; C_2 is the second radiation constant, 1.4388×10^8 (µm K); λ is the radiation wavelength (µm); and T is the absolute radiant temperature (K).

Moreover, the Stefan–Boltzmann law states that the total amount of energy M emitted by a blackbody is proportional to the fourth power of its absolute temperature T, and consequently M rapidly increases with the T increasing:

$$M = \int_0^\infty M_\lambda d\lambda = \sigma T^4 \left[Wcm^{-2} \right]$$

(2)

where σ is the Stefan–Boltzmann constant ($\sigma = 5.67 \times 10^{-4}$ W cm^{-2} K^{-4}).

However, there is an inverse relationship between the maximum wavelength emission λ_{max} of a blackbody and its temperature when expressed as a function of wavelength (Wien radiation law):

$$\lambda_{max} = \frac{2898}{T} [\mu m]$$

(3)

The Wien law demonstrates that the λ_{max} moves toward lower values at the increasing of the surface temperature T of a body. For example, the maximum emission value for the sun is 0.483 µm (corresponding to the blue-green part of the visible spectrum), 0.966 µm for a incandescent bulb (near infrared [NIR]), and 9.66 µm for the surface of the earth (thermic infrared).

Consequently, for any given material, the amount of solar radiation that it reflects, absorbs, transmits, or emits varies with wavelength. When that amount (usually intensity, as a percent of maximum) coming from the material is plotted over a range of wavelengths, the connected points produce a curve called the material's spectral signature (spectral response curve) (Figure 1). For example, at some wavelengths,

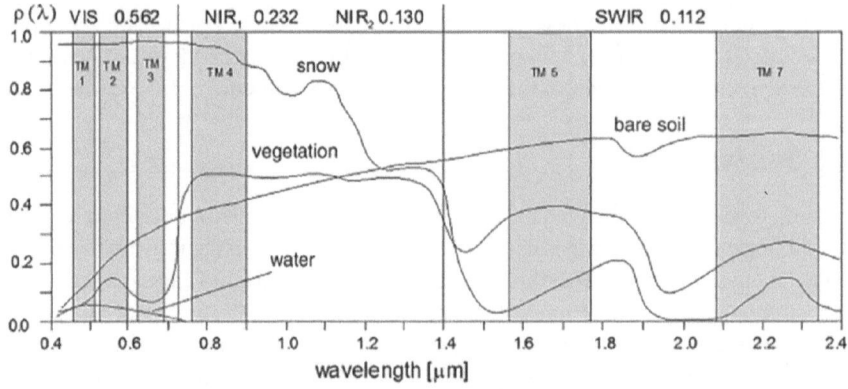

FIGURE 1 Schematization of solar radiation spectral distribution of typical surfaces' spectral reflectivity curves, and of Landsat TM spectral bands.
Source: Brivio et al.[3]

sand reflects more energy than green vegetation; however, at other wavelengths, it absorbs more (reflects less) than the vegetation.

There are two main types of RS using different sensors to register the data: passive RS and active RS. A sensor can be defined as a device that measures a physical quantity and converts it into a signal that can be read by an observer or by an instrument. This physical quantity is the radiance (W m^{-2} sr^{-1}) or the spectral radiance (W m^{-2} sr^{-1} λ^{-1}); they are radiometric measures that describe the amount of electromagnetic energy that passes through or is emitted from a particular area, and falls within a given solid angle in a specified direction. They characterize both emission from diffuse sources and reflection from diffuse surfaces.

Passive sensors detect the natural radiation emitted or reflected by the object. Reflected sunlight is the most common source of radiation measured by passive sensors. Active sensors emit energy in order to scan objects and areas, detecting and measuring the radiation that is reflected or backscattered from the target. RADAR (radio detection and ranging) is an example of active RS where the time delay between emission and return is measured, establishing the location, height, speed, and direction of an object (Figure 2). In most cases, the instruments register a raster image. The image contains a fixed number of rows and columns of pixels, as the smallest individual picture element in the image (Figure 3).

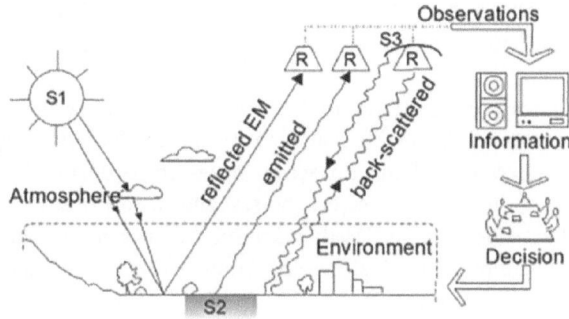

FIGURE 2 Remote sensing basic scheme. Natural energy sources are the sun (S1) and the earth (S2), while the antenna (S3) is an artificial source. Sensor R registers the information carried by electromagnetic waves. This information is stored and elaborated to be used for environmental studies.
Source: Modified from Brivio et al.[3]

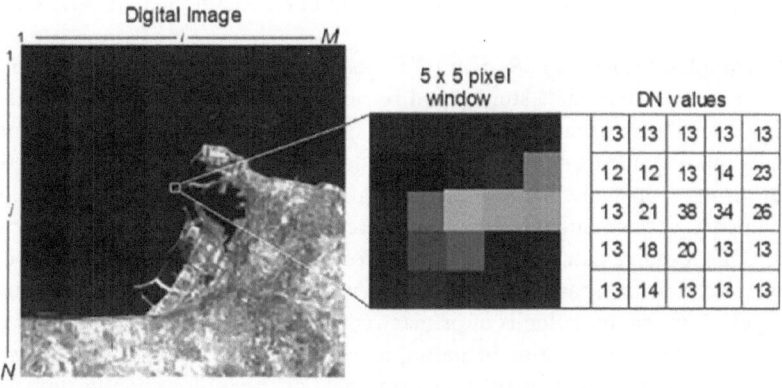

FIGURE 3 Digital image concept and description. A digital image is formed by a matrix of $N \leftrightarrow M$ pixels, easily visible only with an enlargement in the window 5 $\leftrightarrow$ 5. A digital number (DN) is assigned to each pixel in function of the intensity of the electromagnetic energy registered by sensors.
Source: Modified from Brivio et al.[3]

Each pixel holds a quantized value representing the amount of electromagnetic energy registered by instruments at any specific point. Consequently, the quality of RS image data consists of its spatial, spectral, radiometric, and temporal resolutions. The spatial resolution represents the size of the surface area (i.e., in square meters) being measured on the ground, determined by the sensors' instantaneous field of view, from less than 1 m to more than 1 km. A higher resolution means more image details.

The spectral resolution is the wavelength interval size (discreet segment of the electromagnetic spectrum) recorded by the instruments mounted on board, normally a radiometer, able to measure the radiant flux of electromagnetic radiation. A multispectral radiometer collects electromagnetic energy in different spectral bands, normally from the visible (0.4 μm) to the long-wave infrared (15 μm) part of the spectrum (e.g., Enhanced Thematic Mapper Plus [ETM+] mounted on board of the Landsat satellite has 8 bands recorded), while a hyperspectral imaging deals with hundreds of narrow spectral bands, producing a spectra of all pixels in the scene (e.g., Hyperion sensor on Earth Observing-1 has 220 bands with a spectral resolution of 0.10–0.11 μm per band). The radiometric resolution is the number of different intensities of radiation that the sensor is able to distinguish. Typically, this ranges from 8 to 14 bits, corresponding to the 256 levels of the gray scale and up to 16,384 intensities or "shades" of color in each band.

Finally, temporal resolution is defined as the frequency at which images are recorded/captured in a specific place on the earth's surface (National Oceanic and Atmospheric Administration Advanced Very High Resolution Radiometer [NOAA-AVHRR] capture an image every 12 hr, while Landsat ETM+ every 16 days) (Table 1).

Remote sensing data are processed and analyzed with a digital image processing system, using computer algorithms to improve the quality and quantity of the information registered. As an image is formed by an array of numbers, digital image processing uses simple or complex mathematic algorithms to obtain image restoration; to eliminate errors, noises, and distortions; to improve the information content extraction; and to recognize and classify pixels on the basis of their digital signatures.

A simple process, for example, is the "false-color" composition. Three images acquired in different spectral bands could be combined together to obtain an RGB (red–green–blue) color image where each color tone is the combination of the information contained in each band. The term "false color" identifies images whose colors represent measured intensities outside the human eye capacity.

A complex mathematical model is an artificial neural network, inspired by the structure of physically interconnected neurons in biology and able to process information using a connectionist approach through an interconnected group of artificial neurons. It provides an efficient alternative to map complex nonlinear relationships between input and output datasets without requiring a detailed knowledge of underlying physical relationships. **Table 1** List of main satellites with on-board sensors description

Finally, Geographic Information System (GIS) is a database system used for manipulating data, utilizing a data input system, data storage and retrieval, data manipulation and analysis, and data output. The data inputs are usually spatial and non-spatial data derived from a combination of existing maps, aerial photographs, and RS imagery. The data output is used to generate reports in tabular form, digital displays, or maps. Because the input data on which a GIS is based becomes obsolete quickly, it is important to update this data. Remote sensing is often the most cost-effective source for these updates. The resource database should contain as much base inventory data as possible, as well as simple layer selection and spatial and tabular query capabilities, to illustrate how this data is accessible and useful. Data manipulation is of primary concern, including the tracking of metadata that describe the legacy of the information including source, compilations scale, data of last compilation, and other information needed to qualify the usefulness of the information to support a particular application.

What goes beyond a GIS is the Spatial Data Infrastructure, tools that are interactively connected on the web for retrieving, presenting, and processing data in an efficient and flexible way, in a coordinated series of agreements on technology standards, institutional arrangements, and policies.

Remote Sensing Applications

Water

Although not all deleterious processes can be measured directly (e.g., overfishing), many environmental and ecological properties can be measured using RS. These properties include sea surface temperature (SST), chlorophyll-*a* (Chl-*a*) and suspended sediment concentration, salinity, wind speed, algal blooms, etc. Given a robust understanding of the ecosystem responses to these environmental parameters and some in situ field observations, many other biological benchmarks can often be indirectly derived (e.g., fish abundance).[4]

In the open waters, color variations chiefly depend on the presence and abundance of plankton: these waters are defined as Case I. Conversely, coastal and estuarine systems are often characterized by optically complex waters (Case II), with high concentrations of highly reflective water constituents, the color of which is also due to dissolved organic matter and suspended sediments. The constituents, characterized by variation at much smaller geographical scales, are able to modify natural equilibrium.[5]

In recent years, coastal zones, probably more than any other parts of the earth, have been exposed to pressure and processes of change caused by natural processes (waves, currents and storms, etc.), and human activities.[6] As a large percentage of the world's coastlines are poorly mapped or maps have become outdated, satellite imagery can be a useful and economic tool to overcome these constraints, producing maps containing thematic information of relevant interest in coastal environment protection: planimetric mapping of waterlines and beach and coral reef position, shallow water bathymetry, and thematic maps of cover types on adjacent land and islands. This may include mapping of bare sand, mangroves, seagrass, reef flats, wetlands, etc. The number of classes (categories) (e.g., coral reef, seagrass, sand, hard substrate) distinguishable by RS depends on many natural factors (including atmospheric clarity, surface roughness, water clarity, and water depth) and on the type of RS sensor. Very high resolution pixels (0.1–0.8 m) are required to accurately quantify the percentage of bleached corals. The most likely solution to this limitation of pixel size is the application of methods that estimate the cover of substrata within pixels. These methods known as spectral unmixing, even though they have been developed for terrestrial RS, could be adapted to aquatic environment, taking into account the interference of water column.[4]

In shallow water, the water column (including the substances dissolved or in suspension in the medium) and the seafloor contribute to the upwelling signal leaving the water body. An inverse modeling of the radiation transfer can be developed for simultaneous deriving of seafloor reflectance, depth, water-column subsurface reflectance, and diffuse attenuation coefficient from RS of coastal shallow waters.[7] This method relies on a synthetic database of subsurface irradiance reflectance-compiled field and simulated data and the use of a forward shallow-water reflectance model.[8] Uncertainties in the computation of Chl-*a* absolute values can arise primarily owing to the presence, in the water column, of optically active materials other than phytoplankton and related pigments (i.e., dissolved organic matter and suspended inorganic particles), with partially overlapping spectral signatures. Nevertheless, experience has shown that remotely sensed data can provide unique information on phytoplankton growth patterns over a range of space scales (i.e., from a few kilometers to entire basins) and of time scales (i.e., from days to years) not available by any other means.[9]

Ultraviolet (UV) radiation and high levels of photosynthetically active radiation can have a variety of negative impacts on marine phytoplankton, zooplankton, nekton, and benthos. Satellites provide global time series measurements of incident UV radiation.

Understanding the way in which cetaceans interact with the surrounding environment has become a key tool, not only in their subsequent conservation but also in the management of the ecosystems of which they form a part. Due to their high trophic status and vulnerability to anthropogenic effects, marine mammals such as cetaceans are ideally suited to act as indicator species of ecosystem change and anthropogenic threats.[10] Cetacean studies have focused on their relationships with sea temperature.

Thermal infrared SST measurements are derived from radiometric observations at wavelengths of ~3.7 µm and/or near 10 µm. Although the 3.7 µm channel is more sensitive to SST, it is primarily used only for night-time measurements because of relatively strong reflection of solar irradiation in this wavelength region, which contaminates the retrieved radiation. Both bands are sensitive to the presence of clouds and scattering by aerosols and atmospheric water vapor. For this reason, thermal infrared measurements of SST first require atmospheric correction of the retrieved signal and can only be made for cloud-free pixels. Thus, maps of SST compiled from thermal infrared measurements are often weekly or monthly composites, which allow enough time to capture cloud-free pixels over a region. Thermal infrared instruments that have been used for deriving SST include AVHRR on NOAA Polar-Orbiting Operational Environmental Satellites (POES), Along-Track Scanning Radiometer (ATSR) aboard the European Remote Sensing Satellite (ERS-2), the Geostationary Operational Environmental Satellite (GOES) Imager, and Moderate Resolution Imaging Spectroradiometer (MODIS) aboard NASA Earth Observing System (EOS) Terra and Aqua satellites.

Images supplied by the NERC Remote Sensing Data Analysis Service (RSDAS) were atmospherically corrected and provided in a Mercatore projection. Values or digital numbers (DN) associated with each pixel of the AVHRR images supplied were converted into real SST (°C) values using the following equation as supplied by the RSDAS[10]:

$$SST = DN \times 0.1 - 0.3 \tag{4}$$

where SST is the sea surface temperature (°C) and DN is the digital number or the value of each pixel.

Chlorophyll-*a* concentration data could be obtained from the Sea-Viewing Wide Field-of-View Sensor (Sea-WiFS) monthly composite images using the following equation, also supplied by RSDAS:

$$CHL = 10^{(DN \times 0.015 - 2.0)} \tag{5}$$

Where CHL is the Chl-*a* concentration (µg L^{-1}).

Higher numbers of striped dolphins were generally observed in the 21–24°C range in the Mediterranean.[11]

Environmental protection from spills of hydrocarbon compound over the sea surface is currently an important subject of increasing public concern. Oil spills, depending on the exact hydrocarbon content and type, usually involve extensive areas of film on sea surface, a fact that reduces water roughness and can therefore allow the detection by synthetic aperture radar (SAR) images, providing information on location, size, distance from the land, etc.[12,13] The automatic detection of oil slicks in SAR images is not a simple task, as objects resembling oil spills (often called look-alikes) occur frequently in SAR images, especially in low wind conditions. Most frequently, look-alikes are produced by organic film, grease, wind front areas, land, plankton formations, rain cell, current shear zones, and upwelling zones. The probability of each object extracted after image classification to be an oil spill was estimated using an artificial intelligence fuzzy logic modeling system. The fuzzy logic theory has emerged as a useful tool for modeling processes that are too complex for conventional quantitative techniques, or when the available information from the process is qualitative, inexact, or uncertain. Fuzzy logic addresses qualitative information perfectly as it resembles the way humans make inferences and decisions.[14] It fills an important gap in system design methods, which is between purely mathematical approaches (e.g., system design) and purely logic-based approaches (e.g., expert systems). While other approaches require accurate equations to model real-world behaviors, fuzzy design can accommodate the ambiguities of real-world human language and logic. It provides an intuitive method for describing systems in human terms and automates the conversion of those system specifications into effective models. Traditional set theory is based on bivalent logic, where an object is either a member of a set or not. Contrary to that, fuzzy logic allows a number or object to be a member of more than one set and most importantly it introduces the notion of partial membership.[15]

Most of the oil spill detection studies use low-resolution SAR data (quick-looks) with nominal spatial resolution of 100 m×100 m to detect oil spills. Low-resolution data is sufficient for large-scale monitoring; however, small and fresh spills cannot be detected sufficiently as they are represented in very few pixels and present brighter contrast than the bigger and older oil spills. Neural networks are able to detect and classify dark formations. A neural network's ability to successfully handle nonlinearly separable classes is a big advantage against the commonly used statistical approaches.[16] The exploitation of optical satellite images allows more frequent (sometimes daily) information if compared with SAR images. However, good weather conditions and daylight are mandatory conditions for correct detection. The physical and geometrical features of an oil spill are computed by the feature extraction module for an automatic supervised classification: geometrical features (area, perimeter, and complexity of the object) and gray level features (object standard deviation, maximum contrast, mean contrast).[17] Due to its thermal inertia (lower than the sea water), oil polluted areas have a higher brightness temperature in AVHRR thermal infrared (TIR) images collected in daytime, and the opposite during the night. In fact, materials having high thermal inertia show resistance toward a change in temperature, resulting in long-term temperature stability. In daytime, an oil spill should exhibit a relatively higher brightness temperature in AVHRR TIR images as, due to solar radiation, oil becomes warmer than the surrounding water.[18] Finally, GIS can qualitatively and quantitatively characterize not only spatial and temporal distribution of oil spills but also the environmental conditions of the sea basins as a whole. Such an environment can be created by means of integration in a GIS of different databases for seawater quality, nutrient and chemical composition, and contextual information about slick position relative to surrounding objects (ships, ship lanes, rigs, platforms, natural seeps).[19,20]

Groundwater is one of the most valuable natural resources, supporting human health, economic development, and ecological diversity. Nevertheless, it is considered to be vulnerable mainly due to anthropogenic activities such as agriculture and waste disposal. Satellite data provides quick and useful baseline information about factors controlling the occurrence and movement of groundwater, such as geology, lithology, geomorphology, soils, land use/cover, drainage patterns, and lineaments. Remote sensing helps minimize the amount of field data collection in the exploration and assessment of groundwater resources, in estimation of natural recharge distribution, and in subsurface flow and pollution modeling.[21]

A pollution risk map can be obtained combined a vulnerability map with a land use map in a GIS. The vulnerability map illustrates the potential decrease of groundwater protection as a result of the flow regime toward the aquifer, analyzing the hydrogeologic and geomorphologic conditions of the basin and the isotopic ratio$^{18}O/^{16}O$ measured in the collected water samples (springs, rivers, lakes). The updated land use map is produced by satellite image data set analysis. The socioeconomic component is introduced to integrate the whole process leading to the assessment of aquifer's pollution risk, a very important element for decision making and planning.[22]

Agricultural activities have been identified as major sources of non-point source pollutants of ground and surface waters, with nitrogen being one of the most important and problematic nutrients. The use of mathematical models is a common approach for analyzing and describing the status quo, identifying interdependencies between agriculture and hydrosphere, and investigating the effects of agricultural environmental reduction measures. Satellite imagery has been used to substitute low-resolution, outdated, and inadequately differentiated land use maps with high-resolution land cover and imperviousness maps. By applying crop-specific classified satellite imagery in a semiautomated way, the nitrogen surpluses can be transferred and spatially located to main crops on the field scale.[23]

Wetlands and lakes are very sensitive ecosystems that play a key role in maintaining ecological equilibrium and biodiversity. They provide a habitat for wildlife, support a rich biodiversity, and aid in floodwater management and water quality improvement besides having esthetic and educational benefits for humans. Moreover, local economies depend on the water and vegetation of wetlands and lakes for fisheries, reed harvesting, grazing, and recreation. Unfortunately, in many countries, the extensive loss of wetlands and trophic pollution of lakes have occurred. Imaging spectrometry is a versatile tool for assessing water quality in small lakes, monitoring the variation of submerged vegetation in coastal zones, and performing ad hoc studies on transitional ecosystems as lagoons. The Chl-a and the TSS maps, obtained by combining MIVIS hyperspectral airborne data with semi-empirical modeling, confirmed the hypereutrophic-dystrophic conditions of the lakes and the elevated load of nutrient and suspended matter transported by the tributary rivers (Figure 4). The surface temperature map also revealed the discharge of warmer waters by urbanized and industrialized areas.[24] Moreover, MIVIS data demonstrates a good capacity in evaluating variations in submerged vegetation (Figure 5). Macrophyte meadows are in fact indicators of water and sediment quality, as their disappearance from many freshwater and coastal marine environments can be attributed mainly to the phenomena of eutrophication.[25] Once the models are calibrated with in situ measurements, bio-optical models may be applicable over selected lakes. They are mathematical equations that relate radiometric variables observed above or below the water surface (e.g., satellite-derived water reflectance) to the inherent optical properties (i.e., absorption and backscattering coefficients of colored dissolved organic matter and particle).[26]

In recent decades, common reeds (*Phragmites australis*) have been monitored for various purposes, including wetland management for nature conservation. Common reeds are an ecosystem with a high value of biodiversity, ideal for fish, amphibian, and bird reproduction; moreover, they function as a cleaning filter for the lake waters as they can remove up to 10%–15% of the sediment nitrogen content and subtract significant amounts of heavy metals. Lake water eutrophication has been identified as the main reason for reed die-back, while in the hotter and drier Mediterranean area, *P. australis* seems to

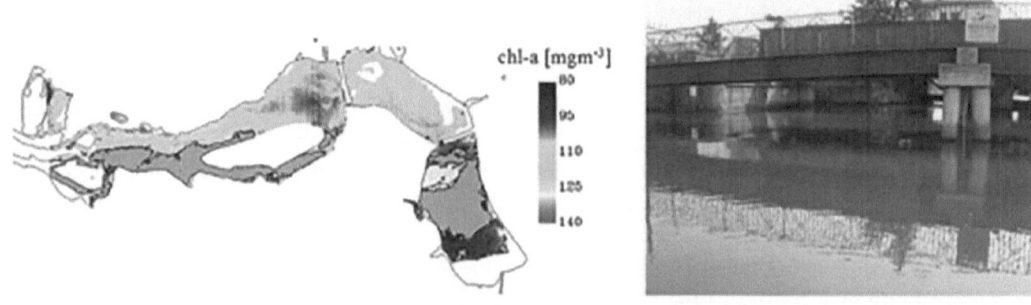

FIGURE 4   MIVIS-derived Chl-*a* maps of three lakes of Mantua, Italy. The high value concentrations are due to eutrophication caused by excessive nutrients from agricultural sources that color of green water.
Source: IREA-CNR, Milan, Italy.

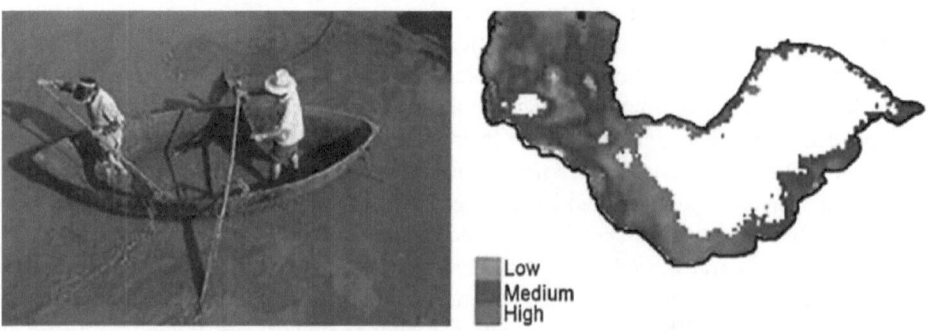

FIGURE 5   Left: Algae bloom situation at Hefei Yicheng, China, on June. Right: Probability of finding phycocyanin pigments in the algal bloom observed for the scene acquired by MERIS on May (in white, the water types in better status).
Source: IREA-CNR, Milan, Italy.

grow vigorously and to expand, even in eutrophicated areas. High normalized difference vegetation index (NDVI) values highlight greater amount of reed biomass that is the result of the higher germinative capacity in optimal environmental conditions.[27]

Land

As for land, ecosystem vegetation plays a fundamental role in the energy and mass exchange between land surface and atmosphere, and can be considered a good indicator of environmental stress. Environmental health conditions, stand structure, damage levels, and other phenomena can be determined using satellite images.[28] The spectral vegetation indices are used to investigate changes in plant stress due to soil contamination. A vegetation index combines two or more spectral bands to enhance the vegetative signal while minimizing background effects; they are commonly used to measure the sensitivity of vegetation to stress. As hyperspectral RS images are increasingly becoming available, further exploration of the increased number of spectral variables is a logical step.[29]

Canopy leaf area index (LAI) is typically defined as the one-sided area of green foliage projected onto a unit area of ground. The LAI is critical in estimating terrestrial carbon sequestration, net primary production, hydrologic watershed budgets, and pollutant deposition. The LAI retrieval method is based on an analytical solution to a multiple scattering equation. The algorithm uses different input data layers to produce a single LAI map: reflectance in the red and NIR bands, satellite zenith view angle, solar zenith altitude angle, sun sensor relative azimuth, and terrain characteristics (elevation, slope, and aspect).[30] A simplified approach uses a semi-empirical regression model between ground measurements and spectral vegetation indices (VIs), such as band ratios or normalized band difference, computed from remotely sensed images.[31,32]

Atmospheric composition affects reflectivity of the spectrum and makes it difficult to compare satellite images from different seasons. To reduce impacts from the atmosphere, the normalized ratio vegetation index (NRVI) has been applied to satellite imagery to study Chl-*a* concentration.

For Landsat ETM/TM, TM3 is a main band for absorbing the spectrum from Chl-*a* and is represented by the red band with a wavelength range varying from 630 to 690 nm. TM4 is an NIR band with a wavelength range varying from 760 to 900 nm. The band is sensitive to the category of green vegetation and is therefore often used for detecting green vegetation. Based on the characteristics of the two bands, three VIs could be selected for RS analysis: NDVI, ratio vegetation index (RVI), and NRVI, all of which are relatively simple in arithmetic and widely used.

The equation of the RVI is

$$RVI = \frac{TM3}{TM4} \tag{6}$$

The equation of the NDVI is

$$NDVI = \frac{TM4 - TM3}{TM4 + TM3} \tag{7}$$

The equation of the NRVI is

$$NRVI = \frac{RVI - 1}{RVI + 1} \tag{8}$$

The NRVI has a positive correlation with Chl-*a* concentration, while NDVI has a negative correlation. Correlative degrees between NRVI and Chl-*a* are also much higher than that of NDVI, showing that the obtained regressions capture the relation between the water quality parameters (nutrient pollution loaded mainly with agricultural wastewater, phosphorous pollution loaded mainly with runoff from forest areas, pollution loaded with a mixture of domestic and industrial wastewater, and agricultural runoff and natural lake water) and the band reflectance values.[33]

Change in leaf spectral reflectance is a symptom of deteriorating forest health. Increase in red reflectance due to reduced chlorophyll absorption, decrease in NIR reflectance due to reduced cell vigor, and shifts in the red edge between these two regions have been commonly used as indicators of leaf stress.[34]

Topography also plays an important role in vegetation sensitivity to air pollution. In general, elevated and wind-ward sites appear to be more affected than depressed and leeward locations. Consequently, wind direction could apparently be more important than topography.[35] Using RS techniques to achieve a detailed analysis of the type and status of vegetation can help us understand the fate of persistent organic pollutants (POPs) in the environment. These POPs are generally bioaccumulated in the food chain and affect living organisms owing to their long persistence and toxicity. Forests play a key role in the environmental distribution of POPs since they act as filters of these chemicals, trapping them in the air compartment and transferring them to forest soils, consequently decreasing their atmospheric half-lives. Therefore, it is necessary to include the vegetation compartment in multimedia models, to understand and predict the fate of these substances in the environment. Multimedia models used to study the fate of POPs and their redistribution in different environmental compartments, from the atmosphere to soil or water via vegetation interface, require foliage biomass information. Foliage biomass (FB) is estimated with the following model:

$$FB_i = LAI_i \times \frac{1}{SLA_i} \times \left(\frac{FW}{DW} \right)_i \tag{9}$$

where FB_i is the foliage biomass of vegetation class i (kg m^{-2}), LAI_i, is the leaf area index of vegetation class i, SLA_i, is the specific leaf area (m^{2}kg^{-1}) of vegetation class i, and $(FW/DW)_i$, is the relative ratio between fresh and dry biomass weight of vegetation class i.[36]

Natural fires and arson greatly affect forests, vegetation, and the environment in general (Figure 6) Mercury (Hg) emissions from fires may be also detected directly using reflectance spectroscopy, field

FIGURE 6 False-color Landsat-7 images taken over Ukraine on July 2001, illustrating the different possibilities in identifying agriculture fires. Left: RGB natural color composition of visible band highlights fires smoke, its intensity and direction, and burned areas. Right: RGB false color composition of visible, and infrared bands, identifies active fires and burned areas.
Source: IREA-CNR, Milan, Italy.

sampling, and laboratory analysis; however, many contaminated regions are covered by vegetation. The Hg released to the atmosphere from vegetation is primarily related to Hg concentrations in foliage. It varies with tissues of plants—predominantly leaves, bark, and root—foliar age, forest typologies, and soil characteristics.

Satellite observations were used to assess Hg emissions through the assessment of spatial and temporal distributions of forest phytomass content and forest burnt areas.[37] Plants with the greatest amount of foliar Hg showed symptoms of stress earlier than plants of other treatments. Foliar Hg appears to influence spectral reflectance at several specific regions of the electromagnetic spectrum. The red edge position is the wavelength of the maximum value of the first derivative of the spectra in the 680–740 nm region and is positively related to chlorophyll concentration and biomass of leaves and canopies.[38] Shifts in vegetation spectra, occurring in both the visible and the NIR part of the spectrum, could be metal induced due to geochemical stress or the presence of old waste deposit sites.

Mining activities, such as coal, are often responsible for environmental pollution. Consequently, the forest patches near the mining sites have been badly degraded, whereas patches away from the mining sites registered relatively less impact. The degree of degradation was based on density, canopy structure, ground flora, and soil conditions (presence of ash and pollutants). The prime reason of these changes can be attributed to the discharge of noxious/toxic gases and particulate matter during drilling, blasting, and loading/unloading operations. It has been recorded that the areas near the industrial area and affected areas hold fewer plant species than that of areas far away from these site. Variation in number of herbs per unit area was found to decrease significantly when approaching mining sites.[39]

Urban green spaces provide a variety of functions essential for improving the quality of citizen life, while urban heat island tends to aggravate the negative effects of climate over urban areas and their surroundings.

Using NDVI multitemporal image differencing it is possible to detect subtle changes and to discern "from–to" changes using a postclassification comparison.[40]

Land surface temperature (LST) retrieved by RS is an important parameter in characterizing surface energy balance and a key parameter in monitoring farming drought, and very significant in studies of global change, meteorology, hydrology, ecology, agricultural, etc. Results reveal that LST is significantly different in various types of land use, the highest is bare land; the second is built-up land; followed by cropland, grass, and forest; and at the lowest in water.[41]

City growth is increasing population and the change of land cover types from permeable to anthropogenic impervious surfaces. Impervious surface area (ISA) is defined as constructed surfaces—roofs, roads, parking lots, driveways, and sidewalks. Impervious surfaces can alter the natural hydrological condition by increasing the volume and rate of surface runoff and decreasing groundwater recharge and base flow. Pollutants either dissolved or suspended in water, or associated with sediment including nutrients, heavy metals, and oil and grease, can accumulate and wash away from ISAs. The percentage of impervious surface estimated from satellite RS data can be used to assess the spatial extent of urban land use, as well as urban development density.[42]

Building energy saving is defined as the rational use of energy and constant improvement of energy efficiency under conditions that guarantee and improve the comfort of the building. Using Landsat TM thermal data, surface radiance values for each scene pixel were calculated with the use of calibration coefficients to identify build structure, building material, vegetation, and heating methods.[43]

Night light emissions that originate mainly from large urban areas are among the main elements of environmental pollution: disturbance of biological rhythms, psychological effects, and environmental degradation. The Defense Meteorological Satellite Program, run by the NOAA, has the capability to detect faint sources of visible near infrared (VNIR) emissions on the earth's surface, making it possible to detect cities and towns. Modern GIS packages offer advanced tools for visibility analysis, including line-of-sight estimations and viewshed mapping. The viewshed identifies the cells in an input raster that can be seen from one or more observation points. Moreover, it is possible to identify areas affected with indirect light pollution.[44]

Unlike the United States, where the use of asbestos is still legal but tightly controlled, on January 1, 2005 (following the directive 76/769/CEE), the marketing and use of products containing asbestos was banned through-out the European Union. Prohibition is due to the possible diffusion of asbestos fibers into the environment constituting a health hazard leading to asbestosis (mesothelioma) cancer. Hyperspectral sensors, characterized by a high spectral and/or spatial resolution, allow reliable quantitative measurements of specific absorption features of urban materials. Spectral analyses can distinguish the asbestos spectrum shape from other roofing materials and backgrounds. Moreover, the classification of the TIR emissivity dataset permits good discrimination between buildings and open spaces and among different roof types, even though residual false positives were still present due to difficulties in the temperature/emissivity separation procedure.[45]

Around oil fields, a considerable part of lighter hydrocarbon fraction normally evaporates, while the remaining fraction in liquid form could be partially recovered. The scars of the damage caused by the spillage of oil remains as soot, tarmats, and tarcrete, which are unrecovered denser hydrocarbon fractions. The surface area covered by the "oil lakes" continues to shrink visibly, leading to the suggestion that there is a disappearance of oil-polluted surfaces. However, thermal RS data provides evidence that the spatial spread of hydrocarbon-contaminated surfaces is much more than previously reported. The addition of hydrocarbons alters soil composition, and these compositional changes are reflected as temperature variations. Landsat TM data has been processed to obtain the LST. Emissivity affects the apparent temperature due to changes in the thermal properties of materials (conductivity, density, capacity, and inertia). The effect of darkness will not affect the LST observations, so waste dumping during night-time can be easily identified for pollution control and environmental management.[46]

Kuwait's oil lakes and oil-polluted surfaces were an act of sabotage imposed on the desert environment during the 1990 to 1991 Arabian Gulf War. The oil formed networks of oil rivers and lakes that accumulated around oil wells and in relatively low areas. The majority of Kuwait's producing oil fields have high reservoir pressures that maintain natural flowing oil from the reserves to the surface. During the 10 months of oil fires, some of the oil fields, especially the Greater Burgan, were more or less inundated with crude oil. The interpretation of high-resolution satellite imagery (Landsat TM, Spot, and Indian Remote Sensing IRSI) shows that contaminated areas demonstrated remarkable resilience, in terms of vegetation growth, seven years after the burning of the oil wells, with increased rainfall from 1991 to 1995. The stabilization of the top sandy soil with tarmat and soot prevented the blowing away of plant seeds, and therefore increased the chances of vegetation growth. The soot, which later turned into black soil, did not appear to be a hindrance to vegetation growth. Oil lakes, tarmats, and black soil continue to decrease and are not always readily observed on the surface. However, harmful chemicals remain in the soil.[47]

In understanding environmental contamination/pollution, one must understand the trajectory of heavy metals in different environmental compartments, which can be seen as different morphological regions (relief forms) shown in RS images. A digital terrain model, generated from the available stereo-pair bands of the ASTER image in the VNIR system, helps aid visualization and interpretation through 3-D observation of different rock types, as well as geomorphometric information (elevation, declivity, and aspect) of the slopes, which potentially reside in, feed, and conduct the landfill wastes.[48] The GIS and RS have an important role in the linkage and analysis of changes in land use-land change (LULC) patterns data, in particular for detection (direct and indirect), extrapolation and interpretation, area calculation, and monitoring. Information on LULC changes is required to achieve environmentally sound management and decision making for future developments.[49]

Air

Over recent years, Earth observation (EO) satellite sensors sent into orbit with improved temporal and spatial resolution, together with enhanced radiometric accuracy, has led to the possibility of using satellite data within the framework of air quality analysis. In this context, the synoptic view and the daily/hourly repetition cycle of satellite observations can provide the monitoring of transboundary air pollution. Moreover, they can allow estimates of air pollutant concentration fields characterized by a more homogeneous spatial distribution and with a more complete coverage of the domain of interest with respect to those carried out by sparse in situ samplings. The processing of the radiance measured by satellite sensors in the spectral range from 0.4 to 2.1 μm supplies information on the atmospheric aerosol loading in terms of columnar aerosol optical depth (AOD) or aerosol optical thickness (AOT) as a measure of light extinction by aerosol in the atmosphere (typically at 0.55 μm), during their overpass time.[50]

Low and moderate spatial resolution satellite sensors have already shown their capability in tracking air pollution in general and aerosols particularly at a global scale. Sensors with moderate to high spatial resolution (such as MODIS and MERIS) seem also to be appropriate for aerosol retrieval at a regional scale by using the differential textural analysis (DTA) code. The DTA code quantifies the contrast reduction as local "textural degradation" on geometrically and radiometrically corrected satellite images.

The code, applied to a set of geo-corrected images, is able to retrieve and map AOT values relative to a reference image assumed to be clean of pollution with a homogeneous atmosphere.[51] As the total atmospheric optical thickness depends on molecular structure (Rayleigh scattering) and aerosols, it is possible to separate the contribution of the "natural" atmospheric molecules from "contaminants" such as aerosols. The AOT can be defined as the integrated extinction coefficient over a vertical column of unit cross section of atmosphere.

Aerosols smaller than the wavelength of satellite sensors cannot be detected, so only coarse particles more than a few micrometers can be detected by MODIS bands. Band combinations of 1, 3, and 7 of MODIS image were used to extract maps of aerosols. Bands 1 and 3, covering the optical region of the electromagnetic spectrum (0.459–0.876 μm), are used to collect information on aerosols and particulate matters, while band 7 covering the infrared region (2.105– 2.55 μm) is used for calibration purposes only.[52]

Aerosol optical thickness in the visible (or atmospheric turbidity) can be considered as an overall air pollution indicator in urban areas, among others, because during photochemical pollution episodes, light extinction is due to particles, while only the yellow-brownish coloration of smog is due to NO_2. Moreover, the presence of particles in the atmosphere always causes an increase of the extinction coefficient, which is strongly correlated with small particle concentration.

Atmospheric turbidity, which expresses overall particulate pollution levels in urban areas, can be extracted by comparing multitemporal high-resolution data. The main prerequisite to carry out this comparison is that no change has arisen in the ground intrinsic reflectance between the data sets used. The noise introduced by changes in ground reflectance is reduced by the combination of two complementary procedures based on physically independent optical effects that are induced by the aerosols in different spectral areas.[53]

Vegetation fire is a global phenomenon that affects large areas and a variety of biomes of the world, from circumboreal forests to the tropical belt through the woodland and shrubland of temperate regions.[54] Since the late 1970s, prescribed and wild vegetation fires have been recognized as a major source of atmospheric trace gases and aerosol particles that affect the composition of the atmosphere, the global climate, and air pollution.[55] In an exceptional event, large vegetation fires raged throughout the Indonesian archipelago in 1997 and 1998, causing a smog blanket that covered more than 3 million square kilometers, with economic losses estimated at over U.S. $4.5 billion.[56] Both active fires and the extent of burned areas can be observed from satellite. Fire radiative power is measured over active fires, and is provided by the MODIS sensor and by the SEVIRI (Spinning Enhanced Visible and InfraRed Imager) sensor onboard the Meteosat Second Generation satellite. Fire radiative power is a very good candidate for any assessment of emission from fires, although it relies on active fires observed by the EO systems, which are a temporal sampling of burning activity. It is consequently possible to highlight similarities and differences in the seasonality and geographical distribution of emission at the global and continental levels.[57] In addition to direct damage on vegetation, devastating fires produced large quantities of gaseous air pollutants and particulate matter (PM_{10} and $PM_{2.5}$, airborne particles smaller than 10 and 2.5 mm in size, respectively) dispersed over the surrounding region, increasing respiratory diseases, asthma, bronchitis, and eye irritation.

The MODIS sensor can measure aerosol abundance and size over land and water with nearly global coverage at moderate spatial resolutions. In addition, MISR (Multiangle Imaging SpectroRadiometer), on board the EOS satellites, is able to provide information on aerosol type and plume top heights. Particle information retrieved by satellite sensors may be suitable for monitoring the spatial and temporal trends of particle concentrations over large geographical areas.[58] The MODIS aerosol optical properties were used in a semi-empirical approach to estimate $PM_{2.5}$ content at ground level. Comparison with daily $PM_{2.5}$ sampled on the ground showed good agreement, with the satellite-based concentrations tending to underestimate the values by at most 20%.

Substantive results suggest that changes in AOD have a significant impact on infant mortality due to respiratory diseases, providing evidence that air pollution's adverse effects, although nonlinear, are not only present in large cities, but also in lower pollution settings that lack ground measures of pollution. A lack of reliable measures of air pollution across wide geographic areas hampers research on pollution's effects on health.[59] Meteorological parameters (air pressure, air temperature, relative humidity, and wind velocity) can influence the estimation of PM from MODIS AOT data. Air quality is usually monitored at fixed ground stations. Although ground measurements are able to indicate the concentration level of air pollutants and their temporal variations precisely, this method of monitoring is limited by its huge expense and sparse spatial coverage. To complement these sparse precise measurements with the estimates derived from satellite, it is more reliable to monitor PM10 from MODIS AOT data at a high temperature and relative humidity, but at low pressure and wind velocity. There is an inverse relationship between the pollution level and the accuracy at which it is monitored from the MODIS data. These findings should serve as useful guidance in selecting the appropriate meteorological conditions under which air pollution can be monitored reliably from MODIS AOT data.[60]

The problem of particulate pollution in the atmosphere has attracted new interest given the recent scientific evidence of the ill-health effects of small particles. Aerosol optical thickness in the visible (or atmospheric turbidity), which is defined as the linear integral of the extinction coefficient due to small airborne particles, can be considered an overall air pollution indicator in urban areas. Reflectance in the visible bands measured by satellite can be used to derive the PM10. In fact, the reflectance at the TOA (top of atmosphere) measured from the satellite is subtracted by the amount given by the surface reflectance to obtain the atmospheric reflectance. The atmospheric reflectance derived from Landsat TM signals can be used as independent variables in a calibration regression analysis to measure the PM10 values.[61]

Using meteorological and other ancillary datasets, an empirical relationship between AOT and $PM_{2.5}$ mass can be obtained, assessing the effects of wind speed, cloud cover, and mixing height on particulate matter air quality. The analysis shows that the $PM_{2.5}$–AOT relationship strongly depends on aerosol concentrations, ambient relative humidity, fractional cloud cover, and height of the mixing layer.[62]

Ground level observations indicated that PM concentration varies widely across different regions, which was mainly due to the difference in weather conditions and anthropogenic emissions. Results showed that MODIS AOD had a better positive correlation with the coincident hourly average PM concentration than with daily average, due to diurnal variation in PM mass measurements. After correcting AOD for relative humidity, the correlation did not improve significantly, suggesting that the humidity was not the main factor affecting the correlation of PM with AOD. The statistical regression analysis between MODIS AOD and PM mass suggested that the satellite-derived AOD is a useful tool for mapping PM distribution over large spatial domains.[63]

A recent method for multi-objective optimization of air quality monitoring systems based on satellite RS of the troposphere uses atmospheric turbidity as a surrogate for air pollution loading. The values of AOD have been extracted by images obtained by sensors on board in-polar-orbit Earth satellites (specifically SPOT 5 and Landsat 7) at the time of their daily overpass. Through an image-processing algorithm, the spatial distribution of optical depth over the whole area is obtained. Assimilating the optical depth field with the field of the mixing layer height, the scattering coefficient of the lowermost part of tropospheric aerosol was reckoned. A physico-chemical and multiphase thermodynamic equilibrium model was used to estimate the secondary aerosol formation and primary aerosol. Using non-linear multiple regression, the experiential relationship between the scattering coefficient of primary and secondary aerosol, its ambient air concentration, and relative humidity can be calculated.[64]

Ozone (O_3) is one of the most dangerous of the phytotoxic air pollutants. Its effects on plants include reductions in photosynthesis, visible leaf injury, growth limitation, and accelerated senescence. Advanced RS techniques using hyperspectral sensors demonstrated the feasibility of detecting the stress in its early phase by monitoring excess energy dissipation pathways, such as chlorophyll fluorescence and non-photochemical quenching.[65] The total ozone mapping spectrometer (TOMS) measures the reflected spectrum from the earth to estimate total column ozone thickness, including maps of erythemal (biologically damaging) UV reaching the earth's surface. Recently, TOMS has been replaced by the ozone monitoring instrument (OMI), which is on board the NASA satellite Aura. The OMI can distinguish between aerosol types such as smoke, dust, and sulfates, and can measure cloud pressure and coverage, providing data to derive tropospheric ozone.[4]

An important indicator for volcanic activity is the emission of trace gases such as sulfur dioxide (SO_2). During an eruption, SO_2 is the third most abundant gas found in volcanic plumes, after H_2O and CO_2. Changes in SO_2 flux can be a precursor for the onset of volcanic activity. In addition, SO_2 is also produced by anthropogenic activities such as power plants, refineries, metal smelting, and the burning of fossil fuels; however, its atmospheric background level is usually very low. Satellite-based instruments operating in the UV spectral region have played an important role in monitoring and quantifying volcanic SO_2 emissions. The TOMS was the first satellite instrument to detect volcanic SO_2 released during the El Chichon eruption in 1982. The detection limit to measure volcanic and anthropogenic SO_2 greatly improved for the Global Ozone Monitoring Experiment (GOME), on board the ERS-2 satellite, and the Scanning Imaging Spectrometer for Atmospheric Cartography (SCIAMACHY), on board the ENVISAT satellite. Last but not least, the Infrared Atmospheric Sounding Interferometer (IASI) on MetOp-A is also able to detect SO_2 with an excellent global coverage in combination with small footprints. Satellite-based observations provide valuable information for detecting and tracking eruption plumes, and therefore minimize the risk of aircraft encounter with hazardous volcanic clouds. GOME-2 has higher sensitivity, especially for SO_2 at lower altitudes, whereas IASI also offers night-time observations, a higher spatial resolution, and an estimation of the altitude of the plume. With commercial and freight air traffic growing globally, the risk of aircraft encounter with hazardous volcanic clouds is increasing, as many volcanoes are not regularly monitored and atmospheric winds can rapidly distribute ash and gas.[66] In fact, once aloft, winds can transport the ash and gases rapidly and in multiple directions, depending on the wind speed and wind shear (the change in wind direction with height). The ash can cause extensive damage to aircraft, stalling engines, abrading windscreens and damaging sensitive avionics equipment. Volcanic gases, specifically SO_2, may also pose a hazard to aircraft, and because gas and ash have different specific gravities they may separate and travel at different speeds and heights in a sheared atmosphere.[67] The 2010 eruptions of the volcano Eyjafjallajökull in Iceland, which, although relatively small for volcanic eruptions, caused enormous disruption to air travel across western and northern Europe. About 20 countries closed their airspace (a condition known as ATC Zero), and it affected hundreds of thousands of travelers.

Conclusion

Remote sensing is nowadays the most applicable tool not only for detection, monitoring, and tracking of pollutants, but also for damage assessment where damage is demonstrable by its effect on the environment.

There is a physical limitation that can never be overcome. Taking marine pollution as an example, it is only possible to differentiate between surfaces (i.e., oil slick and clear water) if they reflect or emit radiation in different proportions; that is, if they have different spectral signatures. This physical limitation means that toxins in water, for example, cannot be detected unless they have an effect on something that is "visible" to the sensor, such as phytoplankton. Another limitation is that electromagnetic radiation has very little power to penetrate objects; for example, it is almost entirely absorbed at the water surface for wavelengths of NIR and longer, and can only penetrate a matter of meters (approximately 15–30 m for very clear water) in the visible wavelengths. This means that RS can only operate on the surface skin of oceans and seas. Fortunately, this is where most marine pollution is focused.[68]

Recent developments in microwave RS, theory, and sensor availability have resulted in new potential and capabilities, such as the ability to extract/detect subsurface parameters and features using these techniques. More and more research is required to refine and implement these approaches. Consequently, there is a need to develop an optimal sensor system including both active and passive microwave techniques for more effective monitoring of the environment. It will allow a range of applications and the synergism of the two types of measurements to provide more useful and new information. Such studies will not only enhance and refine RS applications in environment studies, but will also significantly contribute to the sensor development program.

Finally, the integration of expert systems and spatial decision support systems with GIS is a very interesting area of research, aiding effective and timely decision making concerning the planning, design, analysis, operation, and maintenance of environmental resources systems. Moreover, these tools can greatly reduce the time and effort required in traditional approaches, automating the process of solving pollution problems and helping the selection of cost-effective management alternatives.

References

1. Sabins, F.F. Fundamental considerations. In *Remote Sensing Principles and Interpretation*; W.H. Freeman and Co.; San Francisco, 1978; 1–16.
2. Campbell, J.B. History and scope of remote sensing. In *Introduction to Remote Sensing*, 2nd Ed.; Taylor & Francis; London, 1966; 1–21.
3. Brivio, P.A.; Lechi, G.; Zilioli, E. *Principi e Metodi di Telerilevamento;* Città Studi Ed.; Milan, Italy, 2006.
4. Mumby, P.J.; Skirving, W.; Strong, A.E.; Hardy, J.T.; LeDrew, E.F.; Hochberg, E.J.; Stump, R.P.; David, L.T. Remote sensing of coral reefs and their physical environment. Mar. Pollut. Bull. **2004**, *48*, 219–228.
5. Kabbara, N.; Benkhelilb, J.; Awadc, M.; Barale, V. Monitoring water quality in the coastal area of Tripoli (Lebanon) using high-resolution satellite data. ISPRS J. Photogramm. Remote Sens. **2008**, *63*, 488–495.
6. Sesli, F.A.; Karsli, F.; Colkesen, I.; Akyol, N. Monitoring the changing position of coastlines using aerial and satellite image data: An example from the eastern coast of Trabzon, Turkey. Environ. Monit. Assess. **2009**, *153*, 391–403.
7. Dekker, A.; Brando, V.; Anstee, J.M. Retrospective sea-grass change detection in a shallow coastal tidal Australian lake. Remote Sens. Environ. **2005**, *97*, 415–433.
8. Durand, D.; Bijaoui, J.; Cauneau, F. Optical remote sensing of shallow-water environmental parameters: A feasibility study. Remote Sens. Environ. **2000**, *73*, 152–161.
9. Barale, V.; Jaquet, J.M.; Ndiaye, M. Algal blooming patterns and anomalies in the Mediterranean Sea as derived from the SeaWiFS data set (1998–2003). Remote Sens. Environ. **2008**, *112*, 3300–3313.
10. Tetley, M.J.; Mitchelson-Jacob, E.G.; Robinson, K.P. The summer distribution of coastal minke whales (*Balaenoptera acutorostrata*) in the southern outer Moray Firth, NE Scotland, in relation to co-occurring mesoscale oceanographic features. Remote Sens. Environ. **2008**, *112*, 3449–3454.
11. Panigada, S.; Zanardelli, M.; MacKenzie, M.; Donovan, C.; Mélin, F.; Hammond, P.S. Modelling habitat preferences for fin whales and striped dolphins in the Pelagos Sanctuary (Western Mediterranean Sea) with physiographic and remote sensing variables. Remote Sens. Environ. **2008**, *112*, 3400–3412.

12. Li, Y.; Li, J. Oil spill detection from SAR intensity imagery using a marked point process. Remote Sens. Environ. **2011**, *114,* 1590–1601.

13. Cheng, Y.; Li, X.; Xu, Q.; Garcia-Pineda, O.; Andersen, O.B. SAR observation and model tracking of an oil spill event in coastal waters. Mar. Pollut. Bull. **2011**, *62,* 350–363.

14. Krohling, R.A.; Campanharo, V.C. Fuzzy TOPSIS for group decision making: A case study for accidents with oil spill in the sea. Expert Syst. Appl. **2011**, *38,* 4190–4197.

15. Keramitsoglou, I.; Cartalis, C.; Kiranoudis, C.T. Automatic identification of oil spills on satellite images. Environ. Model. Softw. **2006**, *21,* 640–652.

16. Topouzelis, K.; Karathanassi, V.; Pavlakis, P.; Rokos, D. Detection and discrimination between oil spills and look-alike phenomena through neural networks. ISPRS J. Photogramm. Remote Sens. **2007**, *62,* 264–270.

17. Cococcioni, M.; Corucci, L.; Lazzerini, B. *Issues and Preliminary Results in Oil Spill Detection Using Optical Remotely Sensed Images,* Proceedings of International Workshop on the Analysis of Multi-Temporal Remote Sensing Images, 2007. MultiTemp 2007, Leuven, Belgium, 18–20 July 2007; Eds.; IEEE: New York, 2007.

18. Casciello, D.; Lacavat, T.; Pergolat, N.; Tramutoli, V. *Robust Satellite Techniques (RST) for Oil Spill Detection and Monitoring,* Proceedings of International Conference on Geoinformatics IEEE, Fairfax, USA, August 12–14, 2009; Eds.; IEEE: New York, 2009.

19. Yu, I.A.; Zatyagalova, V.V. A GIS approach to mapping oil spills in a marine environment. Int. J. Remote Sens. **2008**, *29* (21), 6297–6313.

20. Dahish, A.S.; Ahmad, A. In *An Application of Geographical Information System and Remote Sensing Techniques for Detection of Oil Spill,* Proceedings of International Conference on Geoinformatics IEEE, Fairfax, USA, August 12–14, 2009; Eds.; IEEE: New York, 2009.

21. Jha, M.K.; Chowdhury, A.; Chowdary, V.M.; Peiffer, S. Groundwater management and development by integrated remote sensing and geographic information systems: Prospects and constraints. Water Resour. Manage. **2007**, *21,* 427–467.

22. Dimitriou, E.; Zacharias, I. Groundwater vulnerability and risk mapping in a geologically complex area by using stable isotopes, remote sensing and GIS techniques. Environ. Geol. **2006**, *51,* 309–323.

23. Montzka, C.; Canty, M.; Kreins, P.; Kunkel, R.; Menz, G.; Vereecken, H.; Wendland, F. Multispectral remotely sensed data in modelling the annual variability of nitrate concentrations in the leachate. Environ. Model. Softw. **2008**, *23,* 1070–1081.

24. Bresciani, M.; Giardino, C.; Longhi, D.; Pinardi, M.; Bartoli, M; Vascellari, M. Imaging spectrometry of productive inland waters. Application to the lakes of Mantua. Ital. J. Remote Sens. **2009**, *41* (2), 147–156.

25. Giardino, C.; Bartoli, M.; Candiania, G.; Bresciani, M.; Pellegrini, L. Recent changes in macrophyte colonisation patterns: An imaging spectrometry-based evaluation of southern Lake Garda (northern Italy). J. Appl. Remote Sens. **2007**, *1,* 1–17.

26. Giardino, C.; Oggioni, A.; Bresciani, M.; Yan, H. Remote sensing of suspended particulate matter in Himalayan lakes. A case study of Alpine lakes in the Mount Everest region. Mt. Res. Dev. **2010**, *30* (2), 157–168.

27. Bresciani, M.; Stroppiana, D.; Fila, G.; Montagna, M.; Giardino, C. Monitoring reed vegetation in environmentally sensitive areas in Italy. Ital. J. Remote Sens. **2009**, *41* (2), 125–137.

28. Bochenek, Z.; Ciolkosz, A.; Iracka, M. Deteriorations of forests in the Sudety Mountains, Poland, detected on satellite images. Environ. Pollut. **1998**, *101,* 163–168.

29. Kooistraa, L.; Salasc, E.A.L.; Cleversc, G.P.W; Wehrensb, R.; Leuvena, R.S.E.W.; Nienhuisa, P.H.; Buydensb, L.M.C. Exploring field vegetation reflectance as an indicator of soil contamination in river floodplains. Environ. Pollut. **2007**, *127,* 281–290.

30. Nikolov, N.; Zeller, K. Efficient retrieval of vegetation leaf area index and canopy clumping factor from satellite data to support pollutant deposition assessments. Environ. Pollut. **2006**, *141,* 539–549.

31. Chen, J.M.; Cihlar, J. Retrieving leaf area index of boreal conifer forests using Landsat TM images. Remote Sens. Environ. **1996**, *55*, 153–162.

32. Boschetti, M.; Brivio, P.A.; Carnesale, D.; Di Guardo, A. The contribution of hyperspectral remote sensing to identify vegetation characteristics necessary to assess the fate of persistent organic pollutants (POPs) in the environment. Ann. Geophys. **2006**, *49*, 177–186.

33. Zhengjun, W.; Jianming, H.; Guisen, D. Use of satellite imagery to assess the trophic state of Miyun Reservoir, Beijing, China. Environ. Pollut. **2008**, *155*, 13–19.

34. Tuominen, J.; Lipping, T.; Kuosmanen, V. *Assessment of ENVI Forest heath Tool in Detection of Dust and Seepage*, Proceedings of IEEE International Geoscience & Remote Sensing Symposium, Boston, USA, July 06–11, 2008; Eds.; IEEE: New York, 2008.

35. Rastmanesh, F.; Moore, F.; Kharrati-Kopaei, M.; Behrouz, M. Monitoring deterioration of vegetation cover in the vicinity of smelting industry, using statistical methods and TM and ETM+ imageries, Sarcheshmeh copper complex, Central Iran. Environ. Monit. Assess. **2010**, *163*, 397–410.

36. Boschetti, M.; Brivio, P.A.; Carnesale, D.; Di Guardo, A. The contribution of hyperspectral remote sensing to identify vegetation characteristics necessary to assess the fate of Persistent Organic Pollutants (POPs) in the environment. Annals of Geophysics, **2006**, *49*, 177–186.

37. Cinnirella, S.; Pirrone, N. Spatial and temporal distributions of mercury emissions from forest fires in Mediterranean region and Russian federation. Atmos. Environ. **2006**, *40*, 7346–7361.

38. Dunagan, S.C.; Gilmore, M.S.; Varekamp, J.C. Effects of mercury on visible/near-infrared reflectance spectra of mustard spinach plants (*Brassica rapa* P.). Environ. Pollut. **2007**, *148*, 301–311.

39. Joshi, P.K.; Kumar, M.; Paliwal, A.; Midha, N.; Dash, P.P. Assessing impact of industrialization in terms of LULC in a dry tropical region (Chhattisgarh), India using remote sensing data and GIS over a period of 30 years. Environ. Monit. Assess. **2009**, *149*, 371–376.

40. Rafiee, R.; Mahiny, A.S.; Khorasani, N. Assessment of changes in urban green spaces of Mashad city using satellite data. Int. J. Appl. Earth Obs. Geoinf. **2009**, *11*, 431–438.

41. Meng, D.; Gong, H.; Li, X.; Zhao, W.; Gong, Z.; Zhu, L.; Hu, D. *Study of Urban Heat Island Based on Remote Sensing in Beijing-Capital Zone*, Proceedings of Urban Remote Sensing Joint Event, Shanghai, China, May 20–22, 2008; Eds.; IEEE: New York, 2009.

42. Xian, G.; Crane, M.; Su, J. An analysis of urban development and its environmental impact on the Tampa Bay watershed. J. Environ. Manage. **2007**, *85*, 965–976.

43. Huiping, H.; Bingfang, W.; Jingjing, Z.; Yuemin, Z. *The Method and Applications of Remote Sensing in Urban Building Heating-loss Monitoring*, Proceedings of Urban Remote Sensing Joint Event, Shanghai, China, May 20–22, 2008; Eds.; IEEE: New York, 2009.

44. Chalkias, C.; Petrakis, M.; Psiloglou, B.; Lianou B. Modelling of light pollution in suburban areas using remotely sensed imagery and GIS. J. Environ. Manage. **2006**, *79*, 57–63.

45. Bassani, C.; Cavalli, R.M.; Cavalcante, F.; Cuomo, V.; Palombo, A.; Pascucci, S.; Pignatti, S. Deterioration status of asbestos-cement roofing sheets assessed by analyzing hyperspectral data. Remote Sens. Environ. **2007**, *109*, 361–378.

46. Saif ud din; Al Dousari, A.; Literathy, P. Evidence of hydrocarbon contamination from the Burgan oil field, Kuwait—Interpretations from thermal remote sensing data. J. Environ. Manage. **2008**, *86*, 605–615.

47. Kwarteng, A.Y. Remote sensing assessment of oil lakes and oil-polluted surfaces at the Greater Burgan oil field, Kuwait. Int. J. Appl. Earth Obs. Geoinf. **1999**, *1*, 36–47.

48. de Oliveira, M.T.G.; Rolim, S.B.A; de Mello-Farias, P.C; Meneguzzi, A.; Lutckmeier, C. Industrial pollution of environmental compartments in the Sinos River Valley, RS, Brazil: Geochemical-biogeochemical characterization and remote sensing. Water Air Soil Pollut. **2008**, *192*, 183–198.

49. Doygun, H.; Alphan, H. Monitoring urbanization of Iskenderun, Turkey, and its negative implications. Environ. Monit. Assess. **2006**, *114*, 145–155.

50. Di Nicolantonio, W.; Cacciari, A.; Tomasi, C. Particulate matter at surface: Northern Italy monitoring based on satellite remote sensing, meteorological fields, and in-situ samplings. IEEE J. Select. Top. Appl. Earth Obs. Remote Sens. **2009**, *2*, 284–292.
51. Retalis, A.; Sifakis, N. Urban aerosol mapping over Athens using the differential textural analysis (DTA) algorithm on MERIS-ENVISAT data. ISPRS J. Photogramm. Remote Sens. **2010**, *65*, 17–25.
52. Sohrabiniaa, M.; Khorshiddoust, A.M. Application of satellite data and GIS in studying air pollutants in Tehran. Habitat Int. **2007**, *31*, 268–275.
53. Sifakis, N.I. Quantitative mapping of air pollution density using Earth observations: A new processing method and application to an urban area. Int. J. Remote Sens. **1998**, *19* (17), 3289–3300.
54. Brivio, P.A.; Maggi, M.; Binaghi, E.; Gallo, I. Mapping burned surfaces in Sub-Saharan Africa based on multi-temporal neural classification. Int. J. Remote Sens. **2003**, *24*, 4003–4018.
55. Crutzen, P.J.; Andreae, M.O. Biomass burning in the tropics: Image on atmospheric chemistry and biogeochemical cycles. Science **1990**, *250*, 1669–1678.
56. Stolle, F.; Chomitz, K.M.; Lambin, E.F.; Tomich, T.P. Land use and vegetation fires in Jambi Province, Sumatra, Indonesia. For. Ecol. Manage. **2003**, *179*, 277–292.
57. Stroppiana, D.; Brivio, P.A; Gregoire, G.M.; Liousse, C.; Guillaume, B.; Granier, C.; Mieville, A.; Chin, M.; Petron, G. Comparison of global inventories of monthly CO emissions derived from remotely sensed data. Atmos. Chem. Phys. Discuss. **2010**, *10*, 17657–17697.
58. Liu, Y.; Kahn, R.A.; Chaloulakou, A.; Koutrakis, P. Analysis of the impact of the forest fires in August 2007 on air quality of Athens using multi-sensor aerosol remote sensing data, meteorology and surface observations. Atmos. Environ. **2009**, *43*, 3310–3318.
59. Gutierrez, E. Using satellite imagery to measure the relationship between air quality and infant mortality: An empirical study for Mexico. Popul. Environ. **2010**, *31*, 203–222.
60. Zha, Y.; Gao, J.; Jiang, J.; Lu, H.; Huang, J. Monitoring of urban air pollution from MODIS aerosol data: Effect of meteorological parameters. Tellus **2010**, *62B*, 109–116.
61. Lim, H.S.; MatJafri, M.Z.; Abdullah, K.; Saleh, N.M.; Wong, J. *Extracting Spatial Data From Satellite Sensor to Support Air Pollution Determination Using Remote Sensing Technique,* Proceedings of International Geoscience and Remote Sensing Symposium, Barcelona, Spain, July 23–27, 2007; Eds.; IEEE: New York, 2007.
62. Gupta, P; Christophera, S.A.; Wang, J.; Gehrig, R.; Leed, Y.; Kumar, N. Satellite remote sensing of particulate matter and air quality assessment over global cities. Atmos. Environ. **2006**, *40*, 5880–5892.
63. Guo, J.P.; Zhang, X.Y.; Che, H.Z.; Gong, S.L.; An, X.; Cao, C.X.; Guang, J.; Zhang, H.; Wang, Y.Q.; Zhang, X.C.; Xue, M.; Li, X.W. Correlation between PM concentrations and aerosol optical depth in eastern China. Atmos. Environ. **2009**, *43*, 5876–5886.
64. Sarigiannis, D.A.; Saisana, M. Multi-objective optimization of air quality monitoring. Environ. Monit. Assess. **2008**, *136*, 87–99.
65. Meroni, M.; Panigada, C.; Rossini, M.; Picchi, V.; Cogliati, S.; Colombo, R. Using optical remote sensing techniques to track the development of ozone-induced stress. Environ. Pollut. **2009**, *157*, 1413–1420.
66. Rix, M.; Valks, P.; Hao, N.; van Geffen, J.; Clerbaux, C.; Clarisse, C.; Coheur, P.F.; Loyola D.G.; Erbertseder, T.; Zimmer, Z.; Emmadi, S. Satellite monitoring of volcanic sulfur dioxide emissions for early warning of volcanic hazards. IEEE J. Select. Topics Appl. Earth Obs. Remote Sens. **2009**, *2* (3), 196–206.
67. Prata, A.J. Satellite detection of hazardous volcanic clouds and the risk to global air traffic. Nat. Hazards **2009**, *51*, 303–324.
68. Clark, C.D. Satellite remote sensing of marine pollution. Int. J. Remote Sens. **1993**, *14* (16), 2985–3004.

Solid Waste Management: Life Cycle Assessment

Ni-Bin Chang,
Ana Pires, and
Graça Martinho

Introduction

In the end of 2005, the European Commission (EC) published a proposal for a directive on waste [COM (2005) 667,[1] which resulted in the publishing of the new Waste Framework Directive[2]] that is currently under implementation in European institutions. One element of the proposal is the establishment of a hierarchy of principles (i.e., waste hierarchy principle hereafter) to be applied for waste management. The commission proposal includes the following reference to the hierarchy:

> The Member States are to take measures, as a matter of priority, for the prevention or reduction of waste production and its harmfulness and, secondly, for the recovery of waste by means of re-use, recycling and other recovery operations.

It provides for a flexible application of the hierarchy and does not explicitly make a suggestion between the d...l/l/l/l/k,ifferent recovery options such as reuse, recycling, and recovery of energy, making the policy blurred. Later on, with the publishing of the new Waste Framework Directive 2008/98/EC,[1] resulting from COM (2005) 667, a five-step hierarchy was proposed to encourage the member states in the EC to take measures, as a matter of priority, for the following: 1) prevention; 2) preparing for reuse; 3) recycling; 4) other recovery, e.g., energy recovery; and 5) disposal. The waste hierarchy principle for prioritizing waste management options has touched the base of pollution prevention and resource conservation. Yet recent climate change impacts and resource scarcity made some of these options for solid waste management (SWM) no longer risk informed and forward looking. As a consequence, another amendment was introduced that seeks to support the flexibility of the waste hierarchy principle by stating the following:

> When applying the waste hierarchy [...], Member States shall take measures to encourage the options that deliver the best overall environmental outcome. This may require specific waste streams departing from the hierarchy where this is justified by life-cycle thinking on the overall impacts of the generation and management of such waste."

Given that extensive studies are required to deviate from the hierarchy, modern sustainable SWM strategies must be tied to an all-inclusive assessment metrics covering socioeconomic aspects, cost–benefit analysis, technical feasibility, public health requirements, environmental impacts, ecological footprint, carbon footprint, and even sociopolitical considerations. Hence, sustainable SWM has been deemed as a renewed paradigm, in which understanding how sustainable decisions in association with life cycle assessment (LCA) can be reached in a societal context becomes a critical task.

Multicriteria decision making (MCDM) processes have been helpful in translating the features of real-world problems to decision-making processes. The problems of MCDM can be broadly classified into two categories: multiple-attribute decision making (MADM) and multiple-objective decision making (MODM). The main features that differentiate both methods are the type of criteria (attributes or objectives) and the number of alternatives. Whereas MADM has a finite number of alternatives, MODM has an infinite number of alternatives. With the prescribed waste management alternatives in our study by which the salient features for sustainable SWM can be retrieved distinctly, MADM was thus selected to aid in our decision analysis. To explore the decision making with respect to a suite of criteria, this entry presents a comparative study through two different MADM approaches for the purpose of demonstration. They include the simple additive weight (SAW) and the technique for order preference by similarity to ideal solution (TOPSIS). The practical implementation was assessed by a case study in Setúbal peninsula, Portugal, where there is an acute need to improve the quality of SWM to achieve a sustainable solution.

Life Cycle Assessment for Solid Waste Management

Life cycle assessment is a compilation and evaluation of the inputs, outputs, and potential environmental impacts of a product system throughout its life cycle.[3] Thus, LCA is a tool for the analysis of the environmental burden of products at all stages in their life cycle—from the extraction of resources, through the production of materials, product parts, and the product itself, and the use of the product to the management after it is discarded, either by reuse, recycling, or final disposal (in effect, therefore, "from cradle to grave").[4]

The concept of LCA was produced in the 1960s as a way to cumulatively account for energy use leading to the projection of future resource supplies and demand.[5] Since the 1990s, LCA has been applied in a broad range of different fields including SWM. The ISO 14040 standard[3] points out the main areas of application of LCA, including the identification of improvement possibilities, decision making, selection of relevant environmental performance indicators, and marketing. The most influential momentum in the context of ISO 14040 applied for SWM is to explore how the decision-making processes in a social context can make the final option different.

The salient use of LCA for SWM was conducted in Denmark, aiming to meet the need to properly manage the packaging waste embedded in the packaging materials.[6–10] Life cycle assessment may also be applicable in conducting intercomparisons among waste treatment alternatives for specific waste streams, such as recycling of cardboard[11] and recycling vs. incineration of scrap paper.[12] Earlier applications of LCA for SWM focused on the evaluation of waste hierarchy principle in some industrialized countries during 1990.[13] In the literature,[14] it is clear that waste hierarchy principle cannot be used to substitute thorough assessment in most cases. Once LCA is capable of promoting a holistic view of

SWM, several combinations of different technologies can be meaningfully applied to support essential SWM against the contradictory suggestions in early waste hierarchy paradigm. For this reason, LCA has been recommended by the European Union Waste Framework Directive[2] to verify if waste hierarchy principle is the best solution, particularly from the environmental point of view.

The applications of LCA tools for waste management appeared mainly in Europe and United States for the screening of technologies under the umbrella of the integrated solid waste management (ISWM) systems. Several LCA models were tailored specifically for ISWM systems, and examples may include but are not limited to WISARD,[15] IWM,[16] WASTED,[17] and EASEWASTE.[18] Specifically, the IWM versions 1 and 2[19,20] for SWM systems provide life cycle inventory (LCI). The models enabled decision makers and waste managers to use an LCA for assessing their specific waste management configurations without in-depth knowledge of the theory and methodology and allowed them to learn how changes in the system could affect the environmental impacts through scenario analysis.[21] However, generic models, like UMBERTO, Gabi, and SimaPro, were applied to SWM, with raising potentials concerning SWM specificities.[22] Besides, extended tools focusing specially on the possible impact of economic features on decision-making process were developed. In United States, for example, the municipal solid waste – decision support tool (MSW-DST) developed by the Research Triangle Institute and United States Environmental Protection Agency,[23,24] where the environmental methodology is based on the use of LCA and the cost methodology with respect to the full-cost accounting, became available in the 2000s.

Application of LCA in SWM systems has also been promoted through the combination with other systems analysis tools to reach a sustainable decision. In this field, LCA was combined with site-dependent and qualitative approaches to evaluate waste-to-energy taxation.[25] Life cycle assessment and strategic environmental assessment were integrated to assess economic and environmental impacts for weight-based tax in waste incineration.[26] The ORWARE model, developed by Dalemo et al.,[27] Björklund, Dalemo, and Sonesson,[28] and Eriksson et al.[29,30] combines LCA with a simulation tool and material flow analysis. Solano et al.[31,32] developed a model for ISWM to obtain the best solution through LCA and an optimization model for balancing economic and environmental considerations. From the previous review,[22] it can be summarized that LCA results have been capable of changing packaging and packaging waste management, influencing the selection of waste treatment technologies, affecting regulation assessment, challenging waste hierarchy paradigm, increasing knowledge, and developing tools and methods for decision support. Those systems analysis models can be flexibly woven to deal with SWM issues with varying features, and more discussion can be seen in the work of Harrison et al.[33] and Chang, Pires, and Martinho.[22]

Regardless of these cases above where LCA was applied in a variety of SWM systems, it is important to understand how LCA could influence the decision making in many SWM systems and, especially, the decision-making process. The methodology capable of linking different criteria is MADM, such as the applications in the work of Kijak and Moy,[34] Contreras et al.,[35] and Skordilis,[36] all of which can promote the understanding in decision making for SWM.

Study Area and System Description

The Setúbal peninsula is located in the district of Setúbal, covers an area of 1522 km², and has 714,589 inhabitants.[37] The area is divided into nine municipalities, as shown in Figure 1. The nine municipalities have associated to manage their municipal solid waste (MSW), on a regional scale. AMARSUL company is owned by the local municipalities and is responsible for managing the MSW system since 1997. The SWM system is composed of nine recycling centers, two material recovery facilities (MRFs), two landfills, one transfer station, and one aerobic mechanical biological treatment (MBT).

Nowadays, this SWM system keeps on promoting the separation of paper/cardboard, glass, and light packaging (plastics, metals, and composite packaging) waste by means of curbside recycling systems. Each type of waste is collected separately in three specific containers and then directly sent to the MRF

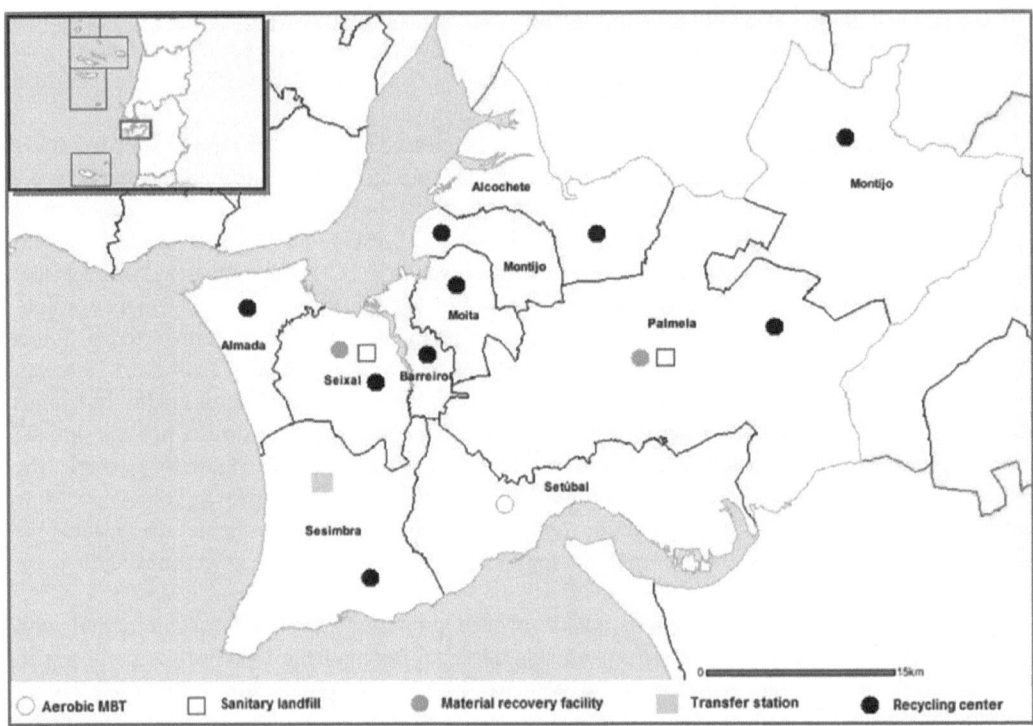

FIGURE 1   The geographical location of the Setúbal peninsula SWM system.

for recycling, recovery, and reuse. The remaining waste fractions are then collected through a door-to-door and/or bin collection scheme, which is destined for final disposal at landfills. In the case of the Sesimbra municipality, the waste stream is first sent to the transfer station, which is then followed by the final disposal at sanitary landfills. Yet the residual waste after waste separation and recycling collected from the Setúbal municipality is transported to an aerobic MBT plant, where the "stabilized residue" can be produced as fertilizer to be applied as agriculture soil-amendment materials.

Within this MSW system, there is a recent need to make some changes in order to comply with the Packaging and Packaging Waste Directive[38] and Landfill Directive.[39] The National Plan for MSW (i.e., designated as PERSU II) has decided to pursue the construction of several more MBT units. An anaerobic digestion (AD) MBT unit, with a mechanical treatment to separate recyclables and high-calorific material to produce refuse derived fuel (RDF), is predicted. It is expected that the unit will work with two separate lines, in which one is related to the biodegradable municipal solid wastes (BMW) and the other is for the residual waste streams. The RDF may be combusted in an incinerator to generate electricity. The existing aerobic MBT plant will be maintained as usual. It is expected that both MRF plants (manual sorting) will be substituted by two automated units.

The schematic of the SWM to be analyzed can be shown in Figure 2, which generally covers all stages of SWM involved, from raw waste pickup to the delivery to bins, to some intermediate processing units, and to the final disposal at landfills. Both AD MBT lines are represented as two separated processes.

These SWM processes include collection and transportation of residual waste and recyclables, waste treatment, waste transport from waste treatment facilities to final destination, energy-from-waste or waste energetic recovery (ER), and landfilling. Several final destinations for recyclables are located in Spain rather than Portugal, specifically for composite packaging and ferrous and nonferrous metal packaging materials.

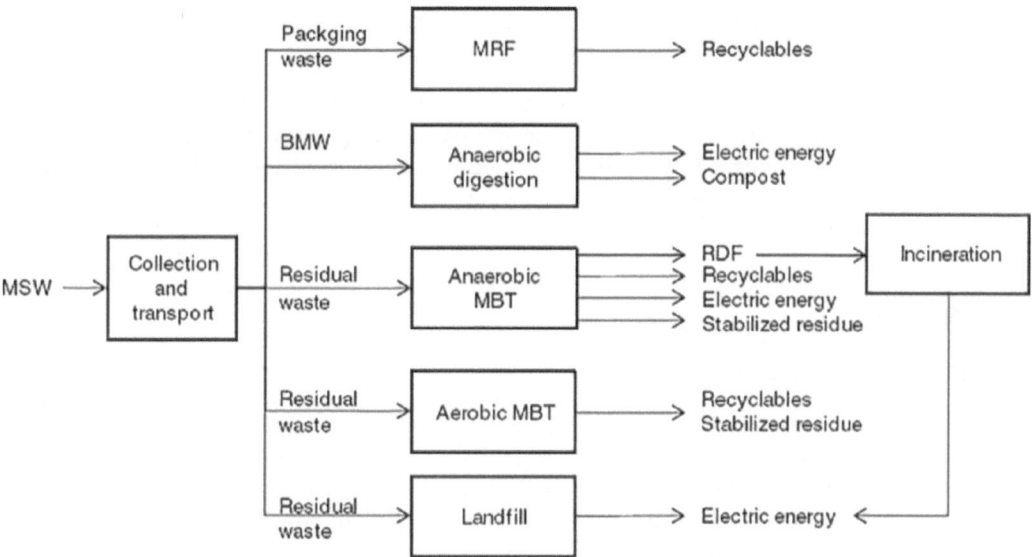

FIGURE 2   The schematic of the SWM system at Setúbal Peninsula.

Based on this system, Table 1 presents the 18 management alternatives for assessment plus the actual situation (base scenario). Those alternatives include waste collection and separate recycling of the three packaging materials through bin systems, which handle 12.4% of the current MSW in the study area. This MRF system is responsible for the compliance with the prescribed target in the Packaging Waste Directive.

Alternative 0 refers to the predicted change that will take place in the Setúbal peninsula waste management system. The remaining alternatives were designed to examine some special options for complying with the Landfill Directive. For example, alternative 1 emphasizes the inclusion of aerobic MBT; alternative 4 signifies the use of AD MBT; alternative 6 examines the specific case of using a BMW AD line. In general, alternatives 0, 3, and 5 are options for differing intermediate processing. Separation of the high-calorific fraction of waste for ER was considered through the production of RDF and the direct burning of the high-calorific fraction in municipal incinerators.

The 18 alternatives of SWM were assessed by considering two scenarios:

- *Baseline scenario*: Targets may be reached without systematic involvement and evolution, meaning that targets fulfillment can be promoted by several external agents such as government, the Green Dot Society (Sociedade Ponto Verde), and promotion campaigns that motivate a better environmental consciousness. The system may be financed by using water consumption tax for waste management to be included in the water billing system.

TABLE 1   The Distribution of Waste Streams Associated with Each Alternative in the SWM System

Fraction (%) Option	Alternatives							
	0/0ᵃ/0ᵇ	1/1ᵃ	2/2ᵃ/2ᵇ	3/3ᵃ	4/4ᵃ/4ᵇ	5/5ᵃ/5ᵇ	6/6ᵃ	Base
MRF	12.4	12.4	12.4	12.4	12.4	12.4	12.4	4.8
Anaerobic digestion BMW	5.4	0	0	13.3	0	7.5	28.7	0
Anaerobic digestion MBT	28.2	0	33.9	0	49.6	38.9	0	0
Aerobic MBT	13.2	49.7	15.8	32.6	0	0	0	13.8
Landfill with ER	40.8	37.9	37.9	41.7	38.0	41.2	58.9	81.4

ᵃ Alternatives considering RDF production plus incineration of high-calorific fraction.
ᵇ Alternatives not considering RDF production but considering incineration of high-calorific fraction from MBT.

- *Pay-as-you-throw (PAYT) scenario*: Targets can be reached by imposing an economic instrument— PAYT—to be implemented by various levels of MSW system managers.

The renewed interest in studying both instruments rests upon the actual billing system in Portugal, which does not comply with the philosophy of the new Waste Framework Directive due to the fact that the water billing system is dependent on the water consumption rate rather than the waste production rate. The PAYT system, being designed based on a pay-per-bag system, would fully comply with the polluter-pays principle.

Methodology

Multiattributive Decision Making

As we know, MADM is one type of multicriteria decision-making category for problem solving in decision making, by which a finite number of alternatives identified by a group of experts can be analyzed. Several MADM methods have been applied in waste management, including ELECTRE,[40] PROMETHEE[41] and GAIA,[42] AHP,[43] TOPSIS,[44] and SAW.[45] To proceed in this study, SAW and TOPSIS were selected for demonstration, both of which are well-proven techniques and make it possible to rank the solutions via a preference order for SWM with a logical procedure and a set of criteria allowing a better appliance for nonexperts, according to Cheng et al.[45] Good performance when compared with more sophisticated methods may be anticipated.[46,47]

Even though criteria of different natures might have different units, this would not be a problem in implementing SAW and TOPSIS methods, once a normalization procedure may be used to perform the criteria aggregation. In detail, SAW is a well-defined method, which is intuitive and easy to understand.[48] In SAW, the role of the decision maker is to assign weights to each attribute. The decision maker can then obtain a total score for each alternative simply by multiplying the scale rating for each attribute value by the weight assigned and summing these products over the attributes. The alternative with the highest score is the one prescribed to the decision maker. The mathematical formula of the SAW is given by Cheng et al.:[46]

$$U_j = \sum_{i=1}^{n} w_i r_{ij}, \, j = 1, 2, \ldots, m \tag{1}$$

where w_i is the weight of the attribute and r_{ij} is the normalized impact matrix, by which the normalization may be provided through a linear scale transformation.

TOPSIS, developed by Hwang and Yoon[44] based upon the concept that the chosen alternative should have the shortest distance from the positive ideal solution and the farthest from the negative ideal solution, was applied in this study too. A utility value $D(i)$ for each alternative i can be obtained by calculating the relative distance from i to the ideal solution, which can be described as follows:[45]

Step 1. Calculate the normalized decision matrix. The normalized value n_{ij} is calculated as

$$n_{ij} = x_{ij} / \sqrt{\sum_{j=1}^{m} x_{ij}^2}, \quad j = 1, \ldots, m, \, i = 1, \ldots, n \tag{2}$$

Step 2. Calculate the weighted normalized decision matrix. The weighted normalized value v_{ij} is calculated as

$$v_{ij} = w_i n_{ij}, \quad j = 1, \ldots m, I = 1, \ldots, n \tag{3}$$

where w_i is the weight of the ith attribute or criterion, and

$$\sum_{i=1}^{n} w_i = 1c$$

Step 3. Determine the positive ideal and negative ideal solution:

$$A^+ = \left\{ v_1^+, \ldots, v_n^+ \right\} = \left\{ \left(\max_j v_{ij} | i \in I \right), \left(\min_j v_{ij} | i \in J \right) \right\}$$

$$A^- = \left\{ v_1^-, \ldots, v_n^- \right\} = \left\{ \left(\min_j v_{ij} | i \in I \right), \left(\max_j v_{ij} | i \in J \right) \right\} \quad (4)$$

where i is associated with benefit criteria, and j is associated with cost criteria.

Step 4. Calculate the separation measures, using the n-dimensional Euclidean distance. The separation of each alternative from the ideal solution and for the negative ideal solution are given as, respectively,

$$d_j^+ \left\{ \sum_{i=1}^{n} \left(v_{ij} - v_i^+ \right)^2 \right\}^{\frac{1}{2}}$$

$$d_j^- \left\{ \sum_{i=1}^{n} \left(v_{ij} - v_i^- \right)^2 \right\}^{\frac{1}{2}} \quad j = 1, \ldots m \quad (5)$$

Step 5. Calculate the relative closeness to the ideal solution. The relative closeness of the alternative j with respect to ideal solution is defined as

$$R_j = d_j^- / \left(d_j^+ + d_j^- \right), \quad j = 1, \ldots, m \quad (6)$$

Since $d_j^- \geq 0$ and $d_j^+ \geq 0$, then $R_j \in [0,1]$.

Step 6. Rank the preference in descending order.

Life Cycle Assessment

According to ISO 14040,[3] LCA consists of four major stages: goal and scope definition, LCI, life cycle analysis, and interpretation of the results. Below is a detailed description of each stage to enhance comprehension.

Goal and Scope Definition

The purpose of the LCA is to assess the 18 alternatives mentioned in Table 1. The same framework was applied to both scenarios as defined by the water billing system and PAYT, respectively. The LCA provided in this study is of an attributional type. We applied the "zero burden assumption," which suggests that waste management carries none of the upstream environmental burdens into the SWM system.[49]

In an LCA system with multiple products, which is the case in our study, it is necessary to set up the methodological framework. According to ISO 14044,[50] the system boundary should be geared toward expanding the product system to include the additional functions related to the coproducts to avoid allocation. In this LCA, the material recycling, energy recovery, and fertilizer application (i.e., stabilized residue waste) of MSW were included in the LCA, which collectively resulted in an expansion of the system boundary. In this LCA, the emissions resulting from the referred operations were included as the baseline information as the emissions of those competing products and energy recovery potential resulting from those alternative operations were also included for the purpose of comparison. In this context, the system can be expanded to include additional burdens of coproduct processing and the avoided burdens of any avoided processes (i.e., substitution or avoided burden method).[4,51–53]

To ensure correct implementation of the avoided burden method through successful MSW recycling and reuse, the expanded system products should have the same function as the raw products. The substitution ratios were then applied considering closed-loop and open-loop procedures. Table 2 presents the substitution ratios for recovered materials and energy consumed. In the cases where substitution ratio is 1:1, they were considered as a closed-loop procedure with a hypothesis that no changes occurred in the inherent properties of the recycled material.[54] For example, 1 kg of recycled glass can replace 1 kg of virgin glass without considering degradation of the material during the recycling so that the quality of the secondary material may not be worse than that of the primary material.[54] The materials included in this situation in our study were glass, metals, polyethylene plastics, plastic wood, fertilizers, and

TABLE 2 Criteria Applied in Decision Making

Evaluation Criteria	Description
Technical criteria	
Landfill deviation rate	Ratio between waste not landfilled and total waste generated in a year (in percentage).
Environmental criteria	
Abiotic depletion (AbD)	Extraction of natural non-living resources. It is the difference between resources consumed during waste life cycle and resources consumption avoided from materials and energy substituted (in kg Sb eq).
Acidification	Referent to acidifying pollutants emitted during waste life cycle. The calculation is the difference between impacts from waste life cycle and the avoided impact from substituted materials and energy (in kg SO_2 eq).
Eutrophication	It is the consequence of high levels of macronutrients, such as nitrogen and phosphorus. It is the difference between eutrophication substances' potential impact during waste life cycle and avoided impacts from substituted materials and energy (in kg PO_4^{3-} eq).
Global warming potential (GWP)	Represents the impact of greenhouse gas (GHG) emissions on the radiative forcing of the atmosphere, inducing climate change. It is obtained from GHG potential impact from waste life cycle less the GHG impact from substituted materials (in kg CO_2 eq).
Human toxicity (HT)	It is the difference between impacts on human health of toxic substances emitted and the impacts from substituted materials and energy life cycle [in kg para-dichlorobenzene (p-DCB) eq].
Photochemical oxidation (PO)	Represents the formation of reactive chemical compounds, such as ozone, by action of sunlight on certain primary air pollutants. The calculation is provided from impact difference between waste life cycle and materials and energy substituted life cycles (in kg C_2H_2 eq).
Gross energy requirement (GER)	Amount of commercial energy that is required directly and indirectly by the process of making a good or service. It is the difference between energy consumed and energy produced (in kJ).
Economic criteria	
Investment costs	Represents the amount to be expended to implement the alternative (in infrastructure, equipment, vehicles, and land) (in millions €).
Operational costs	Related to the amount to be expended during alternative operation (in material, electricity, maintenance, and labor) and to financial costs like annuity (in €).
Operational revenues	The amount related to the profit obtained from selling products (energy, recyclables, compost) or
Social criteria	with the avoidance of landfilling products (RDF, recyclables) (in €).
Fee	It is the amount paid by the population to finance MSW management system (in €/t).

electricity. Specifically, 15% of the electricity consumed in Portugal was purchased from Spain, and the ratio can be taken into account too, with a proportion of 85:15, for carrying out the LCA. Polyethylene, expandable polystyrene, and plastic wood are specific cases having a 1:1 substitution ratio, since they appeared only one time in the market, such that degradation of the material was not considered.

In the cases where substitution ratio is less than 1, an open-loop allocation procedure was applied since degradation of the material should be considered, such as the cases of polyethylene terephthalate (PET), paper/cardboard, and paper from composite packaging. The calculation of substitution ratios was based on the limit number of times that a specific material can be recycled and reused repeatedly.[55] For PET, the limit number of recycling with respect to losing physical properties considered was five times.[56] Concerning paper from composite packaging, the same limit may be applied given that the proportion of paper in the packaging (0.75%) may be assumed, and the calculation procedure adopted by Rigamonti et al.[54] was applied. The substitution ratio adopted for PET was collected from the Institute for Prospective Technological Studies.[57]

Life Cycle Inventory

The LCI is the second phase of the LCA. It is an inventory of input/output data related to the SWM system that is being studied. It involves the collection of the data, which is necessary to meet the goals of the defined study.[50] In accordance with the scope of the study, an LCI was prepared for the waste management activities specified in Figure 2. The Umberto 5.5 software package was used to support the LCA.

Concerning each operational unit analyzed in the AMARSUL system, a short description of the data and assumptions considered for prescribed scenarios were provided. First of all, some of the information applied for our systems analysis was provided by the Empresa Geral do Fomento, co-owner of the SWM system at AMARSUL, which is responsible for the management of this MSW system, and the Portuguese Environment Agency. The rest of information was drawn from the Umberto software library and from the selected data sources such as the vendors of machinery features.

Life Cycle Analysis and Interpretation

Our LCA was then carried out using the Umberto 5.5 software package with the aid of the entire LCI as described in the previous section. Following the methodology suggested by the ISO 14040-44 standard,[3,50] environmental indicators were obtained for covering different impact categories. The impact categories being studied include abiotic depletion, acidification, eutrophication, global warming, human toxicity (HT), and photochemical oxidation. The characterization factors applied to each impact category are proposed by the CML 2000 method.[4]

Determination of Weighting Factors

The criteria applied to assess the alternatives in this study are partially based on some traditional criteria, which concern investment costs, operational costs, and possible revenues. Besides, the fee for waste disposal to be applied to the general public is an important criterion, which is considered as a social concern. Such a fee system is designed to increase the landfill life span, though. Due to this reason, the target of landfill space saving was considered as a technical criterion. Environmental criteria related to environmental impact categories in the context of LCA are those to be particularly tested for sensitivity analysis in this study (Table 2).

To determine the appropriate weighting factors for this application, all relevant criteria associated with different stakeholders involved in the decision-making processes were taken into account. The stakeholders involved in such decision-making processes of SWM are composed of governmental agencies, municipalities, technical groups, academics, and environmentalists. In general, governmental agencies would favor these economic criteria, whereas academia might prefer to weigh all criteria

TABLE 3

Criteria	Weights Options			
	Overall with Equal Weight	Technical	Economic	Social
Investment costs, operational costs, operational revenues	0.2	0.01	0.33	0.01
Fee	0.2	0.01	0.01	0.96
Landfill deviation rate	0.2	0.96	0.01	0.01

TABLE 4   Weight Sets for Traditional and LCA Criteria

	Weights Options				
	Overall with Equal Weight	Technical	Environmental	Economic	Social
Criteria					
AbD, Acid., Eutr., GWP, HT, PO, GER	0.08	0.01	0.14	0.01	0.01
Investment costs, operational costs, operational revenues	0.08	0.01	0.01	0.30	0.01
Fee	0.08	0.01	0.01	0.01	0.89
Landfill deviation rate	0.08	0.89	0.01	0.01	0.01
LCA criteria selected					
AbD/GWP/HT[a] or Acid./Eutr./PO[a] or AD/Acid./GER[a]	0.125	0.01	0.317	0.01	0.01
Investment costs, operational costs, operational revenues	0.125	0.01	0.01	0.317	0.01
Fee	0.125		0.01		0.93
Landfill deviation rate	0.125	0.93	0.01		

[a] The remaining environmental impact categories will have zero weight; their contribution to decision making will not be considered.

equally. Yet municipalities would have a closer look into a delicate balance between economic and social criteria simultaneously, whereas technical groups may focus on technical criteria. With such observations, a regional survey waste conducted to obtain stakeholders' weighting factors, being the results summarized in Table 3.

As a few more environmentalists were brought into the discussion to investigate the differential views, the LCA criteria were configured, signified, and summarized in such a decision-making process. To assess in detail the effect due to the inclusion of those LCA criteria in the SWM decision-making processes, three combinations of impact categories from LCA were tested. This combination will result in the application of three environmental impact categories (Table 4).

Alternative Screening and Ranking

Ranking Based on Traditional Criteria

When choosing the traditional criteria in SAW, the best option to be applied is conformed to the current one. That is mainly due to the revenues from selling the electric power generated at the sanitary landfill since the investment made for hardware construction has been present already, resulting in a reduced fee. However, there is a need to implement some essential measures to com-ply with Landfill and Packaging Directives in the future, which drives us to look at some more management options. Alternative P.A3* is the best when considering the same weight to all criteria. From a technical point of view, the option of treating MSW in MBT plants with AD treatment and ER of high-calorific fraction is the best option. In regard to the consideration of economic criteria only, the option of collection and treatment of separate BMW streams in an AD plant, including RDF production, is the best option. In having only the social criteria, the best option is to maximize the aerobic MBT plant. The results are collectively presented in Table 5. When applying the same practices based on the TOPSIS method, the best options associated

TABLE 5 Complete Ranking of the Alternatives from SAW Methodology

Ranking	Overall with Equal Weight		Technical		Economic		Social	
	Alt.	SAW	Alt.	SAW	Alt.	SAW	Alt.	SAW
1	Base	0.786	P.A4[b]	0.989	Base	0.878	Base	0.989
2	P.A3[a]	0.781	A4[b]	0.989	P.A6[a]	0.831	P.A1	0.595
3	A3[a]	0.780	P.A5[b]	0.968	A6[a]	0.831	P.A1[a]	0.593
4	P.A1[a]	0.776	A5[b]	0.968	P.A6	0.829	A1[a]	0.591
5	A1[a]	0.776	P.A4[a]	0.963	A6	0.828	A1	0.588
6	P.A3	0.774	A4[a]	0.963	P.A3[a]	0.801	P.A4[b]	0.564
7	A3	0.774	P.A3[a]	0.952	A3[a]	0.801	A4[b]	0.563
8	P.A1	0.773	A3a	0.952	P.A3	0.800	P.A4[a]	0.560
9	P.A4[b]	0.772	P.A2[b]	0.946	A3	0.800	A4[a]	0.555
10	A4[b]	0.772	A2[b]	0.946	P.A1	0.790	P.A3[a]	0.548
11	A1	0.771	A0[b]	0.946	P.A1[a]	0.789	P.A3	0.546
12	P.A6[a]	0.770	P.A0[b]	0.945	A1[a]	0.788	A3[a]	0.546
13	A6[a]	0.769	P.A5[a]	0.937	A1	0.788	A3	0.545
14	P.A4[a]	0.765	A5[a]	0.937	P.A5[b]	0.775	P.A4	0.542
15	P.A5[b]	0.765	P.A0[a]	0.925	A5b	0.775	A4	0.536
16	A5[b]	0.764	A0[a]	0.925	P.A5[a]	0.773	P.A5[b]	0.532
17	P.A6	0.764	P.A3	0.925	A5[a]	0.772	A5[b]	0.531
18	A4[a]	0.763	A3	0.925	P.A4[b]	0.769	P.A5[a]	0.529
19	A6	0.763	P.A1[a]	0.922	A4[b]	0.769	A5[a]	0.528
20	P.A5[a]	0.757	A1[a]	0.922	P.A4[a]	0.767	P.A6[a]	0.521
21	A5[a]	0.756	P.A2[a]	0.920	A4[a]	0.766	A6[a]	0.518
22	P.A2[b]	0.741	A2[a]	0.920	P.A4	0.756	P.A6	0.518
23	A2[b]	0.740	P.A1	0.902	A4	0.754	A6	0.514
24	A0[b]	0.736	A1	0.902	P.A5	0.754	P.A2[b]	0.510
25	P.A2[a]	0.734	P.A6[a]	0.834	A5	0.754	A2[b]	0.509
26	P.A0[b]	0.733	A6[a]	0.834	A0[b]	0.750	P.A2[a]	0.508
27	A2[a]	0.733	P.A6	0.813	P.A2[b]	0.750	A2[a]	0.504
28	P.A0[a]	0.731	A6	0.813	A2[b]	0.749	P.A5	0.502
29	A0[a]	0.731	P.A0	0.644	P.A0[a]	0.749	A5	0.501
30	P.A5	0.663	A0	0.644	A0[a]	0.749	P.A2	0.488
31	A5	0.663	P.A5	0.552	P.A2[a]	0.748	A2	0.487
32	P.A0	0.656	A5	0.552	A2[a]	0.747	P.A0[b]	0.486
33	P.A4	0.654	P.A2	0.502	P.A0[b]	0.745	A0[b]	0.485
34	A4	0.652	A2	0.502	P.A2	0.733	P.A0[a]	0.484
35	P.A2	0.638	P.A4	0.459	A2	0.733	A0[a]	0.483
36	A2	0.638	A4	0.459	P.A0	0.726	P.A0	0.457
37	A0	0.071	Base	0.308	A0	0.726	A0	0.456

Note: Alt., alternatives.

[a] Alternatives considering RDF production plus incineration of high-calorific fraction.

[b] Alternatives not considering RDF production but considering incineration of high-calorific fraction from MBT.

with the prescribed weighted criteria selected are similar to those obtained from the case of SAW except when all the criteria present the same weight. These results are collectively presented in Table 6. Comparing the scenarios with PAYT (denoted as P in the context) and with water billing indicates that the best is the PAYT system except when the base alternative is taken into account.

TABLE 6   Complete Ranking of the Alternatives from TOPSIS Methodology

Ranking	Overall with Equal Weight		Technical		Economic		Social	
	Alt.	TOPSIS	Alt.	TOPSIS	Alt.	TOPSIS	Alt.	TOPSIS
1	P.A1[a]	0.623	P.A4[b]	0.997	Base	0.600	P.A1	0.992
2	A1[a]	0.621	A4[b]	0.997	P.A6[a]	0.581	P.A1[a]	0.986
3	P.A1	0.617	P.A5[b]	0.960	A6[a]	0.579	A1[a]	0.977
4	A1	0.611	A5[b]	0.960	P.A6	0.577	A1	0.962
5	P.A3[a]	0.610	P.A4[a]	0.952	A6	0.575	P.A4[b]	0.813
6	A3[a]	0.609	A4[a]	0.952	P.A3[a]	0.507	A4[b]	0.808
7	P.A3	0.599	P.A3[a]	0.929	A3[a]	0.506	P.A4[a]	0.794
8	A3	0.598	A3[a]	0.929	P.A3	0.504	A4[a]	0.760
9	P.A4[b]	0.596	P.A2[b]	0.921	A3	0.503	P.A4	0.705
10	A4[b]	0.596	A2[b]	0.921	P.A1	0.474	P.A3[a]	0.703
11	P.A4[a]	0.586	A0[b]	0.921	P.A1[a]	0.471	P.A3	0.694
12	A4[a]	0.581	P.A0[b]	0.921	A1[a]	0.470	A3[a]	0.692
13	P.A5[b]	0.573	P.A5[a]	0.904	A1	0.469	A3	0.684
14	A5[b]	0.572	A5[a]	0.904	P.A5[b]	0.434	A4	0.665
15	P.A6[a]	0.562	P.A0[a]	0.883	A5[b]	0.433	Base	0.637
16	P.A5[a]	0.560	A0[a]	0.883	P.A5[a]	0.430	P.A5[b]	0.599
17	A6[a]	0.559	P.A3	0.878	A5[a]	0.429	A5[b]	0.591
18	A5[a]	0.559	A3	0.878	P.A4[b]	0.418	P.A5[a]	0.580
19	P.A6	0.549	P.A2[a]	0.873	A4[b]	0.418	A5[a]	0.570
20	A6	0.546	A2[a]	0.873	P.A4[a]	0.414	P.A6[a]	0.515
21	P.A2[b]	0.538	P.A1[a]	0.873	A4[a]	0.411	P.A6	0.491
22	A2[b]	0.537	A1[a]	0.873	P.A4	0.406	A6[a]	0.491
23	P.A2[a]	0.526	P.A1	0.835	A4	0.403	A6	0.465
24	A2[a]	0.523	A1	0.835	P.A5	0.402	P.A2[b]	0.439
25	A0[b]	0.522	P.A6[a]	0.704	A5	0.402	A2[b]	0.430
26	P.A0[b]	0.519	A6[a]	0.704	A0[b]	0.376	P.A2[a]	0.425
27	P.A0[a]	0.513	P.A6	0.666	P.A0[a]	0.374	P.A5	0.409
28	A0[a]	0.512	A6	0.666	A0[a]	0.373	A5	0.401
29	Base	0.481	P.A0	0.353	P.A2[b]	0.369	A2[a]	0.397
30	P.A0	0.347	A0	0.353	P.A0[b]	0.369	P.A2	0.304
31	A0	0.347	Base	0.188	A2[b]	0.369	A2	0.295
32	P.A5	0.345	P.A5	0.177	P.A2[a]	0.366	P.A0[b]	0.240
33	A5	0.345	A5	0.177	A2[a]	0.364	A0[b]	0.232
34	P.A4	0.324	P.A2	0.084	P.A2	0.351	P.A0[a]	0.226
35	A4	0.320	A2	0.084	A2	0.350	A0[a]	0.214
36	P.A2	0.288	P.A4	0.004	P.A0	0.328	P.A0	0.014
37	A2	0.288	A4	0.004	A0	0.328	A0	0.009

Note: Alt., alternatives.

[a] Alternatives considering RDF production plus incineration of high-calorific fraction.

[b] Alternatives not considering RDF production but considering incineration of high-calorific fraction from MBT.

Ranking Based on Traditional and LCA Criteria Together

By looking into the results obtained through SAW and TOPSIS with the inclusion of LCA criteria, the results vary considerably. In most cases, the option of treating MSW through AD MBT followed by ER of waste is the best option, as shown in Table 7. Only when the economic criteria were taken into

TABLE 7   Complete Ranking of the Alternatives from SAW Methodology—All Impact Categories

Ranking	Overall with Equal Weight Alt.	SAW	Technical Alt.	SAW	Environmental Alt.	SAW	Economic Alt.	SAW	Social Alt.	SAW
1	P.A4[b]	0.734	P.A4[b]	0.968	P.A4[b]	0.709	Base	0.805	Base	0.908
2	A4[b]	0.733	A4[b]	0.968	A4[b]	0.709	P.A6[a]	0.790	P.A4[b]	0.575
3	P.A4[a]	0.697	P.A4[a]	0.940	P.A4[a]	0.654	A6[a]	0.789	A4[b]	0.574
4	A4[a]	0.696	A4[a]	0.940	A4[a]	0.654	P.A6	0.786	P.A4[a]	0.567
5	P.A5[b]	0.661	P.A5[b]	0.940	P.A5[b]	0.596	A6	0.786	A4[a]	0.562
6	A5[b]	0.661	A5[b]	0.940	A5[b]	0.596	P.A4[b]	0.765	P.A1[a]	0.552
7	P.A5[a]	0.634	P.A2[b]	0.909	P.A5[a]	0.556	A4[b]	0.765	P.A1	0.552
8	A5[a]	0.634	A2[b]	0.909	A5[a]	0.556	P.A5[b]	0.762	A1[a]	0.550
9	P.A2[b]	0.564	P.A5[a]	0.909	P.A2[b]	0.453	A5[b]	0.761	A1	0.545
10	P.A0[b]	0.564	A5[a]	0.909	A2[b]	0.453	P.A4[a]	0.759	P.A5[b]	0.537
11	P.A2[a]	0.546	A0[b]	0.907	P.A0[b]	0.426	A4[a]	0.758	A5[b]	0.536
12	A2[b]	0.545	P.A0[b]	0.907	P.A2[a]	0.426	P.A5[a]	0.757	P.A5[a]	0.531
13	A0[b]	0.545	P.A3[a]	0.888	A0[b]	0.426	A5[a]	0.756	A5[a]	0.530
14	P.A0[a]	0.545	A3[a]	0.888	A2[a]	0.426	P.A3[a]	0.748	P.A3[a]	0.513
15	A2[a]	0.517	P.A0[a]	0.885	P.A0[a]	0.382	A3[a]	0.748	A3[a]	0.511
16	A0[a]	0.516	A0[a]	0.885	A0[a]	0.382	P.A3	0.745	P.A3	0.510
17	P.A6[a]	0.463	P.A2[a]	0.884	P.A6[a]	0.270	A3	0.745	P.A4	0.509
18	A6[a]	0.463	A2[a]	0.884	A6[a]	0.270	P.A1[a]	0.733	A3	0.508
19	P.A6	0.446	P.A3	0.861	P.A6	0.247	A1[a]	0.733	P.A2[b]	0.506
20	A6	0.446	A3	0.860	A6	0.247	P.A1	0.732	A2[b]	0.505
21	P.A3[a]	0.350	P.A1[a]	0.857	P.A4	0.100	A1	0.731	A4	0.504
22	P.A3	0.349	A1[a]	0.857	A4	0.100	P.A2[b]	0.728	P.A6[a]	0.503
23	A3[a]	0.331	P.A1	0.837	P.A5	0.083	A2[b]	0.728	P.A2[a]	0.502
24	P.A1[a]	0.331	A1	0.837	P.A3[a]	0.083	A0[b]	0.726	A6[a]	0.500
25	A3	0.324	P.A6[a]	0.792	A3[a]	0.079	P.A2[a]	0.724	A2[a]	0.498
26	P.A1	0.324	A6[a]	0.792	A5	0.079	A2[a]	0.723	P.A6	0.498
27	A1[a]	0.314	P.A6	0.772	P.A3	0.054	P.A0[b]	0.722	A6	0.495
28	P.A4	0.313	A6	0.772	P.A0	0.054	P.A0[a]	0.722	P.A0[b]	0.481
29	P.A5	0.307	P.A0	0.601	A3	0.053	A0[a]	0.722	A0[b]	0.481
30	P.A0	0.307	A0	0.601	A0	0.053	P.A4	0.707	P.A0[a]	0.476
31	A1	0.306	P.A5	0.517	P.A1[a]	0.040	A4	0.706	A0[a]	0.475
32	A4	0.305	A5	0.517	A1[a]	0.040	P.A5	0.705	P.A5	0.471
33	A5	0.286	P.A2	0.467	P.A2	0.030	A5	0.705	A5	0.470
34	P.A2	0.286	A2	0.467	A2	0.030	P.A2	0.681	P.A2	0.455
35	A0	0.265	P.A4	0.432	P.A1	0.012	A2	0.681	A2	0.454
36	A2	0.265	A4	0.432	A1	0.012	P.A0	0.676	P.A0	0.427
37	Base	0.231	Base	0.276	Base	−0.118	A0	0.676	A0	0.426

Note: Alt., alternatives.

[a] Alternatives considering RDF production plus incineration of high-calorific fraction.

[b] Alternatives not considering RDF production but considering incineration of high-calorific fraction from MBT.

account, applying the AD to treat a source separated from BMW followed by RDF production remains the best option, if the base case is not considered. When the TOPSIS was applied, the results, as presented in Table 8, have changed, being preferable options with AD and incineration of high-calorific fraction. However, concerning social criteria, the option of maximizing aerobic MBT is considered as the best solution. Also, at almost criteria cases, the option of applying PAYT is better than the same technological option with the water billing system.

TABLE 8 Complete Ranking of the Alternatives from TOPSIS Methodology—All Impact Categories

Ranking	Overall with Equal Weight		Technical		Environmental		Economic		Social	
	Alt.	TOPSPS	Alt.	TOPSPS	Alt.	TOPSPS	Alt.	TOPSPS	Alt.	TOPSPS
1	P.A4[b]	0.795	P.A4[b]	0.975	P.A4[b]	0.807	Base	0.575	P.A1[a]	0.881
2	A4[b]	0.795	A4[b]	0.975	A4[b]	0.807	P.A6[a]	0.574	P.A1	0.879
3	P.A4[a]	0.789	P.A5[b]	0.953	P.A4[a]	0.801	A6[a]	0.572	A1[a]	0.879
4	A4[a]	0.789	A5[b]	0.953	A4[a]	0.801	P.A6	0.570	A1	0.871
5	P.A5[b]	0.777	P.A4[a]	0.946	P.A5[b]	0.791	A6	0.568	P.A4[b]	0.809
6	A5[b]	0.777	A4[a]	0.946	A5[b]	0.791	P.A3[a]	0.511	A4[b]	0.805
7	P.A5[a]	0.765	P.A2[b]	0.915	P.A5[a]	0.779	A3[a]	0.510	P.A4[a]	0.791
8	A5[a]	0.765	A2[b]	0.915	A5[a]	0.779	P.A3	0.507	A4[a]	0.759
9	P.A2[b]	0.732	A0[b]	0.915	P.A2[b]	0.743	A3	0.506	P.A4	0.689
10	A2[b]	0.732	P.A0[b]	0.915	A2[b]	0.743	P.A1	0.485	P.A3[a]	0.683
11	P.A2[a]	0.724	P.A3[a]	0.909	P.A2[a]	0.735	P.A1[a]	0.484	P.A3	0.673
12	A2[a]	0.724	A3[a]	0.909	A2[a]	0.735	A1[a]	0.483	A3[a]	0.672
13	A0[b]	0.717	P.A5[a]	0.900	A0[b]	0.729	A1	0.480	A3	0.664
14	P.A0[b]	0.716	A5[a]	0.900	P.A0[b]	0.729	P.A5[b]	0.460	A4	0.652
15	P.A0[a]	0.703	P.A0[a]	0.879	P.A0[a]	0.715	A5[b]	0.459	Base	0.634
16	A0[a]	0.703	A0[a]	0.879	A0[a]	0.715	P.A5[a]	0.455	P.A5[b]	0.604
17	P.A1[a]	0.578	P.A2[a]	0.870	P.A1[a]	0.577	A5[a]	0.455	A5[b]	0.596
18	A1[a]	0.578	A2[a]	0.870	A1[a]	0.577	P.A4[b]	0.454	P.A5[a]	0.585
19	P.A1	0.567	P.A3	0.865	P.A4	0.570	A4[b]	0.454	A5[a]	0.575
20	A1	0.566	A3	0.865	A4	0.570	P.A4[a]	0.449	P.A6[a]	0.506
21	P.A4	0.560	P.A1[a]	0.863	P.A1	0.565	A4[a]	0.446	A6[a]	0.484
22	A4	0.560	A1[a]	0.863	A1	0.565	P.A4	0.417	P.A6	0.483
23	P.A3[a]	0.547	P.A1	0.827	P.A2	0.545	A4	0.415	A6	0.459
24	A3[a]	0.547	A1	0.827	A2	0.545	P.A5	0.411	P.A2[b]	0.450
25	P.A2	0.536	P.A6[a]	0.699	P.A3[a]	0.544	A5	0.410	A2[b]	0.442
26	A2	0.536	A6[a]	0.699	A3[a]	0.544	A0[b]	0.405	P.A2[a]	0.436
27	P.A0	0.535	P.A6	0.662	P.A0	0.543	P.A2[b]	0.405	A2a	0.410
28	A0	0.535	A6	0.662	A0	0.542	A2[b]	0.404	P.A5	0.406
29	P.A3	0.534	P.A0	0.354	P.A3	0.532	P.A0[a]	0.402	A5	0.399
30	A3	0.534	A0	0.354	A3	0.532	A0[a]	0.401	P.A2	0.309
31	P.A6[a]	0.528	Base	0.192	P.A5	0.529	P.A2[a]	0.401	A2	0.299
32	A6[a]	0.528	P.A5	0.179	A5	0.529	A2[a]	0.399	P.A0[b]	0.265
33	P.A5	0.522	A5	0.179	P.A6[a]	0.527	P.A0[b]	0.398	A0[b]	0.257
34	A5	0.522	P.A2	0.091	A6[a]	0.527	P.A2	0.366	P.A0[a]	0.251
35	P.A6	0.519	A2	0.091	P.A6	0.518	A2	0.366	A0[a]	0.241
36	A6	0.519	P.A4	0.037	A6	0.518	P.A0	0.346	P.A0	0.072
37	Base	0.290	A4	0.037	Base	0.278	A0	0.346	A0	0.071

Note: Alt., alternatives.

[a] Alternatives considering RDF production plus incineration of high-calorific fraction.

[b] Alternatives not considering RDF production but considering incineration of high-calorific fraction from MBT.

TABLE 9 Best-Five Ranking for the Three Different LCA Impact Categories from SAW

	Overall with Equal Weight		Technical		Environmental		Economic		Social	
	Alt.	SAW	Alt.	SAW	Alt.1	SAW	Alt.	SAW	Alt.	SAW
				Ranking of AbD/GWP/HT						
1	P.A6[a]	0.701	P.A4[b]	0.968	P.A6[a]	0.595	Base	0.833	Base	0.941
2	A6[a]	0.700	A4[b]	0.968	A6[a]	0.595	P.A6[a]	0.824	P.A1	0.560
3	P.A6	0.689	P.A5[b]	0.948	P.A6	0.575	A6[a]	0.823	P.A1[a]	0.559
4	A6	0.688	A5[b]	0.948	A6	0.575	P.A6	0.821	A1[a]	0.557
5	P.A4[b]	0.600	P.A4[a]	0.941	P.A5[b]	0.348	A6	0.820	P.A4[b]	0.557
				Ranking of Acid./Eutr./PO						
1	P.A4[b]	0.858	P.A4[b]	0.989	P.A4[b]	0.989	Base	0.855	Base	0.962
2	A4[b]	0.857	A4[b]	0.989	A4[b]	0.989	P.A6[a]	0.801	P.A1	0.588
3	P.A4[a]	0.832	P.A4[a]	0.962	P.A4[a]	0.934	A6[a]	0.800	P.A1[a]	0.587
4	A4[a]	0.831	A4[a]	0.962	A4[a]	0.934	P.A6	0.798	A1[a]	0.585
5	P.A2[b]	0.759	P.A5[b]	0.960	P.A2[b]	0.786	A6	0.797	A1	0.582
				Ranking of AbD/ Acid./GER						
1	P.A4[b]	0 858	P.A4[b]	0.989	P.A4[b]	0.989	Base	0.858	Base	0.965
2	A4[b]	0.857	A4[b]	0.989	A4[b]	0.989	P.A6[a]	0.816	P.A1	0.586
3	P.A4[a]	0.841	P.A5[b]	0.965	P.A4[a]	0.958	A6[a]	0.815	P.A1[a]	0.586
4	A4[a]	0.840	A5[b]	0.965	A4[a]	0.958	P.A6	0.813	A1[a]	0.584
5	P.A5[b]	0.809	P.A4[a]	0.963	P.A5[b]	0.877	A6	0.812	P.A4[a]	0.580

Note: Alt., alternatives.

[a] Alternatives considering RDF production plus incineration of high-calorific fraction.

[b] Alternatives not considering RDF production but considering incineration of high-calorific fraction from MBT.

Another concern when applying LCA criteria is that the total number and type of impact category used in the SAW and TOPSIS could impact the final options. If we reduce the consideration from seven to three criteria, the results would be changed in both SAW and TOPSIS. When looking into the case of SAW, such a change is salient as only AbD/global warming potential (GWP)/HT impact categories were applied, as shown in Table 9. In this case, the AD of BMW separated from source with the inclusion of RDF ER in an incineration plant would be the best option for more criteria. For the remaining environmental impacts the results are mostly in accordance with the total LCA criteria applied (Table 7). Such a finding suggests that there are some LCA criteria that would have a bigger influence on the ranking outcome.

The phenomenon mentioned before has also been verified via TOPSIS (Table 10). When looking into the case of TOPSIS, for LCA criteria AbD/GWP/HT, the results were changed even more than those in the counterpart. When only the technical criterion was taken into account, the option favored the use of an anaerobic MBT plant with ER from the high-calorific fraction, without producing RDF. Such an option is always advantageous no matter how many LCA criteria were included for screening and ranking. Options like the adoption of a full aerobic MBT plant to divert organic waste from landfill as well as selective collection of BMW plus RDF production with burning in an incineration plant were selected as the best options when considering the other combinations of environmental impact categories in LCA (Table 10). Yet the inclusion of a varying number of criteria in LCA might impact the preference for PAYT. It obviously leads to the conclusion that the PAYT can be a better instrument to meet the environmental targets from a sustainable point of view.

TABLE 10   Best-Five Ranking for the Three Different LCA Impact Categories from TOPSIS

	Overall with Equal Weight		Technical		Environmental		Economic		Social	
	Alt.	TOPSIS	Alt.	TOPSIS	Alt.	TOPSIS	Alt.	TOPSIS	Alt.	TOPSIS
					Ranking of AbD/GWP/HT					
1	P.A6[a]	0.735	P.A4[b]	0.976	P.A6[a]	0.753	P.A6[a]	0.587	P.A1	0.903
2	A6[a]	0.734	A4[b]	0.976	A6[a]	0.753	Base	0.586	P.A1[a]	0.903
3	P.A6	0.723	P.A5[b]	0.956	P.A5[b]	0.742	A6[a]	0.585	A1[a]	0.901
4	A6	0.723	A5[b]	0.956	A5[b]	0.742	P.A6	0.583	A1	0.894
5	P.A5[b]	0.722	P.A4[a]	0.946	P.A6	0.741	A6	0.581	P.A4[b]	0.807
					Ranking of Acid./Eutr./PO					
1	P.A4[b]	0.887	P.A4[b]	0.997	P.A4[b]	0.996	Base	0.591	P.A1[a]	0.938
2	A4[b]	0.887	A4[b]	0.997	A4[b]	0.996	P.A6[a]	0.569	P.A1	0.937
3	P.A4[a]	0.882	P.A5[b]	0.958	P.A4[a]	0.977	A6[a]	0.567	A1[a]	0.935
4	A4[a]	0.881	A5[b]	0.958	A4[a]	0.977	P.A6	0.565	A1	0.926
5	P.A2[b]	0.833	P.A4[a]	0.952	P.A2[b]	0.893	A6	0.563	P.A4[b]	0.816
					Ranking of AbD/Acid./GER					
1	P.A4[b]	0.799	P.A4[b]	0.983	P.A4[b]	0.992	Base	0.595	P.A1[a]	0.930
2	A4[b]	0.799	A4[b]	0.983	A4[b]	0.992	P.A6[a]	0.576	P.A1	0.928
3	P.A4[a]	0.789	P.A5[b]	0.957	P.A4[a]	0.961	A6[a]	0.574	A1[a]	0.928
4	A4[a]	0.787	A5[b]	0.957	A4[a]	0.961	P.A6	0.572	A1	0.918
5	P.A5[b]	0.740	P.A4[a]	0.950	P.A5[b]	0.847	A6	0.570	P.A4[b]	0.814

Note: Alt., alternatives.

[a] Alternatives considering RDF production plus incineration of high-calorific fraction.

[b] Alternatives not considering RDF production but considering incineration of high-calorific fraction from MBT.

Final Remarks on the Importance of LCA

The inclusion of environmental criteria into the MADM process can result in a change of the best option when compared with the cases where only all or part of the traditional criteria are taken into account. This proves that a sustainable decision must include all elements of sustainability, or else the decision would end up a biased option. Even though LCA impact category can be considered, the results would vary when differing numbers of LCA impact category were assessed in the MADM process. Impact categories are mutually exclusive, however. When all impact categories were included, the normalization process in the MADM process might dilute the relative importance or effect of some impact categories and, therefore, distort the outcome to some extent. In any circumstance, it is not possible to implement weighted criteria to restore such an outcome. To reach a sociopolitical sustainability, the selection of an impact category should be the stakeholders' responsibility. Regardless of some discrepancies present in the selection of impact categories and normalization, the final results obtained from both SAW and TOPSIS allow improvements for SWM in Portugal complying with the new guidelines and mandatory items with respect to those sustainable principles reinforced by European policies. Overall, the option of applying PAYT has been shown to be a good option relative to the water billing system.

Conclusions

In this entry, we analyze how LCA could have an impact on the decision-making processes of SWM. A case study in Portugal clearly indicates that the inclusion of LCA for screening and ranking has shown the potential of helping SWM decision making. The inclusion of LCA impact categories, together with traditional criteria associated with economic considerations, is capable of producing relatively sustainable decisions, which otherwise would not be reached. Within the analysis, both SAW and TOPSIS were proven capable of showing which alternatives should be favored. Comparing traditional criteria, these two methods have quite-consistent outputs in terms of several weighted criteria applied. When considering only the traditional criteria, the base case is the best option, without regard to environmental regulations. With the inclusion of LCA criteria, the best solution is to adopt the AD MBT plant followed by the energy recovery at incineration plant. Overall, the option of applying PAYT has been shown to be a good option relative to the water billing system. The selection of the number and type of LCA criteria (environmental impact categories) to be included in the MADM process is flexible, depending on the stakeholders' preference during the decision analysis. Such selection must be careful, however, since it ends up making an obvious difference in the final decision analysis results.

References

1. Council of European Parliament. *Proposal for a Directive of the European Parliament and of the Council on Waste— COM(2005) 667;* European Union: Brussels, Belgium, 2005.
2. Council of European Parliament. *Directive 2008/98/EC of the European Parliament and of the Council of 19 November 2008 on Waste and Repealing Certain Directives,* Official Journal L312/3; European Union: Strasbourg, France, 2008; 3–30.
3. ISO. *ISO 14040—Environmental Management—Life Cycle Assessment—Principles and Framework;* International Standard Organization: Switzerland, 2006.
4. Guinée, J.B.; Gorree, M.; Heijungs, R.; Huppes, G.; Kleijn, R.; van Oers, L.; Wegener Sleeswijk, A.; Suh, S.; Udo de Haes, H.A.; de Bruijn, J.A.; van Duin, R.; Huijbregts, M.A.J., Eds. *Handbook on Life Cycle Assessment—Operational Guide to the ISO Standards;* Kluwer Academic Publisher: Dordrecht, Netherlands, 2002.
5. Available at http://www.epa.gov/nrmrl/lcaccess/pdfs/600r06060.pdf (accessed December 2010).
6. Available at http://www2.mst.dk/Udgiv/publications/1998/87-7909-026-5/pdf/87-7909-026-5. pdf (accessed January 2010).
7. Available at http://www2.mst.dk/Udgiv/publications/1998/87-7909-025-7/pdf/87-7909-025-7. pdf (accessed January 2010).
8. Available at http://www2.mst.dk/Udgiv/publications/1998/87-7909-024-9/pdf/87-7909-024-9. pdf (accessed January 2010).
9. Available at http://www2.mst.dk/Udgiv/publications/1998/87-7909-023-0/pdf/87-7909-023-0. pdf (accessed January 2010).
10. Available at http://www2.mst.dk/Udgiv/publications/1998/87-7909-014-1/pdf/87-7909-014-1.pdf (accessed January 2010).
11. Finnveden, G.; Person, L.; Steen, B. *Förpackningar i kret-sloppet: Atervinning av mjölkkartong— En LCA-studie av skillnader i miljöbelastning (Packaging in Circulation: Recycling of Cardboard for Milk—An LCA Study of the Differences in Environmental Load),* Report 4301; Swedish Environmental Protection Agency: Stockholm, Sweden, 1994.
12. Finnveden, G.; Ekvall, T. *Energieller materialåtervinning av pappersförpakningar? (Energy Recovery or Material Recycling of Paper Packaging?);* Svensk Kartongåtervinning AB: Stockholm, Sweden, 1998.
13. Sakai, S.; Sawell, S.E.; Chandler, A.J.; Eighmy, T.T.; Kos-son, D.S.; Vehlow, J.; van der Sloot, H.A.; Hartlén, J.; Hjel-mar, O. World trends in municipal solid waste management. Waste Manage. **1996**, *16* (5/6), 341–350.
14. Klöpffer, W. Conference reports: LCA in Brighton. Int. J. Life Cycle Assess. **2000**, *5* (4), 249.
15. Available at http://www.ecobilan.com/wisard/index.php (accessed December 2008).
16. Available at http://www.iwm-model.uwaterloo.ca/iswm_booklet.pdf (accessed December 2009).
17. Diaz, R.; Warith, M. Life-cycle assessment of municipal solid wastes: Development of the WASTED model. Waste Manage. **2006**, *26* (8), 886–901.

18. Kirkeby, J.; Christensen, T.; Bhander, G.; Hansen, T.; Birgisdottir, H. LCA modelling of MSW management system: Approach and case study. In *Proceedings Sardinia 2005*, Tenth International Waste Management and Landfill Symposium, S. Margherita di Pula, Cagliari, Italy, October 3–7, 2005; Cossu, R., Stegmann, R. Eds.; CISA, Environmental Sanitary Engineering Centre, 2005.

19. McDougall, F.; White, P.; Franke, M.; Hindle, P. *Integrated Solid Waste Management: A Life Cycle Inventory*; Blackwell Science Ltd.: Oxford, 2001.

20. White, P.; Franke, M.; Hindle, P. *Integrated Solid Waste Management: A Life-Cycle Inventory*; Blackie Academic & Professional: Glasgow, U.K., 1995.

21. Winkler, J.; Bilitewski, B. Comparative evaluation of life cycle assessment models for solid waste management. Waste Manage. **2007**, *27* (8), 1021–1031.

22. Chang, N.B.; Pires, A.; Martinho, G. Empowering systems analysis for solid waste management: The tools and perspectives in European countries. Crit. Rev. Environ. Sci. Technol. **2011**, *41* (16), 1449–1530.

23. Weitz, K.; Barlaz, M.; Ranji, R.; Brill, D.; Thorneloe, S.; Ham, R. Life cycle management of municipal solid waste. Int. J. Life Cycle Assess. **1999**, *4* (4), 195–201.

24. Thorneloe, S.; Weitz, K.; Jambeck, J. Application of the US decision support tool for materials and waste management. Waste Manage. **2007**, *27* (8), 1006–1020.

25. Nilsson, M.; Bjorklund, A.; Finnveden, G.; Johansson, J. Testing a SEA methodology for the energy sector: A waste incineration tax proposal. Environ. Impact Assess. Rev. **2005**, *25* (1), 1–32.

26. Björklund, A.; Finnveden, G. Life cycle assessment of a national policy proposal—The case of a Swedish waste incineration tax. Waste Manage. **2007**, *27* (8), 1046–1058.

27. Dalemo, M.; Sonesson, U.; Bjorklund, A.; Mingarini, K.; Frostell, B.; Nybrant, T.; Jonsson, H.; Sundqvist, J.-O.; Thyselius, L. ORWARE—A simulation model for organic waste handling systems. Part 1: Model description. Resour., Conserv. Recycl. **1997**, *21* (1), 17–37.

28. Björklund, A.; Dalemo, M.; Sonesson, U. Evaluating a municipal waste management plan using ORWARE. J. Cleaner Prod. **1999**, *7* (4), 271–280.

29. Eriksson, O.; Frostell, B.; Bjorklund, A.; Assefa, G.; Sundwvist, J.O.; Granath, J.; Carlsoon, M.; Baky, A.; Thyselius, L. ORWARE—A simulation tool for waste. Resour., Conserv. Recycl. **2002**, *36* (4), 287–307.

30. Eriksson, O.; Reich, M.; Frostell, B.; Bjorklund, A.; Assefa, G.; Sundqvist, J.O.; Granath, J.; Baky, A.; Thyselius, L. Municipal solid waste management from a systems perspective. J. Cleaner Prod. **2005**, *13* (3), 241–252.

31. Solano, E.; Dumas, R.; Harrison, K.; Ranjithan, S.; Barlaz, M.; Brill, E. Life-cycle–based solid waste management. II: Illustrative applications. J. Environ. Eng. **2002**, *128* (10), 993–1005.

32. Solano, E.; Ranjithan, S.; Barlaz, M.; Brill, E. Life-cycle–based solid waste management. I: Model development. J. Environ. Eng. **2002**, *128* (10), 981–992.

33. Harrison, K.; Dumas, R.; Solano, E.; Barlaz, M.; Brill, E.; Ranjithan, S. Decision support tool for life-cycle–based solid waste management. J. Comput. Civ. Eng. **2001**, *15* (1), 44–58.

34. Kijak, R.; Moy, D. A decision support framework for sustainable waste management. J. Ind. Ecol. **2004**, *8* (3), 33–50.

35. Contreras, F.; Hanaki, K.; Aramaki, T.; Connors, S. Application of analytical hierarchy process to analyze stakeholders preferences for municipal solid waste management plans, Boston, USA. Resour., Conserv. Recycl. **2008**, *52* (7), 979–991.

36. Skordilis, A. Modelling of integrated solid waste management systems in an island. Resour., Conserv. Recycl. **2004**, *41* (3), 243–254.

37. Available at http://www.amarsul.pt/relatorios_e_contas.php (accessed July 2009).

38. Council of European Parliament. *Directive 2004/12/EC of the European Parliament and of the Council of 11 February 2004 Amending Directive 94/62/EC on Packaging and Packaging Waste*, Official Journal L047; European Union: Strasbourg, France, 2004; 26–32.

39. Council. *Council Directive 99/31/EC of April 1999 on the Landfill of Waste,* Official Journal L182; European Union: Luxembourg, 1999; 1–19.
40. Roy, B. The outranking approach and the foundations of ELECTRE methods. Decis. Theory **1991**, *31* (1), 49–73.
41. Brans, J.P.; Mareschal, B.; Vincke, P.H. PROMETHEE: A new family of outranking methods in multicriteria analysis. Oper. Res. **1984**, *84,* 447–490.
42. Mareschal, B. Geometrical representations for MCDA: The GAIA procedure. Eur. J. Oper. Res. **1988**, *34,* 69–77.
43. Saaty, T.L. *The Analytic Hierarchy Process;* McGraw-Hill: New York, 1980.
44. Yoon, K.; Hwang, C.L. Manufacturing plant location analysis by multiple attribute decision making: Part I—Singleplant strategy. Int. J. Prod. Res. **1985**, *23* (2), 345–359.
45. Zanakis, S.H.; Solomon, A.; Wishart, N.; Dublish, S. Multiattribute decision making: A simulation comparison of select methods. Eur. J. Oper. Res. **1998**, *107* (3), 507–529.
46. Cheng, S.; Chan, C.W.; Huang, G.H. Using multiple criteria decision analysis for supporting decision of solid waste management. J. Environ. Sci. Health A **2002**, *37* (6), 975–990.
47. Chang, Y.H.; Yeh, C.H. Evaluating airline competitiveness using multiattribute decision making. Omega **2001**, *29* (5), 405–415.
48. Hwang, C.L.; Yoon, K. *Multiple Attribute Decision Making—Lecture Notes in Economics and Mathematical Systems;* Springer: Berlin, 1981.
49. Ekvall, T.; Assefa, G.; Björklund, A.; Eriksson, O.; Finnveden, G. What life-cycle assessment does and does not do in assessments of waste management. Waste Manage. **2007**, *27* (8), 989–996.
50. ISO. *ISO 14044—Environmental Management—Life Cycle Assessment—Requirements and Guidelines;* International Standard Organization: Switzerland, 2006.
51. Tillman, A.M.; Ekvall, T.; Baumann, H.; Rydberg, T. Choice of system boundaries in life cycle assessment. J. Cleaner Prod. **1994**, *2* (1), 21–29.
52. Thomassen, M.A; Dalgaard, R.; Heijungs, R.; de Boer, I. Attributional and consequential LCA of milk production. Int. J. Life Cycle Assess. **2008**, *13* (4), 339–349.
53. Finnveden, G.; Hauschild, M.Z.; Ekvall, T.; Guinée, J.; Heijungs, R.; Hellweg, S.; Koehler, A.; Pennington, D.; Suh, S. Recent developments in life cycle assessment. J. Environ. Manage. **2009**, *91* (1), 1–21.
54. Rigamonti, L.; Grosso, M.; Sunseri, M.C. Influence of assumptions about selection and recycling efficiencies on the LCA of integrated waste management systems. Int. J. Life Cycle Assess. **2009**, *14* (5), 411–419.
55. Rigamonti, L.; Grosso, M.; Giugliano, M. Life cycle assessment for optimising the level of separated collection in integrated MSW management systems. Waste Manage. **2009**, *29* (2), 934–944.
56. Comieco. Carta: Introduzione (Cardbord: Introduction). Matrec. 2008. Available at: http://www.matrec.it/ (accessed January 2010).
57. Delgado, C.; Barruetabeña, L.; Salas O. *Assessment of the Environmental Advantages and Drawbacks of Existing and Emerging Polymers Recovery Processes,* JRC Scientific and Technical Reports; European Communities: Luxembourg, 2007; 1–278.

27

Sustainable Development: Ecological Footprint in Accounting

Simone Bastianoni,
Valentina
Niccolucci, Elena
Neri, Gemma
Cranston,
Alessandro Galli,
and Mathis
Wackernagel

Introduction

Since the onset, sustainable development has called for a set of suitable tools and sustainability indicators that are able to monitor the relevant aspects of systems under scrutiny. In general, sustainability indicators are required to highlight the major "hot spots" of a system and provide the most relevant elements upon which to carry out a "sustainability diagnosis." This is necessary to determine the most appropriate "sustainability therapy," which may include future action plans, programs, and strategy, to support the development of environmental management policies and the decision-making process.

In recent decades, the global increase in resource consumption, waste production, and environmental pollution has become a core aspect within the sustainable development framework. The mismatch between the earth's resource regeneration rates and humanity's consumption rates is fundamental to understanding sustainability; the importance of this discrepancy was highlighted by H. Daly through the formulation of his famous "sustainability principles." Such principles can be expressed as follows: 1) renewable resources must be used no faster than the rate at which they regenerate; 2) nonrenewable resources must be used no faster than the time it takes for renewable substitutes to be implemented; and 3) pollution and wastes must be emitted no faster than the speed with which natural systems can absorb them, recycle them, or render them harmless. Assessments are therefore needed to account for both the ecosystem's capacity to provide ecological assets to support human activities and the subsequent effects human use has on the environment. A solution could be to focus upon careful management of natural capital with the long-term aim to preserve the integrity of ecosystems and improve human well-being.

The "life-supporting" natural capital upon which humans depend is generated by the negentropic capacity of the planet to convert, via photosynthesis, low-quality forms of energy (e.g., solar energy) into high-quality forms of energy that can be used by living organisms among which are humans. As such, the Ecological Footprint Analysis (EFA) can be considered a resource accounting framework, with a biophysical and thermodynamic basis, providing reliable measures of both (biosphere) supply of and (human) demand for the photosynthetic capacity of the planet.

Ecological Footprint analysis is not able to answer all aspects of sustainability but instead provides an account of the flows and overuse of resources by humanity. The imbalance between human demand and nature's capacity is one part of the sustainability issue, and the Ecological Footprint is able to answer the key question of how much of the earth's regenerative capacity does humanity demand.

The Ecological Footprint is a highly versatile tool with a wide range of applications and functions ranging from territorial levels (cities, regions, nations, etc.), to a wide set of systems. Ecological Footprint analyses help governments, businesses, and individuals to manage resources available and plan on how to use them in a sustainable way.

This entry consists of seven main sections. After a general introduction, the first section details the conceptual background underpinning the Ecological Footprint methodology. There are two fundamental aspects to the methodology—Ecological Footprint and biocapacity—both of which are introduced and described. In the section "The Mathematics of the Ecological Footprint," the mathematics behind Ecological Footprint and biocapacity calculations is explored. This section also discusses the topic of ecological balances (the comparison between Ecological Footprint and biocapacity) and focuses on the concept of biocapacity remainder and deficit. The section "Toward a Multiindicator Approach" offers a joint discussion of the Ecological Footprint with other indicators like human development index (HDI), gross domestic product (GDP), and index of sustainable economic welfare (ISEW). The section "New Insight in Footprint Theory: Toward a ThreeDimensional Ecological Footprint Geography" presents a new theoretical insight: a three dimensional representation of the Ecological Footprint based on the distinction between depletion of natural capital stocks and use of natural capital flows. An overview of existing applications of the Ecological Footprint to environmental management practices is then provided in the section "Applications of Ecological Footprint," from territorial to products and services. The section "Weakness and Limitations of EFA" acknowledges the limitations of the methodology and provides final conclusions.

Fundamentals of EFA

The EFA was introduced as an environmental accounting and management tool able to provide either static snapshots or temporal trends of human demand on ecological assets (i.e., natural resources and ecosystem services).[1,2] The EFA introduced two relevant spatial parameters:[3–5] Ecological Footprint and biocapacity. The Ecological Footprint tracks the amount of biologically productive area directly and indirectly demanded on a continuous basis by humanity to provide the energy and material resource flows used and to assimilate the wastes generated (i.e., CO_2 emissions), given prevailing technologies and resource management practices.

Biocapacity represents the maximum regenerative capacity of the biosphere, which is annually available for human use. Therefore, biocapacity is an intrinsic property for any given area (where the population lives or the product is produced). Ecological Footprint and biocapacity have some common properties—both can be divided into six major land-use categories according to the World Conservation Union classification:[6] cropland, pasture land, forest, fishing ground, built-up land, and carbon uptake land. The first four land types are required to make animal-, plant-, and fish-based food and fiber and wood products available for human consumption. The built-up land is required for shelter and other physical infrastructure such as roads and cities. The last land type represents a sink, the bioproductive space required to absorb the wastes emitted from human economies. Given global data limitations, national footprint assessments currently consider only carbon dioxide emissions from fossil fuel burning as a waste. Other waste streams would also be included, if the data became available. CO_2 emissions from fossil fuel burning are converted into an equivalent forested area needed to remove the excess CO_2 from the atmosphere, given forest sequestration rates. On the demand side, this area is called the carbon footprint; on the biocapacity side, land for CO_2 sequestration is forest (forest can serve either timber and fuel wood production, or carbon sequestration—but these services are typically in competition for space). Due to data limitation about forests with long-term carbon sequestration commitments, current Ecological Footprint accounts do not distinguish between forests for forest products, long-term carbon uptake (or "carbon uptake land"), or biodiversity reserves.

It should be noted that the term "carbon footprint" used in the context of the Ecological Footprint methodology is not to be confused with the "Carbon Footprint" methodology itself. The latter term refers to a different methodology and aims to measure the total amount of all greenhouse gas emissions that are directly and indirectly caused by an activity or are accumulated over the life stages of a product or an activity. Further information on the differences between the carbon footprint component of the Ecological Footprint methodology and the Carbon Footprint methodology can be found in the work of Galli et al.[7]

Ecological Footprint and biocapacity are both expressed in a standardized unit, global hectares (gha);[3,8] the global hectare represents a hectare with world-average biological productivity.[9] Global hectares are normalized so that the number of actual hectares of biologically productive land and sea on the planet is equal to the total worldwide budget of global hectares in any given year.[10] Because of international trade and the dispersion of carbon dioxide wastes, hectares demanded can be physically located anywhere in the world.[11] While the global hectare is the unit recommended by Global Footprint Network, world-average hectares (wha) and nation-specific hectares (nha) are also possible units,[9,12] and results can be easily converted from one unit of measure to the other. Global hectares are needed to measure bioproductivity rather than surface area. Each global hectare has world-average productivity for all land types and provides more information than simply weight (which does not capture the extent of land and sea area used) or physical area (which does not capture how much ecological production is associated with that land). Global hectares are particularly useful for ranking different products based on their total ecological demands. World-average hectares are areas of a specific land type with world-average productivity for that land type (e.g., 1 ha of forestland with the ecological production of the average forest hectare globally). Nation-specific actual hectares are physical areas of a specific land type located within a specific country and characterized by the bioproductivity of that country.

The choice of the unit of measure depends on the posed research question. The main advantage of using global hectares lies in the possibility to compare the ecological demands of nations and/or products across the globe. A productivity-based normalization is necessary to convert actual hectares into global hectares: yield and equivalence factors (EQFs) are used for this purpose. The yield factor (YF) captures the difference between local and global (world-average) productivity.[9,10] It is calculated as the ratio between the yield for the production of each product in a specific nation and the average yield for the production of the same product in the world. In each year, every country has its own set (one for each of the six land types) of YFs.

The EQF captures the difference between the productivity of a given land type and the world-average productivity of all biologically productive land types.[9,10] Equivalence factors are currently calculated by Global Footprint Network by using the Global Agro-Ecological Zones model, which provides data on potential agricultural suitability. The EQF for a land type depends on its level of potential agricultural suitability relative to world-average suitability. A set of EQFs is calculated on a yearly basis. The EQF for marine area is calculated such that a single global hectare of pasture will produce an amount of calories of beef equal to the amount of calories of salmon that can be produced by a single global hectare of marine area. The EQF for inland water is set equal to the EQF for marine area. The set of YFs and EQFs is annually updated and released by Global Footprint Network.

Mathematics of the Ecological Footprint

The EFA may have various purposes and be applied in different contexts. Here we refer to the Ecological Footprint within a geographically delineated area, e.g., a nation and the Ecological Footprint of a product.

Ecological Footprint of a Nation

The Ecological Footprint of a nation, EF_N, is a function of the number of inhabitants, Pop, and the footprint of consumption of a single inhabitant, $EF_{C, i}$, associated with the ith product and/or waste:

$$EF_N = Pop^* \sum_{i=1}^{n} EF_{C, i} \tag{1}$$

where the per capita footprint of consumption is the sum of the per capita footprint of production ($EF_{P, i}$) and imports ($EF_{I, i}$) minus the footprint of exports ($EF_{E, i}$) for the ith product and/or waste.

$$EF_{C, i} = EF_{P, i} + EF_{I, i} - EF_{E, i} \tag{2}$$

In calculating footprints of production, imports, and exports for each ith product and/or waste. the reference formula is

$$EF_i = \sum_{j=1}^{6} \frac{T_i}{Y_{N, i}} * YF_{N, j} * EQF_j = \sum_{j=1}^{6} \frac{T_i}{Y_{W, i}} * EQF_j \tag{3}$$

where

- T_i is the produced, imported, or exported amount of each product i (in tonnes).
- $Y_{N, i}$ is the annual yield (t ha^{-1} yr^{-1}) for the production of each product, i, in the nation, N. This is calculated as the tonnes of product, i, produced annually divided by all areas in the nation on which this product is grown.
- $Y_{W, i}$ is the world-average (W) annual yield (t ha^{-1} yr^{-1}) for the production of each product, i, given by the tonnes of product, i, produced annually across the world divided by all areas in the world on which this product is grown.
- $YF_{N, j}$ is the YF specific to nation, N, and land type, j.
- EQF_j is the EQF for land type, j.

From these equations, it can be derived that the Ecological Footprint of a population is a function of four main factors:[9,10] 1) population size; 2) average standard of living; 3) the average productivity of land and water ecosystems; and 4) the efficiency of resourcing, harvesting, processing, and use.

Biocapacity is an aggregate measure of the amount of bioproductive land available, which is weighted by the productivity of that land. It represents the ability of the biosphere to produce crops, livestock (pasture), timber products (forest), and fish as well as to take up carbon dioxide in forests. It is calculated as reported in Eq. 4:

$$BC_N = \sum_{j=1}^{6} A_{N, j} * YF_{N, j} * EQF_j \tag{4}$$

where $A_{N, j}$ represents the estimated bioproductive area expressed in nation-specific hectares that is available. $YF_{N, j}$ and EQF_j have been defined above.

Biocapacity represents the maximum theoretical rate of resource supply that can be sustained by a territory under prevailing technology and management schemes. Thus, biocapacity depends on natural conditions as well as on prevailing land-use practices in, for example, farming and forestry.[10]

The Ecological Footprint of a population can be directly compared with the biocapacity of the area where the population resides in the same way that expenditure is compared against income in financial terms.[4,10] The resulting "biocapacity balance" reveals whether that population is living, in net terms, within or beyond its local ecological means. This ecological balance has significance at both the global and national level, though it has to be noted that living within one's ecological means does not yet guarantee ecological sustainability.[13,14] Additional criteria such as biodiversity conservation, soil preservation, decreased release of toxic elements, and the like should be tracked in a comprehensive and multidisciplinary sustainability assessment. Further, countries with a biocapacity reserve (where biocapacity>Ecological Footprint) may use their available biocapacity to satisfy their own domestic consumptions or export ecological resources to other nations. This is generally the case for low-income countries that use only a fraction of their locally available resources, like some African and Latin America countries.[13,15,16]

Conversely, countries with a biocapacity deficit (where Ecological Footprint>biocapacity) must rely on biocapacity from outside their own borders or draw down their own natural budget. Many high-income countries (like the United States, Canada, and some western European countries such as Italy, the United Kingdom, and France) have footprints several times larger than their domestic biocapacity.[13,15,16] Such a biocapacity deficit is becoming an increasing economic risk for countries, particularly in a world of growing global overshoot. It highlights a country's dependence on additional external goods and ecological services, which are provided through one or more of the three following mechanisms:[10,13] 1) the biocapacity trade deficit, which consists of net import of biocapacity from other regions of the world; 2) the biocapacity deficit due to depletion, due to an overuse of local resources; and 3) the demand on biocapacity due to occupation of global commons, such as emissions of greenhouse gases into the global atmosphere (rather than domestic absorption) or fishing in international waters.[17]

The analysis of biocapacity and Ecological Footprint trends reveals how human consumption is changing over time. At the global level, the latest data released[13,16] show that humanity is currently operating in a state of overshoot: in other words, demand for natural resources exceeds the regenerative capacity of existing natural capital by at least 50% according to calculations of Global Footprint Network.[15] Furthermore, the gap between Ecological Footprint and biocapacity globally has been continuously increasing since the mid-1970s (Figure 1).

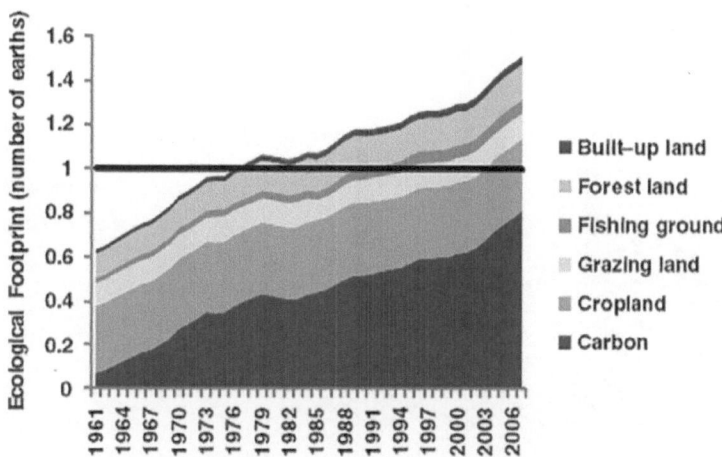

FIGURE 1 Humanity's Ecological Footprint, 1961–2007.

In 2007, humanity's total Ecological Footprint worldwide was 18 billion gha; with world population at 6.7 billion people, the average person's footprint was 2.7 gha.[13,16] However, there was only 11.9 billion gha of biocapacity available that year, equivalent to 1.8 gha per person. This overshoot of approximately 50% means that in 2007, humanity used the equivalent of 1.5 earths to support its consumption. In other words, it would have taken the earth approximately a year and a half to regenerate the resources used by humanity in that year. The largest Ecological Footprint component was the carbon footprint. This has increased by 35% since 1961 and currently accounts for more than half of the global Ecological Footprint.[13,16]

Even if the earth has a high resilience, prolonged biocapacity deficit is not possible since vital ecosystems and nonrenewable stocks would be depleted due to insuperable ecological and thermodynamic constraints. Also, it is not obvious that high-input agriculture can maintain its yields in the long run, particularly in the face of soil loss and potential phosphate limitations. It has become an urgent task to reduce our consumption levels back within the limits of our ecological budget.[14]

The growing global trends, however, hide significant regional variation (Figure 2).

Both demand on and supply of biocapacity are unevenly distributed across the world. Ecological Footprint and biocapacity values can therefore be used to develop new criteria for distinguishing among world nations. For instance, an alternative approach can be used to look at countries, based on their "biocapacity balances" (Figure 3), which helps to identify where resources are located and who uses what and to what extent. While in 1961, approximately 80% of the world population was living in countries characterized by a biocapacity remainder, in 2007, most of the world population was living in countries running a biocapacity deficit situation.

The total Ecological Footprint demanded by a country is strongly related to GDP[18,19] and changes accordingly, in both its extent and its composition, among high-, middle-, and low-income countries (Figure 4).[13]

Generally, high-income countries have per-capita Ecological Footprint values nearly 3 times higher than the world average, the majority of which is from the carbon footprint (approximately 65% of the total value). Conversely, middle- and low-income countries have average Ecological Footprint values that are equal to and lower than the world average, respectively. These countries are frequently characterized by transition economies, in which the carbon footprint component, although increased over the last decades, still constitutes less than 50% of the overall demand (Figure 5).

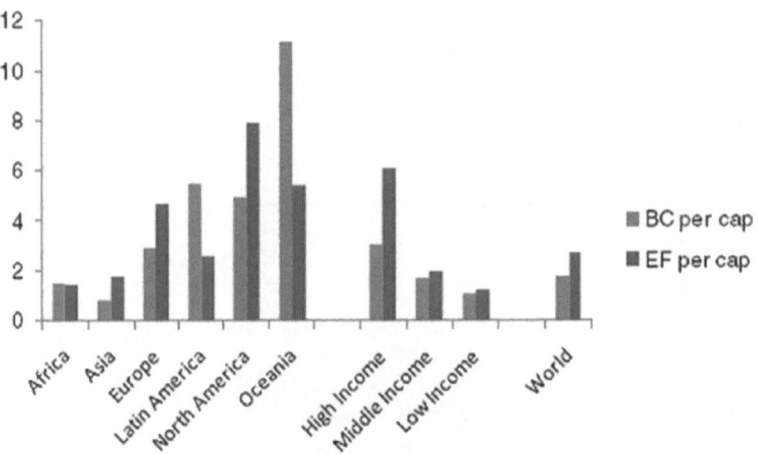

FIGURE 2   Ecological footprint (EF) and biocapacity (BC) by income level and country. The unequal distribution of human demand on bioproductive lands was also investigated by White.[20]

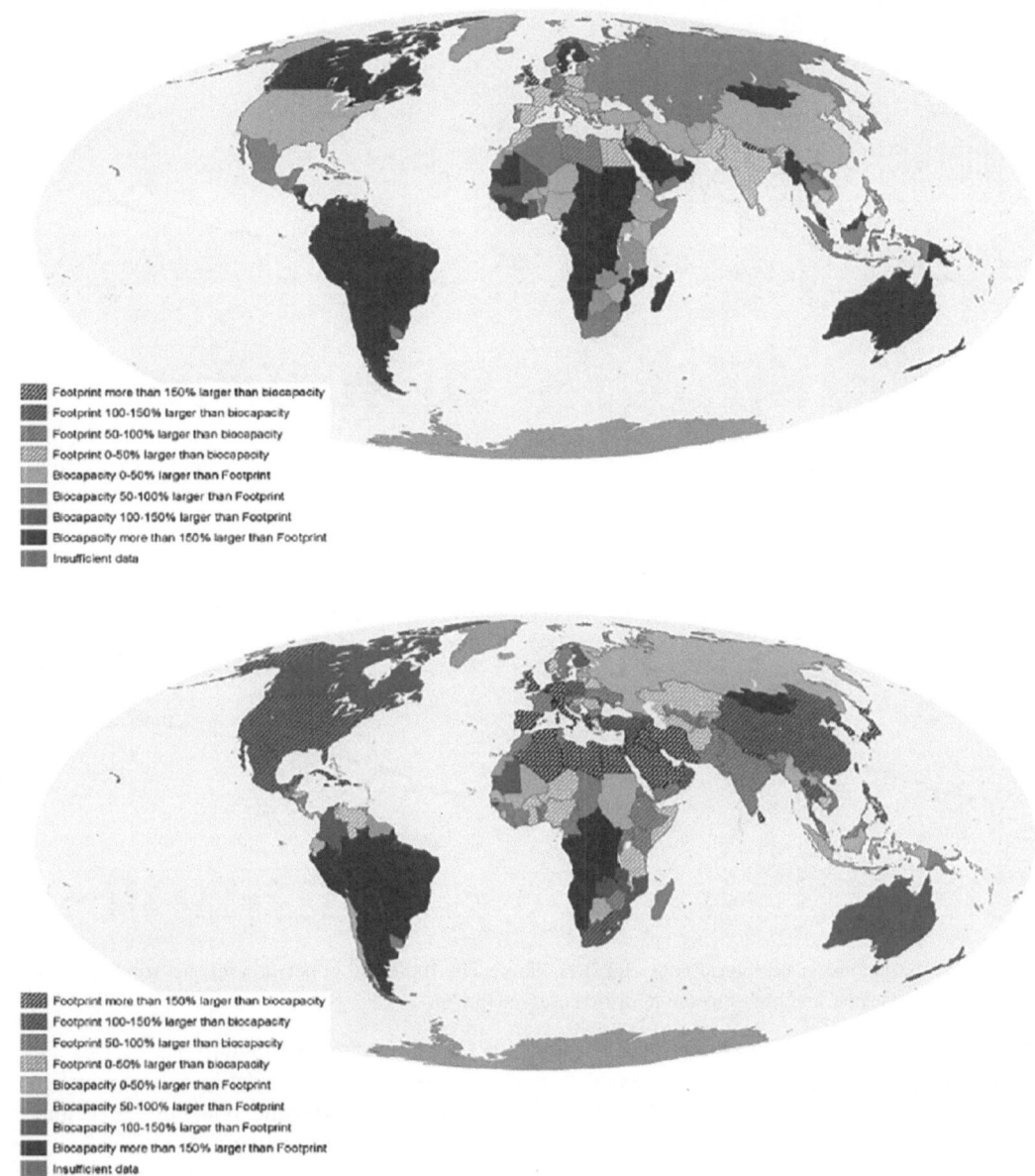

FIGURE 3 Biocapacity reminder/deficit status for world countries in 1961 and 2007. Green nations represent countries where the local biocapacity is greater than their residents' footprint (biocapacity reminder countries); red represents countries where the footprint is greater than local biocapacity (biocapacity deficit countries).

Ecological Footprint of a Product

The Ecological Footprint of a final or intermediate product is defined as the total amount of resources and waste assimilation capacity required in each of the phases required to produce, use, and/or dispose of that product.[8] The lifecycle boundaries can be flexible and changed according to the aim and scope of the analysis i.e., from cradle to gate (production to distribution) or cradle to grave (production to destruction). The Ecological Footprint is evaluated by considering all the direct and indirect inputs that are associated with the analyzed system for its entire life cycle. Each of these inputs is converted in terms

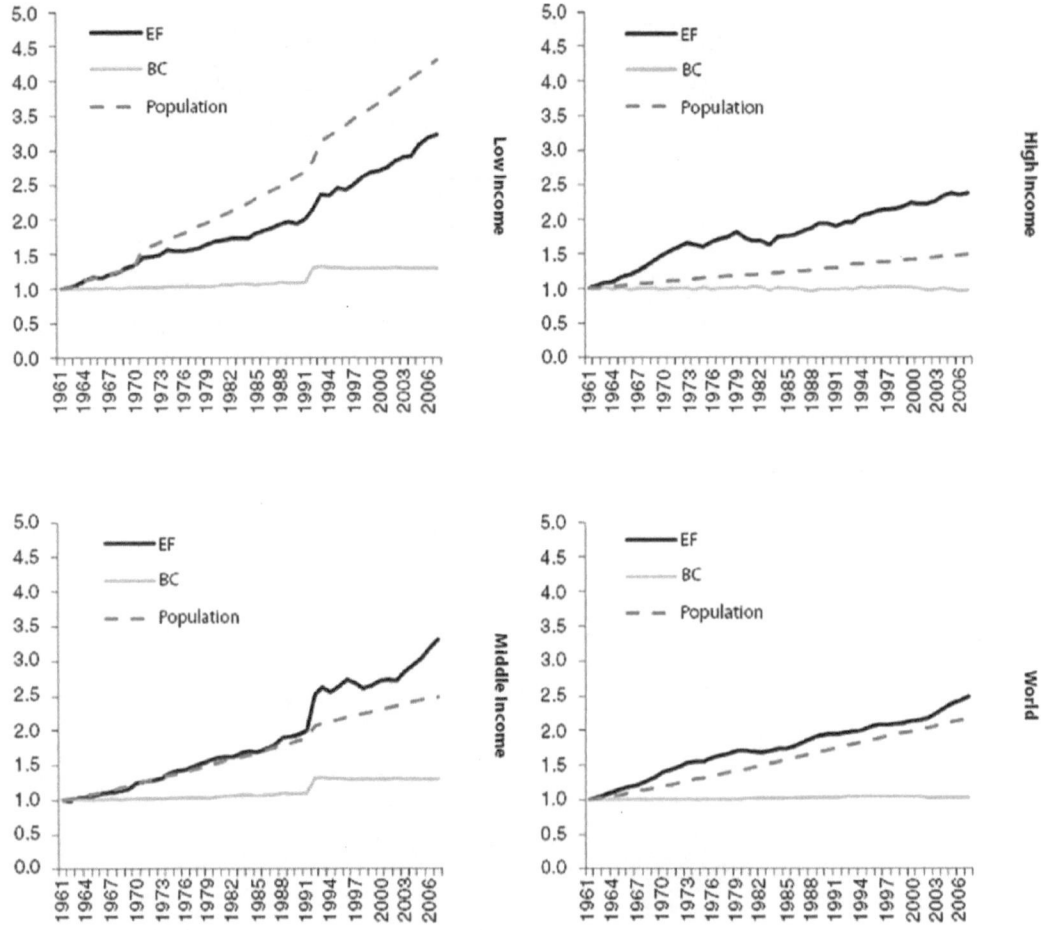

FIGURE 4 Variations on Ecological Footprint (EF), biocapacity (BC), and population for the world and low-income, middle-income, and high-income countries, indexed to 1961.

of the global hectares needed to support their production The EFp is expressed in units of global hectare years (gha yr), not just global hectares.[8] As the Ecological Footprint is strictly related to the production system, the way a product is produced should be clear and identifiable. The functional unit for the analyzed system should be also defined as well as the temporal and spatial boundaries.

There are two widely used approaches for calculating the Ecological Footprint of a product, both standards compliant: process-based life-cycle assessment (P-LCA) and environmentally extended input-output life-cycle assessment.[8] Process-based life-cycle assessment has the advantage of a large amount of detail, as individual product types and even brands can be analyzed, with the general disadvantage of lacking complete upstream coverage of the production chain (e.g., truncation error). Extended input-output life-cycle assessment has the advantage of full upstream coverage but the disadvantage of generality, as input-output tables typically do not disaggregate down to the level of individual product types (e.g., homogeneity assumption).

Following the P-LCA, Ecological Footprint of product (EFp) is given by the sum of the footprints of each input consumed and disposed within the life cycle of the production process as reported below.

$$\mathrm{EF_p} = \sum_{j=1}^{6} \sum_{i=1}^{n} \frac{T_i}{Y_{N,i}} * \mathrm{YF}_j * \mathrm{EQF}_j$$

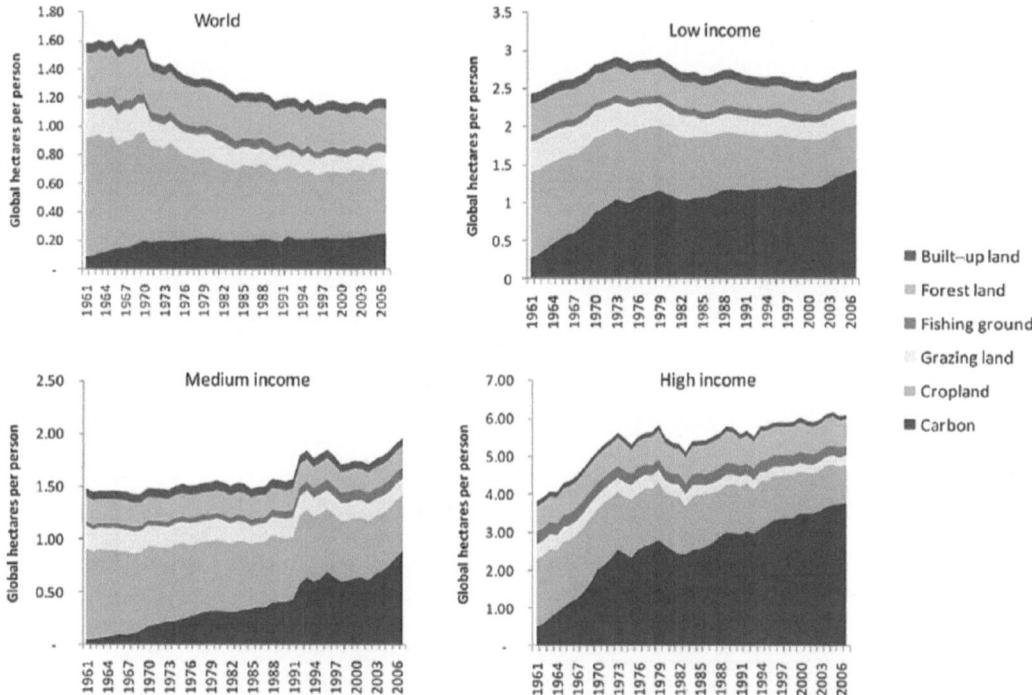

FIGURE 5 Per-capita Ecological Footprint for the world and high-, middle-, and low-income countries by land type, 1961–2007.

$$= \sum_{j=1}^{6} \sum_{i=1}^{n} \frac{T_i}{Y_{W,i}} * \mathrm{EQF}_j \qquad (5)$$

where the variables are as follows:

- i refers to the n-input needed
- j refers to the six different land-use types (cropland, grazing land, fishing grounds, forest area, built-up land, and carbon uptake land)
- YF_j is the YF of the jth land type
- EQF_j is the EQF of the jth land type

Toward a Multi-Indicator Approach

Building on the premise that no single indicator is able to provide a full sustainability diagnosis and indicators should rather be used and interpreted jointly (i.e., the joint use of more than one indicator), this section reports some of the most interesting applications.

For instance, the HDI[21] can be used together with the Ecological Footprint to provide important insights on whether a high level of consumption is necessary for a high level of human development.[15] The HDI is a composite indicator used to rank countries by level of "human development" and then of well-being. It is a comparative measure of life expectancy, literacy, education, and standards of living for countries worldwide. The relationship between Ecological Footprint and HDI has two different categories (Figure 6).

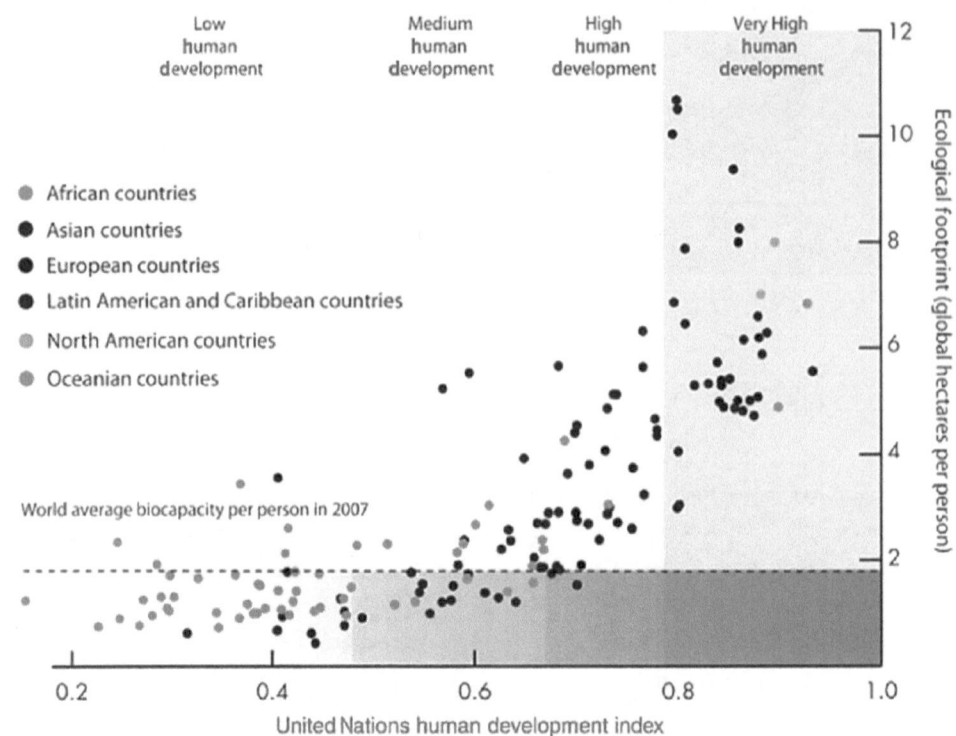

FIGURE 6   Human development index vs. Ecological Footprint, 2007.

While countries with a low level of development (HDI <0.8) report small variations in Ecological Footprint per capita, countries with a higher HDI value than the threshold show significant variations in Ecological Footprint per capita (for example, from Peru with 1.5 gha per person to Luxembourg with over than 9 gha per person). However, as development increases beyond a certain level, so does per-person Ecological Footprint; this is up to a point where small gains in HDI come at the cost of very large Ecological Footprint increases. Moreover, several countries with a high level of development have a similar per-capita Ecological Footprint to countries with a much lower level of development. Together with the breakdown in connection between wealth and well-being above a certain level of GDP per capita, this indicates that a high level of consumption is not necessarily required for a high level of development or well-being.

Ecological Footprint and biocapacity temporal trends have been compared with economic indicators such as GDP and the ISEW.[22] In particular, it has been shown that some (western) countries are characterized by increased economic wealth (measured by GDP) and growing environmental pressure (documented by Ecological Footprint trend) at the expense of a decrease in environmental sustainability (as shown by the increasing gap of Ecological Footprint and biocapacity) and a stagnation or even decrease in welfare (as ISEW demonstrates; Figure 7).[22]

A comparison between Ecological Footprint and life satisfaction—used as a subjective measure of well-being— was performed to understand the nature of their relationship and to test whether it could be possible for humans to reduce their levels of resource consumption without compromising well-being and life satisfaction and, at the same time, protecting the environment.[23,24] The relationship between Ecological Footprint and life satisfaction has two different paths. While countries with an Ecological Footprint of less than 4 gha per capita report significant variations in life satisfaction, countries with higher values (Ecological Footprint >4 gha per capita) report small variations in life satisfaction (Figure 8).

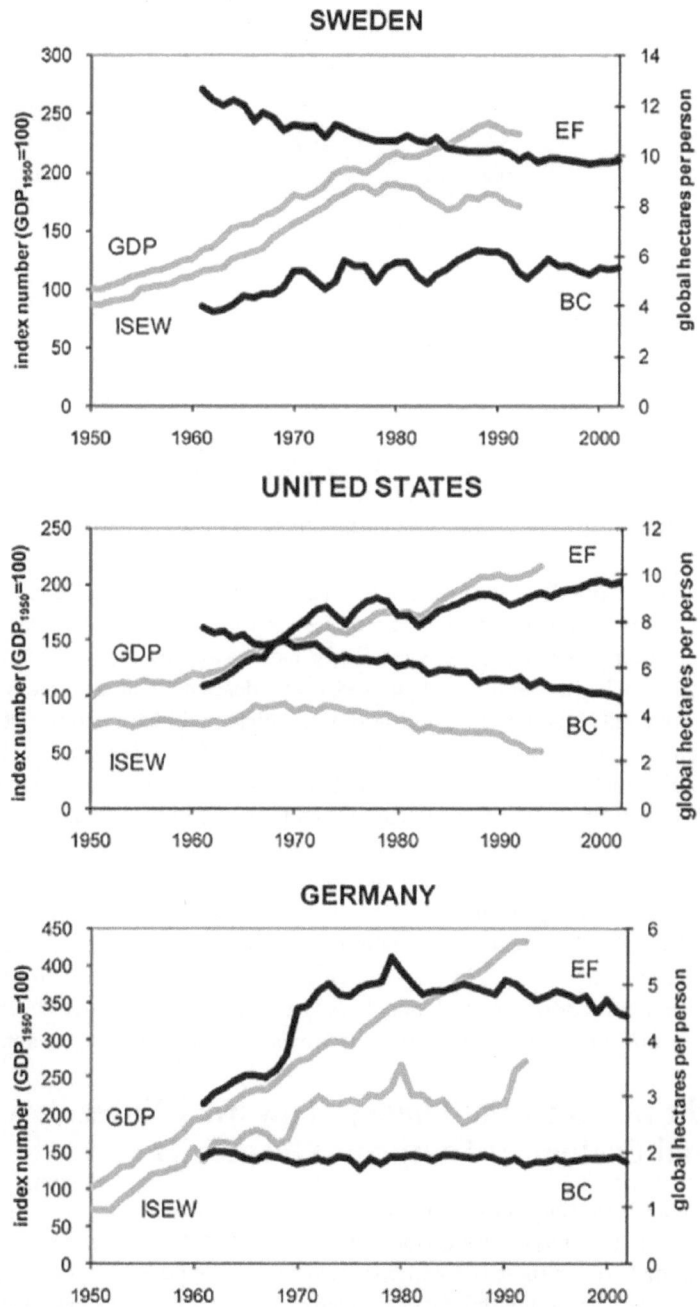

FIGURE 7 The comparison among temporal trends of GDP vs. ISEW (left scale) and Ecological Footprint (EF) vs. biocapacity (BC) (right scale) for Sweden, United States, and Germany.

In general, people with high consumption and income levels are more satisfied with life than people with lower consumption and income levels despite the differing levels and patterns of consumption that are necessary to obtain the same level of satisfaction. On the contrary, people with low consumption and income levels are less satisfied, but the same unit of consumption produces different perceptions of satisfaction (life satisfaction values are more unpaired).

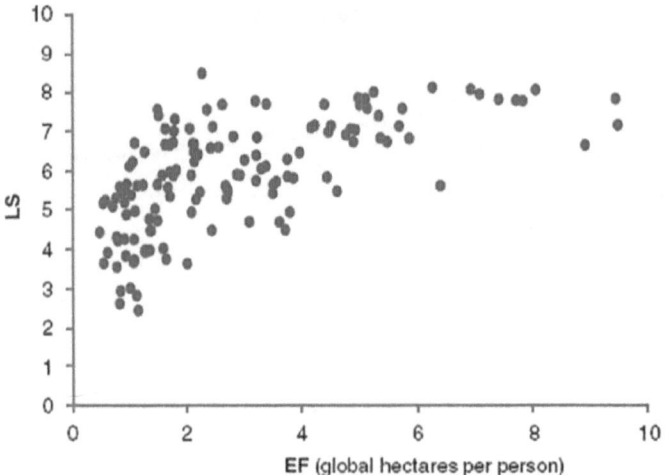

FIGURE 8   Life satisfaction (LS) vs. Ecological Footprint (EF) for 130 world countries.

Biocapacity and Ecological Footprint temporal trends have recently been used to develop alternative interpretations of the geopolitical context of nations around the world.[25,26] Based on their development paths, Ecological Footprint and biocapacity trends for nations around the world can be grouped into four main dynamic typologies: parallel, scissor, wedge, and descent.[25] Each typology corresponds to a particular environmental situation, the implications of which could have extreme relevance for environmental management, economic, and social prosperity as well as the development of sustainability policies. In particular, the role of biocapacity is highlighted in maintaining healthy economies, in offering an acceptable quality of life, and as an essential asset to ensure national competitiveness.[27]

Finally, under the recent EU-funded One Planet Economy Network Europe (OPEN:EU) project, three indicators have been identified as useful and complementary in assessing environmental issues—ecological, carbon, and water footprints—and therefore grouped together to form a suite of indicators called "footprint family."[7] Although not yet comprehensive, this suite provides a quantifiable platform for discussions regarding the limits to biotic resource and freshwater consumption and greenhouse gas emissions, as well as how to address the sustainability of natural capital use across the globe, thus enabling decision makers to more easily understand the environmental consequences of economic activities.

New Insight in Footprint Theory: Toward a Three-Dimensional Ecological Footprint Geography

The presence of global overshoot proves that the current human economy partially relies on natural capital depletion rather than just on sustainable flow consumption.[4] Considering natural capital and its limits, differentiating between these two components is fundamental for environmental planning and management. Recently, a variant of the classical Ecological Footprint model has been proposed, where a distinction between depletion of natural capital stocks and use of natural capital flows is operated.[28] The Ecological Footprint was redesigned as a three-dimensional model (^{3D}EF) with two relevant components, called size (EF_{size}) and depth (EF_{depth}), related to the two different uses of natural capital. The EF_{size} regards the appropriation of the so-called "income" of natural capital, i.e., the consumption of flows of resource yearly regenerated by natural cycles. It is the spatial component expressed in global hectares and plotted in the (x, y) plane. By definition, its value ranges from zero to biocapacity.

On the other side, the EF_{depth} regards the depletion of natural capital, which is the use of stocks of resources requiring a regeneration time longer than the flows. It is the intensity component plotted on the *z*-axis. In particular, the EF_{depth} component arises when overshoot is present and expresses the number of years it would theoretically take to regenerate the natural capital used in one year. By definition, its value ranges from 1 (the reference value called "natural depth") upward (additional depth).

It should be remarked that the two approaches, classical and three-dimensional, are simply two different ways of representing the same footprint values. The ^{3D}EF originates from the fact that flow and stock are technically incommensurable and cannot be summed up because the former is consumed each year and regenerated the following year, whereas the latter represents the irreversible erosion of natural stocks that add up from year to year into an accumulated "environmental debt."

Footprint size and depth have been characterized by opposite trends:[17] 1) EF_{size} grew continuously until mid-1970s, when it reached the asymptote (i.e., the earth biocapacity), and has remained constant ever since; and 2) EF_{depth} has remained equal to the natural depth until the appearance of overshoot, and it has subsequently been growing. Recently, the ^{3D}EF model has been theoretically applied to national case studies with the aim of enhancing the significance and potential usefulness of the Ecological Footprint in tracking relevant issues in the sustainability debate, such as the differentiation between resource stocks and flows.[17] Moreover, EF_{size} can be used as a proxy to highlight the existing intragenerational (in)equity in the appropriation of resources and ecological services by the residents of different nations. At the same time, EF_{depth} enables the relationships between current and future generations to be examined.

Although several questions remain to be addressed, the implementation of both a multilateral trade framework and the ^{3D}EF model in the National Footprint Accounts could form the basis for a new Ecological Footprint-based geography able to differentiate pressures on flows and stocks and identify the spatial/geographical location of such pressures.[4]

Applications of Ecological Footprint

Territorial Systems

Examples of national Ecological Footprint studies can be found in the literature,[29,42] although the most comprehensive set of national Ecological Footprint assessments is represented by Global Footprint Network's National Footprint Accounts.[16]

As of today, more than 35 nations have engaged with the Global Footprint Network directly, 17 nations have completed reviews of the Ecological Footprint, and a few have formally adopted it. Wales has adopted the Ecological Footprint as its headline indicator for sustainability. The Swiss government has incorporated the footprint into its national sustainable development plan. Japan includes the footprint as a measure in its environmental plan, the United Arab Emirates is using the Ecological Footprint as a tool to recommend and assist in the development of longterm science-based policies,[43] and Ecuador has set official footprint reduction targets in its 2009–2013 National Development Plan. There are several other countries that are currently collaborating with Global Footprint Network.[44]

Among nongovernmental organizations, WWF (World Wide Fund for Nature) International, one of the world's most influential conservation organizations, uses the Ecological Footprint in its communication and policy work for advancing conservation and sustainability. WWF has recently established a target of bringing humanity out of overshoot by 2050 and is actively pursuing this goal through its One Planet programs.

Numerous applications have also been performed at various subnational scales.[45–53] Currently, there are two suggested methodologies for subnational Ecological Footprint evaluations: component (or bottom-up) and compound (or top-down) approaches. The component method starts from specific individual consumption and waste production data to then calculate the total Ecological Footprint. While the method is detailed and flexible, several problems, including double counting, the lack of detailed data, and specific conversion factors, make this approach less acceptable. The compound approach evaluates the subnational Ecological Footprint by scaling the national Ecological Footprint value according to differences in consumption and life style. This is the widest used and more complete method.[54,8]

Products and Services

Despite its diffusion and popularity, product Ecological Footprint applications are still scarce, though a few interesting case studies exist. Studies on cultivation of tomatoes,[55] conventional vs. organic wine farming,[56] nectarine production,[57] shrimp and tilapia aquaculture,[58] marine aquaculture of reef fish,[59] and fisheries products[60] have been performed to highlight appropriation of natural capital, efficiency of natural resource use, and environmental pressure. Evaluations of the environmental impact of farms[61] and dairy production[62] as well as assessment of economic and ecological carrying capacity of crops[63] have been proposed via the combined use of the Ecological Footprint with other methods, such as life-cycle assessment, emergy analysis and economic cost, and return estimation.

In the context of product, Ecological Footprint has been used as a basis for the elaboration of the model of the double pyramid, which flanked the food pyramid with the environmental pyramid.[64] This model highlights that, in general, modern lifestyles produce a growing impact on the planet; it tries to promote eco-sustainable life and eating styles.

In the context of industrial processes, the potential of using EFA as an environmental indicator for the textile sector has been considered, although the contribution of wastes other than carbon dioxide should be included in the footprint methodology.[65]

The EFA has also been applied to the tourism sector. Starting from some pioneering studies,[66,67] the EFA has received attention as a key environmental indicator of sustainable tourism,[68] and several studies have been published at different scales.[69–72]

Over time, several tools and software have been developed for Ecological Footprint assessment for both territorial systems and products, although only a few are standard compliant. Most of them are freely downloadable from Web sites such as the footprint calculator (available at http://www.footprint-network.org/en/index.php/GFN/page/personal_footprint/). The REAP (Resource and Energy Analysis Programme) software (available at http://www.sei.se/reap) is a scenario based, integrated resource–environment modeling tool developed by the Stockholm Environment Institute to help local authorities in the U.K. make decisions about how to reduce their Ecological Footprint.

The Footprinter software (available at http://www.foot-printer.com), developed by Best Foot Forward, is comprehensive and powerful analytical software for carbon and Ecological Footprint assessment for products and organizations. It is based on the use of EcoIndex database.

Role of Business

The Ecological Footprint has been used to evaluate the environmental pressure of production processes; this type of investigation is becoming increasingly important to integrate sustainability issues (as natural capital consumption) into industrial and business decision-making processes. In a recent World Business Council for Sustainable Development project (WBCSD), several companies and industries came together to assess their role in helping to shape a future sustainable society for mankind.[73] With the help of Global Footprint Network, the consequences, in Ecological Footprint and biocapacity terms, of the hypothetical scenarios (up to the year 2050) envisioned by WBCSD were estimated. Results from this study showed that humanity will likely require the equivalent of 2.3 planets' worth of resources upon following a business-as-usual scenario or, conversely, 1.1 planets' worth of resources by implementing all the envisioned actions and activities, such as 50% reduction in CO_2 emissions compared with 2005 levels; enhanced forest productivity using better management techniques and an extension of their acreage between 2030 and 2050; increased crop productivity (+2% each year over past trends) due to technological advances and the diffusion of best practices; and changes to the average global nutritional regime, in terms of both diet and calorie content.[74]

Weakness and Limitations of EFA

Despite its popularity, EFA, like most indicators, is not exempt from criticisms regarding philosophical as well as methodological issues. In this section, a brief summary of the main weaknesses and limitations is offered. For further details on this topic see, for example, Best et al.,[75] Fiala,[76] Kitzes et al.,[77] and van der Bergh and Verbruggen.[78]

The first criticism is that the EFA cannot be fully defined as a measure of sustainability. The EFA research question is limited to identify the extent to which humanity is consuming bioproductive land compared with the available biocapacity.[21] We believe consuming resources within the capacity of the planet is a first necessary although not sufficient criterion for sustainability; as such, in order to depict a comprehensive picture of the system analyzed, it is strongly recommended to combine the Ecological Footprint with other complementary indicators (environmental as well as social and economic).

The use of a spatial unit makes some impacts difficult to determine.[75] Ecological Footprint is not able to directly account for all resources that cannot be referred to in spatial terms. This is the case for the depletion of nonrenewable deposits, such as metals, minerals, or fossil fuel reserves. For processes of extraction and refining, only the CO_2 emissions related to these processes are accounted for. The use of fossil fuels is evaluated in an indirect way, considering the amount of forestland required for the absorption of the CO_2 that is emitted. To date, carbon dioxide is the only greenhouse gas accounted for, and its associated footprint relies on the assumption that all emissions are absorbed only by forests and the oceans, neglecting carbon uptake by other biomes.

Other missing elements in EFA are freshwater consumption and soil erosion, even if the latter could be accounted for, at least theoretically. A possible way to include the overexploitation phenomena possible in agricultural land as well as in other land types into the classical Ecological Footprint framework was proposed by Bastianoni et al.[79] As EFA is unable to show unsustainable practices and their consequences, when agricultural EFA is performed, there can be some misunderstanding and misinterpretation of the results. Sometimes, it seems that EFA encourages more intensive farming, as this increases agricultural intensities, resulting in a higher biocapacity. The Ecological Footprint shows pressures that could lead to degradation of natural capital (e.g., reduced quality of land or reduced biodiversity) but does not predict this degradation.

Furthermore, multifunctional land-use patterns are not considered, in order to prevent double counting.[79] Each hectare is counted only once, even though it might provide multiple services. Counting them multiple times would produce an overestimation on Ecological Footprint.

Finally, EFA also is not able to capture the impacts due to the release of long-life toxic materials (e.g., pollution in terms of waste generation, toxicity, eutrophication, etc.), for which no regenerative capacity exists.

Conclusion

This entry offers a comprehensive insight to the EFA. Based on its simple logic and unit of measure, the Ecological Footprint has become a very popular sustainability indicator preserving scientific rigor on an ecological and thermodynamic basis.

By tracking a wide range of human activities, the Ecological Footprint is able to monitor the combined impact of anthropogenic pressures that are more typically evaluated independently and can thus be used to understand, from multiple angles, the environmental consequences of human activities.

The main strength of this methodology is its ability to explain, in simple terms, the concept of ecological limits, thus helping to safeguard the long-term capacity of the biosphere to support mankind and understand how resource issues are linked with economic and social issues.

One of the positive characteristics of this methodology is its ability to make visible aspects that are traditionally invisible for conventional economic analyses. For instance, the choice of an area as a unit of measure reflects the fact that many basic ecosystem services and ecological resources are provided by surfaces where photosynthesis takes place. Unfortunately, these surfaces are limited by physical and planetary constraints: reporting results in terms of an area helps to better communicate the existence of physical limits to the growth of human economies.

Currently, the Ecological Footprint is a robust method widely used to give a measure of the (un)sustainability of consumption patterns at different scales as well as to establish the natural capital requirement of products, services, and activities.

Due to the growing number of applications, Global Footprint Network has released the Ecological Footprint Standards[8] in order to enhance the consistency and quality of footprint assessments. This standard contains a list of mandatory requirements for standards compliance. The document also suggests the best way to present the results avoiding distortion and misinterpretations. The National Footprint Accounts are updated and improved every year, and the annual release of the newest version ensures that the method is more robust, reliable, and detailed than previous versions, though some shortcomings still exist and remain to be addressed. The Ecological Footprint Standards are also periodically updated. The information derived from Ecological Footprint assessments could be included in the environmental management and future planning of territories to promote more competitive lifestyles, resource-efficient strategy, and a more effective management of our ecological assets.

References

1. Rees, W.E. Ecological Footprints and appropriated carrying capacity: what urban economics leaves out. Environ. Urban. **1992**, *4* (2), 121–130.
2. Wackernagel, M. Ecological Footprint and Appropriated Carrying Capacity: A Tool for Planning Toward Sustainability. Ph.D. Thesis. School of Community and Regional Planning, The University of British Columbia, 1994.
3. Wackernagel, M.; Rees, W.E. *Our Ecological Footprint: Reducing Human Impact on the Earth,* 1st Ed.; New Society Publishers: Gabriola Island, Canada, 1996.
4. Wackernagel, M.; Rees, W.E. Perceptual and structural barriers to investing in natural capital: Economics from an Ecological Footprint perspective. Ecol. Econ. **1997**, *20*, 3–24.
5. Wackernagel, M.; Schulz, N.B.; Deumling, D.; Linares, A.C.; Jenkins, M.; Kapos, V.; Monfreda, C.; Loh, J.; Myers, N.; Norgaard, R.; Randers, J. Tracking the ecological overshoot of the human economy. Proc. Natl. Acad. Sci. **2002**, *99*, 9266–9271.
6. The World Conservation Union, United Nations Environment Programme and the World Wide Fund for Nature. *Caring for the Earth: A Strategy for Living Sustainably;* IUCN, UNEP and WWF: Gland, Switzerland, 1991.
7. Galli, A.; Wiedmann, T.; Ercin, E.A.; Knoblauch, D.; Ewing, B.; Giljum, S. Integrating ecological, carbon, and water footprint into a "footprint family" of indicators: Definition and role in tracking human pressure on the planet. Ecol. Indic. **2012**, *16*, 100–112.
8. Global Footprint Network, 2009. *Ecological Footprint Standards 2009;* Global Footprint Network: Oakland, available at http://www.footprintstandards.org. (accessed July 2011).
9. Galli, A.; Kitzes, J.; Wermer P.; Wackernagel M.; Niccolucci V.; Tiezzi E. An exploration of the mathematics behind the Ecological Footprint. Int. J. Ecodyn. **2007**, *2* (4), 250–257.
10. Monfreda, C.; Wackernagel, M.; Deumling, D. Establishing national natural capital accounts based on detailed Ecological Footprint and biological capacity assessments. Land Use Policy **2004**, *21*, 231–246.
11. Kitzes, J.; Peller, A.; Goldfinger, S.; Wackernagel, M. Current methods for calculating national Ecological Footprint accounts. Sci. Environ. Sustainability Soc. **2007**, *4* (1), 1–9.
12. Wiedmann, T.; Lenzen, M. On the conversion between local and global hectares in Ecological Footprint analysis, Ecol. Econ., **2007**, *60* (4), 673–677.
13. Ewing, B.; Reed, A; Galli, A.; Kitzes, J.; Wackernagel, M. *Calculation Methodology for the National Footprint Accounts,* 2010 Ed.; Global Footprint Network: Oakland, CA, 2010.

14. Kitzes, J.; Wackernagel, M.; Loh, J.; Peller, A.; Goldfinger, S.; Cheng, D.; Tea, K. Shrink and share: Humanity's present and future Ecological Footprint. Phil. Trans. R. Soc. Lond. B **2008**, *363* (1491), 467–475.

15. WWF International; Global Footprint Network; ZSL. *Living Planet Report 2010.* WWF: Gland, Switzerland, 2010, available at http://www.footprintnetwork.com. (accessed July 2011).

16. Ewing, B.; Goldfinger, S.; Oursler, A.; Reed, A.; Moore, D.; Wackernagel. M. The *Ecological Footprint Atlas 2010*; Global Footprint Network: Oakland, 2010.

17. Niccolucci, V.; Galli, A.; Reed, A.; Neri, E.; Wackernagel, M.; Bastianoni S. Towards a 3D national Ecological Footprint geography. Ecol. Modell. **2011**, *222* (16), 2939–2944.

18. Jorgenson, A.K. Unpacking international power and the Ecological Footprint of nations: A quantitative cross national study. Sociol. Perspect. **2005**, *48*, 383–402.

19. Jorgenson, A.K.; Burns, T.J. The political–economic causes of change in the Ecological Footprints of nations, 1991–2001: A quantitative investigation. Soc. Sci. Res. **2007**, *36*, 834–853.

20. White, T.J. Sharing resources: The global distribution of the Ecological Footprint. Ecol. Econ. **2007**, *64*, 402–410.

21. UNDP (United Nation Development Programme). *Annual Report 2009;* New York, 2009.

22. Niccolucci, V.; Pulselli, F.M.; Tiezzi, E. Strengthening the threshold hypothesis: Economic and biophysical limits to growth. Ecol. Econ. **2007**, *60* (4), 667–672.

23. Abdallah, S.; Thompson, S.; Michaelson, J.; Marks, N.; Steuer, N. *The (Un)Happy Planet Index 2.0: Why Good Lives Don't Have to Cost the Earth;* New Economics Foundation: London, 2009; 61 pp.

24. Patrizi, N.; Capineri, C.; Rugani, B.; Niccolucci, V. *"SocioEconomic Design and Nature": A Possible Representation through Ecological Footprint. Design and Nature* v. Brebbia, C.A., Carpi, A., Eds.; WIT Press: Southampton, U.K. WIT Trans. Ecol. Environ. **2009**, 527–534.

25. Niccolucci, V.; Tiezzi, E.; Pulselli, F.M.; Capineri, C. Biocapacity vs. Ecological Footprint of world regions: A geopolitical interpretation. Ecol. Indic. **2012**, *16*, 23–30.

26. Global Footprint Network. *The Ecological Wealth of Nations. Earth's Biocapacity as a New Framework for International Cooperation,* 1st Ed.; Goldfinger, S., Poblete, P., Eds.; Global Footprint Network: Oakland, CA, 2010.

27. Moore, D., Brooks, N., Cranston, G., Galli, A. *The Future of the Mediterranean. Tracking Ecological Footprint Trends.* Galli, A., Ed.; Global Footprint Network, available at http://www.footprintnetwork.org. (accessed July 2011).

28. Niccolucci, V.; Bastianoni, S.; Tiezzi, E.B.P.; Wackernagel, M.; Marchettini, N. How deep is the footprint? A 3D representation. Ecol. Modell., **2009**, *220*, 2819–2823.

29. Barrett, J.; Simmons, C. *An Ecological Footprint of the U.K.: Providing a Tool to Measure the Sustainability of Local Authorities;* Stockholm Environment Institute (SEI): Stockholm, Sweden, 2003, available at http://www.lande-con.cam.ac.uk/up211/EP04/readings/footprint_UK.pdf. (accessed July 2011).

30. Galli, A.; Kitzes, J.; Niccolucci, V.; Wackernagel, M.; Wada, Y.; Marchettini, N. Assessing the global environmental consequences of economic growth through the Ecological Footprint: A focus on China and India. Ecol. Indic. **2012**, *17*, 99–107.

31. Wackernagel, M., Lewan, L., Borgström Hansson, C. Evaluating the use of natural capital with the Ecological Footprint. Ambio, **1999**, *28*, 604–612.

32. Bicknell, K.B.; Ball, R.J.; Cullen, R.; Bigsby, H. New methodology for the Ecological Footprint with an application to the New Zealand economy. Ecol. Econ. **1998**, *27*, 149–160.

33. Fricker, A. The Ecological Footprint of New Zealand as a step towards sustainability. Futures **1998**, *30*, 559–567.

34. Van Vuuren, D.P.; Smeets, E.M.W. Ecological Footprints of Benin, Bhutan, Costa Rica and the Netherlands. Ecol. Econ. **2000**, *34*, 115–130.

35. Simpson, R.W.; Petroeschevsky, A.; Lowe I. An Ecological Footprint analysis for Australia. Aust. J. Environ. Manage. **2000**, *7*, 11–18.

36. Ferng, J.J. Using composition of land multiplier to estimate Ecological Footprints associated with production activity. Ecol. Econ. **2001**, *37*, 159–172.

37. Haberl, H.; Erb, K.H.; Krausmann, F.; Loibl, W.; Schulz, N.B.; Weisz. H. Changes in ecosystem processes induced by land use: Human appropriation of net primary production and its influence on standing crop in Austria. Global Biogeochem. Cycles **2001**, *15*, 929–942.

38. Lenzen, M.; Murray, S.A.; A modified Ecological Footprint method and its application to Aust. Ecol. Econ. **2001**, *37*, 229–255.

39. Lenzen, M., Murray, S.A. The Ecological Footprint—Issue and Trends. ISA Research Paper 01–03. The University of Sidney, 2003, available at http://www.isa.org.usyd.edu.au/publications/documents/Ecological_Footprint_Issues_and_Trends.pdf. (accessed July 2011).

40. McDonald, G.W.; Patterson, M.G. Ecological Footprint and interdependencies of New Zealand regions. Ecol. Econ. **2004**, *50*, 49–67.

41. Erb, K.H. Actual land demand of Austria 1926–2000: A variation on Ecological Footprint assessments. Land Use Policy, **2004**, *21*, 247–259.

42. Medved, S. Present and future Ecological Footprint of Slovenia: The influence of energy demand scenarios. Ecol. Modell. **2006**, *192*, 25–36.

43. Abdullatif, L., Alam, T., 2011. *The UAE Ecological Footprint Initiative. Summary Report 2007–2010.* Available at: http://awsassets.panda.org/downloads/en_final_report_ecological_footprint.pdf (accessed July 2011).

44. Global Footprint Network (GFN), 2010. 2010 Annual Report, available at http://www.footprint-network.org/im-ages/uploads/2010_Annual_Report.pdf. (latest access: July 2011). (accessed July 2011).

45. Folke, C.; Jansson, A.; Larsson, J.; Costanza, R. Ecosystem appropriation by cities. Ambio **1997**, *26* (3), 167–72.

46. Bagliani, M.; Galli, A.; Niccolucci, V.; Marchettini, N. Ecological Footprint analysis applied to a sub-national area: The case of the Province of Siena (Italy). J. Environ. Manage. **2008**, *86*, 354–364.

47. Niccolucci, V.; Galli A.; Bastianoni, S. Deriving environmental management practices with the Ecological Footprint analysis: A case study for the Abruzzo Region. In *Ecosystems And Sustainable Development 7*; Brebbia, C.A., Tiezzi, E., Eds.; WIT press, 2009; 195–204.

48. Bagliani, M.; Ferlaino, F.; Procopio, S. The analysis of the environmental sustainability of the economic sectors of the Piedmont Region (Italy). In *Ecosystems and Sustainable Development*; Tiezzi, E., Brebbia, C.A., Uso, J.L., Eds.; WIT Press: Southampton, U.K., **2003**; 613–622.

49. Barrett, J.; Vallack, H.; Jones, A.; Haq, G. A material flow analysis and Ecological Footprint of York. Technical Report. Stockholm Environment Institute: Stockholm, Sweden, 2002.

50. Lenzen, M.; Lundie, S.; Bransgrove, G.; Charet, L.; Sack, F. Assessing the Ecological Footprint of a large metropolitan water supplier: Lessons for water management and planning towards sustainability. J. Environ. Plann. Manage. **2003**, *46*, 113–141.

51. Birch, R.; Wiedmann, T.; Barret, J. *The Ecological Footprint of Greater Nottingham and Nottinghamshire—Results and Scenarios.* Stockholm Environment Institute (SEI), University of York: York, U.K., 2005.

52. Vergoulas, G.; Simmons, C. An Ecological Footprint analysis of Essex-East England. Best Foot Forward Ltd., Commissioned by Essex County Council, 2004.

53. Collins, A.; Flynn, A.; Wiedmann, T.; Barrett, J. The environmental impacts of consumption at a sub-national level: The Ecological Footprint of Cardiff. J. Ind. Ecol. **2009**, *10*, 9–24.

54. Kitzes, J.; Galli, A.; Bagliani, M.; Barrett, J.; Dige, G.; Ede, S.; Erb, K-H.; Giljum, S.; Haberl, H.; Hails, C.; Jungwirth, S.; Lenzen, M.; Lewis, K.; Loh, J.; Marchettini, N.; Messinger, H.; Milne, K.; Moles, R.; Monfreda, C.; Moran, D.; Nakano, K.; Pyhälä, A.; Rees, W.; Simmons, C.; Wackernagel, M.; Wada, Y.; Walsh, C.; Wiedmann, T. A research agenda for improving national Ecological Footprint accounts. Ecol. Econ. **2009**, *68* (7), 1991–2007.

55. Wada, Y. The appropriated carrying capacity of tomato production: The Ecological Footprint of hydroponic greenhouse versus mechanized open field operations. M.A. Thesis. School of Community and Regional Planning, University of British Columbia: Vancouver, Canada, 1993.

56. Niccolucci, V.; Galli, A.; Kitzes, J.; Pulselli, R.M.; Borsa, S.; Marchettini, N. Ecological Footprint analysis applied to the production of two Italian wines. Agric. Ecosyst. Environ. **2008**, *128*, 162–166.

57. Cerutti, A.K.; Bagliani, M.; Beccaro, G.L.; Bounous, G. Application of Ecological Footprint analysis on nectarine production: methodological issues and results from a case study in Italy. J. Cleaner Prod. **2010**, *18* (8), 771–776.

58. Kautsky, N.; Berg, H.; Folke, C.; Larsson, J.; Troell, M. Ecological Footprint for assessment of resource use and development limitations in shrimp and tilapia aquaculture. Aquacult. Res. **1997**, *28*, 753–766.

59. Warren-Rhodes, K.; Sadovy, Y.; Cesar, H. Marine ecosystem appropriation in the Indo-Pacific: A case study of the live reef fish food trade. Ambio **2003**, *32*, 481–488.

60. Parker, R.; Tyedmers P. Uncertainty and natural variability in the Ecological Footprint of fisheries: A case study of reduction fisheries for meal and oil. Ecol. Indic. **2012**, *16*, 76–83.

61. Van der Werf, H.M.G.; Tzilivakis, J.; Lewis, K.; Basset-Mens, C. Environmental impacts of farm scenarios according to five assessment methods. Agric. Ecosyst. Environ. **2006**, *118* (1–4), 327–338.

62. Thomassen, M.A.; de Boer, I.J.M. Evaluation of indicators to assess the environmental impact of dairy production systems. Agric. Ecosyst. Environ. **2005**, *111*, 185–199.

63. Cuandra, M.; Bjorklund, J. Assessment of economic and ecological carrying capacity of agricultural crops in Nicaragua. Ecol. Indic. **2007**, *7* (1), 133–149.

64. Barilla Center for Food and Nutrition. Double Pyramid 2011: Healthy Food for people, Sustainable for the Planet; Parma, IT, 2011, available at http://www.barillacfn.com/en/dp-doppiapyramid/position-paper. (accessed July 2011).

65. Herva, M.; Franco, A.; Ferreiro, S.; Alvarez, A.; Roca, E. An approach for the application of the Ecological Footprint as environmental indicator in the textile sector. J. Hazard. Mater. **2008**, *156* (1–3), 478–487.

66. Gossling, S.; Borgstrom, C.; Horstmeier, H.O.; Saggel, S. Ecological Footprint analysis as a tool to assess tourism sustainability. Ecol. Econ. **2002**, *43* (2–3), 199–211.

67. Hunter, C. Sustainable tourism and the tourist Ecological Footprint. Environ. Dev. Sustainability **2002**, *4*, 7–20.

68. Hunter, C.; Shaw, J. The Ecological Footprint as a key indicator of sustainable tourism. Tourism Manage. **2007**, *28* (1), 46–57.

69. Peeters, P.; Schouten, F. Reducing the Ecological Footprint of inbound tourism and transport to Amsterdam. J. Sustainable Tourism **2006**, *14*, 157–171.

70. Patterson, T.M.; Niccolucci, V.; Bastianoni, S. Beyond "more is better": Ecological Footprint accounting for tourism and consumption in Val di Merse, Italy. Ecol. Econ. **2007**, *62* (3–4), 747–756.

71. Patterson, T.M.; Niccolucci, V.; Marchettini, N. Adaptive environmental management of tourism in the Province of Siena, Italy using the Ecological Footprint. J. Environ. Manage. **2008**, *86* (2), 407–418.

72. Castellani, V.; Sala, S. Ecological Footprint and life cycle assessment in the sustainability assessment of tourism activities. Ecol. Indic. **2012**, *16*, 135–147.

73. WBCSD. *Vision 2050: The New Agenda for Business.* WBCSD: Geneva, Switzerland, 2010.

74. Moore, D.; Cranston, G.; Reed, A.; Galli, A. Projecting future human demand on the Earth's regenerative capacity. Ecol. Indic. **2012**, *16*, 3–10.

75. Best, A.; Giljum, S.; Simmons, C.; Blobel, D.; Lewis, K.; Hammer, M.; Cavalieri, S.; Lutter S.; Maguire C. Potential of the Ecological Footprint for monitoring environmental impacts from natural resource use: Analysis of the potential of the Ecological Footprint and related assessment tools for use in the EU's Thematic Strategy on the Sustainable Use of Natural Resources. Report to the European Commission, DG Environment, 2008.

76. Fiala, N. Measuring sustainability: Why the Ecological Footprint is bad economics and bad environmental science. Ecol. Econ. **2008**, *67* (4), 519–525.
77. Kitzes, J.; Moran, D.; Galli, A.; Wada, Y.; Wackernagel, M. Interpretation and application of the Ecological Footprint: A reply to Fiala. Ecol. Econ. **2009**, *68* (4), 929–930.
78. van den Bergh, J.C.J.M.; Verbruggen, H. Spatial sustainability, trade and indicators: An evaluation of the 'Ecological Footprint'. Ecol. Econ. **1999**, *29*, 61–72.
79. Bastianoni, S.; Niccolucci, V.; Pulselli, R.M.; Marchettini N. Indicator and indicandum: "Sustainable way" vs. "prevailing conditions" in Ecological Footprint definition. Ecol. Indic. **2012**, *16*, 47–50.

28

Environmental Legislation: Asia

Wanpen
Wirojanagud

Introduction

The 21st century has been dubbed "The Asian Century." The region's economic growth is unprecedented, with the rise of China and India as the next economic superpowers.[1] China is the second largest economy and is predicted to take over after the United States as the largest economy in the world by 2020. India is one of the fastest-growing economies and is expected to become the world's third largest economy in the near future.[2–3] More than 58% of the world's population live in Asia and the Pacific region, and Asia is categorized both as the factory of the world as well as the booming market with tremendous growth potential. As a result, natural resources, which are the capital for development of both economic and social sectors, have been remarkably exploited, with a consequent substantial increase in environmental pollution. Without proper environmental management, the ecosystem continues to suffer, resulting in loss of biodiversity, depletion of ecosystem services, desertification, loss of fertile land, atmospheric pollution, aquatic and marine pollution, etc. The consequences of environmental degradation have become more and more acute and chronic over the years. Moreover, natural disasters are more frequent and more severe. As indicated in the Asia-Pacific Disaster 2010 Report of the United Nations (UN), "while the region generated one quarter of the world's GDP, it accounted for a staggering 85 per cent of deaths and 38 per cent of global economic losses due to natural disasters over the last three decades." The report concludes that Asia is the most disaster-prone region in the world.[4] The change in climate patterns has also become more severe.

Environmental status in Southeast Asia, as summarized in Table 1, evidently indicates that over the last decades pressure mostly from anthropogenic activities (rapid population growth, urbanization, economic growth, and consumptive lifestyle) has had a substantial impact on the plentiful natural resources and the environment. Natural resources depletion results in impact on agricultural

TABLE 1 Key Environmental Issues and Causes in ASEAN

Country	Shared Issues	Key Causes
Brunei	Seasonal smoke and haze; solid wastes	Transboundary pollution from land and forest fires; inadequate waste management facilities and practices
Cambodia	Soil erosion; sedimentation; water pollution; deforestation; loss of biodiversity; and threats to natural fisheries	Unmanaged waste and effluent discharge into Tonle Sap Lake; destruction of mangrove wetlands through extensive industrial and aquaculture development
Indonesia	Deforestation; loss of biodiversity; water pollution; air pollution in urban areas; national and transboundary seasonal smoke and haze; land degradation; pollution of Malacca Straits	Deficiencies in urban infrastructure—unmanaged industrial wastes and municipal effluents and waste; vehicle congestion and emissions; extensive land clearance and forest fires for pulp wood and oil palm production; extensive and unmanaged mining activities; national and transboundary industrial pollution; tourist developments in coastal regions beyond carrying capacity
Laos	Deforestation; loss of biodiversity; soil erosion; limited access to potable water; water-borne diseases	Land clearance; shifting cultivation; inadequate water supply and sanitation infrastructure
Malaysia	Urban air pollution; water pollution; deforestation; loss of biodiversity; loss of mangrove habitats; national and transboundary smoke/haze	Vehicle congestion and emissions; deficiencies in urban infrastructure industrial and municipal effluents; extensive land clearance and forest fires for pulp wood and oil palm production; unmanaged coastal developments; tourist developments in coastal regions beyond existing carrying capacity
Myanmar	Deforestation; loss of biodiversity; urban air pollution; soil erosion; water contamination and water-borne diseases	Land clearance; excessive mineral extraction; vehicle congestion and emissions; deficiencies in urban infrastructure—unmanaged industrial and municipal effluents
Philippines	Deforestation in watershed areas; loss of biodiversity; soil erosion; air and water pollution in Manila leading to waterborne diseases; pollution of coastal mangrove habitats; natural disasters (earthquakes, floods)	Illegal forest cutting; land clearance; rapid urbanization and deficiencies in urban infrastructure—unmanaged industrial and municipal effluents, inadequate water supply and sanitation; tourist developments in coastal regions beyond existing carrying capacity
Singapore	Industrial pollution; water shortages; waste disposal problems	Seasonal smoke/haze; limited natural fresh water resources; limited land available for waste disposal
Thailand	Deforestation; loss of biodiversity; land degradation and soil erosion; shortage of water resources in dry season and flooding in rainy season; conflict of water users; coastal degradation and loss of mangrove habitat; urban air pollution; pollution from solid waste, hazardous materials and hazardous waste	Sporadic development and destruction of watersheds; unmanaged aquaculture; tourist growth exceeding growth in carrying capacity; deficiencies in urban and rural infrastructure; freshwater resources polluted by domestic/industrial wastes, sewage, and contaminated runoff
Vietnam	Deforestation and soil degradation; loss of biodiversity; loss of mangrove habitat; water pollution and threats to marine life; groundwater contamination; limited potable water supply; natural disasters (e.g., floods)	Land clearance for industry; forest clearance and chronic impact of Agent Orange; extensive aquaculture and overfishing; growing urbanization and infrastructure deficiencies; inadequate water supply and sanitation (particularly in Hanoi and Ho Chi Minh City)

Source: Data from Nguyen.[5]

productivity; increased frequency of disasters such as floods, landslides, and soil erosion[5]; depletion of aquifers; deterioration of diversity; increased loads from municipal, industrial, and hazardous wastes; atmospheric pollution; and marine and aquatic pollution.[6]

Hence, the key policy question facing Asian governments is how to reconcile economic development and environmental protection. In other words, how to make a transition from the "grow first, clean later" approach to the policy of "sustainable development"—a holistic integration of economic, social, and environmental dimensions.[7] More importantly, the main driving forces in the development of environmental management tools in Asia are international agreements, the occurrence of environmental disasters, and non-harmonized legal framework. Asian countries have shown increasing interest in environmental management by which environmental legislation has been used as one of the important tools in this complex task of integrating environmental protection and economic development. Environmental legislation is among the most determined elements in environmental management toward sustainable development.[8] "Legislation is an important element of the institutional framework for environmental management. The role of legislation is to implement and enforce policy and to provide effective administrative and regulatory mechanisms."[9]

Relevantly, development and implementation of environmental law may involve interaction with legislation and administrative practices and institutions. Thus, environmental laws are undeniably increasingly important in Asia. Environmental law is defined as a body of state and federal statutes intended to protect the environment, wildlife, land, and beauty; prevent pollution and overcutting of forests; save endangered species; conserve water; develop and follow general plans; and prevent damaging practices.[9] The particular law gives individuals and groups the right to bring legal actions or seek court orders to enforce the protection, or demand revisions of private and public activity that may have detrimental effects on the environment.

In terms of the evolution of environmental legislation, two different types of statutes can be distinguished, referred as to the "first generation" and the "second generation" of environmental laws. Although there is no total agreement about how to characterize each generation, generally, the first generation of environmental legislation refers to "command-and-control statutes and regulations administered with technology-based standards and enforced by rule-of-law litigation."[10] Then arrived the "second generation" in the late 1980s with the new concept of "sustainable development" aimed at reconciling environmental protection and economic development.[11] The role of the people, particularly local communities, is highlighted, and compliance incentives and market-based mechanisms have been developed to encourage compliance and provide flexibility.[12]

In the Asia-Pacific region, environmental legislation began to emerge in the 1970s, the same period as the growing global interest in environmental protection. Examples of the early framework laws and regulations for environmental protection are presented in Table 2. Noticeably, there is no lack of environmental legislation in Asia. However, there is a wide gap between law and practice. The countries have to put an emphasis on ways and means to increase the effectiveness of the implementation and enforcement of environmental law.

This entry aims to provide a comparative overview of environmental law and legislation in Asia. It should be noted, however, that the development of environmental legislation in Asia has been uneven and reflects the equally uneven socioeconomic development in the region. In some cases, differences may not be significant in written laws, but lie in the effectiveness of their implementation and enforcement. Due to a wide range of environmental issues and the number of countries in Asia, it is impossible to cover all of them. The entry spotlights four themes—biodiversity conservation, electronic waste (e-waste) management, environmental assessment, and climate change—because of their significance in the Asian context. The choice of countries as case studies aims to demonstrate the uneven development of environmental legislation. Thus, Japanese law serves as a reference of a well-developed body of environmental law, with the second-generation of legislation. China, Korea, Malaysia, Indonesia, Singapore, Hong Kong, Cambodia, Thailand, Laos, and Vietnam are at various stages of development ranging from the first-generation statutes, to development of second-generation environmental laws, and implementation and enforcement.

TABLE 2   Early Laws and Regulations on Environment in the Region

Country	Law or Regulation	Year
Cambodia	Environmental Protection and Natural Resource Management Law	1996
China	Environmental Protection Law	1978
Hong Kong	Water Pollution Control Ordinance	1980
Indonesia	Environmental Management Act No. 4	1982
Japan	Cabinet Directive	1972
Korea	Environmental Preservation Act	1977
Lao PDR	Lao PDR Constitution	1991
Mongolia	Environmental Protection Law	1996
Philippines	Environmental Policy Presidential Decree No. 1151	1977
Singapore	Environmental (Public Health) Act	1969
Thailand	Enhancement and Conservation of the National Environmental Quality Act	1992
Vietnam	Environmental Protection Law	1994

Source: Data from World Bank.[42]

Law on Biodiversity Conservation

Asia-Pacific is one of the richest regions in terms of biodiversity and 60% of the world's species are found in this region. As of 2008, the Asian and Pacific regions had the highest number of threatened species in any of the world's regions, almost one-third of all threatened plants and more than one-third of all threatened animal species.[13] The first generation of biodiversity law consisted mainly in establishing protected areas where human intervention is curtailed or prohibited. However, this exclusionary approach has caused notable socioeconomic impacts, in particular to local communities who used to live in the areas before the establishment of national parks. Therefore, more inclusive approaches have been proposed, such as community-based conservation (comanagement of a protected area by the government and local communities) and payments-based conservation (individuals are paid for their activities to conserve biodiversity). Biodiversity conservation capability is thus a function of a sound legislation in both international and national levels.

Recent Development in International Law and Its Potential Consequences on National Legislations

The international legal instruments, together with agreements that initiate the legislative framework of individual countries for biodiversity conservation, include The Convention on Biological Diversity (CBD) that entered into force on December 29, 1993, the Convention on Wetlands of International Importance Especially as Waterfowl Habitat (Ramsar 1971), the Convention Concerning the Protection of the World Cultural and Natural Heritage (Paris 1972), the Convention on International Trade in Endangered Species of Wild Fauna and Flora (Washington 1973), the Convention on the Conservation of Migratory Species of Wild Animals (Bonn 1979), etc.[14] However, an effective implementation for the international legal agreements is largely dependent on the actions taken by individual countries. Consequently, the most important of all is to establish the legal systems for biodiversity conservation at state level.

Very recently, on February 2, 2011, the Nagoya Protocol on Access to Genetic Resources and the Fair and Equitable Sharing of Benefits Arising from their Utilization was opened for signature by Parties to the Convention of Biological Diversity. Many Asian countries are in the process of amending their legislation to implement the provisions under this new protocol. As most Asian countries are resource-providing countries, the Nagoya Protocol will allow better control of their genetic resources and ensure fair financial compensation among various stakeholders, including local communities.[15]

As for forestry, the new market-based mechanism of "Reducing Emission from Deforestation and Degradation (REDD)," which is under discussion in the UN Climate Change negotiations, should provide additional impetus to the improvement of forest law in Asian developing countries, particularly in the areas of monitoring, reporting, and data collection. Cambodia, Indonesia, Papua New Guinea, the Philippines, and Vietnam are among the first countries to receive support from the UN REDD Program.[16]

Development of Biodiversity Laws at National Level

At present, the biodiversity conservative laws of Asian countries like Japan, Thailand, Indonesia, and Laos contain more or less elements of public participation that characterize the second generation of legislations. Except for Japan, however, communities' rights are still very limited and difficult to enforce. The payment-based conservation approach has generally only been implemented through small-scale projects and not adopted by Asian legislators yet.

China. China covers an enormous land area of 9,600,000 km^2, including complex and varied geomorphology, climate, and natural conditions. That creates a country rich in ecosystems, which can be categorized into five types, namely forest, grassland, desert, inland wetland and other freshwater ecosystems, and ocean and coast. Due to a massive and distinct diversity of flora and fauna, China is regarded as one of the most important biodiversity countries. It is ranked among the top 10 nations in the world diversity of its mammal, bird, amphibian, and plant species. China has been considered to have one of the most important stocks of genetic diversity in the world. It is very important to protect and conserve this biodiversity for the national and international heritage.

China has promulgated a series of laws and regulations. The main domestic laws are Forest Law (1984), Grassland Law (1985), Fishery Law (1986), and Wild Animal Conservation Law (1988). Examples of regulations related to biodiversity include Reproduction and Conservation of Aquatic Resources (1979), Regulation on Forest and Wild Animal Nature Reserves Management (1985), Regulation on Forest Fire Prevention and Control (1988), Regulation on Seed Management (1989), Regulation on Conservation of Terrestrial Wild Animals (1992), Regulation on Nature Reserves (1994), etc.

Regarding the enforcement of the statutes, great progress for in situ and ex situ biodiversity conservation has been achieved. There are, however, still some gaps in the legislation. Based on the current status of conservation legislation in China and in accordance with the Convention on Biological Diversity, more attention should be paid to the conservation legislation for genetic resources, wild plant species, and various natural ecosystems.

Korea. Ecosystems in Korea comprise forest, mountain, freshwater, coastal and marine, and agriculture ecosystems. The total forest area covers 6.394 million hectares, estimated as about 64% of the country's land area. Forests are mainly coniferous, deciduous, and mixed forests. The variety of habitats creates a rich biodiversity of plants, animals, and other living organisms (fungi, protista, prokaryotes, etc.). Some species are considered to be extinct, such as the tiger and Siberian leopard, fox, wolf, and sitka deer. The decline of biodiversity in Korea is associated with its economic development. The main threats to biodiversity include overexploitation of land and biological resources, and environmental pollution.

Under the guiding principles of the Framework Act on Environmental Policy 1990 and the Constitution, the Natural Environment Conservation Act 1991 administered by the Ministry of Environment is Korea's basic law for biodiversity and nature protection. It defines categories of protected areas and provides for species and habitat protection. The law serves as a common framework for nature conservation and strengthens the provisions of other nature laws administered by government agencies. Several government agencies share the responsibility of conservation and sustainable use of biodiversity, in accordance with various laws. The Ministry of Environment is responsible for general biodiversity conservation under the Law of Natural Environment Conservation, Law of Wildlife Protection and Hunting, Law of Wetland Conservation, Law of Natural Parks, and Law of Ecosystem Conservation for Uninhabited Islands. The Ministry of Environment is also responsible for preventing inappropriate uses of natural resources through the Environmental Impact Assessment (EIA) process by the Law of Environmental Impact Assessment. The Forestry Administration, part of the Ministry of Agriculture and Forests, which manages forests under the Law of Forests.

Japan. The country is 3000 km long in the north–south direction, with a vertical range from coasts to mountains, with thousands of islands, and a geological history of intermittent connection to and separation from the continent, and various disturbances such as eruption of volcanoes, flooding of precipitous rivers, and typhoons. Those geological characteristics, together with four definite seasons due to the monsoon climate, create diverse habitats. It is such rich biodiversity that makes Japan one of the 34 biodiversity hotspots identified worldwide. A biodiversity hotspot refers to a region that is originally rich in biological diversity and endemic species but is now exposed to a serious threat of loss of such diversity. Besides the original local geohistorical and natural conditions, the tradition of wet-paddy rice agriculture and the rural lifestyle, which rely on a secondary natural environment known as "satochi-satoyama," or simply "satoyama," as well as the way the land has been used for agricultural purpose, have also contributed to the area's biodiversity richness. As in other countries, biodiversity loss in Japan is due to high economic growth with industrial development, and also natural disasters. It can, however, be said that the Japanese law on biodiversity is regarded as one of the best laws in the field. It goes beyond the protected area–based approach by mainstreaming biodiversity conservation into the daily life of the people. The law defines not only the responsibilities of national government and local governments but also those of businesses, citizens, and private bodies. The elaboration of the National Biodiversity Strategy and regional biodiversity strategies are mandatory. The 4th National Biodiversity Strategy, which was adopted in 2010, sets a long-term goal of 100 years, mid-and short-term targets for 2020 and 2050, and indicates about 720 measures with 35 targets.[17] The results of the implementation have to be reported to the Diet every year in the Annual Report on the State of Biodiversity. The law puts emphasis on preventive and adaptive approaches, including land use planning, research and technology development, EIA, and prevention of global warming. The role of the public is also highlighted with mandatory public consultation before formulation of policies and support of voluntary activities by businesses and citizens for the conservation of biodiversity.

Other relevant legislations are the Nature Conservation Law, Natural Parks Law, Law for the Promotion of Nature Restoration, Law for the Promotion of Biodiversity Conservation Activities, Law for the Conservation of Endangered Species of Wild Fauna and Flora, Wildlife Protection and Proper Hunting Law, Invasive Alien Species Act, and Law Concerning the Conservation and Sustainable Use of Biological Diversity through Regulations on the Use of Living Modified Organisms.

Malaysia. Malaysia has been identified as one of the world's mega-diversity areas with extremely rich biodiversity. Covering much of the country are the tropical forests, the oldest and most biologically diverse ecosystem on Earth. With ratification of the CBD, Malaysia is working toward incorporating into its national policies and planning a set of commitments under the treaty as well as setting the goal to become a world leader in conservation, research, and sustainable utilization of tropical biodiversity by 2020. To accomplish the ratification and goal, Malaysia has enacted a spectrum of legislation aimed at protecting biodiversity. Examples of law relevant to biodiversity conservation are as follows: Environment Quality Act 1974, Fisheries Act 1985, Pesticides Act 1974, Plant Quarantine Act 1976, Protection of Wildlife Act 1972, National Parks Act 1980, National Forestry Act 1984, Parks Enactment 1984, Forest Enactment 1992, Fauna Conservation Ordinance 1963, etc.[18,19]

Malaysia has enacted a number of laws and regulations to protect the nation's environment but, while adequate, there was no single overarching statute (or policy?) that relates to biodiversity conservation and management until the just approved National Policy on Biodiversity. Much of the present legislation is sector based. As stated above, the attainment of biodiversity conservation is significantly dependent on implementation and enforcement of the legislation. The effectiveness of legislation can be accomplished by the dedication of government agencies as well as public participation for accountability.[18,19]

Indonesia. Indonesia is a rich and diversified archipelagic nation. With the topographical characteristics of approximately 13,500 islands and extensive reef system, Indonesia has a wide range of natural habitats, with a wealth of fauna and flora, corals, fish, and other reefs; thus, it is recognized as a major world center for biodiversity. However, that substantial biodiversity is decreasing in the country owing to illegal clearing and deforestation, including large-scale burning in oil palm plantations and small-scale slash and burn for shifting farming, as well as illegal logging and trade in timber. Indonesia has one of the world's worst deforestation rates. In addition, illegal poaching, trade in protected species, and illegal and unsustainable fishing have threatened the country's biodiversity.[20,21]

Regarding the domestic legal framework, as the result of 1972 UN Stockholm Conference on the Environment, Indonesia promptly established of the Office of the State Minister for the Environment and enacted the Environmental Management Act (EMA) No. 4 of 1982 replaced in 1997 by the EMANo. 23 of 1997. This act and its implementing regulations are set in the broader context of the state policies passed every 5 years by the People's Consultative Assembly, which have since the early 1970s progressively entrenched the concepts of sustainable development and natural resources management. The EMA must also be read in the context of other natural resources management acts, such as the Forestry Act No. 41 of 1999, and the Fisheries Act No. 31 of 2004, and associated regulations and decrees transferred to the local governments' agencies. Laws and regulations related to biodiversity conservation are exampled as follows: Act on the Conservation of Biological Resources and their Ecosystems (Act No. 5 of 1990), Decree No. 1 of the Minister of Agriculture on the Conservation of the Riches of the Fish Resources of Indonesia, and Decree of the Ministry of Forestry No. 424/ KPTS-VI/1994 on the Guidelines on Crocodile Management in Indonesia, Fisheries Law (No. 9 of 1985). Similarly to other countries, law enforcement is still not an effective function. Indonesia has no specialized environmental law courts. Environmental cases are heard by the general and administrative courts as well as the Supreme Court on appeal.[21]

Furthermore, a multitude of biodiversity laws in Indonesia tend to be conflicting and uncoordinated. Also, in the forestry sector, the power to manage forests is shared between the central government and regional governments with no clear division of powers and responsibilities. It is believed that the Regional Autonomy Law, which entered into force in 1999, has been abusively used by local authorities to issue their own logging concessions, thus resulting in massive deforestation and forest fires.[22] The Indonesian case shows that a legislation that transfers power from center to periphery has to be carefully drafted to ensure that the power is put in the hands of local communities whose livelihood depends on the forests and biodiversity, and not in the hands of local officials known for corruption and nepotism. Some main legislations of Indonesia relating to biodiversity are the Conservation of Biodiversity and Ecosystems Law (1990) and the Basic Forestry Law (2000). These laws include provisions on participatory forestry planning, people's economic empowerment, partial transfer of authority to regional governments, and community-based forest monitoring.

Philippines. The Philippines is a tropical archipelago of 7100 islands located off the southeast coast of mainland Asia. It occupies a land area of 299,400 km², and territorial waters covering around 2,200,000 km², that create precious terrestrial and aquatic ecosystems and habitat types. As a consequence of rapid loss of biodiversity, as well as widespread destruction of the country's environment, a strong effort has been put into biodiversity conservation, including the formation of the multisector Philippine Council for Sustainable Development (1992), ratification of the Convention on Biological Diversity (1993), and preparation of the Philippine Biodiversity Assessment Report and the National Biodiversity Strategy and Action Plan (NBSAP) (1995–1997). The NBSAP proposes a wide range of strategies and actions, including information generation, in situ and ex situ conservation, legislative and policy development, institutional capability building, information, education and communication, and strengthened international cooperation.[23]

Regarding ecosystem and habitat conservation in the Philippines, it was innovated through the National Integrated Protected Areas System Act of 1992, a landmark piece of legislation that provides the framework for a decentralized, community-based reserve management strategy. The legislation relevant to biodiversity conservation include the following: Act No. 2590 (1916), An Act for the Protection of Game and Fish; RA 7586 National Integrated Protected Areas System Act of 1992; RA 7900,

High-Value Crops Development Act of 1995; PD 1433, Plant Quarantine Decree of 1978; Proc. No. 926, Establishing Subic Watershed Forest Reserve; DAO 20, s 1996, Implementing Rules and Regulations on the Prospecting of Biological and Genetic Resources; DAO 24, s 1991, Shift in Logging from the Old Growth (Virgin) Forests to the Second Growth (Residual) Forests; DAO 20, s 1996, Implementing Rules and Regulations on the Prospecting of Biological and Genetic Resources; etc.[24]

Singapore. The Republic of Singapore, situated off the southern tip of the Malay Peninsula, comprises one major and more than 50 adjacent islands, with a total area of 648 km². The main island is separated from Malaysia by the narrow Johor Strait on the north, and from Indonesia's Riau Archipelago by the Singapore Strait on the south. It is a small country with an urbanized character. Forest land and coastal areas have been cleared to provide land for residential and commercial sites and other developments, resulting in a substantial depletion of flora/fauna and deterioration of natural habitats. At present, the environmental policy states that 5% of the land area should be set aside for nature reserves, national parks, catchment areas, bird sanctuaries, and gardens. Some of the environmental laws related to biodiversity conservation are Fisheries Act 1966, Wild Animals and Birds Act 198, Parks and Trees Act (for parks not gazetted as national parks) 1985, Endangered Species Act 1989, National Parks Act 1990, Animals and Birds Act (Revised) 2002.[25]

To accomplish the vision of *Singapore Today,* which is to be "A Garden City, A Haven for Biodiversity," the Singapore Green Plan (SGP) 2012 was established to provide the direction for protected area management in the next decade. One of the SGP objectives is to ensure the quality of the living environment, including the enhancement of the country's environmental heritage. SGP 2012 also recaps the state's commitment to maintain the 5% of land set aside for nature areas, and provides the direction.[26]

Hong Kong Special Administration Region. Like Singapore, Hong Kong SAR (or Hong Kong in short) experienced very rapid growth and development, to become urbanized and industrial in character. This would have jeopardized the biodiversity of the area, which is recognized as part of the natural heritage of Hong Kong. The most common causes of biodiversity loss in Hong Kong now are habitat destruction associated with infrastructure development, population growth, hillside fires, unsustainable exploitation of wild species, introduction of alien species, pollution, and global environmental change. Possible relevant factors contributing to such causes are lack of a comprehensive legal framework to protect areas of high conservation value and poor enforcement of existing conservation, environmental, and planning laws.[27]

Environmental protection in Hong Kong is constituted by 16 ordinances. Nonetheless, overall environmental laws in Hong Kong are still short of an unambiguous conservation objective. Only the 1995 Marine Parks Ordinance (Cap. 476) and the 1997 Protection of the Harbour Ordinance (Cap. 531) enclose evidently expressed conservation principles. Other legislations protecting flora and fauna are focused on conserving particular species; however, they fail to address the values and principles that lie beneath these objectives. Regarding the Country Parks Ordinance (Cap. 208) originally enacted in 1976, a revision is needed, "To provide for the designation, control and management of country parks and special areas for the purposes of conservation of biological diversity, countryside recreation and education." Additionally, the Environmental Impact Assessment Ordinance (EIAO) is the newest piece of legislation that seeks to protect the environment. The EIAO requires certain designated projects (generally major infrastructure projects) to undertake an EIA before they can be granted an environmental permit for development to proceed. A Technical Memorandum contains guidance on the criteria and guidelines to use for ecological, fisheries, landscape, and visual impact assessment.[27]

As well as domestic laws, Hong Kong is obligated to protect its natural and cultural heritage by international treaties, including the 1973 Convention on International Trade in Endangered Species of Wild Fauna and Flora, the 1979 Convention on the Conservation of Migratory Species of Wild Animals, the 1971 (Ramsar) Convention on Wetlands of International Importance, and the 1972 Convention for the Protection of World Cultural and Natural Heritage. Seemingly, application of the 1992 Convention on Biodiversity in Hong Kong remains limited as there is no sign that the treaty will be formally applied to accomplish the CBD's biodiversity objectives after the government's endorsement.[27,28]

Thailand. Thailand has a total land area of 513,115 km², lying in a hot and humid climatic zone in the middle of Southeast Asia. Much of Thailand is situated in the Mekong River basin, and it is one of the Greater Mekong Subregion (GMS) countries. Thailand also has an extensive coastline. With such a location, it is enriched with biodiversity associated with terrestrial and aquatic (both freshwater and marine) ecosystems. It also covers agricultural ecosystems (about one-fifth of the country), which include biodiversity components of rice, farm crops, and livestock. Similarly to other countries, the major threat to biodiversity is human disturbance through overexploitation of natural resources/ habitat, illegal logging and trading of animals, overhunting of wildlife, deforestation, urban expansion and pollution, etc. Such disturbances cause an adverse reduction of bio-diversity.[29]

Biodiversity in Thailand is safeguarded by a number of laws and regulations. Some of the important ones are the National Park Act 1961, National Forest Reserve Act 1964, Wild Animal Reservation and Protection Act 1992, Plant Quarantine Act of 1964 and Plant Quarantine Act (second issue) 1994, Animal Species Maintenance Act 1966, Importing and Exporting of Goods Act 1979, Enhancement and Conservation of National Environmental Quality Act 1992, and Plant Varieties Protection Act 1999.[30]

In addition, Section 66 Paragraph 1 of the Thai Constitution of 2007 guarantees the right of an individual and communities to participate in the preservation and exploitation of natural resources and biological diversity. However, out of 35 pieces of legislation relating to natural resources, biodiversity and environment, only 4 provides for public participation, namely, the National Promotion and Conservation of Environment Quality Act, B.E. 2535 (1992); the Private Irrigation Act, B.E. 2482 (1939); the Plant Varieties Protection Act, B.E. 2542 (1999); and the Protection and Promotion of the Thai Traditional Medicine Act, B.E. 2542 (1999). Moreover, the rights given to individuals are still very limited, and the authorities retain most of the control in natural resources management.

Lao People's Democratic Republic. Lao PDR, situated in the Mekong River basin, is counted as one of the GMS countries. For Southeast Asia, Lao PDR is one of the countries with a large proportion of land covered with undisturbed forest. It covers a land area of 236,800 km², where the topography is largely mountainous, with elevations above 180 m typically characterized by steep terrain, narrow river valleys, and low agricultural potential. Lao PDR is rich with natural resources such as forestry, minerals, and hydroelectric power. Due to the country's still abundant forestry resources, it thus generates a prosperous biodiversity.[31,32] Similarly to other GMS countries, Lao PDR has enacted laws for biodiversity conservation, but the enforcement is ineffective. Illegal logging and wildlife hunting have been frequently found. Mining and hydropower developments might also cause loss of biodiversity.[31,32]

In Lao PDR, biodiversity conservation is provided for in the Forest Law of 1996, a 1993 decree that designated the first national biodiversity conservation areas, logging ban, decree on (PM Decree No. 67/PM, 1991) protecting trees by logging, decision on adoption of (PM Decree No. 66/PM, 1991) forest conservation, and seed material import regulation (quarantine regulation) that controls importation of plant material.[32]

Cambodia. Cambodia is a tropical country in mainland Southeast Asia with a territory of 181,035 km². It is adjacent to the Gulf of Thailand on the south and shares borders with Thailand (west and north), Laos (north), and Vietnam (east). Cambodia is in the Mekong River basin. The country is dominated by the Mekong River, known as the Tonle Thom or "great river," and the Tonle Sap or "fresh water lake." The Mekong River basin is one of the most biodiverse regions in the world. As situated in the Mekong River Basin, Cambodia has an abundance of terrestrial and aquatic ecosystems that are significant habitat for plants and aquatic organism, fishes in particular. Such ecosystems are invaluable resources for economic development and human well-being.[33,34]

To protect and sustain the country's biodiversity, Cambodia enacted the framework for the Law on Environmental Protection and Natural Resources Management in 1996. Subsequently, the Ministry of Environment was created in 1998, which manages natural resources along with the Ministry of Water Resources and Meteorology and the Ministry of Land Use Management, Urbanization, and Construction. Cambodia has continued to enact more environmental and conservation laws. Examples of the relevant laws are as follows: Environmental Protection and Natural Resources Law 1996, Law on Commune

Administration (part of the decentralization process) 2001, Land Law 2001, Forestry Law 2002, and Wildlife Law 2002. Important subdecrees include Subdecree on Concession Management (moratorium on logging and log transport), Subdecree on Community Forestry, Subdecree on Environmental Impact Assessment, Subdecree on Industrial Agricultural Concessions, Subdecree on Social Concessions, and Royal Decree on Protected Areas 1993.

Cambodia is party to a number of important international conventions of which those stated here are relevant to biodiversity conservation. The Ramsar Convention on Wetlands has been ratified, and Cambodia has identified three wetland sites for recognition: BoengChhmar in the Tonle Sap floodplain, KohKapik on the coast, and a portion of the middle Mekong river north of Stung Treng. In addition, the Tonle Sap is recognized as a UNESCO Biosphere Reserve. Cambodia ratified the Convention on Biological Diversity in 1995. In 1997, the government prepared a biodiversity prospectus and in 2002 completed a National Biodiversity Strategy and Action plan.[19,33]

Vietnam. Vietnam is situated at the eastern side of a peninsula that protrudes into the Eastern Sea, which is a bay of the Pacific Ocean. Within the Mekong River Basin, Vietnam is at the most downstream length of the Mekong River. Due to the country occupied with land, river, and sea, Vietnam is enriched with a variety of ecosystems, including tropical rainforests and monsoon savannah, marine life, and mountainous subalpine scrubland. Additionally, a specific feature of Vietnam is its length of more than a thousand miles from north to south, with a width of only 30 miles from east to west at its narrowest point, thus generating an abundance of natural resource along the coast and sea, and a richness of biodiversity. The threats tobiodiversity are mainly due to transformation of forest and wetland areas to other uses, infrastructure construction, urbanization, industrialization, and environmental pollution.

Vietnam's environmental law is based on its constitution. The Law on Environmental Protection was initially promulgated in 1993; subsequently, Vietnam has enacted a variety of laws and decrees on conservation issues. These laws affect, directly or indirectly, the conservation of biodiversity. Examples of the direct laws are the Decree on the Conservation and Development of Wetlands, the Decree on Protection of the Environment (which details rare and precious flora and fauna), and a related decree that determines methods for regulating their protection and management. The indirect laws include decrees regulating wastewater, controls on businesses creating environmental damage, the 2003 Land Law, etc. Regarding the protected areas, Vietnam has two laws concerning the establishment and management of protected areas. The two statutes are the Law on Forest Protection and Development of December 2004 (revised from the 1991 original), and the Law on Biodiversity that became effective starting July 2009. The Law on Forest Protection and Development provides the guiding principles for the development and use of special-use forests, while the Law on Biodiversity focuses on protected area concerns such as categorization and decentralization of protected area management.

Challenges to Effective Implementation and Enforcement of Biodiversity Law

Evidently, the countries presented here have enough laws and regulations to protect their environment; however, much of the present legislation is sector based. Moreover, the achievement of biodiversity conservation is not a function of the number of laws and regulations but of the implementation and enforcement of such legislation. The effectiveness of legislation can be accomplished by the dedication of government agencies as well as public participation for accountability.[19]

As partly demonstrated through the case studies mentioned above, challenges facing Asian governments in the implementation and enforcement of biodiversity law lie in corruption, weak institutional capacity, lack of reliable data and budget, high demand for alternative land use, and traditional beliefs.

With regard to weak institutional capacity, in most Asian countries, forestry staffs are not adequately trained and are underpaid. To address this shortcoming, governments such as China and Vietnam have raised the pay and living conditions for forest personnel.[19] Inter-agency coordination is another serious institutional problem. There are usually many agencies involved in forest management with overlapping areas of responsibilities or, on the contrary, areas where no agency is in charge.

Economic development has put pressure on forest protection. The construction of hydropower dams and roads, and agricultural activities have often encroached on protected lands. In an effort to reconcile forest protection and economic development, Lao PDR has piloted hydropower levies that support the management of protected areas affected.[34]

Forests and wildlife statistics are often out of date, inaccurate, and incomplete. The remoteness of the forest areas and the reluctance of officials to report forest crimes (as they may be seen as a sign of failed forest management on their part) are among factors that contribute to the lack of reliable data and information sharing among relevant agencies. The use of satellite imagery has helped improve the data collection system to a certain extent.

Traditional beliefs may sometimes come in the way of forest and wildlife protection. In many Asian countries such as China, Vietnam, Thailand, Laos, and Cambodia, the consumption of rare wildlife, such as bear's paw and monkey brains, is seen as status symbols or medicines.

Regional Cooperation

Regional cooperation has played an important role in promoting sustainable forest management. Examples of regional initiatives are the East Asia Forest Law Enforcement and Governance process (EAFLEG) and the Asia Forest Partnership (AFP). Both the EAFLEG and the AFP aim to bring together various stakeholders in forest management, including governments, non-governmental organizations (NGOs), and the private sector. They serve as an informal forum for information sharing, dialogue, and joint action. Within the Southeast Asian Nations Association (ASEAN), the ASEAN Senior Officials on Forestry (ASOF) was entrusted with the task of policy coordination and decision making on regional cooperation in the forest sector.

Law on E-Waste Management

The quantity of e-wastes in Asia has exploded in recent years with the exponential growth in the use of electronic equipment (computers, mobile phones, televisions, refrigerators, etc.) coupled with the consumers' behavior of regularly replacing their devices in order to stay up-to-date with the latest technology. E-wastes often end up in incinerators and/or landfills. Toxic substances such as mercury and lead that are commonly used in electronic products can contaminate the environment, including land, water, and air. E-waste is commonly characterized as hazardous waste. As of July 2008, among 46 countries in the Asia-Pacific region, there are 32 countries that have ratified the Basel Convention on the Control of Transboundary Movements of Hazardous Wastes and Their Disposal and some of them have also ratified the Ban amendment.[35] However, there are some countries in Asia that still lack regulations for controlling hazardous wastes including e-waste. An effective regulatory framework on the management of such wastes is urgently needed. A new legal concept created to deal with e-waste management is the Producer Extended Responsibility (PER) where producers are held liable for the costs of managing their products to end of life of the product. In this way, producers are encouraged to design environmentally friendly products to reduce disposal costs. Only a few countries, such as Japan, Korea, and Taiwan, have a well-established legislation on e-waste management based on the concept of PER and many years of experience in implementation. Some Asian countries such as China have recently enacted a law on the matter; however, the effectiveness of the implementation remains to be seen. Other countries have already passed similar laws, but the laws have yet to enter into force (Thailand and India) or are still at the drafting stage. An adaptation period is usually granted for businesses before the entry into force of the law. Other actors in a product's life cycle such as distributors, repair and customer service providers, consumers, and recyclers may also be required to bear some of the treatment and disposal costs. E-waste discussed herein includes used and waste electrical and electronic equipment (UEEE/WEEE).

China. The Ministry of Environment and Protection is responsible for regulating and controlling e-waste management. At the end of February 2006, China promulgated the law entitled "Administration on the Control of Pollution Caused by Electronic Information Products," which is simply called as China RoHS in the industry.[36] Import of WEEE has taken place since 2002 under the existing laws, including Law of the People's Republic of China on Prevention of Environmental Pollution Caused by Solid Waste, Interim Provisions on Administration of Environmental Protection on Import of Waste and its supplementary provisions, List of Wastes Prohibited against Import (Notice No. 25, 2002), and Catalogue of Solid Waste Forbidden to Import in China (Announcement No. 11, 2008). Regarding UEEE, it is allowed except used TVs; UEEE requires 3C certification. Applicable laws for UEEE are as follows: Administrative Method on Inspection and Supervision of Imported Used Mechanical and Electrical Products, Measures for Administration of Import of Specified Used Mechanical and Electrical Products (Order No. 5, 2008), and Catalogue of Import of Specified Used Mechanical and Electrical Products (Announcement No. 37, 2008). A new law on the management of e-waste has established higher standards for recycling processes and allows only certified recyclers to engage in the e-waste recycling business. To support recyclers to improve their equipment and facilities, a centralized mandatory fund has been established with contributions from domestic producers and sellers of imported electronic devices. The law also places responsibility on manufacturers, distributors, repair and consumer service providers, and recycling companiesto collect and responsibly handle e-waste. One shortcoming of the law, however, is that the scope of their responsibility and the penalizing measures for non-compliance remains vague.[35]

Hong Kong, China. The regulation of e-waste management is under the responsibility of the Environment Protection Department. Hong Kong has begun its waste import and export control through the Waste Disposal Ordinance in 1996. For the purpose of import, WEEE and UEEE (that is classified as WEEE) have been controlled through a permit system in accordance with guidelines on "import and export of hazardous waste including electrical and electronic appliances containing hazardous constituents or components."[35]

Republic of Korea. The Resource Recirculation Policy Division, Ministry of Environment, is responsible for regulating and controlling e-waste management. Specific regulations applicable to UEEE and WEEE do not currently exist; however, in general, e-waste management falls under the "Waste Control Act" of December 1986 and later amendments. Import control of WEEE is performed through application for a license from the Ministry of Environment in accordance with the Act on the Control of Trans-Boundary Movement of Hazardous Wastes and Their Disposal (Basel Convention) and Act on Resource Recycling of Electrical and Electronic Equipment and Vehicles. Import of UEEE is allowed; no specific law is applicable.[35]

Japan. The Ministry of Environment is responsible for regulating e-waste management. Since 2001, Japan has enforced the Fundamental Law for Establishing a Sound Material-Cycle Society to promote comprehensively and systematically the policies for realizing a Sound Material-Cycle Society, providing an umbrella framework for the relevant waste management laws of the country. Regarding the import and export control of WEEE, the Law for the Control of Export, Import, and Others of Specified Hazardous Wastes and Other Wastes was entered into force in 1993. This law stipulates the necessary import/export procedures of hazardous waste to comply with the requirements of the Basel Convention. The Waste Management and Public Cleansing Law of 1970 was amended in 1993 to regulate import and export of waste. None of these laws is applicable for the management of UEEE. With the specific purpose to complement to the Basic Act on Establishing a Sound Material-Cycle Society (2000), the following two specific e-waste recycling legislations have been established: the Law for the Promotion of Effective Utilization of Resources (LPUR), and the Law for Recycling of Specified Home Appliances (LRHA) (1991). The LPUR applies to used computers and small-sized secondary batteries and encourages manufacturers' voluntary efforts to take part in collection and recycling. Recycling costs are borne by both manufacturers and consumers. The LRHA is a stricter regulation. It covers television sets, refrigerators, washing machines, and air conditioners. Manufacturers and retailers have an obligation to take back used products and recycle them. Consumers are required to pay for the cost of transportation and recycling. The LRHA also sets up a procedure that allows tracking a product from the beginning until the end of its life cycle.[35]

Malaysia. The Department of Environment, Ministry of Natural Resources and Environment, is responsible for regulating e-waste management. Under the Environmental Quality Act, 1974, several regulations for the control of scheduled wastes (hazardous wastes) management in Malaysia have been enacted. The principal regulation on e-waste management is the Environmental Quality (Schedule Waste) Regulation of 2005, enforced by Department of Environment, and in which specific categories of e-waste are defined and coded. The Guidelines for the Classification of Used Electrical and Electronic Equipment entered into force in January 2008, which prohibits the import of WEEE and export for the purpose of disposal. Waste generators are allowed to export waste for recycling, recovery, or treatment provided prior written consent are obtained from the importing state. The Ministry of Local Government and Housing has the jurisdiction over households and business entities/institutions and has enacted the Solid Waste Management and Public Cleansing Act of 2007. The Royal Malaysian Customs enforces transboundary movements of hazardous waste under the Customs Act 1967, Customs (Prohibition of Import) Order 2008, and Customs (Prohibition of Export) Order 2008.[35]

Indonesia. The Environmental Impact Management Agency, Ministry of Environment, is responsible for regulating e-waste management; however, neither specific criteria on e-waste nor specific regulations on e-waste management have been established. The existing laws have been employed for WEEE and UEEE. WEEE is only allowed for export, but prohibited for import to Indonesia under Act No. 23 of 1997 on Environmental Management, Articles 20 and 21, Presidential Decree No. 61/1993 Basel Convention Ratification, and Ministerial Decree No. 231/ MPP/Kp/07/1997 Regarding Import Procedure of Waste. Import of UEEE and e-waste for direct consumption by consumers is prohibited under Decree No. 756/MPP/ Kep/12/2003 on Import of Non-New Capital Goods and Decree No. 610/MPP/Kep/10/2004 Regarding Amendment of No. 756/MPP/Kep/12/2003.[35]

Philippines. The Department of Environment and Natural Resources (DENR) is responsible for regulating e-waste management. Import of WEEE and UEEE requires permit. Laws applicable to WEEE are the Toxic Substances and Hazardous and Nuclear Wastes Control Act of 1990 (Republic Act No. 6969), DENR Administrative Order 200–6 (Implementing Rules and Regulations for RA 6969), DENR Administrative Order 1994-28 (Interim Guidelines for the Importation of Recyclable Materials Containing Hazardous Substances), DENR Administrative Order 1997-28 (Amending Annex A of DAO 1994-28), and DENR Administrative Order 2004–27 (Amending Annex A of DAO 1994-28). The law applicable to UEEE is DENR Administrative Order 1994-28 (Interim Guidelines for the Importation of Recyclable Materials Containing Hazardous Substances). DAO-94-28 allows the import of electronic assemblies and scrap with the condition that residuals from recycling of materials that contain hazardous substances without any acceptable method of disposal in the Philippines must be shipped back.[35]

Singapore. Export, import, or transit waste requires a permit from the Pollution Control Department of Singapore in accordance with the Hazardous Waste (Control of Export, Import, and Transit) Act. Import/export of UEEE are allowed if there are documents to support that the appliances for import/export are in working condition and suitable for reuse. Export of UEEE that are not suitable for reuse is prohibited. Import of UEEE for the purpose of dismantling and re-export of the dismantled components are prohibited.[35]

Thailand. The Ministry of Natural Resources and Environment and the Ministry of Industry are the administrative authorities for hazardous waste and e-waste management. Both UEEE and WEEE are controlled under the Hazardous Substance Act B.E. 2535 (AD 1992) in Thailand. UEEE can be imported only under a subordinate law for import control of UEEE. Import of UEEE is allowed only for reuse, repair/refurnish back to its original purposes, disassembly and recycle/recovery under certain conditions. Thirty-two UEEE items require import permits from the Ministry of Industry. WEEE can be imported and exported under a subordinate law and following Basel Convention procedures. The e-waste management act will only enter into force in 2014 and will seek to regulate the entire life cycle of an electronic product. At the beginning of its life cycle, importers and producers will be taxed, and the tax money collected will be used for e-waste management. At the end of the product's life cycle, the law will promote construction of an integrated waste management facility at production sites and other areas throughout the country.[35]

Lao People's Democratic Republic. Lao PDR does not have legislation discretely mentioning e-wastes; however, e-waste is considered hazardous waste by defining that hazardous and toxic wastes include batteries, old paint cans, aerosol, and other refuse. Such wastes are mixed with municipal solid wastes that are disposed at landfills. Accordingly, there are no specific laws or regulations directed to e-wastes. For solid waste disposal itself, a decree on waste management is planned in connection with the finalization of the revision of the Environmental Protection Law of 1999. Nonetheless, some laws have implications for solid waste management in Lao PDR, for example, the Environmental Protection Law (Article 22), the Decree on Implementing the Environmental Protection Law (Article 9.4), and in addition some of the provinces and the Capital City of Vientiane have issued specific regulations on urban environmental management including solid waste management. Thus, Lao PDR is encountering solid waste problems due to several reasons, as follows: inadequate legal framework, ambiguous institutional responsibilities and lack allocation of responsibilities on solid waste management to specific institutions, insufficient budget allocation to carry out functions in accordance with the law, etc.[37]

Cambodia. At present, Cambodia does not produce EEE products. Regulation on e-wastes is the responsibility of the Ministry of Environment. Specific law and/or regulation to properly manage, recycle, and dispose EEW does not yet exist. Cambodia, however, uses the following existing laws for WEEE management: Subdecree on Solid Waste Management and the Inter-Ministerial Declaration on SWM in Cities and Provinces. Import of WEEE to Cambodia is banned, while import of both new EEE and UEEE is allowed for domestic consumption. Export of household waste and hazardous waste from Cambodia requires approval from the Ministry of Environment, export license from the Ministry of Trade, and permit from the import country (BCRC China, 2009).

Vietnam. Regulating e-waste management is the responsibility of Hazardous Waste Management Division, Waste Management and Environment Promotion Agency, Vietnam Environment Administration, Ministry of Natural Resources and Environment. In January 2006, Vietnam promulgated the Implementation Rules for the Law on Trade (No. 12/2006/ND CP), which bans import of waste materials (both WEEE and UEEE), toxic chemical substances, and second-hand commodities, including electronic, cooling, and home appliances. Other applicable laws are the Regulation of Management of Hazardous Waste (Decision No. 155/1999/QD-TTg), and Decision No. 23/2006/QD-BTNMT on the List of Hazardous Waste. In Circular No. 12/2006/TT-BTNMT, export of hazardous waste shall follow Basel Convention procedures.[35]

Transboundary Movements of Hazardous Waste

Since the entry into force of the Basel Convention on the Control of Transboundary Movements of Hazardous Wastes in 1992, most Asian countries have acceded to the Convention and enacted national laws banning import of waste into their territories. However, despite the legal prohibition, China and India continue to be the world's largest waste dumping yards. One of the reasons is that there are loopholes in the laws that provide for some exceptions to the ban. For example, in India, the law allows for imports of secondhand computers and laptops if they are intended for donations to educational institutions or NGOs. Due to several arrests for e-waste smuggling recently, the government is considering a complete ban on e-waste.[38] In China, some recyclable wastes such as poly-silicon and artificial fiber continue to be allowed into the country. To prevent the smuggling of prohibited waste, all solid waste imports are required to undergo electronic inspections.[39]

Challenges to Effective Implementation and Enforcement of E-Waste Management Law

Barriers to effective implementation and enforcement of e-waste management laws, like other pollution-control laws, are as follows: 1) lack of specific legal frameworks; 2) complexity of institutional arrangements; and 3) lack of technological, financial, and human resources.

Lack of specific legal frameworks. As the above-mentioned case studies show, many Asian countries have yet to enact a specific e-waste management law.

Complexity of institutional arrangements. Agencies involved in pollution control are often numerous with overlapping areas of responsibilities as well as gray areas where the institutional responsibilities are unclear or lacking. Lack of coordination is not only an issue between ministries but also between central governments and local authorities, as well as between the administration and the industrial sector.

Lack of technological, financial, and human resources. E-waste management requires considerable investments in planning, staff training, purchasing technology, and building new facilities. The laws need to provide funding or economic/fiscal incentives to manufacturers and local authorities in order to allow them to comply with environmental standards. Or else, many would prefer to pay a little extra money to officials who would turn a blind eye on their polluting practices. How to raise sufficient resources for pollution control remains a big challenge for Asian governments. Thus far, taxation is the most commonly adopted method in resource raising (property taxes, sewerage charges, and vehicle taxes, for instance). However, taxation is never popular. International assistance by means of finance, technology transfer, and capacity building is also essential to help improve the capacity of developing countries in addressing pollution problems.[40]

Law on Environmental Assessment

Environmental assessment considered herein includes EIA and strategic environmental assessment (SEA). EIA and SEA are related methods with the purpose to prevent, mitigate, and compensate adverse environmental impacts that may be caused by a proposed activity. EIA focuses on projects such as construction of a dam, industry, mining, etc., whereas SEA is applied to policies, plans, programs, and macro projects (in some countries) such as a transportation development plan, energy development plan, international airport project in Hong Kong, etc. EIA regulations were established in most Asian countries in the 1980s and 1990s. In terms of legislation, they are varied ranging from none (Myanmar), to very recent and not widely applied legislation (Laos and Cambodia), to moderately to highly applied legislation (Thailand, Malaysia, the Philippines, Indonesia), and to extensively applied EIA regulation within a broader planning framework (Japan, Hong Kong, South Korea, China).[41,42] Accordingly, EIA has been practiced throughout Asia with varying degrees of rigor and effectiveness. However, SEA is a relatively new method and, thus far, few Asian countries have incorporated SEA into their legislation.[42]

In 2006, the World Bank released a report called "Environmental Impact Assessment Regulations and Strategic Environmental Assessment Requirements: Practices and Lessons Learned in East and Southeast Asia." The report divides countries in the region into three categories: 1) Hong Kong SAR, Japan, and Korea, which have the most advanced EIA/SEA legislations and effective implementation; 2) China, the Philippines, Indonesia, and Thailand, which have less strict EIA/SEA legislations and have encountered difficulties in their implementation; and 3) Vietnam, Mongolia, Lao PDR, and Cambodia, which have recently established EIA/SEA legislations and are at an early stage of implementation. Some countries are exampled for discussion herein with the content is mostly drawn from such mentioned report.

China. The EIA system has been in place since 1979, which was the year that China enacted the Environmental Protection Law. This law contained broad elements requiring EIA, particularly for construction projects. Later in 1986, the first legal document on EIA in China was issued by the

Ordinance of Environmental Protection of Construction Projection (1986). A series of regulations on construction projects were issued, including Environmental Protection Procedures for Construction Project (SEPA, 1990), Regulation of Environmental Protection of Construction Projects (State Council No. 253, 1998), and Environmental Management Catalogue for Construction Projects (SEPA, 1999). To broaden environmental assessment, the current EIA law has been modified and extended to cover plans, and has become SEA inclusive. A new law on EIA was approved by the National People's Congress in 2002, and has functioned since September 1, 2003. The new EIA law incorporates the concept of SEA for plans and programs, but not for policies. Subsequently, the EIA law covers two large areas: plans and construction projects. Plans are divided into two categories: 1) plans for land use, regional, watershed, and offshore development; and 2) "specific plans," which include agriculture, industry, livestock breeding, forestry, natural resources, cities, energy, transportation, tourism, etc. However, enforcement in legislation, public participation, and capacity building should be undertaken for applications of policy-and plan-based SEA.[42]

Hong Kong SAR. To solve environmental problems encountered in the country, the Hong Kong government has put forth effort to mitigate, control, and prevent such problems. Environmental assessment is considered an important tool in preventing environmental pollution. It is applied not only to individual projects but also to strategic policies and proposal that facilitate the country moving toward sustainability. The EIA process has been applied to projects since 1986, to plans since 1988, and to strategies and policies since 1992. Hong Kong's EIAO was enacted in 1997 in order to formalize the 15 years experience with EIA, environmental monitoring, and auditing, and came into force in April 1, 1998. The EIA in Hong Kong is considered SEA inclusive. With a successful application of EIA/SEA tools with proven records in legal provision, technical capacity, training and implementation, Hong Kong has become one of the most transparent EIA systems in the world.[42]

Japan. The EIA concept was introduced in Japan in the 1960s and implemented through various administrative guidelines, sector legislation (such as the Public Water Area Reclamation Law), and ordinances and guidelines issued by local authorities. The unified law called "the Environmental Impact Assessment Law" was finally adopted in 1997 and took effect in 1999. The law adopts a listing method by scale to identify projects for which environmental impact statement (EIS) is required.[42] Legal requirements that make the Japanese EIA system more strict and comprehensive than those of many other countries in the region are, for instance, EIA requirements for small-scale projects with potential adverse impacts on the environment and emphasis on public participation. Therefore, public opinion is requested at both the scoping stage and the EIA conduction stage, and a period of 100 days is provided for public hearings and information display before submission of EIA report.

Although Japan has yet to make SEA into law, there is a strong political will at both the national and regional level to integrate SEA in the policy-making process. The Ministry of Environment and other ministries have adopted SEA guidelines such as the Ministry of Environment's preliminary guideline on SEA in the formulation of municipal waste management plans. Local governments are also active in implementing SEA.

Republic of Korea. The development of the EIA system was initiated in 1977 through the Environmental Preservation Act, and put into effect by the legislation of "Regulations on the Preparation of EIA" enacted in February 1981. After the Environmental Administration was upgraded to the ministerial level in 1990, the previous Environmental Preservation Act was divided into a number of separate laws. One of those is the Basic Environmental Policy Act enacted in August 1990. The Environmental Impact Assessment Act was enacted as a separate law on June 11, 1993, and was put into effect on December 12, 1993. To increase efficiency of the system, the EIA Act was further revised in 1997, and became the EIA law of 1997. The current EIA system is considered as SEA exclusive. Nonetheless, an SEA type of system was applied in the late 1990s known as the Prior Environmental Review System (PERS), which is mainly implemented for various plans and programs. The current PERS has been amended as an SEA type in general, but not to cover policy level. Thus, EIA in Korea includes two types, the PERS conducted at a planning stage and EIA carried out at the project-development stage. In this system, a decision on whether to execute a development project will be made at the planning stage, taking into account environmental concerns.[42]

Malaysia. The Environmental Quality Act (EQA) was enacted in 1974 as the major federal environmental statute. It was not until 1987 that EIA procedures were introduced under the EQA as a control preventative mechanism. The EIA is well established in Malaysia under the responsibility of the federal government.[43] The situation is currently changing by delegation of EIA powers to the state level. The states of Sarawak and Sabah have adopted independent impact assessment procedures for natural resource management, and it is possible that other states may follow. However, the EIA at state level has faced some problems due to insufficiency of skilled staffs, low institutional capacity, and an absence of effective monitoring of mitigation measures.[43] Therefore, there is a need to strengthen the state capability on EIA implementation. EIA in Malaysia is considered SEA exclusive. However, there is now evidence of an up-and-coming commitment to SEA in the country. Government objectives in environmental protection and management are moving forward, although the regulatory framework to achieve these objectives is not, as yet, fully developed.[44]

Some major infrastructure projects, such as roads and power facility development, could be subject to SEA procedures in the future. Integration of SEA into policy, plan, and programs is necessary to secure a more environmentally sustainable development in the country.

Indonesia. The development of the EIA system was initiated by Government Regulation No. 29 (1986) in compilation with the provisions of Article 15 of the former Environmental Management Act No. 4/1982. Later, Government Regulation No. 51 (1993) concerning EIA imposed significant revisions to the assessment system. Currently, Regulation No. 27/1999 is a revision of EIA regulation No. 51/1993. The new regulation is expected to be improved and provide a more democratic basis. The EIA system is the responsibility of the Environmental Impact Management Agency. The EIA in Indonesia is project based and SEA exclusive. However, the government has realized the importance of SEA in the decision making process, but its application is not compulsory. The Ministry of Environmental published a book on Strategic Environmental Assessment that provides the fundamentals, procedures, and benefits of applying SEA in the policy, plan, and program process.[42]

Philippines. Originally, the EIS system, which is equivalent to the EIA system, was conceived in Philippines Environmental Policy (P, D, No 1151). The actual establishment of the EIA system began with Presidential Decree (P.D.) No. 1586 in 1978. After issuance of some respective decrees, the EIS was adopted in the document *DAO No. 30 of 2003 Implementing Rules and Regulations (IRR) for the Philippine Environmental Impact Statement (EIS) System,* which was issued in 2003. The EIS system is the responsibility of the DENR. The EIS system is well established in the Philippines, including a legal mandate, administration, procedure, and guidelines. It is regarded as extremely comprehensive and perhaps entails the most stringent requirements in the whole Southeast Asia region.[45] The current EIA system in the Philippines is still project based and SEA exclusive. However, some SEA initiatives have been undertaken. In DAO 30/2003, it was stated that, "The EMB shall study the potential application of EIA to policy-based undertakings as a further step toward integrating and streamlining the EIS system" (Article II, Section 7). The SEA covering policy and plan are being considered to be contained in a new EIA Act.[46]

Singapore. Implementation of EIA in Singapore has been operated through the Environmental Pollution Control Act 2000 and the Land Planning process. Regarding EIA through Pollution Control, an EIA may be required for particular projects specified by the Ministry of Environment and Water Resource that have potential to cause pollution affecting public health; for example, petrochemical works, gasworks, and refuse–incineration plants; foreign investment projects using or storing large quantities of hazardous substances; etc. This is likely a project-based EIA. By EIA through Land Planning, Singapore established a document called the *Concept Plan,* which broadly outlines land-use policies in the country, of which the policies are translated into detailed proposals for local areas called "Development Guide Plans" (DGPs). The basic environmental concerns are considered in the DGPs. This is an SEA-like approach for spatial planning. With the provision of relatively effective laws and efficient centralized planning mechanisms, the lack of an EIA law does not appear to obstruct environmental management endeavors. Not only the comprehensive planning and effective pollution control mechanisms but also, more importantly, a stringent enforcement system makes it possible for Singapore to move toward sustainable development.[42]

Thailand. The EIA system was established in Thailand through the Improvement and Conservation of the National Environmental Act (1975), followed by the Enhancement and Conservation of Environmental Quality Act (1992). Although the EIA system is well established in Thailand with a good number of qualified personnel, the law remains vague on many issues and public hearing is optional for some projects. The recent EIA procedures for projects identified as having potential adverse impacts to natural resources, environment, and health legally require public involvement through a strict and specific process. EIA for such listed projects is named as Environmental Health Impact Assessment.[47,48] Regarding SEA implementation by law, Thailand has not made SEA mandatory yet. However, SEA has been performed for some projects specified by line agencies for certain areas of interest, such as SEA for Economic Zone at Border Territory at Chiang Rai Province by the Office of Natural Resource and Environment Policy and Planning (2005).[49] SEA legislation is now under development. At this moment, it can be stated that EIA in Thailand is project based and SEA exclusive.

Lao PDR. Lao PDR enacted the regulations on EIA back in 2000, and these regulations were revised and upgraded to decree with the Prime Minister Decree on Environmental Impact Assessment No. 112 of 2010. The Ministry of Natural Resource and Environment is responsible for administrating the EIA system, approving EIAs, and for issuing Environmental Compliance Certificates Related legislation on social impacts, including the Decree on the Compensation and Resettlement of Development Projects, 192/PM of 2005, and the Regulations for Implementing Decree 192/PM on Compensation and Resettlement of People Affected by Development Projects.

Since the EIA system of Lao PDR is quite recent, it sets relatively high legal standards and requirements, particularly focusing on social aspects. EIA reports in Lao PDR are called Environmental and Social Impact Assessment. Despite a relatively good legal basis, many barriers to effective implementation persist in Lao PDR, particularly the lack of qualified professionals in EIA. Thus, many capacity building programs for EIA personnel have been funded by international organizations and donor countries most prominently the long-term development cooperation with Sweden (Strengthening Environmental Management Phase I and II) and the assistance from the Government of Finland (Environmental Management Support Programme). At this time, EIA is legally implemented as project based and SEA exclusive; however, through the above-mentioned support, the Ministry of Natural Resources and Environment is revising the Environmental Protection Law and developing a decree and guidelines on SEA. This is combined with comprehensive capacity building and case studies on SEA.[42]

Cambodia. In 1996, the National Assembly of Cambodia enacted the Law on Environmental Protection and Natural Resource Management (EPNRM) as a framework law governing environmental protection and natural resources management. The law requires the Royal Government to prepare national and regional environmental plans. In addition, there are subdecrees concerning a wide range of environmental issues, including EIAs, pollution prevention and control, public participation, and access to information (SIDA). Cambodia has subsequently established an EIA system under the EPNRM law through the EIA Subdecree on Environmental Impact Assessment issued in 1999, which mandates general requirements, procedures, and responsibilities. The subdecree instructs the Ministry of Environment to formulate implementing rules and guidelines.[50] This EIA system covers only projects and is SEA exclusive.

Vietnam. EIA was first mentioned in the Law on Environmental Protection (LEP) 1994 Article 18, which stipulates that organizations and individuals must submit EIA reports to be appraised by the state management agency for environmental protection. The *Government Decree on Providing Guidance for the Implementation of the Law on Environmental Protection* (Government Decree No. 175/CP, 1994) is an important legal document on EIA in Vietnam.

The implementation of the EIA system is the responsibility of the Ministry of Natural Resources and Environment. The current EIA system in Vietnam is basically consistent with international practice.[51] SEA has already been adopted conceptually in the Vietnamese legislative framework, for example, in the LEP, GD 175/CP, and Circular No. 490/TT-BKHCNMT, where "EIA not only must be carried out at project level, but also for master plans for development of regions, sectors, provinces, cities and industrial zones." There are several cases of applying SEA in Vietnam in recent years. As plans are covered by the EIA system, the system is conceptually SEA inclusive. The government is considering accommodating SEA in the new environmental legislation.

Conclusion

Development of environmental law in the region began about four decades ago. However, it has been uneven depending on the economic development of each country. According to the level of development of environmental legislation and effectiveness of the implementation, Asian countries may be divided into three categories: 1) countries with advanced economy, comprehensive environmental legislations, capable and well-coordinated institutional framework, and relatively effective implementation, such as Japan, Korea, Hong Kong, and Singapore; 2) countries with a developing economy, relatively well-established legislations but ineffective implementation due to problems such as institutional complexity and lack of qualified personnel, such as China, Malaysia, Indonesia, the Philippines, and Thailand; 3) countries that have recently emerged from conflict and are rebuilding, and therefore are at an early stage of developing environmental legislations and institutions, such as Laos, Cambodia, and Vietnam.

Environmental legislation in Asia is incorporating more and more elements of the "second generation of environmental legislation," which are public participation, compliance incentives, and market-based mechanism. However, regarding the role of the people, in many cases, the law recognizes rights of local communities in natural resource management but does not provide for procedure or institutional framework that would enable the effective exercise of such rights. In addition, environmental legislations in Asia are struggling to keep pace with the rise of the emerging environmental problems and the level of environmental degradation. The problem of e-waste is one good example. For environmental assessment including EIA and SEA, all the countries presented herein have enacted with Environmental Protection Law that include EIA. However, SEA implemented by law is limited. The EIA systems of most countries by law are project based and SEA exclusive. Only some countries such as China, Hong Kong, and Vietnam have environmental assessment system by law for EIA with SEA inclusion.

To ensure effective implementation, a good environmental legislation must be complemented with a capable and coordinated institutional framework, qualified personnel, and an adequate budgetary allocation. Regional and international cooperation are of utmost importance in the development and implementation of environmental legislations in Asia, particularly in the areas of capacity building, technical cooperation, and funding.

Acknowledgments

This entry could not have been completed without the great help of Miss JinjutaManotham, Second Secretary, Ministry of Foreign Affairs, Thailand, for searching some meaningful environmental legislation-related topics; Mr. Brian J D'Arcy, Environmental Expert, Scotland, U.K.; Mr. Peter Gammelgaard Jensen, Chief Technical Advisor, Grontmij, Denmark, for editing the manuscript; and Miss SriratSuwannakom, my research assistant, for formatting the document. I would like to thank the Research Center for Environmental and Hazardous Substance Management, KhonKaen University, National Excellence Center for Environmental and Hazardous Substance Management for supporting me on the relevant research and academic services. I would also like to thank the *Encyclopedia of Environmental Management,* Taylor & Francis Group, for giving me the opportunity to write this entry.

References

1. Kim, M.; Hodges, H.J. Is the 21st century an "Asian Century"? Raising more reservations than hopes. Pac. Focus **2010**, *25* (2), 161–180.
2. Kastner, S.L. The global implications of China's rise. Int. Stud. Rev. **2008**, *10* (4), 786–794.
3. Browne, A. Why China is the REAL master of the universe, available at http://www.dailymail.co.uk/news/article-559133/Why-China-REAL-master-universe.html (accessed January 2012).
4. UNESCAP. *Asia Pacific Disaster Report 2010*; Protecting Development Gains, Information and Communications Technology and Disaster Risk Reduction Division: Bangkok, Thailand, 2010.
5. Nguyen, L.D. *Environmental Indicators for ASEAN: Developing an Integrated Framework*; UNU-IAS Working Paper No. 109; United Nations University Institute of Advanced Studies: Yokohama, Japan, 2004.
6. UNESCAP. *State of the Environment in Asia and the Pacific 2000, ST/ESCAP/2087*; United Nations: New York, 2000.
7. King, P.; Annandale, D.; Bailey, J. A conceptual framework for integrated economic and environmental planning in Asia—A literature review. J. Environ. Assess. Policy Manag. **2000**, *2* (3), 297–315.
8. Hezri, A.A.; Hasan, M.N. Towards sustainable development? The evolution of environmental policy in Malaysia. Nat. Resour. Forum **2006**, *30* (1), 37–50.
9. Chapter 2.3: National Legislation, Module 2: Policy and Legislation, Mekong River Commision's Watershed Management Resource Kit. Available at http://www2.gtz.de/snrd/wmrk/2PolLeg/203_National_Legislation.pdf
10. Arnold, C.A. Fourth-Generation Environmental Law: Integrationist and Multimodal. Express O. Available at http://works.bepress.com/anthony_arnold/1.
11. Kheng-Lian, K. The first generation of environmental laws in Asia. KohKheng-Lian. In *Towards a "Second Generation" in Environmental Laws in the Asian and Pacific Region: Select Trends*; Lin-Heng, L., Manguait, M., Socorro, Z., Eds.; IUCN: Gland, Switzerland, 2003; 15–25.
12. Arnold, C.A. Fourth-generation environmental law: Integrationist and multimodal, http://works.bepress.com/anthony_arnold/1 (accessed May 2011).
13. ESCAP; ADB; UNEP. *Preview Green Growth, Resources and Resilience Environmental sustainability in Asia and the Pacific*; ST/ESCAP/2582,2010; United Nations: Bangkok, Thailand, 2010; 5.
14. Sustainable Hunting Project. *Regional Action Plan for Moving Toward Sustainable Hunting and Conservation of Migratory Birds in Mediterranean Third Countries*; Bird-life International: Cambridge, U.K., 2007.
15. Secretariat of the Convention on Biological Diversity. The Nagoya Protocol on Access and Benefit-Sharing of Benefits Arising from their Utilization to the Convention on Biological Diversity: text and annex, Convention on Biological Diversity; United Nations, Quebec, Canada, 2011.
16. Meridian Institute. *Reducing Emissions from Deforestation and Forest Degradation (REDD): An Options Assessment Report*; Prepared for the Government of Norway; Angelsen, A., Boucher, D., Brown, S., Merckx, V., Streck C., Zarin, D., Eds.; Meridian Institute: Washington, D.C., 2009.
17. O'Neill, H.; Wandel, A. *Future Policy Award 2010: Celebrating the World's Best Biodiversity Policies*; World Future Council: Hamburg, Germany, 2010.
18. UNCTAD. Latiff, A. and A.H. Zakri (UniversitiKebang-saan Malaysia). Protection of Traditional Knowledge, Innovations and Practices: The Malaysian Experience (paper presented at the UNCTAD Expert Meeting on Systems and National Experiences for Protecting Traditional Knowledge, Innovations and Practices). 30 October –1 November 2000. Available at http://r0.unctad.org/trade_env/docs/malaysia.pdf (accessed May 2012).
19. Nagle, J.C. The effectiveness of biodiversity law. J. Land Use Environ. Law **2009**, *24*, 203.

20. Maxim, S.; Hadad, I.; Sitorus, S. Biodiversity conservation in Indonesia: The case of KEHATI, available at http://www.synergos.org/knowledge/03/asiafinancingkehati.htm (accessed June 2011).

21. International Development Law Organization. Strengthening Environmental Law Compliance in Indonesia, Towards Improved Environmental Stringency and Environmental Performance. Development Law Update 2006,6, available at http://www.idlo.int/publications/30.pdf (accessed June 2011).

22. Tan, A.K. Environmental laws and institutions in Southeast Asia: A review of recent developments. SYBIL **2004**, *VIII*, 177–192.

23. Meniado, A.P.; Garcia, J.L.; Madamba, E.J. The Philippines. In *Biodiversity Planning in Asia. A Review of National Biodiversity Strategies and Action Plans*; Carew-Reid, J., Ed.; IUCN: Gland, Switzerland, 2002; 216–236.

24. Senga, R.G. Establishing protected areas in the Philippines: Emerging trends, challenges and prospects. George Wright Forum **2001**, *18* (1), 56–65.

25. Bugna, S.C. A profile of the protected area system in Singapore. Asian Biodivers. **2002**, *2* (2): 30–33.

26. National Parks Board. *Conserving Our Biodiversity: Singapore's National Biodiversity Strategy and Action Plan*; National Parks Board: Singapore, 2009.

27. Wan, J.; Telesetsky, A. *Creating Opportunities: Saving Hong Kong's Natural Heritage*; Civic Exchange: Hong Kong, China, 2002, 1–53.

28. Sharma, C. Enforcement mechanisms for endangered species protection in Hong Kong: A legal perspective. VJEL **2003–2004**, *5* (1), 1–34.

29. Office of Environmental Policy and Planning. *National Report on the Implementation of Convention on Biological Diversity*; Ministry of Science, Technology and Environment: Bangkok, Thailand, 2002; 1–60.

30. Office of Natural Resources and Environmental Policy and Planning. *Thailand: National Report on the Implementation of the Convention on Biological Diversity*; Ministry of Natural Resources and Environment: Bangkok, Thailand, 2009; 1–76.

31. Country Report on the State of Plant Genetic Resources for Food and Agriculture, Lao PDR, available at http://www.fao.org/docrep/013/i1500e/Lao%20Peoples%20Democratic%20Republic.pdf (accessed August 2011).

32. Clarke, J.E. Protected area management planning. Oryx **2000**, *34* (2), 85–89.

33. FAA 118/119 Analysis, Conservation of Tropical Forests and Biological Diversity in Cambodia, available at http://www.oired.vt.edu/sanremcrsp/documents/team-room/usaid-info/USAID-Cambodia-Forest-and-Biodiversity-Report.pdf (accessed July-August 2011).

34. ICEM. Lessons learned in Cambodia, Lao PDR, Thailand and Vietnam: Review of protected areas and development in the Lower Mekong River Region, Indooroopilly, Queensland, Australia. ICEM **2003**, 1–104.

35. Basel Convention Coordinating Center for Asia and the Pacific. *Report of the Project on the Import/Export Management of E-waste and Used EEE, Asia-Pacific Regional Centre for Hazardous Waste Management Training and Technology Transfer;* Department of Environmental Science and Engineering, Tsinghua University: Beijing, China, 2009.

36. Kirschner, M. RoHS in China, available at http://www.conformity.com/A0725 (accessed July 2011).

37. Wittmaier, M.; Langer, S.; Wolff, S.; Bilitewski, B.; Werner, P.; Stefan, C.; Schingnitz, D.; Wiesmeth, H.; Parthasarathy, R.; Wooldridge, C.; Green, J.; Quynh, D.N.; Viet, L.H.; Ngan, N.V.C.; Hoang, N.X.; Trang, N.T.D.; Minh, P.H.; Touch, V.; Samell, K.; Khouangv-ichit, S.; Daladone, P.; Tia, S.; Songkasiri, W.; Commins, T.: Framework conditions for waste management in Lao PDR, Vietnam, Cambodia and Thailand (Ch. 2). In *INVENT—Innovative Education Modules and Tools for the Environmental Sector, Particularly in Integrated Waste Management, Part I Curricula and Modules*; Handbook/e-book; University of Applied Sciences: Bremen, Germany, 2009; 45–87.

38. Thakur, P. Govt may ban import of e-waste, *Times of India,* September 6, 2010.
39. Hurst, C. *China's Rare Earth Elements Industry: What Can the West Learn?*; Institution of the Analysis of Global Security (IAGS): Potomac, MD, 2010; 1–42.
40. UNEP. *Asia-Pacific Environmental Outlook* 2; United Nations Environment Programme, Regional Resource Centre for Asia and the Pacific: Pathumthani, Thailand, 2001.
41. Li, C.J. *Environmental Impact Assessments in Developing Countries: An Opportunity for Greater Environmental Security?* Working Paper No. 4; USAID and Foundation for Environmental Security and Sustainability (FESS): Falls Church, VA, 2008.
42. World Bank. *Environmental Impact Assessment Regulations and Strategic Environmental Assessment Requirements Practices and Lessons Learned in East and Southeast Asia, Safeguard Dissemination Note No.* 2; Environment and Social Development East Asia and Pacific Region: Washington, D.C., 2006.
43. Briffett, C.; Obbard, J.P.; Mackee, J. Towards SEA for the developing nations of Asia. Environ. Impact Assess. Rev. **2003**, *23* (2), 171–196.
44. ADB. Country assistance plan (2000–2002) Malaysia. Asian Development Bank, 1999. Cited by Briffett, C.; Ob-bard, J.P.; Mackee, J. Towards SEA for the developing nations of Asia. Environ. Impact Assess. Rev. **2003**, *23* (2), 171–196.
45. Tan, A.K.J. APCEL Report: Environmental Law (ASEAN-10), Faculty of Law, National University of Singapore, 2000. Cited by World Bank. *Environmental Impact Assessment Regulations and Strategic Environmental Assessment Requirements Practices and Lessons Learned in East and Southeast Asia, Safeguard Dissemination Note No. 2*; Environment and Social Development East Asia and Pacific Region: Washington, D.C., 2006.
46. Villaluz, M.G. Advancing the EIA system in the Philippines, UNEP EIA Training Source Manual, 2003. Cited by World Bank. *Environmental Impact Assessment Regulations and Strategic Environmental Assessment Requirements Practices and Lessons Learned in East and Southeast Asia, Safeguard Dissemination Note No. 2*; Environment and Social Development East Asia and Pacific Region: Washington, D.C., 2006.
47. Section 4. Constitution of the Kingdom of Thailand, B.E. 2550 (2007), Government Gazette, Vol. 124, Part 27a, dated 24th August B.E. 2550 (2007).
48. The Enhancement and Conservation of National Environmental Quality Act B.E. 2535 (NEQA 1992). Government Gazette, Vol. 109, Part 37, dated 4th April B.E. 2550 (2007); Government Gazette, Vol. 119, Part 102, dated 8th October B.E. 2550 (2007).
49. ONEP. *SEA in the Border Area in Chiang Rai Province, Final Report*; Office of Natural Resources and Environmental Policy and Planning: Bangkok, Thailand, 2005.
50. Government of Cambodia. *Cambodia's Report to WSSD—National Assessment of Implementation of Agenda 21—Progress, Challenges and Directions*; Government of Cambodia: Phnom Penh, Cambodia, 2002.
51. Obbard, J.P.; Lai, Y.C.; Briffett C. Environmental assessment in Vietnam: Theory and practice. J. Environ. Assess. Policy Manag. **2002**, *4* (3), 267–295.

Bibliography

1. Aoki-Suzuki, C. *Trade of Second Hand Electrical and Electronic Equipment (SH-EEE) in Asia: Focusing on Actors in Reuse Markets and the Need for Deepened Actor Analysis and Integrated Sustainability Assessment,* ISIE Asia-Pacific Meeting and ISIE MFA-Con Account Meeting, Tokyo, Nov 7–9, 2010; Institute for Global Environmental Strategies: Tokyo, Japan, 2010.
2. Asian Development Bank (ADB). *Emerging Asia: Changes and Challenges;* ADB: Manila, Philippines, 1997.
3. Brandon, C.; Ramesh, R. *Toward an Environmental Strategy for Asia*, World Bank Discussion Paper No. 224; World Bank: Washington, D.C., 1993.

4. Bruch, C.; Mrema, E. *UNEP Guidelines and Manual on Compliance with and Enforcement of Multilateral Environmental Agreements*, Proceedings of the Seventh International Conference on Environmental Compliance and Enforcement, Marrakech, Morocco, Apr 9–15, 2005; Gerardu, J., Jones, D., Markowitz, K., Zaelke, D., Eds.; International Network for Environmental Compliance and Enforcement: Washington, D.C., 2005.

5. Chongrak, P.; Indra, G. Asian environmental status: Emerging issues and future scenarios. In *Environmental Management Tools. A Training Manual.*Routray, J.K., Mohanty, A., Eds.; UNEP: Asian Institute of Technology, Pathumthani, Thailand, 2006.

6. Clive, B. Environmental Impact Assessment in Southeast Asia: Fact and Friction. Geo J. **1999**, *49* (3), 333–338.

7. Daniel, E.; Marie, P. *Globalization and The Environment in Asia, United States–Asia Environment Partnership Framing Paper;* USAID Development Experience Clearinghouse: Washington, D.C., 1999; 1–53.

8. Department for Economic and Social Information and Policy Analysis. *Glossary of Environment Statistics,* Studies in Methods, Series F, No. 67; United Nations: New York, 1997; 1–83.

9. Desai, B.H. Multilateral environmental agreements: Legal status of the secretariats. J. Environ. Law **2010**, *23* (1), 155–157.

10. Evans, P.J. Industry and the environment in Asia. TDRI Q. Rev. **1998**, *13* (3), 9–27.

11. Available at http://ewasteguide.info/stepping-efforts-con (accessed July 2011).

12. Available at http://lawprofessors.typepad.com/environmental_law/2007/01/what_is_environ.html (accessed July 2011).

13. Available at http://timesofindia.indiatimes.com/india/Govt-may-ban-import-of-e-waste/article-show/6501864.cms (accessed August 2011).

14. Available at http://www.cecphils.org/node/55 (accessed August 2011).

15. Available at http://www.chinaenvironmentallaw.com/wp-content/uploads/2009/03/regulations-on-waste-electric-and-electronic-products-chn-eng.pdf (accessed September 2011).

16. Available at http://www.env.go.jp/en/laws/recycle/10.pdf (accessed September 2011).

17. Available at http://www.ide.go.jp/English/Publish/Down-load/Spot/pdf/30/007.pdf (accessed August 2011).

18. Available at http://www.indiaenvironmentportal.org.in/files/DraftE-waste-Rules303.10.pdf (accessed September 2011).

19. Available at http://www2.kankyo.metro.tokyo.jp/anmc21_WM/legislation.htm (accessed September 2011).

20. Iqbal, M.T. Environmental law and multilateral environmental agreements (MEAs). In *Environmental Management Tools, A Training Manual*; Routray, J.K., Mohanty, A., Eds.; School of Environment, Resources and Development, Asian Institute of Technology: Pathumthani, Thailand, 2006; 75–82.

21. Jones, S. Highlights of waste control laws and regulations in China. In *A China Environmental Health Project Fact Sheet*; China Environment Forum's Partnership with Western Kentucky University on the USAID-Supported China Environmental Health Project: Bowling Green, KY, 2007.

22. Kaniaru, D.; Kurukulasuriya, L. UNEP's role in capacity building in environmental law. In *International Capacity Building*, Fourth International Conference on Environmental Compliance and Enforcement. Chiang Mai, Thailand, Apr 22–26, 1996; International Network for Environmental Compliance and Enforcement: Washington, D.C., 1996.

23. Lee, S.; Na, S. E-Waste recycling systems and sound circulative economies in East Asia: A comparative analysis of systems in Japan, South Korea, China and Taiwan. Sustainability **2010**, *2*, 1632–1644.

24. Memon, A. Devolution of environmental regulation: EIA in Malaysia. In *Case Studies for Developing Country*; UNEP EIA Training Resource Manual 2002. Sadler, B., Fuller, K., Ridgway, B., McCabe, M., Baily, J., Saunders, R., Eds.; UNEP: Geneva, Switzerland, 2002; 45–61.

25. Millennium Ecosystem Assessment. *Ecosystems and Human Well-Being: Synthesis*; Island Press: Washington, D.C., 2005.

26. Ministry of the Environment. *Environmental Impact Assessment in Japan*; Environmental Policy Bureau, Ministry of the Environment, Government of Japan: Tokyo, Japan. Available at http://www.env.go.jp/en/policy/assess/pamph.pdf

27. O'Connor, D. *Grow Now/Clean Later, or Pursuit of Sustainable Development?*; Research Programme on Economic Opening, Technology Diffusion, Skills and Earnings, Working Paper No. 111; OECD Development Centre: Paris, France, 1996.

28. Pescott, M.J.; Durst, P.B. Leslie, R.N. Forest law enforcement and governance: Progress in Asia and the Pacific; RAP PUBLICATION 2010/05; FAO: Bangkok, Thailand, 2010; 1–205.

29. Peter, H.A. Asian cultural influences on environmental legal norms: RodaMushkat, International Environmental Law and Asian Values, Toronto, UBC Press. Rev. Québécoise Droit **2004**, *17* (1), 283–286.

30. Pulhin, J.M. Trends in forest policy in the Philippines policy. In *Trend Report 2002*; Inoue, M., Ed.; The Institute for Global Environmental Strategies (IGES); Forest Conservation Project; Soubun Printing Co. Ltd: Tokyo, Japan, 2003; 29–41.

31. Richardson, B.J.; Wood, S., Eds. *Environmental Law for Sustainability*; Hart Publishing: Oxford, U.K., 2006; 1–18.

32. Schluep, M.; Hagelueken, C.; Kuehr, R.; Magalini, F.; Maurer, C.; Meskers, C.; Mueller, E.; Wang, F. *Recycling—From E-Waste to Resources*; Final Report, Sustainable Innovation and Technology Transfer Industrial Sector Studies; UNEP and UNU: Paris, France, 2009.

33. Suzuki, K. Sustainable and environmentally sound land use in rural areas with special attention to land degradation, APFED Third Substantive Meeting (APFED3), Guilin, People's Republic of China, Jan 23, 2003; Asia-Pacific Forum for Environment and Development: APFED3/EM/03/Doc.4.

34. UNEP. *Global Environment Outlook 2000*; Earthscan Publications Ltd.: London, 1999.

35. UNEP. *Global Environment Outlook-3*; Earthscan Publications Ltd.: London, 2002.

36. World Bank. *World Development Indicators 2001*; Development Data Cebter, The World Bank: Washington, D.C., USA, 2001.

37. Yang, H.; Innes, R. Economic incentives and residential waste management in Taiwan: An empirical investigation, environmental and resource economics. Eur. Assoc. Environ. Resour. Econ. **2007**, *37* (3), 489–519.

ELE: Focuses on the Use of Legislation or Policy to Address Environmental Problems

29

Environmental Policy

Sanford V. Berg

Introduction

When economic activity leads to pollution and over-use of common property resources, government intervention can improve social welfare. Pollution involves a market failure in which damages caused by a producer or consumer are imposed on third parties. These damages can involve personal health, the physical deterioration of buildings, and foregone options for the future. Of course, if transaction costs are low, those that are causing pollution damage can be taken to court if the liability rules are clear. Destruction of common property resources such as losing unique ecological habitats, endangering particular species, or destroying valued scenic vistas is another form of market failure affecting the environment. Because there may not be clear property rights to such elements of the environment, these common property resources can be overutilized. Given the lack of well-defined property rights, government enacts environmental laws to address these market failures. However, identifying and quantifying the damage caused by pollution sources or inflicted upon sensitive ecosystems can be difficult, making any determination of the benefits and costs a contentious exercise. Consequently, choosing policies that define the extent of the environmental protection can be both contentious and problematic.

The next section describes the multidisciplinary inputs that are incorporated into environmental policy analysis, selection, and implementation. Other topics addressed here include policy impacts, the burden of proof, economic evaluation, and the strengths and limitations of policy options.

Multidisciplinary Approach to Developing Environmental Policy

Environmental economics is the study of how economic and environmental issues interact. Issues addressed by environmental economists include but are not limited to evaluating ways to reduce pollution, analyzing the trade-offs between using renewable and nonrenewable resources, or estimating monetary values for ecosystems or habitat. While no single field of study contains all the insights needed to develop and implement sound environmental policies, the focus here will be on economics because it provides a system for incorporating many perspectives and it is the framework by which environmental policy is designed and evaluated. Depending on the burden of proof, the resulting policies might be excessively stringent (costly relative to their benefits) or inadequate for the protection and preservation of environmental features that affect human health and welfare and have intrinsic value.

We know from materials balance that human activity does not create matter but only changes its form, concentration, and location, thus there is a need for physical sciences such as chemistry, physics, and biology to help inform environmental policy. While all societies affect natural systems, the scale of potential impacts has grown with economic development. There is evidence that as incomes rise, citizens are willing to devote relatively more resources to controlling environmental impacts. Moreover, many citizens would like to see much more attention given to reducing current damages and limiting the risks for future harm, hence an understanding of societal and political dynamics is also important for informing environmental policy.

The development and implementation of sound environmental policy draws upon information and procedures from many fields of study. Here, economics is utilized as the framework for integrating the concepts, measurements, and values required for the steps:

1. Determine appropriate regulatory objectives (through citizen participation in political processes and community consensus-building)
2. Balance those objectives to determine regulatory priorities
3. Identify and legislate oversight responsibilities for environmental agencies
4. Develop (a) mechanisms for monitoring environmental impacts (such as ambient air and water quality) and (b) methodologies for integrating new scientific understandings of environmental impacts into the policy prioritization process
5. Define the appropriate targets for different types of pollutants and the protection of biodiversity
6. Determine (and then apply) the appropriate policies for meeting objectives
7. Analyze environmental indictors on a regular basis, checking for noncompliance
8. Evaluate the impacts, recognizing potential biases in the measures and the ways impacts are valued
9. Establish an effective process for monitoring and reviewing the framework, including the penalties and sanctions applied when there is noncompliance

These steps require input from a number of disciplines that shape the way we see things. Although technical training allows analysts to delve deeply into subjects in a consistent manner, awareness of other disciplines' perspectives can be important for constructive environmental policy-making, including:

- Engineers look to technology for solutions to environmental problems. They are able to incorporate new (often expensive) control technologies into energy extraction, production, consumption (energy efficiency), and pollutant disposal and storage (as with nuclear waste).
- Meteorologists and hydrologists analyze pollution transport in air and water systems. They have a deep understanding of the impacts of discharges under different conditions. In conjunction with demographers and epidemiologists, they can estimate the doses received by different population groups.
- Medical scientists and toxicologists analyze the dose response relationships for citizen health, conducting exposure and risk assessments.

- Ecologists study the impacts of pollutants on the local and global environment, assess the value of ecosystem services, and track invasive species and biodiversity. Climate scientists help assess the causes and consequences of changes in local and global temperatures and other weather patterns.
- Materials scientists look at damages caused by air and water pollution. The associated impacts include cleaning and painting buildings, treatment costs, and shorter life spans for affected equipment.
- Political scientists focus on issues of power, legitimacy, social cohesion, and the roles of different stakeholder groups in influencing environmental policies. Consensus is critical because ultimately, in a democratic system, there needs to be widespread agreement on the desired outcomes if the system is to avoid instability.
- Economists emphasize the importance of efficiency in resource allocation. They apply benefit-cost analysis and tend to depend on price signals to provide incentives for the adoption of appropriate control technologies and conservation measures.
- Planners deal with land-use and zoning issues, given population growth projections. Planners integrate legal constraints with historical experience, bringing topological, aesthetic, and geographical elements to the analysis.
- Archeologists and anthropologists provide insights on the impacts of dams, mines, and their related economic activities on unique historical sites, local populations, and indigenous groups. Such impacts create difficult valuation issues.[1,2]
- Lawyers spotlight the institutions of policy implementation. For example, rules and regulations attempt to pay significant attention to procedural fairness. Due process contributes to the legitimacy of outcomes. If different parties perceive that there is no transparency and no opportunity for participation, environmental policy will be perceived as unreasonable and the laws will either be changed or they will be disobeyed in a variety of ways.
- Environmentalists advocate sustainability and environmental equity. The by-products of energy production affect public health and have environmental outcomes. Those impacts have economic value, but often that value is nonmonetary or difficult to quantify. For example, generation and transmission siting decisions incorporate impacts on biodiversity and sustainability.
- Ethicists help society understand personal values and notions of stewardship. Humans have a clear responsibility to leave future generations with a legacy of sound institutions and a clean environment, though the best means to this end are often not obvious.

Thus, physical, biological, and social scientists attempt to uncover patterns and identify lessons to help us improve policy. Given the complexity of environmental issues, most environmental problems are managed, not solved.

Impacts

Energy production and consumption impact people and the environment in a number of ways. For example, activities can damage ecosystems in the extraction phase (oil drilling or coal mining) or involve cross-media emissions in the consumption phase that can lead to further ecosystem damage. Emissions can be from a single point or a mobile source. In addition, they can be continuous or intermittent (with exposure and impacts depending on wind and other weather conditions or the presence of other chemicals). The transport mechanism can be complicated and involve multiple jurisdictions (as with SO_2 and NOx—emissions lead to "acid rain" or ozone problems in downwind areas).

Air

Issues range from local concentrations of particulate matter in the atmosphere to concerns over anthropogenic climate change. Consequences for health, ecosystems, agriculture, coastal settlements, species survival, and other impacts make atmospheric change a serious policy issue. For example, long-range transport means pollutants cross national boundaries and require coordination. Other pollutants—such as greenhouse gas emissions of CO_2—require coordination not due to transport, but because the effects are global in nature regardless of where emissions occur.

Water

Effects of contaminants vary in surface waters and groundwater. The United States has primary standards to protect public health (with maximum contamination levels [MCLs] for toxic wastes). Secondary standards and associated MCLs are meant to protect public welfare (for example, ensuring that the taste, odor, and appearance of groundwater do not result in persons discontinuing water use). Other environmental issues include species loss and dealing with nonindigenous, invasive species.

Land Use

Siting is an issue for electricity generators, transmission lines, and distribution systems (other aspects of land use include urban sprawl and availability of land for agriculture. The focus here is on the environmental impact of energy systems. For example, social investments in mass transit affect emissions from mobile sources (autos). However, environmental policy addresses many other issues, such as the use of pesticides and fertilizers by agriculture or deforestation). The problem of not in my back yard (NIMBY) is universal: we like the convenience of electricity but do not want its production or transport to affect our own property. Surface coalmines are an eyesore, but restoration can be costly. Hydroelectric dams can affect fisheries, flood unique canyons (causing a loss of scenic vistas), damage ecosystems (as in the Amazon), or displace human populations (as with China's Three Gorges Project). Solar collection stations and wind generators require space and have impacts on aesthetics. For some, viewing large windmills along the crest of a lovely mountain range is an eyesore. For others, the same scene is a symbol of hope.

Environmental policy-makers must be aware of the relationship between changes in impacts in one medium and changes in impacts in other media. For example, reducing airborne emissions of mercury will also lead to reduced mercury concentrations in rivers and lakes. However, it may also be the case that reducing ozone precursors from auto emissions by using methyl tert-butyl ether (MTBE) leads to increasing harm to bodies of water as the MTBE precipitates out in rain. Finally, there may be policy and impact trade-offs that must be evaluated with reducing CO_2 emissions through a greater use of nuclear energy. The policy reduces greenhouse gases but raises issues and associated risks of waste storage and protection. In all cases, the links between different environmental media and different environmental policy must be understood for society to properly evaluate the trade-offs.

Burden of Proof

Because environmental issues tend to be complex, delays in responding to citizen concerns and new scientific information can lead to negative impacts or a local crisis. What is more problematic: erring on the side of environmental protection or erring on the side of development? When science is unclear or when studies yield conflicting outcomes, the issue of burden of proof arises. Two types of errors are possible. In a Type I error, a hypothesis is rejected when it is in fact true (e.g., deciding that a pollutant causes no health damages when in fact it does). Rejecting the hypothesis of a health link would lead to more emissions (and citizen exposure) than otherwise would be the case.

A Type II error occurs when the decision maker fails to reject a hypothesis that is in fact false (e.g., not rejecting the hypothesis that low doses of a pollutant have no damaging side effects for certain types of citizens, such as asthmatics, who are viewed as potentially sensitive to a particular pollutant). If in fact at low doses the pollutant does not have negative health impacts, environmental regulators might have imposed standards that induced costly compliance strategies that were based on the Type II error.

Dose-response models that do not reject linear functions when the actual relationships are non-linear would fall into this category.

Both types of errors have costs. However, the political implications may depend on the type of error, leading decision-makers to prefer making errors that are difficult to detect. Thus, it can be argued that environmental regulators will tend to avoid making Type I errors. When evidence accumulates and shows conclusively that a pollutant has health impacts, those responsible for environmental policy do not want to be blamed for acting too slowly. Furthermore, citizens might prefer excessive caution (labeled a "precautionary bias"). On the other hand, Type II errors can result in regulators imposing high abatement costs onto polluters (and those purchasing associated products) in a manner that is not cost effective.

A related issue is whether or not the environmental impact is irreversible. If it is not reversible, a case can be made that the burden of proof should be assigned to those who assert that relatively higher levels of pollution are not problematic. On the other hand, if abatement costs are systematically under-estimated and the benefits of pollution reduction are overestimated, it is possible to devote excessive resources to limiting environmental impacts.

Economic Framework

Economists are aware that it is difficult to place monetary values on many impacts of pollution but argue that environmental amenities must be balanced against other valued goods and services.[3] Some view economists as overemphasizing the efficacy of market incentives to the exclusion of other instruments. However, because economics offers a consistent framework for integrating insights from other fields, it will be described here.

Cost-Benefit Analysis (CBA)

The most fundamental economic analysis looks at how pollution impacts (reflected in "external costs") cause excessive consumption of polluting goods in the absence of government intervention. These external costs are the negative spillover effects of production or consumption for which no compensation is paid (e.g., a polluted stream that damages the health of those living along the stream). Producers consider the environment to be a free input; hence they only minimize private costs. If these external costs are added to the private costs (reflected in the supply curve), this is the total social cost.

Figure 1 shows how a competitive product market yields an equilibrium price ($4) and quantity (80 units per week). However, in the absence of public intervention, the price only reflects the private costs of production, not damages imposed on others (amounting to $2 when the 80th unit is produced, but this is assumed to be less if only 65 units of the good are produced. The external costs are higher at

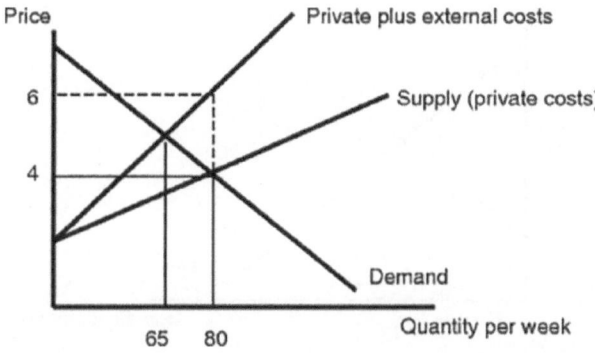

FIGURE 1   Private costs and external costs.

higher levels of output presumably because damages rise dramatically when there are very high concentrations of the pollutant in the atmosphere). Determining the extent of those damages requires some valuation metric.

For now, let us assume that the analysts "got it right" in estimating both benefits and costs. This is a strong assumption because environmental services are notoriously hard to price. This problem can limit the ultimate effectiveness of CBA because the abatement costs tend to be short-term and quantifiable, but the benefits (avoided damages) are often long-term and difficult to quantify. For now, consider the impacts of environmental regulation within the CBA framework. Regulation requires pollution abatement activity, raising production costs but reducing the pollution and associated damages (as shown in Figure 2).

The imposition of environmental regulation raises production costs (shifting the supply curve up) and reduces equilibrium consumption of the polluting good (from 80 to 75 per week) because the price has risen (from $4.00 to $4.40). In addition, external costs are reduced (so the sum of private and external costs is now $5 when 75 units of the good are produced). Emissions are reduced (though this particular figure only indicates the reduction in damages, not the precise reduction in emissions).

The next question is how much pollution abatement makes economic sense, since control costs rise rapidly as emissions are cut back towards zero. Continuing with our illustrative example, Default three depicts the total benefits of abatement and the total cost of abatement. The latter depends on the abatement technology and input prices and the interdependencies among production processes (for retrofitting control technologies). It is relatively easy to compute abatement costs from engineering cost studies, although predicting future control costs is not easy because innovations will create new control technologies. The benefits from abatement (or the reduction of pollution damages—the cost of pollution) depend on the size of the affected population, incomes (indicating an ability to pay for environmental amenities), and citizen preferences (reflecting a willingness to pay). The benefits can be very difficult to estimate. Consider, for example, the health benefits of reduced particulates in the atmosphere, habitat values, and citizen valuations of maintaining a habitat for a particular species. Physical benefits can be found from dose-response studies. Various survey methods and market proxies for computing willingness to pay to avoid experiencing the impacts of pollution have methodological problems. However, if the dollar metric is to be used for determining the benefits of environmental improvements, techniques can at least establish rough benchmarks (as shown in Figure 3) (Some argue strongly against the use of CBA.[4]).

The total benefits and costs to the marginal benefits and costs can be related because, for economists, the issue is not zero emissions versus unlimited emissions. In the former situation, if 80 units of the good results in 80 tn of emissions, then zero emissions reduced would be characterized as having no abatement costs (but also, no benefits from abatement). When the total benefits equal the total cost of abatement (at 65 tn of emissions reduced per week in Figure 3), the last reductions in emissions were very

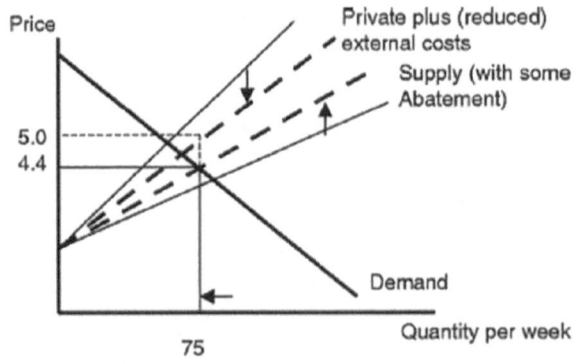

FIGURE 2   Reducing external costs.

costly and the additional benefits were fairly small. Zero abatement activity would also be inefficient in this example because the marginal damages are very high and marginal abatement costs are quite low. Economics tries to determine the optimal amount of emissions. In the hypothetical example, the "optimal" quantity of reduced emissions is about 25, where the marginal abatement cost is just equal to the marginal benefits of $2. These are depicted in Figure 4.

This outcome means that there are still 55 tn of emissions per week. If the estimated benefits and costs of pollution abatement are correct in this illustration, economic efficiency would be violated if additional resources were devoted to abatement activity. For example, if 50 tn of emissions were reduced (so only 30 tn of the pollutant are released), the marginal benefit would be about $1, but the marginal cost would be greater than $5. From the standpoint of economic efficiency, those resources could be used to create greater value in other activities.

Of course, the difficulty of obtaining a common dollar metric for all the impacts of different pollutants means that benefit-cost analysis must incorporate a range of values. The range could be incorporated in Figure 4 as a band around the marginal benefit curve indicating one standard deviation from the calculated values. A conservative approach would recognize that the marginal benefit function could be above that depicted in Figure 4, which would lead to optimal emission reduction of more than 25 tn per week (improving the ambient air quality).

Further complicating the analysis are production and exposure interdependencies. For example, the marginal cost of abatement associated with one type of emission may depend on the level of treatment (or abatement) for another contaminant. A joint optimization problem results, with the basic principles

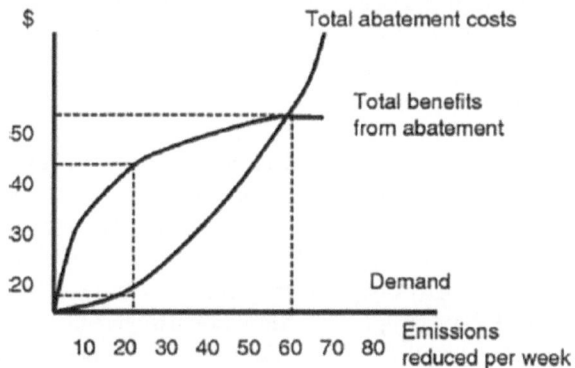

FIGURE 3 Total benefits and total cost of abatement.

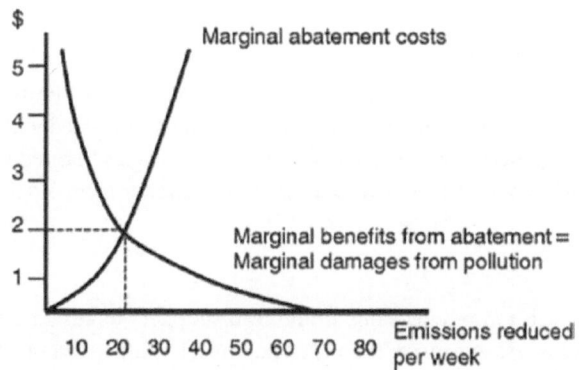

FIGURE 4 Marginal benefits and marginal cost of abatement.

unchanged. Many investments in abatement equipment have this characteristic: once one set of contaminants is being reduced from a discharge flow and the cost of dealing with additional contaminants can be relatively low. For example, in the case of water discharges, if iron or manganese is removed via the precipitation method, total dissolved solids (TDS) is reduced and there may be an improvement in water clarity.

Interdependencies can also arise on the benefit side when the dose-response relationship for a particular contaminant is influenced by the presence of other contaminants. Again, in the case of secondary groundwater standards, perceptions of odor and color will be affected by whether or not they occur in combination. Such considerations must be factored into the analysis when comparing the benefits and costs of different treatment options.

Cost-Effective Analysis

Instead of trying to estimate the dollar benefits of saving a human life (or reducing the incidence of asthmatic attacks), one can compare the number of lives saved per dollar spent in abatement activity across programs. Thus, cost-effective analysis involves finding the least-cost method of achieving a given economic or social objective such as saving lives or retaining unique ecological settings. No dollar value (or explicit measure of avoided damages) is placed on that objective.[5] One advantage of this approach is that the focus is on minimizing the cost of meeting the (politically determined) target. It promotes consistency across a range of programs that might be designed to address a particular problem, whether that involves health impacts or a loss of habitat. Cost-effective analysis facilitates comparisons across programs, leading to reallocations of resources devoted to meeting such targets as new information is gathered over time.

Policy Instruments

Political systems have passed legislation and created agencies to apply laws to improve environmental performance. For example, in the United States, the Water Pollution Control Act of 1956 and the Clean Air Act of 1963 and subsequent amendments to both pieces of legislation have focused on achieving ambient standards. The U.S. Environmental Protection Agency is responsible for implementing these laws, and in other nations agencies have also been established to reduce emissions and improve environmental outcomes. A number of policy options can lead to emission reductions.[6,7] These instruments have different economic efficiency implications. In addition, some of these approaches are difficult to implement (due to being information-intensive), some are not cost effective (in that other approaches achieve the same outcome at lower cost), and the distributional implications can differ across these approaches (tax burdens differ or some groups obtain valuable assets).

Tax on the Polluting Good

An excise tax could be imposed on the good, cutting back consumption to 65 units per week (Figure 1). Of course, the problem is not with the product but with the emissions associated with its production. Thus, this option does not provide incentives for developing new technologies that reduce abatement costs—it represents a static approach to the problem because it does not promote technological innovation.

Tax on Emissions

A penalty or charge for each ton of emissions would lead suppliers to cut back on emissions—to the extent that the abatement is less expensive than the tax. Thus, in Figure 4, a tax of $2/tn would lead to the optimal reduction of pollutants. In addition, it provides incentives for innovation in the control technology industry. Firms will seek ways to reduce abatement costs, thus reducing their pollution taxes. This strategy is likely to be opposed by polluters who will be passing the taxes on to customers (where the ultimate incidence depends on supply and demand elasticities in the product market).

Tradable Emissions Permits

The same result (and incentive) is obtained if "allowances" of 25 tn are allocated to polluting firms, limiting emissions (the situation is not completely identical—a tax has certain costs to firms but yields uncertain overall abatement because regulators will not have precise estimates of abatement costs; the allowances have certainty in terms of overall abatement but uncertain cost. Of course, with monitoring, the tax can be varied over time to achieve a desired ambient condition). This approach provides an incentive for those with low abatement costs to reduce emissions and sell their permits (allowances) to others whose abatement costs would be very high. This places entrants at a disadvantage because incumbent firms are "given" these valuable allowances. The SO_2 regime in the United States has this feature. Of course, the initial allocations raise political issues (because permits represent wealth). In establishing a tradable permit regime, an environmental agency must determine the allowed level of emissions (here, 25 tn) and whether additional constraints might be applied to local areas with particular circumstances. In addition, the energy sector regulator has to make decisions regarding the treatment of cost savings from the regime. For example, savings might be passed on to consumers or retained by firms. The latter situation provides an incentive for utilities to participate in the emissions trade markets. A sharing plan can also be utilized so customers benefit as well.

Tighten Liability Rules

An alternative approach would utilize a court-based system, where fees would be assessed against those responsible for damaging the health of others, for reducing the economic value of assets, or for reducing the amenity values of ecosystems. Of course, this approach requires a well-specified set of property rights and clear causal links between specific emitters and affected parties. The transaction costs of such a system (resources devoted to negotiations and legal activity) could be prohibitive for many types of pollutants.

Emission Reduction Mandates (Quantity-Based Command-and-Control)

Although equal percentage cutbacks sound "fair," this strategy is not cost-effective because abatement costs can differ widely across pollution sources. If there are scale economies to emission reductions, it would be most efficient to have a few firms reduce emissions. The least-cost way to achieve a given overall reduction in emissions will involve differential cutbacks from different firms.

Mandate a Specific Control Technology (Technology-Based Command-and-Control)

This "command and control" strategy is not cost-effective because production conditions and retrofitting production processes differ across firms (based on the age of the plant and other factors). However, this policy option has been utilized in a number of situations as a "technology-forcing" strategy.

Other Policy Issues

The above instruments have been utilized in different circumstances. Additional issues include intrinsic benefits, income distribution, sustainability, and renewable resources.

Intrinsic or Nonuse Benefits

Some people take a more expansive view of environmental amenities as they attempt to separate economic values from inherent values. However, this might be partly accounted for in terms of the perceived benefits to future generations. Intrinsic benefits from environmental programs include option values, existence values, and bequest values.[8] The first value represents a form of insurance so future access to a potential resource is not eliminated due to current consumption. The rationale behind option values is closely related to the "margin for error" argument noted earlier. Existence values reflect a willingness to pay for the knowledge that the amount of contaminant in the environment does not exceed particular levels or that a particular species (or level of biodiversity) is retained. The resource or ecological system is available for others. The bequest values can be interpreted as the willingness to pay for preserving a resource (or a geographic site) for future generations.

Redistributive Effects

It is important to note that citizens being harmed by emissions are not necessarily the same as those who are consuming the polluting good (such as electricity). Even if a particular program has positive net benefits, some parties are likely to be losers. They are seldom compensated and left better off, raising concerns about the distributional consequences of alternative policies. Furthermore, those harmed may have lower incomes (and thus, a lower willingness to pay to avoid damages due to the lower ability to pay). This point underscores the role of fairness as a factor that might outweigh efficiency considerations in some circumstances. Some agencies have been forbidden to use CBA on the grounds that the numbers are too speculative and that social concerns should be given priority. Intergenerational concerns can be interpreted as reflecting redistributive considerations.

Sustainable Development

Some of the issues associated with energy involve the use of nonrenewable resources (irreversibility). Some citizens argue that sustainability requires development that can be supported by the environment into the future. These people wish to ensure that resources are not depleted or permanently damaged. However, since sustainability depends on technology and innovations change resource constraints, defining the term with precision is quite difficult.

Renewable Energy Resources

Generating electricity without fossil fuels (e.g., hydro, wind, solar, biomass) is sometimes referred to as using green options. Green options are often limited in the amount (and reliability) of energy produced in a given time period. Utility applications for renewable resources include bulk electricity generation, on-site electricity generation, distributed electricity generation, and non-grid-connected generation. A number of regulatory commissions have required utilities to meet renewable portfolio standards. Such strategies reduce dependence on a particular energy source (to reduce the region's vulnerability to supply disruptions or rapid price run-ups). In addition, such requirements imply that managers are not making the most efficient investments in long-lived assets. Also, note that demand reduction through energy-efficient technologies is a substitute for energy, whatever the source.

Conclusions

The three main trends in environmental regulation in recent years have been shifting from command-and-control regulation towards a greater use of economic instruments (such as emissions trading), seeking more complete information on the monetary value of environmental costs and benefits, and a tendency for addressing environmental objectives in international meetings, as with the Kyoto Protocol.[9]

The interactions between economic and environmental regulation raise important policy issues. If energy sector regulation and environmental regulation remain separate, some means of harmonization may be necessary to promote improved performance. Collaboration would involve clarifying the economic duties of the environmental regulator and the environmental duties of the economic regulator. To avoid regulatory competition, agencies sometimes establish task forces or other mechanisms for identifying and resolving issues that might arise between jurisdictional boundaries (across states or between state and federal authorities). Such cooperation can serve to clarify the division of responsibilities and identify regulatory instruments that will most effectively meet economic and social objectives.

In summary, policy-makers respond to domestic political pressures by devising institutions and instruments to address pollution and environmental sustainability.[10] Although no single field of study contains all the tools necessary for sound policy formulation, economics does provide a comprehensive framework for evaluating the strengths and limitations of alternative policy options. Because of the pressures brought to bear by powerful stakeholders, adopted policies and mechanisms are not necessarily cost minimizing. The resulting inefficiencies may partly be due to considerations of fairness, which places constraints on whether, when, how, and where environmental impacts are addressed. As emphasized in this survey, citizens want to be good stewards of the land. We appreciate the adage: "The land was not given to us by our parents; it is on loan to us from our children." How to be good stewards— through the development and implementation of sound environmental policies—has no simple answer given the complexity of the issues that need to be addressed.

References

1. Maler, K.-G.; Vincent, J.R. In *Handbook of Environmental Economics: Environmental Degradation and Institutional Responses*; North-Holland: Amsterdam, 2003; Vol. 1.
2. Maler, K.-G.; Vincent, J.R. *Handbook of Environmental Economics: Valuing Environmental Changes*; North-Holland: Amsterdam, 2005; Vol. 2.
3. Viscusi, W.K.; Vernon, J.M.; Harrington, J.E., Jr. *Economics of Regulation and Antitrust*; MIT Press: Cambridge, MA, 2000.
4. Ackerman, F.; Heinzerling, L. *Priceless: On Knowing the Price of Everything and the Value of Nothing*; New Press: New York, 2004.
5. Freeman, A.M., III *The Measurement of Environmental and Resource Values: Theory and Methods*, 2nd Ed. Resources for the Future: Washington, DC, 2003.
6. Portney, P.; Stavins, R. *Public Policies for Environmental Protection*; Resources for the Future: Washington, DC, 2000.
7. Vig, N.; Kraft, M.E. *Environmental Policy and Politics: New Directions for the Twenty-first Century*, 5th Ed.; Congressional Quarterly Press: Washington, DC, 2003.
8. Krutilla, J. Conservation reconsidered. Am. Econ. Rev. **1967**, *57* (4), 777–786.
9. Busch, P.-O.; Jörgens, H.; Tews, K. The global diffusion of regulatory instruments: the making of a new international environmental regime. In *The Annals of the American Academy of Political and Social Science*, Sage Publications: Thousand Oaks, CA, 2005, Vol. 598, 146–167.
10. Kraft, M.E. *Environmental Policy and Politics*, 3rd Ed.; Pearson/Longman: New York, 2004.

30

Environmental
Policy: Innovations

Alka Sapat

Introduction

Over the past three decades, American environmental policy management has often entailed the search for new solutions and innovative ways of managing environmental problems. These solutions have taken the form of environmental policy innovations (EPIs), which have frequently been pioneered and adopted by local and state entities, at times with the help of citizen groups and non-profit organizations. They have become an important means of managing environmental problems, particularly for intractable, multimedia, multijurisdictional environmental issues. While policy innovations have been studied in a number of areas at the state level,[1-19] less attention has been paid to policy innovations in environmental management.[20-24] The focus of this entry is to shed light on these environmental policy initiatives, termed here as EPIs.

To provide an understanding of these policy initiatives, I begin by defining and discussing EPIs and supply some understandings of the concept. Next, I discuss the nature and type of a large majority of EPIs; in particular, I focus on their reliance and inclusion of collaborative forms of environmental management and provide some examples. The fourth section of the entry analyzes some of the main factors affecting the adoption of EPIs, including determinants such as institutional commitment, resources, the severity of the problem, and the role of interest groups. The effect of neighboring entities in spurring innovation adoption and diffusion is also considered in this section. The entry ends with a conclusion that summarizes the main points of the entry, provides some considerations for further research, and discusses the policy implications of adopting environmental policy initiatives as a strategy for managing the environment.

Environmental Policy Innovations

EPIs here are defined in this entry as government-sponsored initiatives to protect the environment. This is based in part on the original definition of a policy innovation originally espoused by Walker[1] who examined the adoption of new policies by states, which he termed as "innovations." Walker's work also differentiated between states that were "leaders" and those that are deemed to be "laggards," along with examining the diffusion or the spread of these innovations from leading states to other states. Thus, he analyzed the amount of time that elapsed between the invention of an innovative program and a state's adoption of it. Based on 88 different state programs, he concluded that larger, wealthier, and more industrialized states adopt new programs more rapidly than do small, less-developed states. He also found strong regional relationships among states that were thought to reflect conditions of emulation and competition among proximate states. Walker's research lead to further investigation into state policy innovations. Gray[3] undertook a more detailed analysis and rejected Walker's contention that leading states lead no matter what the policy area; in a study of education, welfare, and civil rights policies, Gray found that innovativeness was not a pervasive factor. She concluded that based on the issue and the era, different states would comprise the leading, "middle-adopting," and lagging clusters.

During the next two decades, there were numerous other studies of state innovation that were published in a variety of policy areas. These included studies of innovation in juvenile corrections,[12] consumer affairs,[13] technology,[14] energy,[15] tort law,[16] judicial administration,[17] and human services.[18] These studies explored various determinants of innovativeness in different policy areas. However, while they expanded the scope of policy areas subject to innovation analysis, they did not lead to major advances in the conceptualization of state innovations or in the empirical approaches to its investigation.

In the 1990s, however, significant advances were made in the field of state innovations research. Research on state innovation was tested with sophisticated empirical methods in the area of state lottery adoptions[4] and state taxation.[5] In both these studies, the authors extended theories of innovation to different policy areas and made a substantial methodological contribution to the literature by refining the empirical analysis of state innovation with the employment of event history analysis (EHA), a pooled cross-sectional time-series technique. Berry and Berry[4,5] also analyzed the importance of internal determinant explanations (which posit that the factors causing a state government to innovate are the political, economic, and social characteristics of a state) and regional diffusion explanations (which point toward the role of policy adoptions by neighboring states in prompting a state to adopt).

An interesting twist on the diffusion aspect of state innovations was presented in a study of state living will laws.[19] In this study, Glick and Hays argued that as policy innovations diffuse across states, a process of "policy reinvention" occurs.[19] Policy innovations, in their view, are not adopted in their original form; rather, they are changed or selectively adapted as innovations by other states. The contribution of this study lay in refinements that added to earlier understandings of innovation diffusion.

Another variation on state innovation research was the study undertaken by Berry.[6] In this study, Frances Stokes Berry[6] extended the original research undertaken by Berry and Berry[4,5] to state innovations in strategic planning by various regulatory agencies. This extension was interesting as it was one of the first studies to empirically evaluate an innovation adopted by state bureaucracies as opposed to legislatures. This research was followed by more sophisticated understandings of the catalysts and actors behind the adoption of innovations; in his analysis, Mintrom[7,8] focused specifically on the importance of policy entrepreneurs at the state level. Using both secondary data sources and primary data gathered from a 50-state survey of state education officials in an EHA, Mintrom examined both the consideration of school choice and its adoption by states across time. He focuses in particular on the role of policy entrepreneurs by developing a theory of policy entrepreneurship and testing that theory with respect to school choice. Using EHA, he concluded that policy entrepreneurs act as catalysts in the policy innovation process by promoting ideas for dynamic policy change, persuading legislators, and networking with others in government.

In the area of environmental policy, state policy innovations has been the subject of research by Rabe,[20–23] Sapat,[24] and others. For instance, in a comparison of Canada and the United States, Rabe[21] found that despite the relatively decentralized Canadian institutional framework and more centralized and "bureaucratic" framework in the United States, the latter had been much more active and effective in devising innovative policy approaches. Comparing four U.S. states and four Canadian provinces, he concluded that within those sample jurisdictions, the United States was clearly ahead of Canada in cross-media permit integration, pollution prevention, disclosure of information on toxic substances, and achieving greater refinement and use of environmental outcome indicators. Similarly, in later research analyzing state experimentation in the realm of climate change, Rabe[22] analyzed the factors prompting successful state initiatives in reducing greenhouse gases. He finds that some states did succeed in cutting through traditional partisan divides and that agent- based entrepreneurs helped to develop policy ideas and form viable coalitions.

State EPIs adopted by administrative agencies (rather than by legislatures) have also been the subject of past research. In a study of state policy innovations adopted by agencies in the areas of hazardous waste and groundwater contamination, Sapat[24] found that state agencies are more likely to adopt innovations to deal with problems created by hazardous waste contamination than for groundwater contamination. Further, she finds that state environmental managers are not directly influenced by interest groups and that the inclusion of all stakeholders is likely to lead to greater support for new policy initiatives.

EPIs as defined here are drawn from and are based on this past research. An EPI as defined here can be one that is initiated by an agency or government institution (such as a state legislature, county government, or city) or it can be one that is borrowed from another government entity/ institution. Thus, EPIs can be adopted by legislatures or administrative agencies. Also, as defined in prior research by Sapat,[24] EPIs are defined as initiatives that protect the environment. The emphasis on the positive aspects of EPIs is important, because initiatives can have both positive and negative benefits. The term EPI as used in this entry is positive, i.e., it refers to laws, regulations, and policies that are beneficial to the environment. In other words, the goals of these EPIs are to regulate pollution problems and protect the environment.

At this point, a caveat is necessary. Clearly, policy choices can be subjectively interpreted; they may be viewed as being positive for the environment by some and as bad for the environment by others. This is particularly true within the realm of environmental politics.[25–27] For instance, the use of economic incentives, such as fees, pollution taxes, or offsets, is viewed as being beneficial by some and negative by others.[25] Even environmental organizations are divided on this issue. For instance, large environmental interest groups such as the Environmental Defense Fund, which utilize the services of economists and scientists, view economic incentives as being beneficial for the environment.[28] However, a number of smaller environmental interest groups are more suspicious about the environmental benefits of adopting market-based incentives.[29] Moreover, the problem of clearly defining an environmentally "positive" policy solution is also compounded by the fact that for some environmental issues, even scientific experts cannot come to an agreement on what the best technical solution is to a pollution problem.[30] Hence, there is extensive debate about the benefits of particular environmental policies, because of conflicting perspectives regarding policy outcomes and because of incomplete and inconclusive technical knowledge of solutions to environmental problems. To overcome some of these problems, the EPIs discussed here are those that are deemed as being beneficial for the environment by the United States Environmental Protection Agency (USEPA).[31,32] These assessments are arguably less subjective than those carried out by environmental or industry interest groups.

The Nature of EPIs: Conflict to Collaboration

Following the definition given above, EPIs can be categorized in various ways. For instance, they may be categorized according to their function; the USEPA provides a portfolio for instance, in which various environmental innovations are characterized by the core agency function that they address, such as helping in improving service delivery, enhancing regulatory outcomes, supporting superior environmental performance, designing targeted geographic solutions, etc. While it is useful to categorize environmental innovations thus, other ways of classifying environmental innovations may provide a broader and deeper understanding of these initiatives.

In this entry, I contend that rather than categorizing EPIs purely by function, it is also useful to classify such innovations by the underlying regulatory and managerial premises of such policy innovations. More specifically, it is instructive to understand the extent to which such innovations rely on collaborative as compared to non-cooperative, more conflict-based approaches to environmental management. This perspective would be useful because there has been growing evidence in both practice and theory about the usefulness of collaborative approaches to managing the environment and natural resources.

Reasons for the emergence of these collaborative approaches have been many and varied, but the primary reasons for the growth in collaborative forms of environmental management have been ascribed to the problems stemming from command and control measures.[33–39] These traditional forms of regulation have often resulted in failure to achieve key goals and/or protracted conflict. Traditional forms of environmental regulation that rely on bureaucratic, adversarial, and often technology-based regulatory approaches are also seen as being far too rigid and, more importantly, ineffective, in that they fail to address multimedia and multijurisdictional environmental hazards, particularly those stemming from non-point pollution.[33,36] Furthermore, traditional hierarchical approaches do not allow for more democratic forms of participation and often rely on unrealistic models of individual and administrative rationality.[37,40]

To overcome such problems, a search for alternative solutions to managing environmental problems began to evolve and grow. For instance, a study by John in 1994[40] found that states and local governments were increasingly turning away from "top–down command-and-control" regulation to a bottom–up style, which he termed "civic environmentalism." Civic environmentalism relies primarily on the use of non-regulatory tools, which encompass market-based incentives to regulation.

Civic environmentalism or collaborative environmental approaches or similar interchangeable terms that can be used to describe a more cooperative, less adversarial form of environmental management are to be found in a number of EPIs that are undertaken by environmental agencies at different levels of government. Indeed, some EPIs were initially pioneered and adopted because of the failure of conflict-based approaches to manage environmental problems. For instance, environmental permitting processes engender a high level of stakeholder frustration and conflict due to permitting backlogs, long lead times, costs, and uncertainty. To improve these processes and reduce conflict and backlog times, some state and local agencies adopted EPIs to shift away from media-specific permitting for individual facilities and also to improve internal agency permitting processes. For instance, Massachusetts shifted away from a facility permitting approach to a multimedia, sector-based regulatory approach, targeting sectors with large numbers of small sources, as an alternative to facility-specific state permits with industry-wide environmental performance standards and annual self-certifications of compliance.[31,41]

Similarly, other EPIs take a collaborative approach by helping to enhance partnerships to resolve environmental problems that cannot be effectively solved without the participation and collaboration of multiple actors. Leveraging such partnerships enables environmental agencies to solve complex problems by harnessing energies and resources of stakeholders to achieve mutually desirable outcomes. An example of an EPI that does this is the effort undertaken in South Carolina to reduce neighborhood contamination. This initiative enlists the support of numerous community organizations and local businesses to provide education and outreach to reduce community exposure to lead and other hazardous substances as part of the Charleston–North Charleston Community-Based Environmental Partnership (CBEP).[42]

Other examples of collaboration-based EPIs[43] are provided in Table 1.

To summarize, EPIs are initiatives adopted to protect the environment and a number of them have been adopted at the local, state, and federal level by both legislative and administrative agencies. These EPIs are characterized by more collaborative approaches to environmental management. To understand EPIs further, we can try to ascertain which factors are likely to affect the adoption of these innovations. It is to this issue that I turn to in the next section.

TABLE 1   Examples of EPIs

Neighborhood Contamination Reduction—South Carolina
Enlists numerous community organizations and local businesses to support education and outreach to reduce community exposure to lead and other hazardous substances as part of the Charleston–North Charleston Community- Based Environmental Partnership (CBEP). (http://www.epa.gov/Region4)
Northeast Ohio Initiatives—Ohio
Responds to regional economic, sprawl, ecosystem and infrastructure challenges through a 15-county, community-based approach. (http://www.epa.gov/glnpo/lakeerie/leneohio.htm)
Urban Environmental Program—Boston, Massachusetts
Adopts a community-based approach including the city of Boston, Massachusetts DEP, and community organizations to improve the quality of life in urban settings by targeting issues such as asthma and indoor air quality, lead poisoning, vacant lots and green spaces, and pollution prevention. (http://www.epa.gov/boston/eco/uep/boston/)
Pollution Complaint Response—Indiana
Coordinates an agency-wide, multimedia response to citizen inquiries and complaints using Web-based information, enabling the agency to reduce costs and increase public trust. (http://www.in.gov/idem/5274.htm)
Ford Good Neighbor Dialogue—Illinois
Brings together stakeholders, academics, and agency representatives in a collaborative process to periodically discuss a large manufacturing facility's environmental management and performance. (http://www.delta-institute.org/)
Community Environmental Awareness Project—Michigan
Develops an approach to improve the way environmental information is presented and made available to the public; the goal of the CEAP is to improve the public's access to and understanding of how major industries are performing under environmental laws and regulations. (http://www.deq.state.mi.us/ceap)

Resources, Needs, Politics, and Other Determinants

The factors that lead to the adoption of EPIs can be many and varied. Prior research on innovation adoption in other areas, as well as on EPI adoption is useful here. The following factors are some of the most influential determinants of innovation adoption as seen in extant research.

Need/Problem Severity

Keeping in with the adage that "necessity is the mother of invention," researchers have traditionally regarded problem severity as a significant influence on the adoption of innovations.[1,3,44,45] For innovations adopted at the state and local level, the expectation is that states and local governments rather than the federal government are likely to understand local problems more clearly and are hence likely to be responsive to the needs and problems present in their jurisdictions.[27,45] For instance, in a study of state innovation, ranging from state testing on teacher competency and rail passenger service to state regulation on sodomy, Nice[45] found that the problem environment was prominent in explaining five of eight policy innovations analyzed in this study. For EPIs, the severity of the pollution problem itself has been regarded as one of the most important reasons for innovation adoption. To deal with intractable environmental problems, several entities (local, state, non-profit, and even private institutions) have come up with new ways to solve old problems.[20,31]

Institutional Factors: The Importance of Commitment and Capacity

While necessity can often drive innovation, institutional commitment and capacity have also been found to be important for the adoption of policy innovations.[24,40,46] The theoretical origins of the role of institutional actors may be found in ideas related to roles of institutional elites and theories of institutionalism and neo-institutionalism.

    Elites: The importance of elites in the policy-making process is theoretically supported by the elite perspective of policy analysis. Elite theorists argue that power is concentrated in the hands of elites who use the resources of their respective organizations to manage and impose order on society.[47] Societal stability, according to this perspective, rests not on a common political culture and a set of values, but on a forced consensus created and reinforced by the elite. Pluralist politics can coexist with the governmental elite; however, key decisions regarding policy are made by elites. Popular and electoral politics are, for the most part, mainly symbolic and concerned with middle-level policy issues, according to the elite perspective.[47]

The elite theory of the policy process is close to the neopluralist view: neo-pluralists, such as Lindblom,[48] challenged the pluralist notion that power was diffuse and argued instead about the privileged position of business. Similarly, others such as E.E. Schattschneider pointed out, "the flaw in the pluralist heaven is that the heavenly chorus sings with a strong upper-class accent."[49]

Institutionalism and Neo-institutionalism: The motivation of institutional actors and elites is also stressed in theories that emphasize the importance of institutions in policy and governance.[50-54] Institutional theories hold that government actors can act independently of interest group pressures and other factors. According to this view, government actors are not just simply "pawns" of various interest groups; rather, the perceptions and attitudes of these actors shape the way they process information and affect independently the choices they make.[52,54] Institutional theories also recognize that informational constraints and computational limitations of political actors prevent actors from making purely "rational decisions" that are independent of the actor's subjective representation of the decision problem.[52,55] Thus, the attitudes and ideological views of institutional actors can influence their choices in innovation adoption.

Based on the theories discussed above, it is likely that pro-environmental policy actors will likely push for adoption of EPIs. Related to that issue is the importance of resources. Adoptions of EPIs require both the motivation and the commitment of key institutional actors in terms of time, money, and other resources. Typically for most policy adoptions, resources are required. For this reason, wealthier and larger states and local governments are more likely to be innovation adopters.[11,56] Resource provision by third- party institutions, such as non-profits, or by higher levels of government, such as by the federal government to state and local governments, can also help spur innovation adoption and diffusion. For instance, the USEPA provides State Innovation Grants to spur innovation adoption; these grants are typically provided to states with projects in three areas: environmental results programs, environmental management systems and permitting, and environmental leadership programs such as EPA's Performance Track.[31]

However, while institutional resources may be required for innovation adoption, some EPIs are adopted to economize on resources or to raise revenues. For instance, EPIs in pesticide regulation adopted by certain states helped raise fees, adding to revenue sources.[24] The role of resources or the importance of economic factors can also be played out in other ways. That is, objectives of economic development by states could deter their willingness to adopt strict environmental regulations.[57-60] For instance, states with a higher percentage of economic activity relying on manufacturing or industrial activities with negative environmental consequences are less likely to favor the adoption of innovative laws and programs that would negatively affect revenue-generating industries. This raises the issue then of interest group support and its importance for EPI adoption.

Interest Group Support

Interest groups have always played an important role in the American policy process, due in part to constitutional provisions for pluralism. Theoretically, the pluralist perspective of policy making argues that public policy is the product of democratic participation by individuals who are represented by organizations such as interest groups. Theories of interest group influence on legislators and bureaucrats range from narrow views of interest group influence[61-63] to those that posit interest groups as not exercising any more influence over regulatory policy than other actors or bureaucrats.[64,65]

Pluralist theorists, such as Truman[ZS] and Dahl[67] argued that group exchange (pluralism) was the dominant (and desired) method of political decision making. In their view, this was possible because the decentralization of American institutions and the relative openness of the political system guaranteed competing groups access and some degree of power. Pluralism thus presents a view of the political system in which multiple centers of power compete to shape policy. Power is diffuse, fluid, multifaceted, and dispersed among a number of groups. Political participation is largely a goal- oriented activity in which citizens take part in order to obtain some benefit from the government.[68]

Pluralist group theory was questioned by a number of scholars, mainly neo-pluralists, who argued that dominant groups and interests had the ability to control the political agenda, preventing pluralist discourse over a full range of policy options.[47,48,69] Group theory was also evaluated in part in analyses of regulatory policies.[62,70,71] A number of scholars argued that regulatory policies emerged from a political equilibrium produced by coalitions of regulated industries and their customer groups, leading to "agency capture."[62,72,73] This narrow interest group perspective is based on the assumption that certain groups of people, who are organized and powerful in terms of possessing economic resources, will have the capability to dominate policy at the subnational level.

While this view is compatible with the Madisonian perspective of private parochial interests modifying policy outcomes, it has been modified considerably to account for characteristics of interest groups such as size and density that could affect their capabilities to exert influence over state regulations.[74] Moreover, the motivations and actions of other political actors, bureaucrats, and other interest groups themselves have also been found to be important in influencing policy outcomes.[75] Further, the narrow view of interest groups adduced by scholars such as Stigler[62] and Posner[73] have been more successful in explaining economic regulation as opposed to social regulation, such as consumer protection and regulation of the environment. The traditional "iron triangle" or "agency capture" theory does not adequately account for the emergence and power of newly organized and motivated consumer, environmental, and other "public interest" groups, which seek social benefits.

With respect to environmental policy, past studies have found environmental interest groups to be important in influencing policy outcomes. For instance, Ringquist[76] and Hird[27] analyzed the influence of environmental interest groups that countered the regulatory priorities advocated by industry groups. Ringquist[76] found environmental interest groups to have a significant influence on state water quality regulation in a group influence model of state policy influence.

With regard to EPIs, the presence and activities of environmental interest groups are likely to be important as well. However, industry interest group opposition to EPIs may be tempered by the nature of these policy innovations. As discussed above, a number of EPIs rely on collaborative nonadversarial approaches to solve environmental problems. Past research has also shown that some EPIs provide new ways to work with such industries to improve compliance with regulations and do not burden them with any additional regulatory costs.[24] Interest group opposition may also be mitigated by public support for certain EPIs, which provide broader benefits than they do costs to the public at large.

Regional Diffusion

Previous empirical studies on the diffusion of innovations have found that external influences by other neighboring entities are important. For instance, adoption of innovations by neighboring states can affect a state's adoption of a policy innovation.[1,4,11,56,77] Policy adoptions by neighboring states can decrease the information costs regarding the possible consequences, including the electoral consequences, of the adoption of a policy.[4] State officials can view the relative success/failure of programs in these adjacent states and decide whether such programs are suitable for their own state. Moreover, if a policy has been successful in a neighboring state, then state officials, in particular state legislators, can, by using the experiences of neighboring states as an example, boost public and legislative support for similar legislation in their own state. This allows them then to selectively utilize the experiences of neighboring states for their own political gain. The decrease in information costs and political uncertainty can thus increase the motivation of state policy makers to initiate or adopt a policy innovation.[24]

While early research on policy innovation diffusion focused on regional influences primarily with respect to policy innovations adopted by states,[8,11,24] later work has also studied factors affecting diffusion at the country level[78] and the local level.[11] On a related note, there has also been research on examining the mechanisms of policy diffusion,[79–82] which has moved away from looking simply at the effects of neighboring states to analyzing the more complex and multiple factors affecting diffusion, such as the presence of competition, learning, imitation, and coercion; this research has found that policy mimicry is often tempered by the size of the state and by the presence of other third-party institutions that may aid diffusion.[11] Thus, while regional effects or "external determinants" may be important, it is also necessary to understand the mechanisms of diffusion, which may be varied and complex.

There are thus, a number of factors that can influence the adoption and implementation of EPIs; need or the type of problem alone may be a necessary but not a sufficient determinant of innovation adoption. Institutional commitment and capacity, the role played by interest groups, and the effects of innovation adoption by neighboring institutions and entities are also important influences affecting the espousal of EPIs.

Conclusion

Complex environmental problems require innovative solutions for effective management. The focus of this entry has been on such initiatives or EPIs. In doing so, this entry began by defining and explaining the nature of such initiatives, including a discussion on past research on state policy innovations. I contend that a number of environmental initiatives have pioneered and are symptomatic of a shift in environmental management to more collaborative ways of solving environmental problems. Need or the severity of the problem has been one of the reasons affecting the adoption of EPIs, but institutional commitment and capacity are also critical to adoption and implementation of policy innovations in general and for EPIs. Given that interest groups are critical in the American policy process, the support or opposition of key groups is also vitally important to innovation adoption. Since a number of EPIs provide new ways to work in partnership rather than in conflict with industries to improve compliance with regulations and do not burden them with any additional regulatory costs,[24] industry opposition to such EPIs may be lessened. Interest group opposition may also be mitigated by public support for certain EPIs, which provide broader benefits than they do costs, to the public at large. Finally, innovation adoption may also be affected by the presence or adoption of innovations by neighboring entities, the external determinants. However, the mechanisms of such diffusion need further study and research.

Future research could also address the policy implications and long-term outcomes of innovation adoption. For instance, do the adoptions of EPIs lead to better environmental outcomes in terms of solving environmental problems and improve environmental quality? Do EPIs foster better outcomes in terms of increasing stakeholder participation and provide for more democratic and innovative means of achieving environmental goals? Preliminary evidence does suggest that EPIs have positive outcomes, but more research is needed on this issue. Given that the environmental problems facing us in the coming century are only likely to grow and be exacerbated by the slow-moving yet devastating effects of more large-scale problems such as climate change, the need for innovative environmental solutions is likely to be critically important.

Acknowledgments

This research was funded in part by the National Science Foundation, NSF Grant No. SBER-9510308. The author is responsible for any omissions or errors.

References

1. Walker, J. The diffusion of innovation among the American states. Am. Polit. Sci. Rev. **1969**, *63*, 880–899.
2. Mohr, L. Determinants of innovations in organizations. Am. Polit. Sci. Rev. **1969**, *63*, 111–126.

3. Gray, V. Innovation in the States: A diffusion study. Am. Polit. Sci. Rev. **1973**, *67*, 1174–1185.
4. Berry, W.D.; Berry, F.S. State lottery adoptions as policy innovations: An event history analysis. Am. Polit. Sci. Rev. **1990**, *84*, 395–415.
5. Berry, W.D.; Berry, F.S. Tax innovation in the States: Capitalizing on political opportunity. Am. J. Polit. Sci. **1992**, *36*, 715–742.
6. Berry, F.S. Innovations in public management: The adoption of strategic planning. Public Adm. Rev. **1994**, *54*, 322–330.
7. Mintrom, M. Policy entrepreneurs and the diffusion of innovation. Am. J. Polit. Sci. **1997**, *41* (3), 738–770.
8. Mintrom, M. Policy Entrepreneurship in Theory and Practice: A Comparative State Analysis of the Rise of School Choice as a Policy Idea. Unpublished Ph.D. dissertation. State University of New York at Stony Brook, 1994.
9. Mooney, C.Z.; Mei-Hsien, L. Legislating morality in the American states: The case of the pre-Roe abortion regulation reform. Am. J. Polit. Sci. **1995**, *39* (3), 599–628.
10. Berry F.S.; Berry, W.D. Innovation and diffusion models in policy research. In *Theories of the Policy Process;* Sabatier, P., Smith, H.-J., Eds.; Westview Press: CO, 2007; 169–180.
11. Shipan, C.R.; Volden, C. The mechanisms of policy diffusion. Am. J. Polit. Sci. **2008**, *52* (4), 840–857.
12. Downs, G.W., Jr.; Mohr, L.B. Conceptual issues in the study of innovation. Adm. Sci. Q. **1976**, *21* (December), 700–714.
13. Siegelman, L.; Smith, R. Consumer regulation in the American states. Soc. Sci. Q. **1980**, *61*, 58–76.
14. Menzel, D.; Feller, I. Leadership and interaction patterns in the diffusion of innovation among the American states. West. Polit. Sci. Q. **1977**, *30* (4), 528–536.
15. Regens, J.L. State policy responses to the energy issue: An analysis of innovation. Soc. Sci. Q. **1980**, *61* (June), 44–57.
16. Canon, B.C.; Baum, L. Patterns of adoption of tort law innovations: An application of diffusion theory to judicial doctrines. Am. Polit. Sci. Rev. **1981**, *75*, 975–990.
17. Glick, H.R. Innovation in state judicial administration: Effects on court management and organization. Am. Polit. Q. **1981**, *9*, 49–69.
18. Sigelman, L.; Roeder, P.W.; Sigelman, C.K. Social service innovation in the American states: Deinstitutionalization of the mentally retarded. Soc. Sci. Q. **1982**, *62* (3), 503.
19. Glick, H.R.; Hays, S.P. Innovation and reinvention in state policy-making: Theory and the evolution of living will laws. J. Polit. **1991**, *53* (3), 835–850.
20. Rabe, B. Power to the states: The promise and pitfalls of decentralization. In *Environmental Policy in the 1990s*; Vig, N., Kraft, M.E., Eds.; Congressional Quarterly Press: Washington, DC, 1995; 31–52.
21. Rabe, B.G. Federalism and entrepreneurship: Explaining American and Canadian innovation in pollution prevention and regulatory integration. Policy Stud. J. **1999**, *27* (2), 288–306.
22. Rabe, B. *Statehouse and Greenhouse: The Stealth Politics of America Climate Change Policy*; Brooking Institution Press: Washington, DC, 2004.
23. Rabe, B. Racing to the top, the bottom, or the middle of the pack? The evolving state government role in environmental protection. *In Environmental Policy: New Directions for the Twenty-First Century*; Vig, N., Kraft, M.E., Eds.; Congressional Quarterly Press: Washington, DC, **2010**; 27–51.
24. Sapat, A. Devolution and innovation: The adoption of state environmental policy innovations by administrative agencies. Public Adm. Rev. **2004**, *64* (2), 141–151.
25. Davis, C.E. *The Politics of Hazardous Waste;* Prentice-Hall: Englewood Cliffs, NJ, 1993.
26. Lester, J.P.; Bowman, A. Subnational hazardous waste policy implementation: A test of the Sabatier-Mazmanian model. Polity **1989**, *21* (4), 731–753.
27. Hird, J. *Superfund: The Political Economy of Environmental Risk;* John Hopkins University Press: Baltimore, MD, 1994.

28. Environmental Defense Fund. Using Economics to Solve Eco-Challenges, available at http://www.edf.org/approach/markets (accessed November 2011).
29. Ingram, H.; Colnic, D.; Mann, D.E. Interest groups and environmental policy. In *Environmental Politics and Policy: Theories and Evidence,* 2nd Ed.; Lester, J.P., Ed.; Duke University Press: Durham, NC, 1995.
30. Barke, R.P.; Jenkins-Smith, H.C.; Silva, C. Consistency and Controversy in the Translation of Scientific Knowledge into Policy Recommendations. Paper Presented at the Annual Meeting of the Midwest Political Science Association, Chicago, IL, April 6–8, 1995.
31. United States Environmental Protection Agency. *Environmental Innovation Portfolio.* U.S. EPA, National Center for Environmental Innovation, Washington, DC, available at http://www.epa.gov/osem/pdf/portfolio.pdf (accessed October 2011).
32. Morandi, L. *Groundwater Protection Legislation: Survey of State Action 1988–1992;* National Conference of State Legislatures: Washington, DC, 1994.
33. Multi-State Working Group on Environmental Performance and Regulatory Policy Program, Mossavar-Rahmani Center for Business and Government. *Environmental Innovation: A Dialogue on the Role of Government, Law, and Regulatory Approaches.* University of Massachusetts, Lowell and the John F. Kennedy School of Government, Harvard University, January 2006.
34. National Academy of Public Administration. *Resolving the Paradox: EPA and the States Focus on Results;* NAPA: Washington, DC, 1997.
35. O'Leary, R.; Durant, R.F.; Fiorino, D.; Weiland, P.S. *Managing for the Environment: Understanding the Legal, Organizational, and Policy Challenges;* Jossey-Bass: San Francisco, 1999.
36. Durant, R.F.; Fiorino, D.; O'Leary, R., Eds. *Environmental Governance Reconsidered: Challenges, Choices, and Opportunities;* The MIT Press: Cambridge, MA, 2004.
37. Koontz, T.M.; Thomas, C.W. What do we know and need to know about the environmental outcomes of collaborative management? Public Adm. Rev. **2006**, *66*, 111–121.
38. Weber, E.P. *Pluralism by the Rules: Conflict and Cooperation in Environmental Regulation;* Georgetown University Press: Washington, DC, 1998.
39. Rogers, E.; Weber, E.P. Thinking harder about outcomes for collaborative governance arrangements. Am. Rev. Public Adm. **2010**, *40* (5), 546–567.
40. John, D. *Civic Environmentalism: Alternatives to Regulation in States and Communities;* Congressional Quarterly Press: Washington, DC, 1994.
41. Massachusetts Department of Environmental Protection. MassDEP Environmental Results Program, available at http://www.mass.gov/dep/service/envrespr.htm (accessed November 2011).
42. United States Environmental Protection Agency. About EPA Region 4, available at http://www.epa.gov/Region4 (accessed November 2011).
43. United States Environmental Protection Agency. Environmental Innovation Portfolio, available at http://www.epa.gov/osem/portfolio/protection.htm (accessed November 2011).
44. Savage, R.H. When a policy's time has come: Cases of rapid policy diffusion, 1983–1984. Publius **1985**, *15* (3), 111–125.
45. Nice, D.C. *Policy Innovation in State Government;* Iowa State University Press: Ames, Iowa, 1994.
46. Karch, A. National intervention and the diffusion of policy innovations. Am. Polit. Q. **2006**, *34* (4), 403–426.
47. Mills, C.W. *The Power Elite.* Oxford University Press: New York, 1956.
48. Lindblom, C. *Politics and Markets: The World's Political Economic Systems;* Basic Books: New York, 1977.
49. Schattschneider, E.E. *The Semisovereign People; A Realist's View of Democracy in America;* Holt, Rinehart and Winston: New York, 1960.
50. Noll, R.; Owen, B.M. *The Political Economy of Deregulation: Interest Groups in the Regulatory Process;* American Enterprise Institute for Public Policy Research: Washington, DC, 1983.

51. Moe, T. The politics of bureaucratic structure. In *Can the Government Govern?* Chubb, J.E., Petersen, P.E., Eds.; Brookings Institution: Washington, DC, 1989.
52. North, D.C. *Institutions, Institutional Change and Economic Performance*; Cambridge University Press: New York, 1990.
53. Chubb, J.E.; Petersen, P.E. *Can the Government Govern?* Brookings Institution: Washington, DC, 1989.
54. Van Horn, C.E. *The State of the States,* 3rd Ed.; Congressional Quarterly Press: Washington, DC, 1996.
55. Herbert A. Decision Making and Problem Solving. Research Briefings 1986: Report of the Research Briefing Panel on Decision Making and Problem Solving. National Academy Press: Washington, DC, 1986.
56. Shipan, C.R.; Volden, C. Bottom-up federalism: The diffusion of antismoking policies from U.S. cities to states. Am. J. Polit. Sci. **2006**, *50* (4), 825–843.
57. Getz, M.; Walter, B. Environmental policy and competitive structure: Implications for the hazardous waste management program. Policy Stud. J. **1980**, *9* (Winter).
58. Feiock, R.C.; Rowland, C.K. Environmental regulation and economic development: The movement of chemical production among states. West. Polit. Q. **1991**, 561–576.
59. Lowry, W. *The Dimensions of Federalism;* Duke University Press: Durham, NC, 1992.
60. Feiock, R.C.; Davis, C. Can state hazardous waste regulation be reconciled with economic development: A test of the Rowland–Goetze model. Am. Polit. Q. **1992**, 19.
61. Bernstein, M. *Regulating Business by Independent Commission*; Princeton University Press: Princeton, NJ, 1955.
62. Stigler, G. The theory of economic regulation. Bell J. Econ. Manage. Sci. **1971**, *2*, 3–21.
63. Peltzman, S. Toward a more general theory of regulation. J. Law Econ. **1974**, *19*, 211–240.
64. Meier, K.J. *The Political Economy of Regulation: The Case of Insurance*; SUNY Press: Albany, NY, 1988.
65. Derthick, M.; Quirk, P.J. *The Politics of Deregulation;* Brookings Institution: Washington, DC, 1985.
66. Truman, D. *The Governmental Process*; Knopf: New York, 1951.
67. Dahl, R. *A Preface to Democratic Theory*; University of Chicago Press: Chicago, IL, 1956.
68. Verba, S.; Nie, N.; Kim, J. *The Modes of Democratic Participation: A Cross-National Comparison*; Sage Publications: Beverly Hills, CA, 1971.
69. Bachrach, P.; Baratz, M.S. Two faces of power. Am. Polit. Sci. Rev. **1962**, *56* (4), 947–952.
70. Becker, G. A theory of competition among pressure groups for political influence. Q. J. Econ. **1983**, *98* (3), 371–400.
71. Wilson, J.Q. *The Politics of Regulation;* Basic Books: New York, 1980.
72. Bernstein, M. *Regulating Business by Independent Commission*; Princeton University Press: Princeton, NJ, 1955.
73. Posner, R. Theories of economic regulation. Bell J. Econ. Manage. Sci. **1974**, *5* (3), 337–352.
74. Aggarwal, V.K.; Keohane, R.O.; Yoffie, D.B. The dynamics of negotiated protectionism. Am. Polit. Sci. Rev. **1987**, *81* (2).
75. Wilson, J.Q. *Bureaucracy: What Government Agencies Do and Why They Do It*; Basic Books: New York, 1989.
76. Ringquist, E. *Environmental Protection at the State Level: Politics and Progress in Controlling Pollution*; M.E. Sharpe: Armonk, NY, 1993.
77. Rogers, E.M. *Diffusion of Innovations,* 5th Ed.; The Free Press: New York, 2003.
78. Simmons, B.A.; Dobbin, F.; Garrett, G. Introduction: The international diffusion of liberalism. Int. Organ. **2006**, *60* (4), 781–810.
79. Berry, W.D.; Baybeck, B. Using geographic information systems to study interstate competition. Am. Polit. Sci. Rev. **2005**, *99* (4), 505–519.

80. Boehmke, F.J.; Witmer, R. Disentangling diffusion: The effects of social learning and economic competition on state policy innovation and expansion. Polit. Res. Q. **2004**, *57* (1), 39–51.

81. Weyland, K. Theories of policy diffusion: Lessons from Latin American pension reform. World Polit. **2005**, *57*, 262–295.

82. Weyland, K. *Bounded Rationality and Policy Diffusion*; Princeton University Press: Princeton, NJ, 2007.

31

Food Quality
Protection Act

Christina D.
DiFonzo

Tolerances

FQPA fundamentally changed the way EPA sets tolerances for pesticide residues in food. EPA must review all (nearly 10,000) pesticide tolerances under new FQPA guidelines. The tolerance assessment schedule developed by EPA called for examining 33% within 3 years after August 1996, 66% within 6 years, and 100% within 10 years. EPA initially took a "worst-first" approach, to review the pesticides it considered to be of greatest risk, particularly to children, by August 1999. Three major pesticide groups, organophosphates (OPs), carbamates, and probable human carcinogens (B2s), were targeted under the worst-first approach. OPs and carbamates, the majority of which are insecticides, are neurotoxins structurally related to nerve gas. They affect the enzyme acetylcholinesterase in animals, including humans. B2 carcinogens are pesticides classified by EPA as having sufficient evidence for causing cancer in lab animals (usually at very high dose levels), but human evidence is lacking. Several important fungicides, plus a few herbicides and insecticides, are classified in this category.

Before FQPA, a single tolerance was established for each pesticide/crop combination, based only on dietary exposure to pesticide residue. Under FQPA, EPA must consider the combined (aggregate) exposure to a pesticide through dietary, drinking water, and nondietary sources (for example, structural, turf, garden, and pet uses) as well as the cumulative exposure to related pesticides with a common mechanism of toxicity. Furthermore, FQPA directs EPA to consider sensitive subpopulations, especially children, when setting tolerances. To insure that sensitive groups are adequately protected, EPA can require a safety factor of up to 10-fold on existing tolerances.

The Risk Cup

An analogy of a "risk cup" is used by EPA to explain changes in the establishment of tolerances under FQPA. Before FQPA, there was a separate risk cup for each pesticide/crop combination, containing only dietary exposure to residue. FQPA creates a separate risk cup for *each group* of related pesticides with common toxicity. Multiple pesticides, as well as multiple residues from all sources—food, water, and nonfood—of each pesticide, go into the same cup. Under this scenario, the cup gets crowded, and individual tolerances for each pesticide/crop combination in the group must get smaller. Furthermore, safety factors for children may reduce the overall size of each cup, potentially by a factor of 10.

Endocrine Disruption

Under FQPA, all pesticides and pesticide additives must be tested for effects on the endocrine system. This may require in vitro and in vivo screening for three different types of endocrine effects: estrogenic (mimics or blocks estrogen), androgenic (mimics or blocks androgens), and thyroid. Of the potential targets of a screening program, these three hormone groups are important in human development, are fairly well studied, and some laboratory methodology is already available to detect changes in level and function. Estimates are that up to 70,000 pesticides and other chemicals will be screened under FQPA and a second law, the Safe Drinking Water Act.

Consumer Right-to-Know

Another issue addressed in FQPA is consumer right-to-know about pesticide residues in food. FQPA mandated that EPA create a brochure to inform consumers about pesticide risks and benefits, and ways to remove residues from food they purchase. The brochure was completed and distributed to supermarkets in early 1999. However, FQPA did not mandate that stores actually display the publication.

Potential Impacts of FQPA

Pesticides that do not meet FQPA standards must either be mitigated (use patterns changed) or eliminated (some or all uses dropped). Thus, as FQPA is implemented, it potentially will have a tremendous impact on American agriculture.

- Changes in labeling or use patterns (number, frequency, and timing of applications) of pesticides to mitigate residue.
- Loss of critical pesticide uses, particularly for so-called minor (specialty) crops. These commodities represent smaller markets for pesticide manufacturers and thus are often "expendable."
- Increases in production costs. Traditional broad-spectrum products might be replaced by more expensive, reduced-risk alternatives that control a narrower range of pests.
- Increased complexity of production and pest management systems. Broad-spectrum pesticides will be replaced by narrower spectrum tactics that require better knowledge and more intense management of the production system on the part of the producer.
- Potential for pesticide resistance. Loss of certain classes of pesticides could lead to resistance to remaining products, which are being relied on too heavily.

Bibliography

1. Colborn, T.; Dumanoski, D.; Myers, J. P. *Our Stolen Future*; Penguin Books U.S.A.: New York, 1996; 316.
2. *Public Law 104–170 to Amend the Federal Insecticide, Fungicide, and Rodenticide Act and the Federal Food, Drug, and Cosmetic Act, and for Other Purposes*, H.R. 1627; Federal Register: Washington, DC, Aug. 3, 1996.
3. Proceedings National Pesticide Impact Assessment Program Workshop, USDA CSREES NAPIAP Program, Sacramento, CA, May 5–7, 1998; Melnicoe, R., Ed.: Washington, DC, 1998; 76.
4. National Research Council. *Pesticides in the Diets of Infants and Children*; National Academy Press: Washington, DC, 1993; 386.
5. U.S. Environmental Protection Agency, Office of Pesticide Programs. *The Food Quality Protection Act (FQPA) of 1996.* http://www.epa.gov/oppfead1/fqpa/

32

Food: Cosmetic Standards

David Pimentel
and Kelsey Hart

Introduction

The American marketplace features nearly perfect fruits and vegetables. Gone are apples with an occasional blemish and fresh spinach with a leaf miner. This increase in the "cosmetic standards" of fruits and vegetables has resulted from the efforts of the Food and Drug Administration (FDA) and the U.S. Department of Agriculture (USDA) to limit the levels of insects and mites in produce, and new standards established by food wholesalers, processors, and retailers. Meeting more stringent standards has led to significant increases in the amounts and toxicity of pesticides used in crops. Increased pesticide use has negative environmental and public health consequences. In comparison, the health risks from consuming herbivorous insects/insect parts in food do not exist and certainly do not justify the increase in pesticide use and the associated problems. Recent research indicates that pesticide use can be reduced by 35% to 50% without any substantial increase in food prices or loss of crop yields.[1] Surveys suggest that the public would support relaxation of cosmetic standards if it decreases pesticide residues in its food.

History of Cosmetic Standards

The FDA sets defect action levels (DALs) for insects and mites allowed in fruits and vegetables and in products made from them. These DALs were established to reduce insect and mite infestation in foods to a reasonable and safe level, because their presence in food products was thought to indicate that crops had insufficient insect and mite control, were improperly washed, were unsatisfactorily inspected, and contained insects and mites harmful to health. Besides visual prejudice against insects in food, there is the well- placed concern that insects such as nonherbivorous houseflies and cockroaches may transmit disease.

During the past 40 years, the FDA has steadily lowered DALs.[2] For example, a fivefold decrease in the number of leaf miners permitted in spinach occurred from 1930 to 1974. As tolerance levels for insects in food have fallen, wholesalers, processors, and retailers have increased their "cosmetic standards" for produce and other food products so that most marketed U.S. produce is visually perfect. Produce distributors encourage high cosmetic standards because their contracts enable them to visually inspect produce before buying and reject it when the supply is excessive. Growers are motivated to produce cosmetically perfect produce to ensure its sale.

However, to meet these increasingly stringent regulations, farmers have had to use greater amounts of increasingly toxic pesticides and implement other pest control strategies. Synthetic pesticide use in

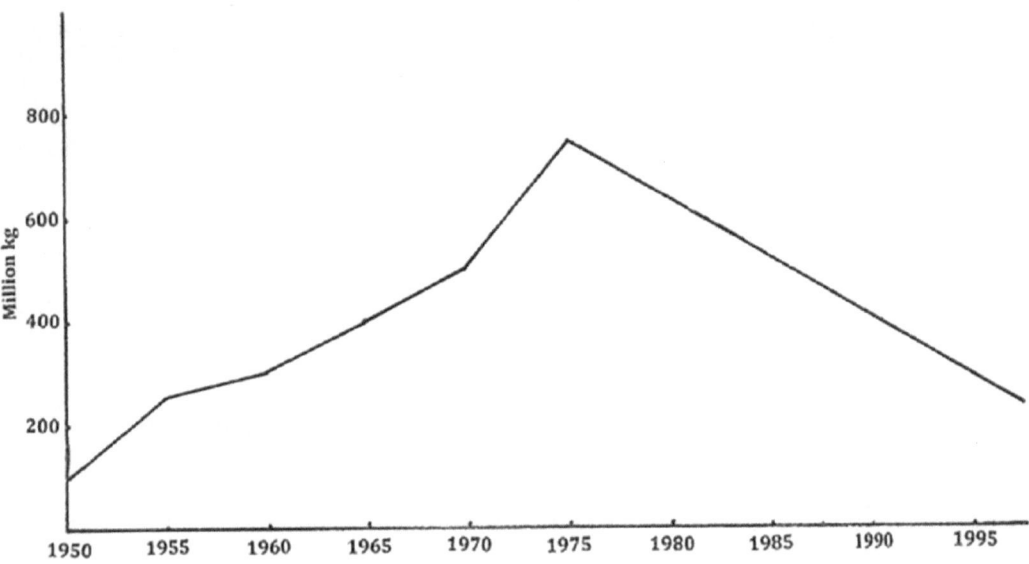

FIGURE 1   The amount of synthetic pesticides produced in the United States. About 90% is sold in the United States. The decline in total amount produced since 1975 is in large part due to the 10- to 100-fold increased toxicity and effectiveness of the newer pesticides.[1,3]

the United States has increased about 33-fold since 1945, and the toxicity of pesticides used has increased 10- to 100-fold in the past 25 years (Figure 1). More pesticides need to be used to produce the blemish-free produce distributors and consumers expect.

There is little evidence that eating herbivorous insects or insect parts is hazardous to human health.[4] However, solid data suggest that the adverse health and environmental impacts of pesticide exposure are substantial.[5] Given the direct correlation between increases in cosmetic standards and increases in pesticide use, why are cosmetic standards and DALs growing increasingly severe and perpetuating further increases in pesticide use?

The increase in cosmetic standards and more stringent DALs are based on the premise that consumers demand unblemished, insect-free food. Clearly, cosmetic appearance of produce is a primary factor consumers use in assessing the quality of produce. Unfortunately, this assessment is often made without more substantive quality information, such as nutritional values or pesticide residue levels. Recent evidence suggests that when consumers are aware of the trade-offs between blemish-free produce and pesticide use, they will purchase produce that is not cosmetically perfect because it has less pesticide residue.[6]

Environmental and Health Effects of Pesticide Exposure

An estimated 617,000 tons of more than 600 different kinds of pesticides are used annually in the United States, at a cost of approximately $9 billion.[7] Still, pests such as insects, plant pathogens, and weeds destroy 37% of all potential food and fiber crops.[8] Typically, each dollar invested in pesticides returns about $4 in crops saved.

However, this economic evaluation does not take into account the impacts of pesticide use on public health and the environment. Approximately 0.1% of applied pesticides reach target pests, leaving 99.9% of the pesticides to impact the environment.[8] Environmental effects of pesticides can be significant: Domestic animals and wildlife can be poisoned or adversely affected by pesticide exposure; beneficial natural enemies of harmful pests can be destroyed by pesticide use; heavy pesticide use can result in pesticide resistance and subsequently even heavier or more toxic pesticide use; and already limited natural resources such as soil, groundwater, and surface water can be contaminated by pesticide residues or drift.[5]

The human health effects of pesticide exposure through food are also diverse and significant. About 35% of foods purchased by American consumers have detectable levels of pesticides, and about 1%–3% of these foods have residue levels over the legal tolerance level 8. These estimates are conservative because detection methods currently detect only about one-third of the pesticides now in use in the United States. The contamination rate is undoubtedly higher for fruits and vegetables because they receive the highest levels of pesticides. One USDA study indicates that some pesticide residue remains in produce even after it is washed, peeled, and cored.[9]

Both the acute and chronic health effects of pesticide exposure are significant. Worldwide, about 26.5 million acute pesticide poisonings occur each year, resulting in about 3 million hospitalizations, approximately 220,000 fatalities, and 750,000 cases of chronic pesticide-related illness.[10] Chronic effects can adversely affect most systems of the human body. U.S. data indicate that 18% of all insecticides and about 90% of all fungicides are carcinogenic.[11] Many pesticides are also estrogenic, linked to increased breast cancer among some women in the United States. Pesticide exposure can also damage the respiratory and reproductive systems, leading to conditions like asthma and infertility.[12–14] In the United States, EPA[15] reports that 300,000 non-fatal pesticide poisonings occur each year.

The negative health effects that pesticides can have are more significant in children. Children have higher metabolic rates than adults, and their ability to detoxify and excrete toxic compounds is different. Also, because of their smaller size, children are typically exposed to higher levels of pesticides than adults. Finally, certain types of pesticides, such as carbamates and organophosphates, are more dangerous for children than adults.[10]

Given the significant environmental and public health impacts that pesticides can have, it appears desirable to limit pesticide exposure to minimize these adverse effects. However, the increasingly stringent DALs and cosmetic standards have resulted in considerable increases in pesticide use. Do the health effects of eating herbivorous insects, insect parts, or blemished produce warrant the risks and the substantial health consequences of increasing pesticide exposure to meet these standards?

Health Effects of Eating Insects/Insect Parts in Food

Even under the current stringent DAL regulatory guidelines, a few insects and mites do remain in or on produce. For instance, the DAL for apple butter is an "average of 5 whole insects or equivalents per 100 grams not counting mites, aphids, thrips, or scale insects." DALs for many other food products are similar. Many insects commonly found in foods and food products are so minute in size that they are practically impossible to eliminate. Although the numbers of insects are strictly limited by FDA regulations, some do remain and are eaten.

This, however, is not a cause for concern. In contrast with the well-documented acute and chronic negative health effects resulting from pesticide exposure, there is not one known case of human illness from ingesting insects and mites in or on foods. In addition, though some insects do carry disease or present health risks—houseflies, for example—all herbivorous insects/mites found on harvested produce are harmless to humans.

While ingesting insects or insect parts in our food may seem distasteful to many Americans, many cultures eat insects by choice. Insects are a substantial source of protein, with digestible protein content ranging from 40% to 65%,[16] and insects, shrimp, lobster, and crawfish are all arthropods; the latter three are often considered food delicacies.

Given that herbivorous insects found on produce are not a health hazard, consumers must decide whether they are willing to tolerate the presence of a few insects rather than insisting on visually "perfect" produce that requires high levels of pesticides.

Conclusions and Future Directions

If the health effects of herbivorous insects found on or in food products are not cause for concern, then the need for strict DALs might be relaxed. Relaxing cosmetic standards for some fruits and vegetables might be feasible. Approximately 10% to 20% of pesticides applied to fruits and vegetables are used only to comply with the current strict cosmetic standards established by the FDA, USDA, wholesalers, and retailers that result in blemish-free pro-duce. Rigorous cosmetic standards are probably unnecessary, since surface blemishes on fruits and vegetables generally do not affect nutritional content, storage life, or flavor. However, will the American public purchase produce that appears less than perfect?

Research on public preferences shows that 97% of Americans prefer food without pesticide residues. In addition, 50%–66% are willing to pay more for food with less pesticide residue.[17] It is estimated that in the United States, pesticide use can be reduced by about 50% without reducing crop yields. The estimated increase in the consumer's food costs would be only 0.6%.[10] This marketplace cost increase does not take into account the positive environmental and health benefits that would be realized if pesticide use were reduced. Sweden, for example, has reduced pesticide use 68% and reduced pesticide poisonings 77%.[18] The small increase in consumer cost would be more than offset by these benefits.

Therefore, given the environmental and health tradeoffs related to high cosmetic standards for produce, it appears that human health and the environment would be best protected by less stringent DALs and relaxed cosmetic standards for produce, to minimize unnecessary pesticide use and related adverse effects.

References

1. Pimentel, D.; Kirby, C.; Shroff, A. The relationship between "cosmetic standards" for foods and pesticide use. In *The Pesticide Question: Environment, Economics, and Ethics*; Pimentel, D., Lehman, H. Eds.; Chapman and Hall: New York, 1993; 85–105.
2. FDA. *The Food Defect Action Levels: Levels of Natural or Unavoidable Defects in Foods That Present No Health Hazards for Humans*; ed.; Department of Health and Human Services, Public Health Service, Food and Drug Administration, Center for Food Safety and Applied Nutrition: Washington, DC, 1995.
3. KRS Network. *U.S. Pesticide Industry Report Executive Summary*. Covington, GA, 2005; 15 pp. available at http://www.knowtify.net/2005USPestIndReptExecSum.pdf.
4. Defoliart, G.R. The human use of insects as food and as animal feed. Bull. Entomol. Soc. Am. 1986, *35*, 22–35.
5. Pimentel, D. Environmental and economic costs of the application of pesticides primarily in the United States. Environ. Dev. Sustainability 2005, *7*, 229–252.
6. Pimentel, D.; Terhune, E.; Dritschilo, W.; Gallahan, D.; Kinner, N.; Nafus, D.; Peterson, R.; Zareh, N.; Misiti, J.; Haber-Schaim, O. Pesticides, insects in food, and cosmetic standards. BioScience 1977, *27*, 178–185.
7. Environmental Protection Agency. Pesticide Market Estimates: Usage, 2000–2001; 3; available at http://www.epa.gov/opp bead1/pestsales/usage2001.htm (accessed February 11, 2010).
8. Pimentel, D.; Greiner, A. Environmental and socioeconomic costs of pesticide use. In *Techniques for Reducing Pesticide Use*; Pimentel, D., Ed.; John Wiley and Sons: Chichester, U.K., 1997; 51–78.
9. Wiles, R.; Campbell, C. *Washed, Peeled—Contaminated: Pesticide Residues in Ready-To-Eat Fruits and Vegetables*; Environmental Working Group: Washington, DC, 1994.
10. Pimentel, D.; Hart, K.A. Pesticide use: ethical, environmental, and public health implications. In *New Dimensions in Bioethics: Science, Ethics and the Formation of Public Policy*; Galston, A.; Shurr, E., Eds.; Kluwer Academic Publishers: Boston, 2001; 79–108.
11. Eastmond, D.A. Mechanisms of Carcinogenesis of the Fungicide and Rat Bladder Carcinogen o-Phenylphenol. EPA Grant No.: R826408 Project Period: May 1, 1998 through April 30, 2001; 5; available at http://cfpub.epa.gov/ncer_abstracts/index.cfm/fuseaction/display.abstractDetail/abstract/876.

12. Weisenburger, D.D. Human health effects of agrichemical use. Hum. Pathol. 1993, *24,* 571–576.
13. Schneider, E.P.; Dickert, K.J. Health costs and benefits of fungicide use in agriculture. J. Agromed. 1993, *1,* 19–37.
14. Yang, L.; Kemadjou, J.R.; Zinsmeister, C.; Bauer, M.; Legradi, J.; Muller, F.; Pankratz, M.; Jakel, J.; Strahle, U. Transcriptional profiling reveals barcode-like toxicoge-nomic responses in the zebrafish embryo. Genome Biol. 2007, *8,* R227.
15. Environmental Protection Agency. *Hired Farm Workers Health and Well-Being at Risk.* United States General Accounting Office Report to Congressional Requesters. February, 1992.
16. Gorham, J.R. Foodborne filth and human disease. J. Food Prot. 1989, *52,* 674–677.
17. Anon. Appearance of produce versus pesticide use. Chem-ecology 1991, *20* (4), 11.
18. Pesticide News. Persistence pays—lower risks from pesticides in Sweden. Pesticide News *34,* 10–11; available at http://www.pan-uk.org/pestnews/Issue/pn54/pn54p10.htm (accessed February 11, 2010).

33

Laws and Regulations: Food

Ike Jeon

Tolerance Limits for Pesticide Residues

The responsibility for ensuring that pesticide residues in foods are not present above the limits is shared by three major government agencies.[1] The Environment Protection Agency (EPA) determines the safety of pesticide products and sets tolerance levels for pesticides. The Food and Drug Administration (FDA) enforces the tolerances in all foods except meat and poultry products. The U.S. Department of Agriculture's Food Safety and Inspection Service (FSIS) regulates commercially processed egg, meat, and poultry products including combination products (e.g., stew, pizza). In addition, any products containing 2% or more poultry or poultry products, or 3% or more red meat or red meat products are also under jurisdiction of the FSIS. The pesticides of concern usually include insecticides, fungicides, herbicides, and other agricultural chemicals. Table 1 illustrates examples of tolerance levels for pesticide residues in several food categories.[2,3] These tolerance levels are extremely low, usually below parts per million, but do not represent permissible levels of contamination where it is avoidable. In addition, blending of a food (or feed) containing a substance in excess of an action level or tolerance with another food (or feed) is not permitted, and the final product from blending is unlawful, regardless of the level of the contaminant.

Regulatory Inspection and Enforcement

The FDA monitors the levels of pesticide residues in processed foods. For imported products, the FDA checks a sample of the food at entry into the United States and can stop shipments at the entry. If illegal residues are found in domestic samples, FDA can take regulatory actions, such as seizure or injunction.

The U.S. Department of Agriculture also monitors pesticide residues in food.[4] The Department was charged in 1991 with implementing a program to collect data on pesticide residues on various food commodities. The program has become a critical component of the Food Quality Protection Act of 1996 and currently is known as the Pesticide Data Program. The data on pesticides in selected commodities are used by the EPA to support its dietary risk assessment process and pesticide registration and by the FDA to refine sampling for enforcement of tolerances.

If a product is in violation of the tolerance limits, it is *adulterated* under the food law. The product may be destroyed or recalled from the market by the manufacturer or shipper. The recall may be initiated voluntarily by the manufacturer (or shipper) or at the request of the regulatory agency. The responsible agency also may seize the product on orders obtained from the Federal courts and may prosecute persons or firms responsible for the violation.

Tolerance Limits for Insect Fragments

Many food materials may contain natural but unwanted debris that cause no health hazards for humans. These debris may include insects, insect fragments, and rodent hairs and are considered unavoidable defects in foods with the current agricultural practices. In fact, the use of chemical substances to control insects, rodent, and other contaminants has little, if any, impact on natural and unavoidable defects in foods. The FDA contends that the use of pesticides does not effectively reduce the presence of these food defects. This has led the regulatory agencies to establish maximum levels of natural or unavoidable defects allowable in foods for human use. The FDA currently lists over 100 products from fruits to fish,[5] and Table 2 shows only several examples. If no defect action level exists for a product, the FDA evaluates and decides on a case-by-case basis using criteria of reported findings such as length of hairs and size of insect fragments.

The FDA sets these action levels under the premise that it is economically impractical to grow, harvest, or process raw products that are totally free of nonhazardous, naturally occurring, unavoidable defects. It is incorrect, however, to assume that because the FDA has an established defect action level for a food, the manufacturer needs only keep defects just below that level. The defect levels do not represent averages of the defects that occur in any of the products. The levels represent limits at which FDA will regard the food product as *adulterated* and, therefore, subject to enforcement action. Like pesticide residues, blending of food with a defect at or above the current defect action level with

TABLE 1 Examples of Tolerance Limits for Pesticide Residues in Human Food

		Action Level	
Substance	Commodity	(Parts per Million)	Remark
Aldrin and dieldrin	Asparagus	0.03	
	Fish	0.3	Edible portion
	Peanuts	0.05	
Chlordane	Carrots	0.1	
	Fish	0.3	Edible portion
	Lettuce	0.1	
	Poultry	0.3	Fat basis
DDT[a]	Carrots	3.0	
	Citrus fruits	0.1	
	Tomatoes	0.05	
Lindane	Beans	0.5	
	Corn	0.1	
	Milk	0.3	Fat basis
	Beef	7.0	Fat basis

Source: FDA[2] and USDA.[3]

[a] Dichlorodiphenyltrichloroethane.

another lot of the same or another food is not permitted. That practice renders the final food unlawful regardless of the defect level of the finished food.

Responsibility of Food Manufacturers

Food manufacturers are required to follow the standard manufacturing procedures under a federal regulation, known as good manufacturing practice (GMP), during food production.[6] The GMP guidelines imply that all food materials used must not exceed the tolerance limits set for pesticide residues or any other poisonous or deleterious substances. The GMP also calls for the same regulatory requirement for natural or unavoidable defects in all food materials. The food materials susceptible to contamination may be tested for compliance or relied on a supplier's guarantee or certification that they are in compliance. In addition, the GMP regulation stipulates that food manufacturers and distributors must utilize at all times quality control operations that reduce natural or unavoidable defects to the lowest level feasible with the current technology.

Potential Consumer Benefits

Through conducting a monitoring program, the federal government agencies work together to improve consumer protection. The EPA will continue to review scientific data on all pesticide products, while the FDA and U.S. Department of Agriculture will closely monitor levels of pesticide residues in all foods including both domestic and imported products. The U.S. Department of Agriculture's data for 1998 suggest that violation of the pesticide tolerance limits was very low in all raw products including fruit and vege, wheat, and milk samples. In 1993, the FDA reported that no pesticide residues were found in infant formulas, and no residues over EPA tolerances or FDA action levels were found in any of the foods that were prepared as consumers normally would prepare them at home.[7]

Acknowledgments

Contribution No. 00–231-B, Kansas Agricultural Experiment Station, Manhattan, Kansas 66506, U.S.A.

TABLE 2   Examples of Tolerance Limits for Natural or Unavoidable Defects in Foods

Product	Defect	Action Level
Sweet corn, canned	Insect larvae	2 or more 3 mm or longer larvae
Macaroni	Insect filth	225 insect fragments or more per 225 g
	Rodent filth	4.5 rodent hairs or more per 225 g
Peaches, canned and frozen	Mold/insect damage	Wormy or moldy on 3% or more fruits
	Insects	1 or more larvae and/or larval fragments whose aggregate length exceeds 5 mm in 12 one-pound cans
Peanut butter	Insect filth	30 or more insect fragments per 100 g
	Rodent filth	1 or more rodent hairs per 100 g
Popcorn	Rodent filth	1 or more rodent excreta pellets or rodent hairs in 1 or more subsamples
Tomato juice	Drosophila fly	10 or more fly eggs per 100 g
	Mold	24% of mold counts in 6 subsamples
Wheat flour	Insect filth	75 or more insect fragments per 50 g
	Rodent filth	1 or more rodent hairs per 50 g

Source: FDA.[5]

References

1. FDA. *FDA's Food and Cosmetic Regulatory Responsibilities*; U.S. Food and Drug Administration: Washington, DC, 1998; 1–5, http://vm.cfsan.fda.gov/~dms/regresp.html (accessed June 2000).
2. FDA. *Action Levels for Poisonous or Deleterious Substances in Human Food and Animal Feed*; U.S. Food and Drug Administration: Washington, DC, 1998; 1–17, http://vm.cfsan.fda.gov/~lrd/fdaact.html (accessed June 2000).
3. USDA. *Domestic Residue Book (Appendix I)*; U.S. Department of Agriculture, Food Safety and Inspection Service: Washington, DC, 1998; 1–30, http://www.fsis.usda.gov:80/OPHS/redbook1/appndx1.htm (accessed June 2000).
4. USDA. *Pesticide Data Program Annual Summary—Calendar Year of 1998*; U.S. Department of Agriculture, Agricultural Marketing Service: Washington, DC, 2000; 1–19.
5. FDA. The Food Defect Action Levels—Levels of Natural or Unavoidable Defects in Foods that Present No Health Hazards for Humans. In *FDA/CFSAN Food Defect Action Level Handbook*; U.S. Food and Drug Administration: Washington, DC, 1998; 1–36, http://vm.cfsan.fda.gov/~dms/dalbook.html (accessed June 2000).
6. CFR. Current good manufacturing practice in Manufacturing, Packing, or Holding Human Food. In *Code of Federal Regulations, Title 21, Part 110*; U.S. Government Printing Office: Washington, DC, 1999; 206–215.
7. FDA. *FDA Reports on Pesticides in Foods*; U.S. Food and Drug Administration: Washington, DC, 1993; 1–5, http://vm.cfsan.fda.gov/~lrd/pesticid.html (accessed June 2000).

34

Laws and Regulations: Pesticides

Praful Suchak

Introduction

Why Regulate Pesticides?

Chemical or biological pesticides have target specific toxicity that controls or eradicates pests falling under different groups. These products, though developed for specific usage, could have adverse effects on living beings and the environment and unchecked use can cause havoc. Regulating pesticides, therefore, would assure reasonable safety in use of these toxic substances and ensure that risks from pesticides to humans and their environment are minimized and are consistent with the benefits achieved by their use in terms of reduced losses.

Regulating pesticides at the international and national level should consider social costs in line with social benefits. Pesticides impose costs on society, such as health risks and environmental degradation, which are not borne by the user. The available policy remedies include bans on individual or classes of chemicals that prohibit the introduction of hazardous compounds into the environment, and economic instruments such as taxes, registration fees, and import duties that work to redistribute and adjust the social costs occurring for pesticide use and also provide the government with revenues that can be used to cover health costs and environmental clean-up activities.

History

The United States in 1910 introduced the Federal Pesticide Act that underwent complete metamorphosis to become the Federal Insecticide, Fungicide and Rodenticide Act (FIFRA) in 1947, which since 1970 is under the auspices of the Environmental Protection Agency.

Australia initiated pesticide legislation with one state in 1925 and by 1945 all states had their individual laws. The Industry Association brought law common to all states in 1995. By the end of 1999 about 95% of the countries in the world had adopted full/partial regulatory systems.

Early in-depth studies were not carried out on the long-term effects of: 1) repeated exposures, 2) residual toxicity, 3) accumulated toxicity, and 4) the impact on environment. With additional knowledge on the cumulative toxicity of chlorinated hydrocarbons such as DDT having come to light, the regulating authorities have started demanding the generation of additional critical toxicological data to assess short-term, long-term, and environmental toxicity of earlier registered pesticides. The European Union has already undertaken reviews of 90 molecules in the first phase by a Commission regulation dated December 11, 1992, to be completed in 12 years, and a further 148 molecules in the second phase effective March 1, 2000. The remaining substances in the European Union would be included in third phase.

Regulatory requirements for pesticides have undergone a change over the past half a century. With the advent of highly sophisticated testing equipment, more knowledge about harmful effects of the toxic chemicals has come to light. Consistent watch by environmentalists and organizations like the Pesticides Action Network (PAN), Greenpeace, Save the Planet groups, and other nongovernmental organizations has resulted in added awareness resulting in hosts of data requirements for registration/reregistration of pesticides.

Although all developed countries and most of the developing countries have their own legislation to regulate pesticides, there have been vast variations in data requirements for registrations between these countries. With globalization it has became imperative to have harmonized data requirements so that the registrant can hope for faster registration in different (pesticide consuming) countries.

Available International Guidelines

1. Agenda 21 of the United Nations Conference on Environment and Development (UNCED)
2. The *Codex Alimentarius*
3. The FAO International Code of Conduct and Prior Informed Consent (PIC)
4. WTO and International Trade with respect to pesticides
5. Agreement on Persistent Organic Pollution (POP)
6. Guidelines of Minor Donor Institutions on the purchase of pesticidess

Implementation Problems

Although FAO took the lead to harmonize data requirements in participating nations for registrations of pesticides, certain problems and practical difficulties have occurred such as

7. The original registrant, having invested huge amounts in data generation, is unable to protect the data
8. Absence of confidentiality assurance by the registering country, creating difficulties in multiple country registrations
9. Recommended uses differ from country to country, resulting in difficulties
10. Unchecked dumping of unsafe or banned pesticides in less-developed countries
11. New registrations by a company other than the original registrant by providing data generated by such a company could not be checked

Steps Undertaken

Although though PIC entry of banned pesticides could be prevented, this instrument has not been fully effective. Once it becomes fully operational legally things should improve.

With the United States implementing the Food Quality Protection Act and fixing maximum residue limits for 3000 toxic compounds, countries worldwide would need to harmonize their registrations on toxic chemicals so as to meet the residue levels in food.

The formation of the European Union with 15 member countries, OECD with 29 members, and the Technical Working Group having EPA, Canada, and Mexico, has accelerated the pace towards harmonization. However, since a vast disparity exists between developed countries on one side and developing countries on the other side, it is rather difficult to have a unified data requirement, particularly in case of risk assessment.

Acceptance of electronic data submission and dossier/ monograph submissions and joint reviews by EU would also pave the way toward harmonization and would address questions in the nondietary exposure area.

Apart from studies related to bioefficiency of the product, the toxicological studies of the toxicant, its analogues, impurities and breakdown products, residual toxicity, etc., as listed in Appendix 1 would help understanding and regulating pesticides.

Present Scenario and Probable Remedies

Substantial evidence exists that pesticides are being applied in a technically and economically inefficient manner. Many developing countries subsidize pesticides and equipment, resulting in excessive use of pesticides.

Also in developing countries, the current legal environment and enforcement capabilities have been inadequate and dysfunctional, thus exerting a significant impact on current levels of pesticide use. This is partly due to lack of resources and partly due to manipulation by vested interests.

The inadequacies of the existing regulatory framework, institutional rigidities, and a bias in favor of pesticide-dependent paths also contribute to improper use of pesticides.

A major problem confronting many countries is the absence of well-established procedural mechanisms for public involvement in the decision making process including crop protection policy. Competing interests with a stake in the process, including farmers, the pesticide industry, and policy makers responsible for food security, argue for a more liberal regulatory stance. On the other hand, environmentalists, public health workers, and consumers demand strict regulation and reduced pesticide volumes.

To be more effective, pesticide regulation and implementation should be handled by a neutral agency like the Ministry of Environment or similar organization and not the Ministry of Agriculture or other interested ministry.

Pesticide policy needs to be integrated into the broader public policy debate concerning the nations' agricultural, environmental, and health strategies.

Nevertheless, two general principles should apply. First, dispassionate analysis of the costs and benefits of pesticide use would provide a useful tool for the formulation of normal policies; and second, the broader and more inclusive the debate, the more likely it is that the outcome will serve the public rather than specific private interests.

Future Global Policy

A uniform global regulatory system needs to ensure

1. Agricultural chemical use increases agricultural output
2. Food supplies are safe from harmful toxicants/ residues
3. Reduced-risk chemical pesticides, biopesticides, and nonchemical alternatives are encouraged
4. Uniform MRLs to eliminate trade barriers
5. Uniform health-based safety standards for pesticide residues
6. Special provisions for certain groups of the population including infants and children

Appendix 1

Toxicological and Other Data Requirements for Pesticide Registration

7. Identity of active substance
    Chemical name
    Empirical and structural formula
    Molecular mass
    Method of manufacture (synthesis pathways)
    Purity
    Identity and content of isomers
    Impurity and additives

8. Physical and chemical properties
    Melting point
    Boiling point and relative density
    Vapor pressure
    Volatility
    Appearance
    Absorption spectra-molecular extinction at relevant wavelength
    Solubility in water/organic solvents
    Partitioning coefficient N-octanol/water
    Stability and hydrolysis rate in water
    Photochemical degradation on surface, in water, and in air
    Thermal stability and stability in air

9. Analytical method
    Analytical method for the determination of the pure active substance in the technical grade.
    For breakdown products and additives in plant products, soil, water, animal body fluids, and tissues.

10. Toxicological and metabolism studies
    Studies on acute toxicity—oral, percutaneous, inhalation, intraperitoneal, skin and, where appropriate, eye irritation, and skin sensitization.
    Short-term toxicity—oral, cumulative toxicity, and other routes inhalation or dermal.
    Chronic toxicity—oral, long-term toxicity, and carcinogenicity.
    Mutagenicity—reproductive toxicity-teratogenicity and multigeneration studies in mammals.
    Metabolism studies in mammals—absorption, distribution, and excretion studies, elucidation of metabolic pathways.
    Supplementary studies—neurotoxicity studies— toxic effects of metabolites from treated plants and toxic effects on livestock and pests.
    Medical data—medical surveillance on manufacturing plant personnel, clinical cases, poisoning incidents from industry and agriculture sensitization/ allergenicity observations, observations on exposure of the general population, and epidemiological studies if appropriate. Diagnosis and specific signs of poisoning, clinical tests, and prognosis of expected effects of poisoning. Proposed treatment: first aid measures, antidotes, and medical treatment. Summary of toxicological studies and conclusions, critical scientific evaluation with regard to all toxicological data, and other information concerning the active substance.

11. Residues in or on treated products, food and feed metabolism in plants and livestock
    In treated plants (distribution, metabolism, binding constituents, etc.).
    In livestock (uptake, distribution, metabolism, binding constituents, etc.).

12. Fate and behavior in the environment
Studies on aerobic and anaerobic degradation under laboratory conditions in different soil types.
Adsorption and desorption in different soil types including metabolites.
Mobility of the active ingredients in different soil types.
Behavior in water and air, rate and route of degradation.

13. Ecotoxicological studies
Effects on birds, fish, aquatic organisms such as Daphnia magna, algae, honeybees, earthworms, other nontarget macroorganisms and microorganisms.

14. Information concerning the labeling including indication of danger and safety measures.

Bibliography

1. Pesticides Policies in Developing Countries—Do They Encourage Excessive Use? In *World Bank Discussions Paper No. 238*; 1994.
2. Asian Development Bank. In *Handbook on the Use of Pesticides in Asia Pacific Region*; ADB: Manila, Philippines, 1987.
3. *Pesticide Policy Project Hannover*; Publication Serial No. 1, January 1995; No. 2, November 1995; No. 3, December 1995; No. 4, December 1996; No. 5, December 1996; No. 6, 1998; No. 7, April 1999; No. 8, April 1999.
4. EC Directives 91/414/EEC and Subsequent Directives Including 1999/80/EC.
5. Proceedings of Asia Pacific Crop Protection Conference 1997 and 1999, PMFAI: Mumbai, India.
6. *Global Pesticides Directory*, 2nd Ed.; Suchak's Consultancy Services: Mumbai, India, 1997. suchakgr@vsnl.net
7. *Pesticides News*; No. 20–47, Pesticides Action Network (PAN): London, 1993 to 1999.
8. Guidelines on the Operation of Prior Informed Consent (PIC) Rome FAO 1990, Guidance to Government in PIC Rome 1991, and Other FAO Publications.
9. U.S. EPA Pesticides Information Network. http://www.cdpr.ca.govt/docs/epa/epachim.htm.

35

Laws and Regulations: Rotterdam Convention

Barbara Dinham

Introduction

When chemical pesticides were introduced 50 years ago, little attention was paid to the environmental and health impacts. With the rapid expansion of use in the 1950s, understanding gradually increased of the consequences of exposure to certain chemicals. Wide-ranging impacts began to be identified, including: environmental persistence and effects on birds and wildlife; residues in soil, water, and air; residues in food; human poisonings from acutely toxic pesticides or long-term health impacts such as cancer; and pest resistance, often leading to dramatic crop losses.

With almost 1000 different pesticides and thousands of formulations on the market to control insects, diseases, weeds, and other pests, action was clearly needed to protect human health and the environment. International standards recommended that governments establish a registration system to authorize each formulation of a pesticide for each specific crop or other use. Concern with some pesticides led governments to ban or restrict them to a limited number of uses. Few developing countries can fully implement a registration scheme, and they are often unaware of bans imposed elsewhere. Recognizing these problems, in the early 1980s, governments, international organizations, and public interest groups began to demand action to provide a warning system to help developing countries regulate or ban the use of hazardous pesticides.

The Rotterdam Convention on Prior Informed Consent Procedure for Certain Hazardous Chemicals and Pesticides in International Trade[1] is the outcome of 15 years of activity on trade in hazardous chemicals. Adopted on 10 September 1998 in Rotterdam, the Netherlands, the Convention was signed by 73 countries[2] and by June 2001 had been ratified by 14 parties. It will become legally binding after 50 countries have ratified.

The Convention takes an important step toward protecting humans and the environment from highly toxic chemicals. For the first time, it will help monitor and control trade in dangerous substances, circulate better information about health and environmental problems of chemicals, and prevent unwanted imports of certain hazardous chemicals.

Central to the Rotterdam Convention is the system of Prior Informed Consent (PIC), a means of obtaining and disseminating decisions of importing countries about their willingness to receive shipments of certain chemicals, and ensuring compliance to these decisions by the exporter. To be included in PIC, a pesticide must be banned or severely restricted for health or environmental reasons by two countries in two different regions of the world—indicating that its adverse effects are a "global concern."

But focusing on banned or severely restricted pesticides may only touch the tip of the iceberg. Industrialized countries rely on trained and informed users able to apply good practice as safeguards: in developing countries where pesticides are often used under conditions of poverty, these measures cannot be applied. Furthermore, older—and often more hazardous—pesticides are often cheaper, making them attractive to poorer farmers. The Convention recognizes that "severely hazardous pesticide formulations" should be included in PIC if they cause health or environmental problems in developing countries or in Eastern Europe—termed "countries with economies in transition"— in the Convention.

History of PIC

A PIC system was first proposed in the early 1980s as part of the International Code of Conduct on the Distribution and Use of Pesticides, negotiated by governments in the Food and Agriculture Organization (FAO) of the UN. Some governments resisted the concept, and the Code was adopted in 1985 without any reference to PIC. But intense pressure from nongovernmental organizations (NGOs) and others won support, and the principle was accepted in 1987. It took until 1989 to establish the wording and issue a revised version of the Code.[3] That same year, the UN Environment Programme (UNEP) included an identical provision in the London Guidelines on the Exchange of Information on Chemicals in International Trade, and a voluntary system was put in place with the FAO acting as the Secretariat for pesticides and UNEP for industrial chemicals. The first pesticides were added in 1991, and by 1995, 22 pesticides and five industrial chemicals were included.

From Voluntary to Legally Binding

The issue of transforming the voluntary scheme into a legally binding international Convention was first mooted in 1992 at the United Nations Conference on Environment and Development (UNCED).[4] In November 1994, the FAO Council meeting agreed to proceed, and this was followed in May 1995 by a decision of the UNEP Governing Council. The two organizations convened an Intergovernmental Negotiating Committee (INC) to draft and agree international legally binding instrument.

Banning Exports of Banned Pesticides

An alternative to PIC strongly advocated at the time was to stop all exports of banned pesticides. However, unless action to limit the market for a banned pesticide could be taken, banning exports could encourage companies to relocate production, possibly in a country with less stringent controls. Preventing the export of banned pesticides would have no effect on severely restricted chemicals. Without a PIC system, a developing country could unwittingly allow the import of banned or severely restricted pesticides, ignorant of action taken by some governments. Many developing countries maintained that an export ban could limit their development, as alternatives were more expensive, and that import decisions should rest with them. PIC does not prevent individual countries from deciding that their banned pesticides should not be exported, but does ensure that regulatory actions are widely shared.

How the Convention Is Operated

In negotiating the text of the Rotterdam Convention, governments built on the experience gained in the voluntary PIC. As a mark of its importance, the Convention began immediately on a voluntary basis, with FAO and UNEP continuing as an interim Joint Secretariat.

Designated National Authorities

To participate in PIC, governments must appoint a Designated National Authority (DNA). By December 2000, 170 governments had appointed a DNA or a focal point. When ratifying the Convention, DNAs must be authorized to carry out administrative functions such as receiving, transmitting, and circulating information.

Notifying Regulatory Actions

When a government bans or severely restricts a pesticide, it must notify the Joint Secretariat within 90 days. Governments need to demonstrate that their action is final and that it was based on a risk evaluation, including a review of scientific data, and the Secretariat will validate the notification. Once two valid notifications from different PIC regions have been received for the same pesticide, it becomes a candidate for PIC.

Chemical Review Committee

The Convention set up a Chemical Review Committee to consider notifications, and advise the Conference of the Parties (CoP—this will replace the INC after ratification). A parallel structure operates in the voluntary phase, with an Interim Chemical Review Committee (ICRC). The Committee will review PIC notifications, and—when they meet the agreed criteria—draft a Decision Guidance Document (DGD).

Two Routes to Be "PIC-ed"

Pesticides in the voluntary PIC were carried forward, and new pesticides continue to be added. By June 2001, the process included 26 pesticides and five industrial chemicals (Table 1).

There are two routes for adding pesticides to the Convention. Under Article 5, a ban or severe restriction in any two regions triggers PIC if the action is taken for health or environmental reasons. Governments have decided that the PIC regions would be: Africa (48 countries), Latin America, and the Caribbean (33 countries), Asia (23 countries), Near East (22 countries), Europe (49 countries), North America (2 countries: Canada and US), Southwest Pacific (16 countries).

The second route is covered in Article 6, and addresses "severely hazardous pesticide formulations." This category applies only to pesticide formulations found to be causing health or environmental problems under conditions of use in developing countries, or countries with economies in transition. These pesticides may not have been banned, but—generally because of high toxicity— cause poisonings and deaths when used without extreme caution. Governments must submit evidence based on a "clear description of incidents related to the problem, including the adverse effects and the way in which the formulation was used." Nevertheless, this kind of evidence is rare, and collecting information is difficult: incidents take place far from medical facilities; many farmers are unaware of the active ingredients of pesticides they use; and it is common to use mixtures of several pesticides. The ICRC is investigating how to deal with these problems.

TABLE 1 Pesticides Covered by the Interim PIC Procedure, November 2000

Banned or severely restricted pesticides[a]

2,4,5-T (dioxin contamination)
Aldrin
Binapacryl (INC6)[a]
Captafol
Chlordane
Chlordimeform
Chlorobenzilate
DDT
Dieldrin
Dinoseb and dinoseb salts
1,2-Dibromoethane (EDB, or ethylene dibromide)
Ethylene dichloride (INC7)[a]
Ethylene oxide (INC7)[a]
Fluoroacetamide
HCH, mixed isomers
Heptachlor
Hexachlorobenzene
Lindane
Mercury compounds
    mercuric oxide
    mercurous chloride, Calomel
    other inorganic mercury compounds
    alkyl mercury compounds
    alkoxyalkyl/aryl mercury compounds
Pentachlorophenol
Toxaphene (INC6)[a]

Severely hazardous pesticide formulations[b]

Monocrotophos
Methamidophos
Phosphamidon
Methyl parathion
Parathion

[a] Indicates that these four pesticides were added to the PIC list at the 6th and 7th International Negotiating Committee meetings.
[b] Only certain formulations of these severely hazardous pesticides are included.
Source: http://www.pic.int/[5]

Import Decisions, Information, and Website

Once a pesticide is included in PIC, the DGD is circulated to all governments who must decide whether to consent to or prohibit its import. Import decisions are posted on the PIC website, and circulated biannually. Governments in exporting countries must ensure that their exporters comply. Of course, many countries are both importers and exporters and under the rules of international trade, a country cannot ban the import of a pesticide that is manufactured and used nationally.

An important tool is the PIC Circular, updated every six months by the Secretariat. Circulated in hard copy and on the website,[5] it includes new bans and severe restrictions, importing country responses, and general progress reports. For the first time, it is easy to access sound information on government regulatory actions, even if these do not meet all the full PIC criteria.

The Convention—More than PIC

Information exchange is an important principle promoted under Article 14 of the Convention. Developing countries lack resources to undertake extensive evaluations of pesticides and governments are encouraged to share scientific, technical, economic, and legal information on chemicals within the scope of the Convention, as well as other information on their regulatory actions.

Building Capacity/Improving Regulations

The process of identifying problem pesticides through PIC will be slow, and there are limitations. In some cases, for example, governments will have no easy substitute, although this may increase the incentive to seek safer and more appropriate alternatives, including Integrated Pest Management strategies.

Financial resources are needed, not only to allow the Secretariat to meet its obligations, but also to ensure that regulators in developing countries can participate in workshops and training sessions. In poorer countries, with competing demands on scarce resources, chemical regulation is not always a priority. The status of an international Convention gives PIC the attention it requires to be effective, and should help attract the necessary funds.

PIC is just one tool, although an important one, in the regulation of pesticides. With good training and additional resources, PIC can play a central role as part of capacity building initiatives to help governments improve their ability to regulate pesticides, and to look for products and strategies that reduce the dependence on hazardous chemicals.

References

1. Rotterdam Convention on the Prior Informed Consent Procedure for Certain Hazardous Chemicals and Pesticides in International Trade, UNEP and FAO, Text and Annexes, January 1999.
2. The signatory countries can be found on the PIC website: http://www.pic.int/. The Convention closed for signatures in September 1999: countries which have not signed accede to, rather than ratify, the Convention, to the same effect.
3. *International Code of Conduct on the Distribution and Use of Pesticides (Amended Version); FAO, 1989. The Code is currently being revised and updated.*
4. United Nations Conference on Environment and Development, Agenda 21, Chapter 19, Environmentally Sound Management of the Toxic Chemicals, Including Prevention of International Illegal Traffic in Toxic and Dangerous Products, UNEP, Nairobi, 1992.
5. Convention text and PIC website (http://www.pic.int/).

36

Laws and Regulations: Soil

Ian Hannam
and Ben Boer

Introduction

At a national level, soil law means a body of law to promote soil conservation enacted by a legislature, e.g., an act, decree, regulation, or other formal legal instrument that is legally enforceable. Soil law, or "soil legislation" as it may also be referred, includes those laws that have primary responsibility for soil conservation, soil and water conservation, and land rehabilitation. They are generally characterized by provisions to mitigate and manage soil erosion and soil degradation and methods to conserve soil resources. Internationally, the legal framework for the conservation of soil can include conventions, protocols, agreements, and covenants, which are expressed to be legally binding. Worldwide, soil law is managed by a variety of legal and institutional systems, which are the individual organizational and operational regimes that have the administrative authority over soil.

Why Law for Soil?

Soil bodies are effectively large ecosystems and comprise fundamental components of the earth's biodiversity. Soil is thus seen as the basis for the conservation of terrestrial biological diversity and the sustenance of all terrestrial organisms, including people. The ongoing and widespread soil degradation as a result of human use of soil provides the imperative for enactment of soil law. The ever-increasing demand for food by rapidly growing populations in many countries in the past few decades has exerted increasing environmental stress on the soil leading to widespread soil degradation.[1] The following definitions provide the context for soil law.

Soil

Soil forms an integral part of the earth's ecosystems and is situated between the earth's surface and bedrock. It is subdivided into successive horizontal layers with specific physical, chemical, and biological characteristics. From the standpoint of history of soil use, and from an ecological and environmental point of view, the concept of soil also embraces porous sedimentary rocks and other permeable materials together with the water that these contain and the reserves of underground water.[2]

Soil Degradation

Soil degradation is a loss or reduction of soil functions or soil uses. It includes aspects of physical, chemical, and biological deterioration, including loss of organic matter, decline in soil fertility, decline in structural condition, erosion, adverse changes in salinity, acidity, or alkalinity, and the effects of toxic chemicals, pollutants, or excessive flooding.[1]

Sustainable Use of Soil

The sustainable use of soils preserves the balance between the processes of soil formation and soil degradation while maintaining the ecological functions and needs of soil. In this context, the use of soil means the role of soil in the conservation of biological diversity and the maintenance of human life.[3]

International Law and Soil

International environmental law is an essential component for setting and implementing global, regional, and national policy on environment and development. There is an increasing recognition of the role of international environmental law to overcome the global problems of soil degradation, including its ability to provide a juridical basis for action by nations and the international community.[4] A number of international and regional instruments introduced in the past 10 years contain elements that can contribute to achieving sustainable use of soil. None are sufficient on their own. Some of the instruments could assist by promoting the management of some of the activities that can control soil degradation. However, this role is not readily apparent except for those that include provisions specifically directed to soil (e.g., see Article IV "Soil"— 1968 *African Convention on the Conservation of Nature and Natural Resources,* final revision text adopted by the African Union Assembly on July 11, 2003).

Declarations

A number of nonbinding declarations and charters draw attention to the fact that soil degradation and desertification are reaching alarming proportions and seriously endangering human survival. They call on states to cooperate and develop the tools to conserve soils. Key declarations relevant to soil include the 1972 *Stockholm Declaration on the Human Environment,* the 1981 FAO World Soil Charter, the 1982 *World Charter for Nature,* the 1982 *Nairobi Declaration,* the 1992 *Rio Declaration on Environment and Development,* and the 2002 *Johannesburg Declaration on Sustainable Development.* Also of relevance is the Programme for the Development and Periodic Review of Environmental Law for the First Decade of the 21st Century, known as the Montevideo Programme; this program includes provisions to improve the conservation, rehabilitation, and sustainable use of soils.[5]

International Conventions, Covenants, Treaties, and Agreements

Many multilateral agreements include provisions that could be used to promote sustainable use of soil, but the provisions are generally tangential to the needs of soil as such. Key global instruments relevant to soil include the 1992 *Convention on Biological Diversity,* the 1992 *United Nations Framework Convention on Climate Change* and the 1997 *Kyoto Protocol,* and the 1994 *United Nations Convention to Combat Desertification.* Relevant regional instruments include the 1968 *African Convention on the Conservation of Nature and Natural Resources* (Revised July 2003), the 1985 *ASEAN Agreement on the Conservation of Nature and Natural Resources,* the 1986 *Convention for the Protection of the Natural Resources and Environment of the South Pacific Region,* the 1986 *European Community Council Directive,* the 1995 *Convention Concerning the Protection of the European Alps,* and the 1998 *Protocol for the Implementation of the Alpine Convention of 1991 in the Area of Soil Protection.*[6]

National Soil Law

Legislation has been used for some 60 years in many countries to control soil degradation problems and to manage soil. A worldwide examination of national legal and institutional frameworks indicates that most countries approach the management of soil in a fragmented manner. The term "soil law" also covers those situations where comprehensive provisions for soil protection and management have been integrated in legislation that protects other aspects of the environment, such as forests, water, biodiversity, and desertification. In general, soil law thus provides for farm planning, implementation of soil erosion control measures, establishing community groups, planning catchment schemes, and compliance and enforcement. Some jurisdictions, such as the United Kingdom, have multiple soil legislation mechanisms that cover a broad range of functions including soil planning, access to sensitive land types, organic farming practices, nitrate sensitive areas, and soil restoration. On the other hand, federally organized countries often have a system where each state or province has its own soil legislation and supportive legal mechanisms. Hybrid situations also exist, such as in the People's Republic of China, which has enacted the *Water and Soil Conservation Law 1991* and the *Desertification Law 2002* at a national level, but causes them to be implemented through a comprehensive provincial system of law and regulations.

There is a wide variety of types of legal mechanisms used to protect and manage soil, including acts, decrees, resolutions, ordinances, codes, regulations, circulars, decisions, orders, and bylaws. Whereas these are generally appropriate, many need to be applied in more inventive ways to effectively manage the soil in an ecosystem context.[3]

Effectiveness of Soil Law

The effectiveness of international and national soil law is generally dependent on two matters: first, the capacity of a legal and institutional framework to manage soil—which is measured by the ability of a legislative and institutional system to achieve sustainable use of soil—and second, by the number and type of essential legal and institutional elements present in a soil statute in a format that enables soil degradation issues to be identified. These need to be backed by the legal, administrative, and technical capability in the particular instrument as a basis of some form of effective action. Capacity is also represented in the form of legal rights, the type of legal mechanisms, and importantly, the number and comprehensiveness of the essential elements and their functional capabilities. Legal and institutional "elements" for soil are the basic, essential components of a legal and institutional system. An individual law can include a number of legal mechanisms in a well thought-out structure that gives an organization the power it needs, through its executive and administrative structure, to address soil degradation. It is also possible that the necessary elements may be distributed among a number of individual laws within a comprehensive national legal and institutional system.[7]

Most key soil management issues are multifactorial (i.e., many include a sociological, a legal, and a scientific component), so it is obvious that generally more than one piece of environmental legislation (along with detailed regulations) and many types of legal and institutional elements will be needed to effectively manage soil degradation issues.[7] Legal and institutional elements can be used to assist in the evaluation of an existing law or legal instrument to determine its capacity to meet certain prescribed standards of performance for the sustainable use of soil. They can also be used to guide the reform of an existing soil law or to develop new legislation for the sustainable use of soil. The manner and degree in which an "essential element" is applied will vary according to the particular type of legal mechanism concerned and its expected role in a particular jurisdiction. For example, an international legal instrument may include a provision for dispute resolution, but the actual implementation of this provision between states might not rely on, or be influenced by, the existence of similar provisions within a law of either of the disputing states.[7]

IUCN Commission on Environmental Law

The Commission on Environmental Law of IUCN (The World Conservation Union) has carried out extensive investigation into the options for a new international instrument focusing on soil. The commission has also identified a variety of ways available for states to approach the task of a detailed legal and institutional analysis and the design of appropriate legal and institutional systems that provides for the effective management of soil. Arising from this work, in which the authors have been centrally involved, two principal strategies can be considered for the development of legal and institutional arrangements for soil. These are:

- A nonregulatory strategy which is characterized by elements for education, participatory approaches, soil management, and incentive schemes
- A regulatory approach that is characterized by statutory soil use plans that prescribe legal limits and targets of soil and land use, issue of licenses or permits to control soil use, and the use of restraining orders and prosecution for failure to follow prescribed standards of sustainable soil use.

These strategies can be approached on a short-term time frame for implementation or a longer-term time frame, which involves substantial reform of existing laws, policies, and institutional and sectoral change.[7]

Conclusions

Soil law in the past has been neglected at the international level and, in many of the world's regions, at the domestic level. However, the growing recognition of soil degradation as a major international environmental issue in the context of the conservation of biological diversity is gradually being addressed, and this is starting to change attitudes toward the benefits of improved international and national legal and institutions for soil.[8] Soil bodies represent complex terrestrial ecosystems. They require careful management of their ecological characteristics through the medium of soil law at a national and international level. This approach is essential for the long-term sustainable use of soil and to meet the food production requirements of the expanding human population of the world, as well as to meet the needs of all flora and fauna that depend on the soil for sustenance.

References

1. Bridges, E.M., Hannam, I.D., Oldeman, L.R., Penning deVries, F., Scherr, S.J., Sombatpanit, S., Eds.; *Response to Land Degradation;* Science Publishers Inc.: Enfield, NH, 2001.
2. Council of Europe European conservation strategy. Recommendations for the 6th European Ministerial Conference on the Environment; Council of Europe: Strasbourg, 1990.
3. Hannam, I.D.; Boer, B.W. *Legal and Institutional Frameworks for Sustainable Soils;* The World Conservation Union: Gland, U.K., 2002.
4. Khan, R. International law of land degradation. In *International Studies. 30:3;* Sage Publications: New Delhi, India, 1993.
5. UNEP. The Montevideo Programme III—The Programme for the Development and Periodic Review of Environmental Law for the First Decade of the 21st Century; 2001 Decision 21/23 of the Governing Council of UNEP, UNEP: Nairobi, February 2001.

6. Sands, P. *Principles of International Environmental Law;* Cambridge University Press: Cambridge, 2003.
7. Boer, B.W.; Hannam, I.D. Legal aspects of sustainable soils: International and National. Rev. Eur. Commun. Int. Environ. Law **2003**, *12* (2), 149–163.
8. WSSD (World Summit on Environment and Development). *A Framework for Action on Agriculture;* WEHAB Working Group: New York, 2002.

37

LEED-EB: Leadership in Energy and Environmental Design for Existing Buildings

Rusty T. Hodapp

Introduction

Existing buildings comprise a significant proportion of the total building stock in the United States and building operations consume large amounts of resources (energy, water, building materials, land, etc.), while generating great amounts of waste. For example, in the United States, commercial and industrial buildings alone are estimated to be responsible for the following[1]:

- 38.9% of primary energy use
- 38% of all CO_2 emissions
- 72% of all electricity use
- 13.6% of all potable water use
- 170 million tons of construction and demolition debris
- Using 40% of raw materials globally

Furthermore, because the average person spends 90% of their time indoors, the quality of a building's interior environment impacts virtually everyone. This suggests a very personal interest in better buildings in addition to the national implications of large-scale resource consumption. Issues such as these have driven government, corporate, and personal interest in sustainability and "green" topics. Applied to the building industry, this interest is forcefully seen in the rise of green building certification programs and, in particular, the Leadership in Energy and Environmental Design (LEED) Green Building Rating System™ of the United States Green Building Council (USGBC). Intended to guide the development and verify the performance of green buildings, LEED rating systems have become well accepted as a national standard. Consisting of a number of products, the LEED Green Building Rating System largely focuses on design and construction of new buildings. One rating system, however, known as LEED for Existing Buildings: Operations and Maintenance (LEED-EB O&M) is oriented at the operation, maintenance, and management of existing buildings. LEED-EB O&M is of particular interest to facility managers, energy managers, owners, or others interested in reducing operating costs, improving indoor environmental quality, and minimizing the environmental impact of buildings as a growing body of case study and research evidence suggests that these outcomes are linked to green building practices.

This entry presents an overview of the LEED-EB O&M Green Building Rating System, its benefits and distinctions from other LEED rating systems, how it is organized and implemented, and the value of high-performance green buildings.

Introduction to LEED

LEED stands for the USGBC's family of standards for rating "green buildings." The LEED Green Building Rating System is USGBC's effort to provide a national standard to define a green building. Used as guideline for design, construction, operation, and maintenance and with third-party certification, LEED provides a consistent, credible means of developing and operating high-performance, environmentally sustainable buildings.

United States Green Building Council

Formed in 1993, the USGBC has become perhaps the most prominent "green building" organization in America. With more than 18,000 member organizations, a network of 78 local affiliates, and more than 140,000 LEED Professional Credential holders, the non-profit organization works through leaders in all sectors of the building industry to advance buildings that are environmentally responsible, profitable, and healthy places to live and work. Driving its mission to transform the building marketplace to sustainability is the Council's LEED Green Building Rating System and related training and professional accreditation programs. USGBC also supports education, research, and advocacy programs as well as strategic alliances with key industry, research organizations, and federal, state, and local government agencies to transform the building market.[2]

LEED Green Building Rating System

The LEED Green Building Rating System is a voluntary, consensus-based, market-driven building rating system. LEED assesses sustainability from a whole-building perspective by evaluating five key areas of a building's performance in terms of economic, environmental, and human health impact: sustainable site development, water savings, energy efficiency, materials selection, and indoor environmental quality. Buildings are awarded different levels of certification (Certified, Silver, Gold, Platinum) based upon the amounts of credits satisfied and points earned. These credits are performance oriented and intended to address specific impacts inherent in the design, construction, operation, and maintenance of buildings.

The initial LEED rating system (referred to as LEED Version 1.0) was released in August 1998. LEED 1.0 was extensively modified and released as Version 2.0 in March 2000 with LEED Version 2.1 following in 2002 and Version 2.2 in 2005. LEED has continued to evolve, undertaking new initiatives and expanding into a family of products.

As of April 2009, the portfolio of LEED rating systems consists of several products targeting specific sectors of the buildings market:

- LEED 2009 for New Construction and Major Renovation
- LEED 2009 for Commercial Interiors
- LEED 2009 for Core and Shell Development
- LEED 2009 for Existing Buildings: Operations and Maintenance
- LEED 2009 for Schools
- LEE 2009 for Retail
- LEED for Healthcare
- LEED for Homes
- LEED 2009 for Neighborhood Development

With new products, technologies, and design innovations coming into the green building marketplace daily, the Rating Systems and Reference Guides will continue to evolve as necessary to stay current and relevant.[3]

Benefits of Green Building Certification

LEED provides a guidebook for the design, construction, operation, and maintenance of green buildings, the general benefits of which will be described in more detail later in this entry. The LEED rating system is flexible in order to provide owners and design teams the ability to accommodate circumstances or goals specific to their project. The rigorous and independent certification process provides firm and compelling proof that the building has achieved the sustainable goals established for it and is performing as intended. The credible assurance that a building is in fact green can be valuable to owners, occupants, investors, and other key stakeholders in the industry as well as the public at large.

LEED and Existing Buildings

With the exception of LEED-EB O&M, the LEED family of products is intended to address the design and construction phases of buildings. The primary users of these products are architects, engineers, construction contractors, and building owners. A building designed and constructed to LEED standards has verifiably incorporated green or sustainable features and, therefore, should perform better in the key impact areas (economic returns, environmental impact, and occupant health and comfort) than a typical building built to basic code standards. Addressing the design and construction phase of a building's life is extremely important because it is in these phases that many irreversible decisions with long-term impacts on the building's performance are made.

Figure 1 depicts graphically what has become well established regarding the life-cycle cost of buildings: 1) the majority (75% or more) of total life-cycle costs occur after construction (i.e., during the O&M phase); and 2) many of the decisions that drive long-term cost occur during programming and design.[4]

However, the fact is that new buildings represent a very small percentage of the total commercial building stock in the United States. According to some sources, new construction amounts to 2% of the total stock of buildings in any one year.[5] No doubt it is safe to assume that the majority of these existing buildings were not designed or are being operated and maintained to green standards. In order to improve the building sector's performance in terms of sustainability, clearly the existing stock must be addressed in addition to new construction. Similarly, since the post-construction phase of a building's life cycle contributes disproportionately to its total cost, resource consumption, and impact on users, standards for operations and maintenance are necessary to maximize the benefits of green practices in the building sector. This is exactly the focus of the LEED-EB O&M Rating System.

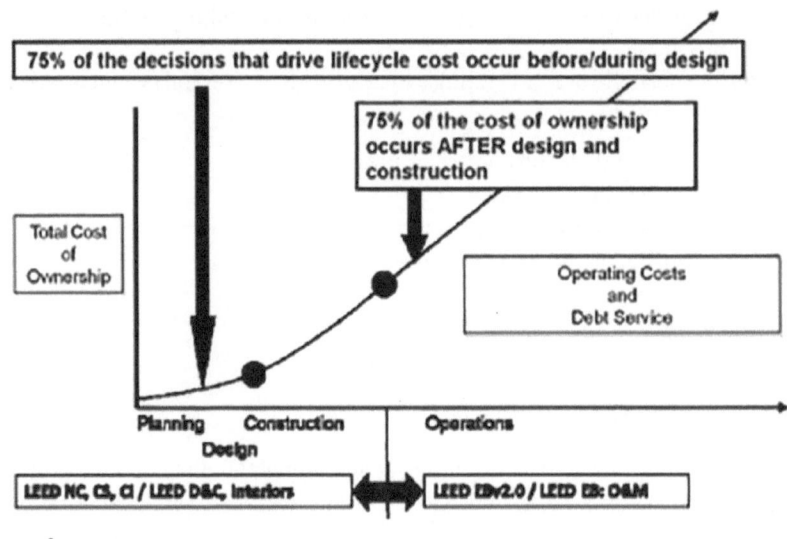

FIGURE 1   Building life-cycle cost curve.

LEED for Existing Buildings: Operations and Maintenance

The LEED-EB O&M Rating System is a set of performance standards for sustainable operations and maintenance of existing buildings of various types and all sizes. It is intended to advance high-performance, healthy, durable, affordable, and environmentally sound practices in existing buildings. LEED-EB O&M provides an entry point into the LEED certification process for the existing building stock. It can be used for buildings new to LEED certification as well as those previously certified under LEED-NC. The USGBC began developing LEED-EB in 2000 and it was tested in a pilot phase involving more than 100 buildings in 2002. The final version (Version 2.0) was released in October 2004. The current version, LEED-EB O&M, was released in April 2009 under the suite of 2009 LEED rating systems and has been further updated in April 2010. The introduction to the LEED 2009 for Existing Buildings: Operations and Maintenance Green Building Rating System states "LEED for Existing Buildings Operations and Maintenance encourages owners and operators of existing buildings to implement sustainable practices and reduce the environmental impacts of their buildings over their functional life cycles." To achieve this, LEED-EB O&M addresses exterior building site maintenance, water and energy use, environmentally preferred products and practices for cleaning and alternations, waste stream management, and ongoing indoor environmental quality.[6]

Issues Addressed by LEED-EB O&M

LEED-EB O&M addresses all the key facets of building operations and maintenance that impact total cost of ownership, the environment, and building occupants. Some examples include the following:

- Energy use
- Water use
- Building operations and maintenance
- Building systems (e.g., mechanical, electrical, plumbing) performance
- Maintenance of building exterior and site
- Ventilation and indoor air quality
- Lighting quality

- Thermal comfort of spaces
- Green cleaning
- Recycling programs
- Green product purchasing programs
- Management of indoor pollutants and toxic substances
- Systems upgrades[7]

LEED-EB O&M seeks to address sustainability on an ongoing basis. This largely falls under the scope of those involved in managing and operating buildings, and clearly, their involvement and expertise are necessary to successfully certify a building under LEED-EB O&M. To the extent the benefits of green buildings and popularity of standards like LEED continue growing, market forces will create new opportunities for facility managers, energy managers, etc., demand for their services, and highlight the overall value of their contributions.

Key Distinctions between LEED-EB O&M and Other LEED Products

Although sharing many common features in terms of structure and process with other LEED products, LEED-EB O&M is fundamentally distinct in three key ways. LEED-NC (and the other new building-oriented products) is essentially a onetime event whereas LEED-EB O&M represents an ongoing process. Second, with LEED-NC, the green building process ends after the design and construction phase. For LEED-EB O&M, the green building phase is a continuous process that deals with ongoing operations, maintenance, and upgrades of a building over its life cycle. Buildings certified under LEED-EB O&M require recertification at least once every 5 years. Finally, given their focus on different phases of a building's life cycle, LEED-NC is primarily a capital budget event while LEED-EB O&M deals with operating budgets.[8]

Overview of LEED-EB O&M

In the same manner as all LEED products, the LEED-EB O&M Rating System is based on evaluations of a building in seven categories:

1. Sustainable sites
2. Water efficiency
3. Energy and atmosphere
4. Materials and resources
5. Indoor environmental quality
6. Innovation (in this case, in operations)
7. Regional priority

Minimum Program Requirements

All projects must meet certain minimum program requirements (MPRs) to be eligible for certification under the LEED 2009 rating systems. MPRs define the minimum characteristics that a project must possess in order to be certified and are intended to: 1) provide clear guidance to users; 2) protect the integrity of LEED program; and 3) reduce challenges during the certification process. The LEED 2009 MPRs for EB O&M are as follows:

1. Must comply with environmental laws
2. Must be a complete, permanent building or space
3. Must use a reasonable site boundary
4. Must comply with minimum floor area requirements

5. Must comply with minimum occupancy rates
6. Must commit to sharing whole-building energy and water usage data
7. Must comply with a minimum building area-to-site area ratio

The ongoing performance data from buildings required as part of the certification will be compiled and used to establish benchmarks for building performance and provide operators an idea of how their building compares on water and energy use. To further its commitment to improving building performance, the USGBC launched the Building Performance Initiative (BPI) in August 2009 to complement the MPR for ongoing performance data. The BPI will make the data collected available to building owners for analysis and feedback.

Prerequisites

Also consistent with other LEED rating systems, LEED-EB O&M requires every project to meet certain prerequisites in order to be considered for certification (see the list of prerequisites by category in Table 1). All prerequisites must be satisfied for a project to be eligible for certification.

The prerequisites include such items as minimum levels of water and energy efficiency, building commissioning, no CFC refrigerants, no-smoking policy, and other basic elements of a high-performance, green building operation. A key prerequisite involves a minimum performance period for the building. LEED-EB O&M requires buildings to be in operation for a minimum of 12 continuous months before certifying (3 months for all prerequisites and credits except Energy and Atmosphere Prerequisite 2 and Credit 1, which require a minimum of 12 months).

Credits and Points

Buildings achieve certification under all LEED products by accumulating a certain number of credit points. Points can be obtained in any combination within and among the credits and categories (see Table 2 for the credit and point breakdown for LEED-EB O&M). All LEED rating systems have 100 base points, and up to 10 bonus points can be earned through Innovation and Regional Priority credits.

TABLE 1   LEED 2009 for EB O&M Prerequisites

Category	Prerequisites
Sustainable sites	0
Water efficiency	1
Energy and atmosphere	3
Materials and resources	2
Indoor environmental quality	3
Total	9

TABLE 2   LEED-EB O&M Credits and Points

Category	Credits	Points
Sustainable sites	9	26
Water efficiency	4	14
Energy and atmosphere	9	35
Materials and resources	9	10
Indoor environmental quality	15	15
Innovation in operations	3	6
Regional priority	1	4
Total	50	110

Each LEED rating system uses the same format for prerequisites and credits. The sections include the following:

- *Intent*—describes the main goal of the prerequisite or credit.
- *Requirements*—specifies the criteria needed to satisfy the prerequisite or credit.
- *Submittals*—specifies the documentation required to demonstrate compliance with the prerequisite or credit.
- *Potential Technologies and Strategies*—identifies means and methods that project teams may consider to achieve the prerequisite or credit.

Certification Levels

LEED rating systems allow buildings to achieve various levels of certification based on points achieved (see Table 3 for the certification levels for LEED-EB O&M).

Registration Process

With the launch of LEED Version 3 in 2009, USGBC implemented a new certification model. LEED v3 consists of three components:

- LEED 2009 rating systems.
- An upgrade to LEED Online to make it faster and easier to use.
- New building certification model—an expanded infrastructure based on ISO standards, administered by the Green Building Certification Institute (GBCI) for improved capacity, speed, and performance.

LEED Online is the primary resource for managing the LEED documentation process. It allows project teams to manage project details, complete documentation requirements for LEED credits and prerequisites, upload supporting files, submit applications for review, receive reviewer feedback, and ultimately earn LEED certification. The GBCI is an independent, third-party organization that has assumed administration of LEED certification for all commercial and institutional projects registered under any of the LEED rating systems.

The process of certifying a building under any of the LEED rating systems is essentially the same—the project is first registered with GBCI using LEED Online. Once a project is registered, access to software tools, supplemental resources, sample documentation, credit interpretation rulings, and other essential information is provided. For LEED-EB O&M, the initial application (application for standard review) must be submitted within 60 calendar days of the performance periods used. The application has to include complete documentation for all prerequisites and enough points for certification. GBCI reviews the application and designates each credit and prerequisite as anticipated pending or denied. This preliminary standard review is targeted (but not guaranteed) for completion within 25 business days of receipt of the application. Within 25 days of receiving GBCI's preliminary standard review, the owner may submit a response including any revised documentation. GBCI will then review and return comments for all credits and prerequisites in response to the preliminary Standard Review and

TABLE 3   LEED-EB O&M Certification Levels

Certification Level	Points Required
Certified	40–49
Silver	50–59
Gold	60–79
Platinum	80+

designate each as awarded or denied. This final Standard Review is targeted for completion within 15 business days of receipt of the completed application. The owner then accepts or appeals the final review. Following acceptance of the final certification review, LEED projects:

1. Will receive a formal certificate of recognition
2. Will receive information on how to order plaque and certificates, photo submissions, and marketing
3. May be included (at the owner's discretion) in the online LEED Project Directory of registered and certified projects
4. May be included (along with photos and other documentation) in the U.S. Department of Energy High Performance Buildings Database[9]

Implementation Process

From a practical standpoint, the process of implementing LEED-EB O&M should generally involve the following steps:

1. Become familiar with the LEED-EB O&M Rating System
2. Gain the support of key decision makers and stakeholders
3. Form a project team
4. Conduct a preliminary building audit and identify corrective actions required to meet prerequisites and/ or opportunities
5. Establish project goals related to target certification level, credits to be pursued, and budget
6. Register the project
7. Create and adopt policies and procedures, implement upgrades, make operational changes, etc., in accordance with the project goals
8. Track performance
9. Assemble and submit required documentation (preliminary and any required corrections or resubmittals)
10. Achieve certification

The minimum performance period for initial certifications under LEED-EB O&M is 12 months. During this period, actual operational performance must be tracked and reported. The performance-tracking period can be as long as 2 years depending upon the project goals and/or implementation strategy.

The USGBC provides project teams with numerous resources including the LEED Online system for managing and preparing the certification application, credit templates that define supporting documentation needed and compliance calculations, credit interpretation rulings that can help answer questions on credits and implementation strategies, and the LEED Reference Guides.[10,11]

Benefits of Green Buildings

The premise inherent in the LEED rating systems is that "green" buildings provide superior value to owners, occupants, and other stakeholders. Typically, the value proposition construct for sustainability, green buildings, etc., is the well-known triple bottom line of economic returns, environmental impact, and social benefit. Green buildings, also known (perhaps more accurately) as high-performance buildings, are premised as providing superior economic returns, reduced environmental impact, and enhanced social benefits. Such buildings in theory:

- Were properly built and/or are well operated and maintained
- Use resources (e.g., energy, water, building materials, O&M supplies) more efficiently
- Provide a safer, more comfortable, and productive working environment for occupants

In fact, there is a robust and growing body of evidence in research and case study that supports these claims. The following examples are illustrative.

A report commissioned by California's Sustainable Building Task Force found that energy savings alone exceeded the average cost increase associated with 33 different LEED buildings studied. When adding the life-cycle cost benefits of water savings, reduced emissions, operations and maintenance efficiencies, and improved occupant productivity and health, the 20 years net present value of the financial benefits of green buildings exceeded the implementation costs by as much as 10–15 times.[12] Case studies on commissioning alone show that construction and operating costs can be reduced from 1 to 70 times the initial cost of commissioning.[13] Improved thermal comfort, reduced indoor pollutants, enhanced ventilation rates, and other characteristics of green buildings have been found to have positive impacts on occupant productivity, student test scores, absenteeism, and incidences of various sicknesses.[14]

Other benefits continue to be demonstrated in case studies and research, including the following:

- Increased building value
- Risk mitigation
- Employee loyalty and recruitment
- Brand image and public relations
- Environmental stewardship[15]

Energy Efficiency Potential of Green Buildings

In LEED 2009, points are allocated among credits based on the potential environmental impacts and human benefits of each credit. As a result, the allocation of points significantly changed in comparison to previous versions of the LEED rating systems. These changes increased the relative importance of reducing energy consumption and building-related greenhouse gas emissions. Reflecting this, one credit in LEED 2009 EB O&M comprises the largest potential amount of points—Energy and Atmosphere Credit 1 Optimize Energy Efficiency Performance provides the opportunity for 18 possible points. Furthermore, with its emphasis (and associated requirements) on demonstrated performance, LEED-EB O&M presents tremendous potential to reduce energy consumption throughout the commercial building sector. When considering the fact that existing buildings comprise some 95% of the commercial building stock in the United States, the magnitude of potential reduction is immense. To put this potential in context, consider U.S. Energy Information Administration (EIA) projections of the impact of energy efficiency on per capita commercial energy consumption. In their *Annual Energy Outlook 2010,* EIA estimates per capita commercial energy consumption in 2035 could be decreased by as much as 17.5% depending upon the degree to which technology-based efficiency improvements are deployed throughout the sector.[16] McKinsey and Company estimates that, by 2020, the United States could reduce annual energy consumption by 9.1 quadrillion British thermal units (BTUs) (23%) of end-use energy (18.4 quadrillion BTUs in primary energy) from a business-as-usual baseline by deploying an array of energy efficiency measures with the commercial sector accounting for 25% of this potential. At full potential, the projected efficiency improvements could reduce greenhouse gas emissions by as much as 1.1 gigatons of CO_2 by 2020 and serve as a bridge to low-carbon energy sources.[17] Similarly, the Electric Power Research Institute (EPRI) estimates that the combination of energy efficiency and demand response programs has the potential to reduce summer peak electric demand in the United States by 157 GW to 218 GW (14% to 20%) by 2030.[18] Finally, consider the potential national impact of LEED energy savings presented in the Green Building Market and Impact Report 2009. This report projects that given the acceleration of LEED adoption, energy savings in the United States could reach 1.75 quadrillion BTUs by 2020 and 3.9 quadrillion BTUs by 2035 (8.3% and 17.3%, respectively, of national annual commercial building energy use). A best-case scenario could see those savings rise to 22.3% by 2030.[19]

Conclusion

The USGBC's LEED Green Building Rating System has become a well-recognized standard for guiding the development of green, high-performance buildings and for verifying that established green building goals have been accomplished. The LEED family of rating systems is focused on the design and construction process for new buildings. However, one rating system—LEED-EB O&M— focuses on operations and maintenance of existing buildings. LEED-EB O&M is of particular interest to facility managers, energy managers, and other professionals involved in building operation and management. LEED-EB O&M provides a guidebook for those interested in "greening" their existing building stock. Implementing these green processes and practices can be an effective means of reducing a building's life-cycle costs, reducing its environmental impact, and improving occupant health and productivity.

References

1. United States Green Building Council. Green Building Facts. Available at http://www.usgbc.org/ (accessed August 2007).
2. United States Green Building Council. About USGBC. Available at http://www.usgbc.org/ (accessed August 2007).
3. United States Green Building Council. *LEED Reference Guide for Green Building Design and Construction 2009 Edition;* USGBC: Washington, 2009; xii.
4. National Research Council. *Investments in Federal Facilities, Asset Management Strategies for the 21st Century;* The National Academies Press: Washington, 2004; 27.
5. Architecture2030. Available at http://architecture2030.org/the_ solution/buildings_solution_how (accessed November 2010).
6. United States Green Building Council. *LEED 2009for Existing Buildings: Operations and Maintenance Rating System (Updated April 2010);* USGBC: Washington, 2010; xvi.
7. Opitz, M. What LEED-EB Is and Why to Use It, available at http://www.fmlink.com/ProfResources/Sustainability/Articles/article.cgi?USBGC:200604-01.html (accessed August 2007).
8. United States Green Building Council. LEED-EB Presentation. Available at http://www.usgbc.org/ (accessed August 2007).
9. Green Building Certification. GBCI LEED Certification Manual. Available at http://www.gbci.org/main-nav/building-certification/leed-certification.aspx (accessed July 2010).
10. Opitz, M. Starting and Managing Your LEED-EB Project, available at http://www.fmlink.com/ProfResources/Sustainability/Articles/article.cgi?USBGC:200609-20.html (accessed September 2007).
11. United States Green Building Council. *LEED 2009 for Existing Buildings: Operations and Maintenance Rating System (Updated April 2010);* USGBC: Washington, 2010; xvi–xxiv.
12. Kats, G.; Alevantis L.; Berman, A.; Mills, E.; Perlman, J. The Costs and Financial Benefits of Green Buildings: A Report to California's Sustainable Building Task Force, available at http://www.cap-e.com/ewebeditpro/items/O59F3259.pdf (accessed September 2006).
13. ASHRAE. *ASHRAE GreenGuide;* Elsevier: Burlington, 2006; 14.
14. Callan, D. Green Building Report: Studies Relate IAQ and Productivity. Build. Operating Manage. **2006,** *52* (11), 6–8.
15. Yudelson, J. *Marketing Green Buildings, Guide for Engineering, Construction and Architecture;* The Fairmont Press: Lilburn, 2006; 5–7.
16. U.S. Energy Information Administration, U.S. Department of Energy. *Annual Energy Outlook 2010,* 2010, 59.
17. Choi Granade, H.; Creyts, J.; Derkach, A.; Farese, P.; Nyquist, S.; Ostrowski, K. Unlocking Energy Efficiency in the U.S. Economy, available at http://www.mckinsey.com/USenergyefficiency (accessed July 2010).

18. Faruqui, A.; Hledik, S.; Rohmund, I.; Sergici, S.; Siddiqui, O.; Smith, K.; Wikler, G.; Yoshida, S. *Assessment of Achievable Potential from Energy Efficiency and Demand Response Programs in the U.S. (2010–2030)*. EPRI: Palo Alto, CA, 2009; xi.
19. Watson, R. Green Building Market and Impact Report, 2009, available at http://www.green-biz.com/business/research/report/2009/11/05/green-building-market-and-impact-report-2009 (accessed July 2010).

38

LEED-NC: Leadership in Energy and Environmental Design for New Construction

Stephen A. Roosa

Introduction

Land development practices have yielded adverse environmental consequences, urban dislocation, and changes in urban infrastructure. Urban development in particular has long been associated with reduced environmental quality and environmental degradation.[2] The rate at which undeveloped land is being consumed for new structures—and the growing appetite of those structures for energy and environmental resources—has contributed to ecosystem disruption and has fostered impetus to rethink how buildings are sited and constructed. While urban developmental patterns have been associated with environmental disruptions at the local and regional scales, the scientific assessments of global impacts have yielded mixed results. In part as a reaction to U.S. development patterns that have traditionally fostered suburbanization and subsidized automobile-biased transportation infrastructure, design alternatives for structures with environmentally friendly and energy efficient attributes have become available.

According to the United Nations Commission on Sustainable Development, "air and water pollution in urban areas are associated with excess morbidity and mortality ... Environmental pollution as a result of energy production, transportation, industry or lifestyle choices adversely affects health. This would include such factors as ambient and indoor air pollution, water pollution, inadequate waste management, noise, pesticides and radiation."[3] It has been demonstrated that a relationship exists between the rates at which certain types of energy policies are adopted at the local level and select indicators of local sustainability.[4] As more urban policies focus on the built environment, buildings continue to be the primary building blocks of urban infrastructure. If buildings can be constructed in a manner that is less environmentally damaging and more energy efficient, then there is greater justification to label them as "green" buildings.

The concept of sustainability has evolved from considerations of land development, population growth, fossil fuel usage, pollution, global warming, availability of water supplies, and the rates of resource use.[5] Thankfully, a vocabulary of technologies and methodologies began to develop in the 1970s and 1980s that responded to such concerns. Driven by ever increasing energy costs, energy engineers began to apply innovative solutions, such as use of alternative energy, more efficient lighting systems and improved electrical motors. Controls engineers developed highly sophisticated digital control systems for heating, ventilating and air conditioning systems. With growing concerns about product safety and liability issues regarding the chemical composition of materials, manufacturers began to mitigate the potential adverse impacts of these materials upon their consumers. Resource availability and waste reduction became issues that began to influence product design. In the span of only 25 years, local governments made curbside recycling programs in larger U.S. cities nearly ubiquitous. Terms and phrases such as "mixed use planning," "brownfield redevelopment," "alternative energy," "micro-climate," "systems approach," "urban heat island effect," "energy assessments," "measurement and verification," and "carrying capacity" created the basis for a new vocabulary which identifies potential solutions. All of these concerns evolved prior to the 1992 U.N. Conference on the Environment and Development, which resulted in the Rio Agenda 21 and clarified the concept sustainability.

In regard to the built environment, architectural designers renewed their emphasis on fundamental design issues, including site orientation, day lighting, shading, landscaping, and more thermally cohesive building shells. Notions of "sick building syndrome" and illnesses like Legionnaires' disease, asthma and asbestosis, jolted architects and engineers into re-establishing the importance of the indoor environmental conditions in general and indoor air quality (IAQ) in particular when designing their buildings.

The decisions as to what sort of buildings to construct and what construction standards to apply are typically made locally. Those in the position to influence decisions in regard to the physical form of a proposed structure include the builder, developer, contractors, architects, engineers, planners, and local zoning agencies. In addition, all involved must abide by regulations that apply to the site and structure being planned. The rule structure may vary from one locale to another. What is alarming is that past professional practice within the U.S. building industry has only rarely gauged the environmental or energy impact of a structure prior to its construction. Prior to the efforts of organizations like the U.S. Green Building Council (USGBC) (established in 1995), the concept of what constituted a "green building" in the United States lacked a credible set of standards.

Concept of Green Buildings

Accepting the notion that sustainable, environmentally appropriate, and energy efficient buildings can be labeled "green," the degree of "greenness" is subject to multiple interpretations. The process of determining which attributes of a structure can be considered "green" or "not green" is inconclusive and subjective. Complicating the process, there are no clearly labeled "red" edifices with diametrically opposing attributes. While it is implied that a green building may be an improvement over current construction practice, the basis of attribute comparison is often unclear, subjective, and confusing. It is often unclear as to what sort of changes in construction practice, if imposed, would lead the way to greener, more sustainable buildings. If determinable, the marketplace must adjust and provide the technologies and means by which materials, components, and products can be provided to construction sites where greener buildings can arise. Since standards are often formative and evolving, gauging the degree of greenness risks the need to quantify subjective concepts.

There are qualities of structures, such as reduced environmental impact and comparatively lower energy usage, which are widely accepted as qualities of green construction practices. For example, use of recycled materials with post-consumer content that originates from a previous use in the consumer market and post-industrial content that would otherwise be diverted to landfills is widely considered an issue addressable by green construction practices. However, evaluation of green building attributes or standards by organizations implies the requirement that decisions be based on stakeholder consensus.

This process involves input to the decision-making processes by an array of representative stakeholders in often widely diverse geographic locations. For these and other reasons, developing a rating system for green buildings is both difficult and challenging.

Rating Systems for Buildings

Rating systems for buildings with sustainable features began to emerge in embryonic form in the 1990s. The most publicized appeared in the United Kingdom, Canada, and the United States. In the United Kingdom, the Building Research Establishment Environmental Assessment Method (BREEAM) was initiated in 1990. BREEAM™ certificates are awarded to developers based on an assessment of performance in regard to climate change, use of resources, impacts on human beings, ecological impact, and management of construction. Credits are assigned based on these and other factors. Overall ratings are assessed according to grades that range from pass to excellent.[6]

The International Initiative for a Sustainable Built Environment, based in Ottawa, Canada, has its Green Building Challenge program with more than 15 countries participating. The collaborative venture is geared toward the creation of an information exchange for sustainable building initiatives and the development of "environmental performance assessment systems for buildings."[7] In the United States, agencies of the central government co-sponsored the development of the Energy Star™ program, which provides "technical information and tools that organizations and consumers need to choose energy-efficient solutions and best management practices."[8] Expanding on their success, Energy Star™ developed a building energy performance rating system which has been used for over 10,000 buildings.

Entering the field at the turn of the new century, the USGBC grew from an organization with just over 200 members in 1999 to 3500 members by 2003.[9] The LEED™ rating system is a consensus-developed and reviewed standard, allowing voluntary participation by diverse groups of stakeholders with interest in the application and use of the standard. According to Boucher, "the value of a sustainable rating system is to condition the marketplace to balance environmental guiding principles and issues, provide a common basis to communicate performance, and to ask the right questions at the start of a project."[10] The first dozen pilot projects using the rating system were certified in 2000.

LEED-NC Rating System

The USGBC's Green Building Rating System is a voluntary, consensus-developed set of criteria and standards. This rating system evolved with a goal of applying standards and definition to the idea of high-performance buildings. The use of sustainable technologies is firmly established within the LEED project development process. LEED loosely defines green structures as those that are "healthier, more environmentally responsible and more profitable."[1]

LEED-NC 2.1 is the USGBC's current standard for new construction and major renovations. It is used primarily for commercial projects such as office buildings, hotels, schools, and institutions. The rating system is based on an assessment of attributes and an evaluation of the use of applied standards. Projects earn points as attributes are achieved and the requirements of the standards are proven. Depending on the total number of points a building achieves upon review, the building is rated as Certified (26–32 points), Silver (33–38 points), Gold (39–51 points) or Platinum (52 or more points).[11] Theoretically, there are a maximum of 69 achievable points. However, in real world applications, gaining certain credits often hinders the potential of successfully meeting the criteria of others. While achieving the rating of Certified is relatively easily accomplished, obtaining a Gold or Platinum rating is rare and requires both creativity and adherence to a broad range of prescriptive and conformance-based criteria.

The LEED process involves project registration, provision of documentation, interpretations of credits, application for certification, technical review, rating designation, award, and appeal. Depending on variables such as project square footage and USGBC membership status, registration fees can range up to $7500 for the process.[12]

LEED Prerequisites Categories and Criteria

To apply for the LEED labeling process, there are prerequisite project requirements which earn no points. For example, in the Sustainable Sites category, certain procedures must be followed to reduce erosion and sedimentation. In the category of Energy and Atmosphere, minimal procedures are required for building systems commissioning. Minimal energy performance standards must be achieved (e.g., adherence to ANSI/ASHRAE/IESNA Standard 90.1–1999, Energy Standard for Buildings Except Low-Rise Residential Buildings, or the local energy code if more stringent), and there must be verification that CFC refrigerants will not be used or will be phased out. In addition, there are prerequisite requirements outlining mandates for storage and collection of recyclable material, minimum IAQ performance (the requirements of ASHRAE Standard 62–1999, Ventilation for Acceptable Indoor Air Quality must be adhered to), and the requirement that non-tobacco smokers not be exposed to smoke.

In addition to the prerequisite requirements, the LEED process assigns points upon achieving certain project criteria or complying with certain standards. The total points are summed to achieve the determined rating. Projects can achieve points from initiatives within the following sets of categories: Sustainable Sites (14 points), Water Efficiency (5 points), Energy and Atmosphere (17 points), Materials and Resources (13 points), and Indoor Environmental Quality (15 points). Use of a LEED Accredited Professional (1 point) to assist with the project[13] earns a single point. Additional points are available for Innovation and Design Process (maximum of 4 points).

Within each category, the specific standards and criteria are designed to meet identified goals. In the category of Sustainable Sites, 20.2% of the total possible points are available. This category focuses on various aspects of site selection, site management, transportation and site planning. The goals of this category involve reducing the environmental impacts of construction, protecting certain types of undeveloped lands and habitats, reducing pollution from development, conserving natural areas and resources, reducing the heat island impacts, and minimizing light pollution. Site selection criteria are designed to direct development away from prime farmland, flood plains, habitat for endangered species and public parkland. A development density point is awarded for projects that are essentially multi-story. If the site has documented environmental contamination or is designated by a governmental body as a brownfield, another point is available. In regard to transportation, four points are available for locating sites near publicly available transportation (e.g., bus lines or light rail), providing bicycle storage and changing rooms, provisions for alternatively fueled vehicles and carefully managing on-site parking. Two points in this category are obtained by limiting site disturbances and by exceeding "the local open space zoning requirement for the site by 25%."[14] In addition, points are available by following certain storm water management procedures, increasing soil permeability, and attempting to eliminate storm water contamination. Potential urban heat island effects are addressed by crediting design attributes such as shading, underground parking, reduced impervious surfaces, high albedo materials, reflective roofing materials, or vegetated roofing. Finally, a point is available for eliminating light trespass.

Water efficiency credits comprise 7.2% of the total possible points. With the goal of maximizing the efficiency of water use and reducing the burden on water municipal systems, points are credited for reducing or eliminating potable water use for site irrigation, capturing and using rainwater for irrigation, and using drought tolerant or indigenous landscaping. This section of the LEED standard also addresses a building's internal water consumption. Points are available for lowering aggregate water consumption and reducing potable water use. Reducing the wastewater quantities or providing on-site tertiary wastewater treatment also earns points.

Energy and Atmosphere is the category that offers the greatest number of points, 24.6% of the total possible. The intents of this category include improving the calibration of equipment, reducing energy costs, supporting alternative energy, reducing the use of substances that cause atmospheric damage, and offering measurement and verification criteria. Optimizing the design energy cost of the regulated energy systems can achieve a maximum of ten points. To assess the result, project designs are modeled against a base case solution which lacks certain energy-saving technologies. Interestingly, the unit of measure for evaluating energy performance to achieve credits is not kilocalories or million Btus, but dollars. Points are awarded in whole units as the percentage of calculated dollar savings increases incrementally. In addition to the ten points for energy cost optimization, a maximum of three additional points is available for buildings that use energy from onsite renewable energy generation. Purchased green power is allocated a single point if 50% of the electrical energy (in kWh) comes from a two year green power purchasing arrangement. This category provides points for additional commissioning and elimination of the use of HCFCs and halon gases. Measurement and Verification (M&V) is allowed a point, but only if M&V options B, C, and D, as outlined in the 2001 edition of the International Measurement and Verification Protocol (IPMVP), are used.

The Materials and Resources category represents 18.8% of the total possible points. This category provides credit for material management; adaptive reuse of structures; construction waste management; resource reuse; use of material with recycled content; plus the use of regionally manufactured materials, certain renewable materials and certified wood products. A point is earned for providing a space in the building for storage and collection of recyclable materials such as paper, cardboard, glass, plastics and metals. A maximum of three points is available for the adaptive reuse of existing on-site structures and building stock. The tally increases with the extent to which the existing walls, floor, roof structure, and external shell components are incorporated into the reconstruction. LEED-NC 2.1 addresses concerns about construction waste by offering a point if 50% of construction wastes (by weight or volume) are diverted from landfills and another point if the total diversion of wastes is increased to 75%. A project that is composed of 10% recycled or refurbished building products, materials, and furnishings gains an additional two points. Another two points are available in increments (one point for 5%, two points for 10%) if post-consumer or post-industrial recycled content (by dollar value) is used in the new construction. To reduce environmental impacts from transportation systems, a point is available if 20% of the materials are manufactured regionally (defined as being within 500 miles or roughly 800 km of the site), and an added point is scored if 50% of the materials are extracted regionally. A point is available if rapidly renewable materials (e.g., plants with a ten year harvest cycle) are incorporated into the project, and yet another point is earned if 50% of the wood products are certified by the Forest Stewardship Council.

The category of Indoor Environmental Quality allows 21.7% of the possible total points available. The goals include improving IAQ, improving occupant comfort, and providing views to the outside. With ASHRAE Standard 62–1999 as a prerequisite, an additional point is available for installing CO_2 monitoring devices in accordance with occupancies referenced in ASHRAE Standard 62–2001, Appendix C. A point is also available for implementing technologies that improve upon industry standards for air change effectiveness or that meet certain requirements for natural ventilation. Systems that provide airflow using both underfloor and ceiling plenums are suggested by LEED documentation as a potential ventilation solution. Points are available for developing and implementing IAQ management plans during construction and prior to occupancy. The requirements include using a Minimum Efficiency Reporting Value (MERV) 13 filter media with 100% outside air flush-out prior to occupancy. There are points available for use of materials that reduce the quantity of indoor air pollutants in construction caused by hazardous chemicals and by volatile organic compounds in adhesives, sealants, paints, coatings, composite wood products, and carpeting. A point is offered for provision of perimeter windows and another for individual control of airflow, temperature, and lighting for half of the non-perimeter spaces. Points are available for complying with ASHRAE Standard 55–1992 (Thermal Environmental Conditions for Human Occupancy), Addenda 1995, and installing permanent temperature and humidity control systems. Finally, points are gained for providing 75% of the spaces in the building with some form of daylighting and for providing direct line of-sight vision for 90% of the regularly occupied spaces.

In the category of Innovation and Design Process, 7.2% of the total possible points are available. The innovation credits offer the opportunity for projects to score points as a result of unusually creative design innovations, such as substantially exceeding goals of a given criteria or standard.

Assessing LEED-NC

The LEED-NC process has numerous strengths. Perhaps the greatest is its ability to focus the owner and design team on addressing select energy and environmental considerations early in the design process. The LEED design process brings architects, planners, energy engineers, environmental engineers, and IAQ professionals into the program at the early stages of design development. The team adopts a targeted LEED rating as a goal for the project. A strategy evolves based on selected criteria. The team members become focused on fundamental green design practices that have often been overlooked when traditional design development processes were employed.

Furthermore, the LEED program identifies the intents of the environmental initiatives. Program requirements are stated and acceptable strategies are suggested. Scoring categories attempt to directly address certain critical environmental concerns. When appropriate, the LEED-NC program defers to engineering and environmental standards developed outside of the USGBC. The components of the program provide accommodation for local regulations. Case study examples, when available and pertinent, are provided and described in the LEED literature. To expedite the process of documenting requirements, letter templates and calculation procedures are available to program users. The educational aspects of the program, which succinctly describe select environmental concerns, cannot be understated. A Web site provides updated information on the program with clarifications of LEED procedures and practice. The training workshops sponsored by the USGBC are instrumental in engaging professionals with a wide range of capabilities.

These considerations bring a high degree of credibility to the LEED process. Advocates of the LEED rating system have hopes of it becoming the pre-eminent U.S. standard for rating new construction that aspires to achieve a "green" label. To its credit, it is becoming a highly regarded standard and continues to gain prestige. Nick Stecky, a LEED accredited professional, firmly believes that the system offers a "measurable, quantifiable way of determining how green a building is."[15]

Despite its strengths, the LEED-NC has observable weaknesses. The LEED-NC registration process can appear to be burdensome, and has been perceived as slowing down the design process and creating added construction cost. Isolated cases support these concerns. Kentucky's first LEED-NC school, seeking a Silver rating, was initially estimated to cost over $200/ft^2 ($2152/m^2) compared to the local standard costs of roughly $120/ft^2 ($1290/m^2) for non-LEED construction. However, there are few comparative studies available to substantially validate claims of statistically significant cost impact. Alternatively, many case studies suggest that there is no cost impact as a result of the LEED certification process. It is also possible that the savings resulting from the use of certain LEED standards (e.g. reduced energy use) can be validated using life-cycle costing procedures. Regardless, LEED-NC fails as a one-size-fits-all rating system. For new construction, Kindergarten to 12th-grade (K-12), school systems in New Jersey, California, and elsewhere have adopted their own sustainable building standards.

There are other valid concerns in regard to the use of LEED-NC. In an era when many standards are under constant review, standards referenced by LEED are at times out of date. The ASHRAE Standard 90.1–1999 (without amendments) is referenced throughout the March 2003 revision of LEED-NC. However, ASHRAE 90.1 was revised, republished in 2001, and the newer version is not used as the referenced standard. Since design energy costs are used to score Energy and Atmosphere points, and energy use comparisons are baselined against similar fuels, cost savings from fuel switching is marginalized. In such cases, the environmental impact of the differential energy use remains unmeasured, since energy units are not the baseline criteria. There is no energy modeling software commercially available that has been specifically designed for assessing LEED buildings. LEED allows most any energy modeling software to be used, and each has its own set of strengths and weaknesses when used for LEED energy modeling purposes. It is possible for projects to comply with only one energy usage prerequisite, applying a standard already widely adopted, and still become LEED certified.

In fact, it is not required that engineers have specialized training or certification to perform the energy models. Finally, LEED documentation lacks System International (SI) unit conversions, reducing its applicability and exportability.

A number of the points offered by the rating system are questionable. While indoor environmental quality is touted as a major LEED concern, indoor mold and fungal mitigation practices, among the most pervasive indoor environmental issues, are not addressed and are not necessarily resolvable using the methodologies prescribed. It would seem that having a LEED-accredited professional on the team would be a prerequisite rather than an optional credit. Projects in locations with abundant rainfall or where site irrigation is unnecessary can earn a point by simply documenting a decision not to install irrigation systems. The ability of the point system to apply equally to projects across varied climate classifications and zones is also questionable and unproven.

While an M&V credit is available, there is no requirement that a credentialed measurement and verification professional be part of the M&V plan development or the review process. Without the rigor of M&V, it is not possible to determine whether or not the predictive preconstruction energy modeling was accurate. The lack of mandates to determine whether or not the building actually behaves and performs as intended from an energy cost standpoint is a fundamental weakness. This risks illusionary energy cost savings. Finally, the M&V procedures in the 2001 IPMVP have undergone revision and were not state-of-the-art at the time that LEED-NC was updated in May 2003. For example, there is no longer a need to exclude Option A as an acceptable M&V alternative.

The LEED process is not warranted and does not necessarily guarantee that in the end, the owner will have a "sustainable" building. While LEED standards are more regionalized in locations where local zoning and building laws apply, local regulations can also preempt certain types of green construction criteria. Of greater concern is that it is possible for a LEED certified building to devolve into a building that would lack the qualities of a certifiable building. For example, the owners of a building may choose to remove bicycle racks, refrain from the purchase of green energy after a couple of years, disengage control systems, abandon their M&V program, and remove recycling centers—yet retain the claim of owning a LEED certified building.

Conclusion

The ideal of developing sustainable buildings is a response to the environmental impacts of buildings and structures. Developing rating systems for structures is problematic due to the often subjective nature of the concepts involved, the ambiguity or lack of certain standards, and the local aspects of construction. While there are a number of assessment systems for sustainable buildings used throughout the developed world, LEED-NC is becoming a widely adopted program for labeling and rating newly constructed "green" buildings in the United States. Using a point-based rating system, whereby projects are credited for their design attributes, use of energy, environmental criteria, and the application of select standards, projects are rated as Certified, Silver, Gold, or Platinum.

The LEED-NC program has broad applicability in the United States and has been proven successful in rating roughly 150 buildings to date. Its popularity is gaining momentum. Perhaps its greatest strength is its ability to focus the owner and design team on energy and environmental considerations early in the design process. Today, there are over 1700 projects that have applied for LEED certification. Due to the program's success in highlighting the importance of energy and environmental concerns in the design of new structures, it is likely that the program will be further refined and updated in the future to more fully adopt regional design solutions, provide means of incorporating updated standards, and offer programs for maintaining certification criteria. It is likely that the LEED program will further expand, perhaps offering a separate rating program for K-12 educational facilities. Future research will hopefully respond to concerns about potential increased construction costs and actual energy and environmental impacts.

References

1. U.S. Green Building Council. *LEED green building rating system.* 2004.
2. Spirn, A.W. *The Granite Garden: Urban Nature and Human Design;* Basic Books: New York, 1984.
3. United Nations. *Indicators of Sustainable Development: Guidelines and Methodologies;* United Nations: New York, 2001; 38.
4. Roosa, S.A. *Energy and Sustainable Development in North American Sunbelt Cities;* RPM Publishing: Louisville, KY, 2004.
5. Koeha, T. What is Sustainability and Why Should I Care? *Proceedings of the 2004 world energy engineering congress,* Austin, TX, Sept 22–24, 2004; AEE: Atlanta 2004.
6. URS. http://www.urseurope.com/services/engineering/engineering-breeam.htm (accessed Feb 2005).
7. iiSBE. International Initiative for a Sustainable Built Environment http://iisbe.org/iisbe/start/iisbe.htm (accessed Feb 2005).
8. United States Environmental Protection Agency. *Join energy star—Improve your energy efficiency.* http://www.energystar.gov. (accessed June 2003).
9. Gonchar, J. Green building industry grows by leaps and bounds; Engineering News-Record; 2003.
10. Boucher, M. Resource efficient buildings—Balancing the bottom line. *Proceedings of the 2004 world energy engineering congress,* Austin, TX, Sept 22–24, 2004; AEE: Atlanta 2004.
11. U.S. Green Building Council. *Green Building Rating System for New Construction and Major Renovations Version 2.1 Reference Guide;* May 2003; 6.
12. U.S. Green Building Council. *LEED—Certification Process.* http://www.usgbc.org/LEED/Project/certprocess.asp. (accessed Oct 2004).
13. U.S. Green Building Council. *Green Building Rating System for New Construction and Major Renovations Version 2.1:* Nov 2002.
14. U.S. Green Building Council. *Green Building Rating System for New Construction and Major Renovations Version 2.1:* Nov 2002; 10.
15. Stecky, N. Introduction to the ASHRAE greenguide for LEED, *Proceedings of the 2004 world energy engineering congress,* Austin, TX, Sept 22–24 2004; AEE: Atlanta 2004.

39

Nanomaterials: Regulation and Risk Assessment

Steffen Foss Hansen,
Khara D. Grieger,
and Anders Baun

Introduction

The topics of regulation and risk assessment of nanomaterials have never been more relevant and controversial in Europe than they are at this point in time. As the first major piece of legislation to be amended in Europe, the cosmetics legislation was adopted in 2009 requiring all nanomaterial-containing cosmetics to be labeled after 2013 and producers to provide a safety assessment of the nanomaterial used.[1,2]

Concurrently with the recasting of various pieces of legislation, such as the Novel Foods Regulation, the European Commission has commissioned an expert-/multistakeholder investigation of whether nanospecific amendments are needed to the current technical guidelines on substance identification and chemical safety assessment, which lie at the core foundations of the European Chemical legislation known as REACH—Registration, Evaluation, Authorization, and Restriction of Chemicals.[3,4] It is the major piece of legislation concerning regulating the manufacturing and applications of nanomaterials, although the text in REACH itself has only been subject to minor changes thus far. A number of other pieces of legislation relevant to the manufacturing, use, and disposal of nanomaterial and products have furthermore not been subject to any nanospecific changes, although they might be revised in the future.

In the following, some of the major pieces of legislation relevant for the regulation of nanomaterials in Europe will be presented. Examples of both horizontal regulation as well as subject-specific legislation will be given. Some of these have yet to take nanospecific issues into consideration, and the focus will therefore be at explaining the limitations of these in handling nanomaterials. For others, nanospecific aspects have recently been taken into consideration and for these the focus will be at explaining how this has been done and what kinds of challenges still need to be addressed.

Chemical risk assessment plays a crucial role in many of these pieces of legislation, and hence a short introduction and discussion of the applicability of chemical risk assessment to nanomaterials will be included.

Registration, Evaluation, and Authorization of Chemicals (Reach)

One of the key pieces of European legislation affecting nanomaterials is the European chemical regulation known as REACH, which went into force in mid-2007.[5] REACH prescribes

1. The registration of chemicals commercialized by manufacturers and importers in Europe as well as the collection of data on their use and toxicity.
2. The evaluation and examination by governments of the need for additional testing and regulation of chemicals.
3. That authorization has to be sought and given to manufacturers in order for them to use chemicals of high concern.
4. European Union (EU)-wide restrictions or complete ban of certain chemicals that cannot be used safely.

The REACH regulation replaced more than 40 other directives and subsequently shifted the responsibility in the registration and authorization process of REACH onto manufacturers and importers (including downstream users of chemicals) to provide data of uses and hazard information. Industry, furthermore, has to show that chemicals of high concern can be used safely. The evaluation and restriction process is still the responsibility of the national authorities, the newly established European Chemical Agency, and the European Commission.

Registration of all the commercialized chemicals in the European market is a tremendous task that is expected to occur gradually. Substances produced or imported in the highest volumes or of the greatest (known) concern are to be registered first. Substances produced or imported in more than 1000 tonnes per year per manufacturer or importer had to be registered by November 30, 2010, by the latest date. This was also the case for substances marketed in 100 tons/yr that have been classified as very toxic to aquatic organisms and for substances produced/imported in more than 1 ton/yr and which have been classified as Category 1 or Category 2 carcinogens, mutagens, or reproductive toxicants. Furthermore, substances entering the European market in yearly quantities above 100 tons, and 10 tons per producers or importers have to be registered by June 1, 2013, and June 1, 2018, respectively.[5]

REACH does not specifically mention nanomaterials, but does cover chemicals in all their physical–chemical states, using the following definition of a substance: "a chemical element and its compounds in the natural state or obtained by any manufacturing process, including any additive necessary to preserve its stability and any impurity deriving from the process used, but excluding any solvent which may be separated without affecting the stability of the substance or changing its composition."[5] Therefore, REACH is formally the relevant legislative frame for industrially used nanomaterials, and the exemption registration of carbon and graphite was redrawn in 2008 to address concerns raised about carbonaceous nanomaterials.[6,7] Companies will now have to register these materials if produced in quantities above 1 ton per producer or manufacturer per year. However, for a number of nanomaterials, it is not evident whether a nanoequivalent of a substance with different physicochemical and (eco)toxicological properties from the bulk substance would be considered as the same or as another substance under REACH.[8]

If a nanomaterial is considered to be a different substance under REACH, hazard information specifically related to the nanoform of the substance would have to be generated for the registration, if produced in more than 1 ton/yr. On the other hand, if a nanomaterial is considered to be the same as a registered bulk material, hazard information data generated for the registration might not be directly relevant for the nanoform of the substance and hence open to discussion.[8,9] In response to these concerns, the European Commission has launched a multistakeholder project on nanomaterials to look into substance identification under REACH in order to get recommendations on whether the nanoform of a substance should be considered different from the bulk form of the substance.[3,4]

If manufacturers and importers produce or import nanomaterials in volumes of more than 10 tons/yr and if it meets the criteria for classification as dangerous or a PBT (persistent, bioaccumulative, and toxic) or vPvB (very persistent and very bioaccumulative), a chemical safety assessment is required that includes information about uses, (eco)toxicological information, exposure assessment, and risk characterization(s).

Thus far, no nanomaterial has been classified as PBT or as vPvB, but if it was to be it is highly unclear how companies should do a chemical safety assessment. Both the Commission of the European Communities[10] as well as the its Scientific Committee on Emerging and Newly Identified Health Risks (SCENIHR)[11] have pointed out that current test guidelines in REACH are based on conventional methodologies for assessing chemical risks and may not be appropriate for assessing risks associated with nanomaterials.

It should be noted that a chemical safety assessment can also be required if a nanomaterial is selected for further evaluation by a member state or by the European Chemicals Agency due to specific concerns; or if a substance is a CMR (carcinogenic, mutagenic, or toxic for reproduction), PBT, vPvB, ED (endocrine disrupting), or substance of equivalent concern.

EU Water Framework Directive

Whereas REACH deals with the manufacturing and import of chemicals, the EU Water Framework Directive (WFD) deals with improving water quality and reducing dangerous chemicals in European river basins. The key aim of the WFD, which was adopted in 2000, is to promote long-term sustainable water use, preventing further deterioration of surface waters, transitional waters, coastal waters, and groundwater, and to protect and enhance the status of aquatic ecosystems with regard to their water needs, terrestrial ecosystems, and wetlands directly depending on the aquatic ecosystems,[12] Article 1.

The WFD establishes water management by a river basin approach with cooperation and joint objective setting across member state borders and even in some cases beyond the EU territory. Geographical and hydrological formation of each river basin determines which member states need to establish and implement a so-called river basin management plan. The river basin management plan, which needs to be updated every 6 years, specifies the measures to be taken to meet the environmental objectives for surface waters, for groundwater, and for protected areas. The WFD prescribes the setting of the environmental quality standards ensuring the general protection of the aquatic ecology, specific protection of unique and valuable habitats, and protection of drinking water resources and bathing water. For instance, for surface waters, member states shall implement necessary measures to prevent deterioration, and promote restoration of artificial and heavily modified water bodies with the aim of achieving "good ecological potential" and "good surface water chemical status" in 2015 by the latest. This has to be done along with a progressive reduction of pollution from a set of "priority substances" and discontinue emissions of priority hazardous substances,[12] Article 4.

For all surface waters, the WFD set a number of "general requirements for ecological protection" as well as a "general minimum chemical standard" and defines "good ecological status" and "good chemical status" in terms of the quality of the biological community, the hydrological characteristics, and the chemical characteristics,[12] Article 4.

The definition of "good chemical status" is especially relevant in regard to nanomaterials as it is defined in terms of compliance with all the quality standards established for chemical substances at the European level. For "priority substances," member states are required to set environmental quality standards (EQSs) to monitor the chemical status of a water body (European Parliament and the Council of the European Union (EP & CEU),[12] Article 16). Thus, the EQS is taken as concentration below which the chemical status is referred to as "good" in the WFD terminology (European Parliament and the Council of the European Union (EP & CEU),[12] Article 2). Even for the so-called priority hazardous substances, only a few EQSs have been set, but more substances will follow with a specific focus at substances that are toxic toward humans and/or aquatic organisms, compounds with a widespread environmental distribution, and those that are discharged in significant quantities.

A key question in regard to WFD is whether nanomaterials are possible candidates as priority substances.[13] In favor of this speaks the widespread and diffuse use of nanoparticles in a range of consumer products along with the hazard characteristics of some nanomaterials such as functionalized carbon nanotubes (CNTs), nanoscale silver, and zinc oxide. Some applications of nanomaterials furthermore involve direct contact with the water cycle, e.g., in relation to their use for water disinfection[14] and wastewater treatment,[15] as well as in regard to the direct use for treating soil and groundwater contamination.[16]

If a given type of nanoparticles is included in the list of priority substances in the future based on environmental occurrence or hazard information, an EQS will have to be defined.[12] To derive an EQS for a priority substance, the WFD outlines that test results from both acute and chronic ecotoxicological standard tests should be used for the "base set" organisms, i.e., algae and/or macrophytes, crustacean, and fish. Estimating EQS for nanoparticles is currently hampered by lack of ecotoxicological data even for the most tested nanoparticles such as C60, CNTs, TiO_2, ZnO, and Ag. For instance, the degradability of C60 and CNTs and their ability to bioaccumulate in the aquatic environment remains to be studied, making it virtually impossible to set an EQS for these two kinds of nanoparticles.[13]

Not only are the number of studies very limited, but the number of tested taxa is also too few to be used in the context of setting an EQS. The reliability and interpretation of the available ecotoxicity data is furthermore impeded as a result of factors such as particle impurities, suspension preparation methods, release of free metal ions, and particle aggregation.[13,17,18]

Besides these issues, mainly related to the lack of relevant data, it is also questionable whether the principles for deriving EQSs for chemicals can be directly transferred to nanoparticles. The setting of EQS is based on a chemical safety assessment similar to the one required under REACH and, as noted above, the European Commission's SCENIHR have pointed out that amendments have to be made to the guidelines for chemical safety assessment.[11,19]

Another manner in which nanomaterials could meet the criteria to be included in a WFD list of priority substances is if there is "evidence from monitoring of widespread environmental contamination".[12] However, when it comes to nanomaterials, monitoring in natural waters represents some profound challenges.[20,21] While applicable methods for in situ monitoring remain to be developed and refined,[22] it is also challenging to set up a reliable monitoring program for nanoparticles since a number of issues still remain to be resolved, e.g., choice of suitable sampling materials, preconcentration/fractionation methods, and analytical methods to characterize and quantify collected particles.[21] Despite significant progress in recent years, reliable methods are not yet available to determine nanoparticle identity, concentrations, and characteristics in complex environmental matrices, such as water, soil, sediment, sewage sludge, and biological specimens.

Pharmaceutical Regulation

Liposomes, polymer–protein conjugates, polymeric substances, or suspensions are examples of well-described and understood medicinal products containing nanoparticles and have been given marketing authorizations within the EU under the existing regulatory framework.[23–26]

As in the case of REACH, nanomedicine and nanomaterials are not specifically mentioned in the EU legislation on medicinal products and devices, tissue engineering, and other advanced therapies. Although the scope of the various EU regulations and directives that constitutes this framework covers nanomedicine, they have been accused of being too general, non-specific, and fraught with difficulties in case of complex drugs.[27,28] Given this, it does seem that it is generally believed that the regulatory framework for medicine covers medical products based on nanotechnology, and that the extensive pre-market safety assessment of medicine in general is sufficient to ensure that the benefits outweigh any identified risks or the adverse side effects.[29,30]

Concerns have, however, been raised that the risk assessment, safety, and quality requirements for medicine may not be designed to address nanomedicine and medical devices based on nanotechnology, as these have to be fulfilled by conformity to established quality systems and published product standards. This might be especially true for novel applications such as nanostructure scaffolds for tissue replacement, nanostructures enabling transport across biological barriers, remote control of nanoprobes, integrated implantable sensory nanoelectronic systems, and chemical structures for drug delivery and targeting of disease.[31]

Currently, the mechanism of action is key to decide whether a product should be regulated as a medicinal product or as a medical device. This could be problematic when it comes to many novel applications of nanomedicinal products as they are likely to span regulatory boundaries between medicinal products and medical devices.[29,31] This is due to the notion that they may exhibit a complex mechanism of action combining mechanical, chemical, pharmacological, and immunological properties, and combining diagnostic and therapeutic functions.

For new marketing authorization applications of pharmaceuticals, an environmental risk assessment has to be provided, which involves a rough calculation of the predicted environmental concentration (PEC) for surface water. Actions have to be initiated if the PEC is predicted to surpass 0.01 ppb.[32] However, this threshold cannot be interpreted as a safe concentration, and it is not based on a scientific evaluation.[32] It could furthermore be problematic when it comes to nanomedicine, as concentration in terms of mass per volume might not be the relevant metric to characterize the environmental hazard of nanomaterials.[33–35]

Nanofood Regulation

Food and food packaging are regulated by a number of directives and regulations in the EU, such as the EU Food Law Regulation and the EU Novel Foods Regulation.[36] As an overarching principle, all food are required to be safe and this overarching principle of safety applies to all foods and food packaging that contain nanomaterials. This has, however, been criticized for being too loose.[37]

During the recent discussion related to the update of regulation regarding food additives, the European Parliament's Committee on Environment, Public Health, and Food Safety stated that it wanted separate limit values for nanotechnologies and that the permitted limits for an additive in nanoparticle form should not be the same as when it is in traditional form.[38] This demand, however, never made into the actual regulation and the final adopted regulation on food additives is limited to requiring that food additives that have been produced via nanotechnology or consist of/or include materials fulfill a number of criteria before it can be include on the list of approved food, food additives, food enzymes, and food flavorings. Nanotechnology and nanomaterials are not defined in the regulation, but these criteria include what use does not pose a safety concern to the health of the consumer at the level of use proposed on the basis of available scientific evidence. Furthermore, there has to be a reasonable technological need that cannot be achieved by other economically and technologically practicable means, and using the food additives should entail consumer advantages and benefits.[39]

Another important piece of legislation in regard to food regulation in the EU is the Novel Foods Regulation. This regulation requires mandatory premarket approval of all new ingredients and products. In 2008, the European Commission adopted a proposal to revise the Novel Foods Regulation with the purpose of improving the access of new and innovative foods to the EU market.[40] The definition of novel foods was broadened to include those modified by new production processes, such as nanotechnology and nanoscience, which might have an impact on the food itself. This proposal is currently being discussed in the European Parliament and is going through what is known as a "third reading," and has to be adopted after a co-decision wherein both the Council of the European Union and the Parliament has to agree on the final text of the regulation. If agreement cannot be reached, it goes to conciliation.

There are a number of areas on which the European Parliament and the Council of the European Union disagree in regard to nanomaterials. The requirement of having mandatory labeling is also controversial and so is the issue of whether to have premarket safety testing of nanotechnology and nanomaterials in food and packaging.[41]

In the first line of revisions suggested to the Novel Foods Regulation, both the Council and the Parliament mention the lack of adequate information and lack of test methods for assessing the risks of nanomaterials.[41] Once the European Commission receives an application for authorization of a novel food, the European Food Safety Authority (EFSA) is responsible for the evaluation of whether a novel food and its use as an ingredient presents a danger to or misleads consumers. By regulation, the EFSA is required to provide assessment on the composition, nutritional value, metabolism, intended use, and the level of microbiological and chemical contaminants. Studies on the toxicology, allergenicity, and details of the manufacturing process may also be considered. No distinction is, however, made in regulation in regard to particle size, and hence nanoparticles will not require new safety assessments if the substance has already been approved in bulk form.

Risk Assessment of Nanomaterials

Three different kinds of limitations have been identified in various independent analysis of the applicability of existing regulatory frameworks when it comes to nanomaterials.

The first category of limitations are related to the limitations of definitions of what qualifies as a "substance," "novel food," etc., when it comes to nanomaterials. For instance, does the definition of a chemical substance cover both the bulk from as well as the nanoform of the substance, and does any given application of nanotechnology to manufacture a given food fall under the definition of a novel food? This issue is currently being discussed in a multistakeholder expert working group; however, this has failed to reach a consensus.[42]

In the second category fall requirements triggered by thresholds values not tailored to the nanoscale, but based on bulk material. For instance, for pharmaceuticals, the environmental concentration of medical products has to be estimated before marketing, and if it is below 0.01 ppb and "no other environmental concerns are apparent," no further actions are to be taken for the medical product in terms of environmental risk assessment.[32] Such a predefined action limit could potentially be problematic since the new properties of nanobased products are expected to also affect their environmental profiles, and this problem has yet to be addressed.[35]

The third category of limitations are related to lack of metrological tools, (eco)toxicological data, and environmental exposure limits as required by, e.g., REACH, the pharmaceuticals regulation, and the recast of the Novel Foods Regulation. The availability of (eco)toxicological data and chemical risk assessments is necessary to support existing legislation.

In regard to REACH, companies are urged to use already existing guidelines when performing chemical risk assessments, despite the fact that both the European Commission[10] and its SCENIHR,[11] as well as others,[9,10] have pointed out that current test guidelines supporting REACH are based on conventional methodologies for assessing chemical risks and may not be appropriate for the assessment of risks associated with nanomaterials.

Chemical risk assessment consists of four elements i.e., hazard identification, dose–response assessment, exposure assessment, and risk characterization. In Europe, legislation for controlling the production, use, and release of chemical substances is based on chemical safety assessment or risk assessment, as described in detail in the "Guidance on Information Requirements and Chemical Safety Assessment".[43] The guidance totals a staggering number of pages and is issued by the ECHA to help companies carry out chemical safety assessments. It includes extensive technical details for conducting hazard identification, dose (concentration)–response (effect) assessment, exposure assessment, and risk characterization in relation to human health and the environment.[43] Each of these four elements holds a number of limitations that are not easily overcome despite the fact that a lot of effort is being put into investigating the applicability of each of these four elements.

Hazard Identification of Nanomaterials

Toxicity and ecotoxicity have been reported on for multiple nanoparticles (metal and metal oxide nanoparticles, carbonaceous nanomaterials, and quantum dots) in scientific studies; however, many of these need further confirmation. Univocal hazard identification is currently impossible as it is hard to systematically link reported nanoparticle properties to the observed effects.

For instance, in regard to multiwalled CNTs (MWCNTs), Poland et al.[44] compared the toxicity of four kinds of MWCNTs of various diameters, lengths, shape, and chemical composition by exposing the mesothelial lining of the body cavity of three mice with 50 mg MWCNT for 24 hr or 7 days. This method was used as a surrogate for the mesothelial lining of the chest cavity. It was found that long MWCNTs "produced length-dependent inflammation, FBGCs, and granulomas that were qualitatively and quantitatively similar to the foreign body inflammatory response caused by long asbestos." Only the long MWCNTs caused significant increase in polymorphonuclear leukocytes or protein exudation. The short MWCNTs failed to cause any significant inflammation at 1 day or giant cell formation at 7 days. Poland et al.[44] also found that the water-soluble components of MWCNT did not produce significant inflammatory effects 24 hr after injection, which rules out that residue metals were the cause of the observed effects, as others previously had speculated on the basis on in vitro studies.[45,46] The findings by Poland et al.[44] have since then been supported by Ma-Hock et al.[47] and Pauluhn et al.[48] in 90-day inhalation toxicity studies.

Less work has been done in regard to exploring the ecotoxicological aspects of nanomaterials, but a number of significant studies have been published.

In 2004, Oberdorster[49] published the first ecotoxicological study and reported observed significant increase in lipid peroxidation of the brain of juvenile largemouth bass after exposure to uncoated fullerenes (99.5%) in concentrations of 0.5 and 1 ppm after exposure for 48 hr. C60 was dissolved in tetrahydrofuran (THF), which have since then led to some discussion about whether C60 or the THF was responsible for the effects observed.[50,51] The use of THF is no longer recommended.[18]

In regard to CNTs, Templeton et al.[52] compared "as prepared" single-walled CNTs (SWCNTs) with electro- phorectically purified SWCNTs and the fluorescent fraction of nanocarbon by-products. They observed an average cumulative life cycle mortality of 13±4%, while mean life cycle mortalities of 12±3%, 19±2%, 21±3%, and 36±11% were observed for 0.58, 0.97, 1.6, and 10 mg/L. Exposure to 10 mg/L showed: 1) significantly increased mortalities for the naupliar stage and cumulative life cycle; 2) a dramatically reduced development success to 51% for the nauplius to copepodite window, 89% for the copepoddite to adult window, and 34% overall for the nauplius to adult period; and 3) a significantly depressed fertilization rate averaging only 64±13%.

A number of studies have furthermore highlighted the need to investigate the potential interactions with existing environmental contaminants or what has become known as the "Trojan horse effect." For instance, Baun et al.[53] found that the toxicity of phenanthrene was increased toward algae and crustaceans following sorption to C60 aggregates. In contrast, Baun et al.[53] found that the toxic effect of pentachlorophenol decreased when C60 was added. After studying the ecotoxicity of cadmium to algae in the presence of 2 mg/L TiO_2 nanoparticles of three different sizes, Hartmann et al.[17] found that the presence of TiO_2 in algal tests reduced the toxicity of cadmium. This is thought to be due to decreased bioavailability of cadmium resulting from sorption/complexation of Cd^{2+} ions to the TiO_2 surface. However, the observed growth inhibition was, however, greater for the 30 nm TiO_2 nanoparticles than could be explained by the concentration of dissolved Cd(II) species alone, which indicates a possible carrier effect, or combined toxic effect of TiO_2 nanoparticles and cadmium.

Dose–Response Relationship in Regard to Nanomaterials

In regard to the second element of chemical risk assessment, it is fundamental that a dose–response relationship can be established so that no-effect concentrations or no effect levels need to be predicted or derived. It is unclear whether a no-effect threshold can be actually be established and what the best hazard descriptor(s) of nanoparticles is, and what the most relevant dose metrics and the what the most sensitive endpoints are. Several studies have reported observing a dose–response relationship. This goes for, especially, in vitro studies on, among others, C60, SWCNTs and MWCNTs, and various forms of nanometals. Normally, dose refers to "dose by mass"; however, based on the experiences gained in biological tests of nanoparticles, it has been suggested that biological activity of nanoparticles might not be mass dependent, but dependent on physical and chemical properties not routinely considered in toxicity studies.[54] For instance, Oberdorster and col-leagues[55,56] and Stoeger and colleagues[57,58] found that the surface area of the nanoparticles is a better descriptor of the toxicity of low-soluble, low-toxicity particles, whereas Wittmaack[59,60] found that the particle number worked best as dose metrics. Warheit et al.[61,62] found that toxicity was related to the number of functional groups in the surface of nanoparticles.

Exposure Assessment

Completing a full exposure assessment requires extensive knowledge about, among others, manufacturing conditions, level of production, industrial applications and uses, consumer products and behavior, and environmental fate and distribution. Such detailed information is not available, and thus far no full exposure assessment has been published for any one or more nanomaterials. This may partly be due to difficulties in monitoring nanomaterial exposure in the workplace and the environment, and partly due to the fact that the biological and environmental pathways of nanomaterials are still largely unexplored.[63] Some efforts have been made to assess occupational, consumer, and environmental exposure, however, both to assess the level of exposure and to assess the applicability of current exposure assessment methods and guidelines.

These are, however, hampered by the paucity of knowledge, lack of access to information, difficulties in monitoring nanomaterial exposure in the workplace and the environment, and by the fact that the biological and environmental pathways of nanomaterials are still largely unexplored. Hence, they should be seen as "proof of principle" rather than actual assessment of the exposure.[64–67]

Risk Characterization

All the information from the first three elements of the risk assessment come together in the fourth and final element of chemical risk assessment, namely risk characterization.[63] In the risk characterization process, exposure levels are compared with quantitative or qualitative hazard information, then suitable predicted no-effect concentrations or derived no-effect levels are determined in order to decide if risks are adequately controlled.[43]

Often, risk characterization boils down to the estimation of a risk quotient. For the environment, this is, for instance, defined as the PEC/predicted no effect concentration (PNEC). If the risk quotient is <1, no further testing or risk reduction measures are needed according to the European Chemical Agency.[43] If it is >1, further testing can be initiated to lower the PEC/PNEC ratio. If that is not possible, risk reduction could be implemented.

A number of studies reported having completed—or attempted to complete—risk assessments of various nanomaterials such as CeO_2, TiO_2, C60, and CNTs.[18,67–69] For instance, in regard to the use of CeO2-based diesel fuel additive in the United Kingdom, Park et al.[67] assessed the risk of CeO_2 causing pulmonary inflammation. First, they estimated an internal dose of $3.8 \times 10^{-7}\,cm^2/cm^2$ by converting the retained dose into surface area units and then dividing by the area of the proximal alveolar region of the lung. Then, they compared this value to the highest noobserved-effect level found in a number of in vitro toxicity studies. This value was $26.75\,cm^2/cm^2$. Assuming that in vitro exposure data can be accurately projected to the in vivo situation, Park et al.[67] concluded that "it is highly unlikely that exposure to cerium oxide at the environmental levels (from both monitored and modeled experimental data) would elicit pulmonary inflammation."

Mueller and Nowack[69] reported having completed the first quantitative risk assessment of nanoparticles in the environment. In a first attempt to derive PEC values, Mueller and Nowack used the threshold concentrations of 20 and 40 mg/L reported in the literature for nano-Ag on *Bacillus subtilis* and *Escherichia coli,* and considered it to be equivalent to a no-observed-effect concentration. For nano-TiO_2 and CNT, the lowest value found in the literature was <1 mg/L for algae, daphnia, and fish.[69] Applying the assessment factor of 1000, the PNEC in water was found to be 0.04, <0.001, and <0.0001 mg/L for nano- Ag, nano-TiO_2, and CNT, respectively. Combining these PNEC values with the predicted exposure, Mueller and Nowack[69] calculated the environmental concentrations in Switzerland for nano-Ag, nano-TiO_2, and CNTs stemming from textiles, cosmetics, coatings, plastics, sports gear, electronics, etc. Assuming worse-case exposure levels, Mueller and Nowack[69] found that the risk quotient for nano-Ag and CNT is less than one-thousandth, and they state that their modeling suggests that currently little or no risk is to be expected from nano-Ag and CNT to organisms in water and air. Nano-TiO_2, on the other hand, might pose a risk to organism in water—according to Mueller and Nowack[69]—with risk quotients ranging from >0.7 to >16. The PNEC for soil could not be determined due to lack of information.

Despite the preliminary risk characterizations by Park et al.,[67] Mueller and Nowack,[69] Shinohara et al.,[68] and Stone et al.,[18] it is important to realize that risk characterization critically involves reflection of the data behind each step and determining what the overall risk will be.[63] As elaborated on previously, each of the three first steps of risk assessment holds a number of challenges, and since risk characterization is the fourth and final step where all the information is to come together, the sum or maybe even the power all of these limitations are conveyed to calculating risk quotients for nanomaterials under REACH.[9]

Revisions of the Technical Guidance of Risk Assessment

The European Commission has commissioned an expert/multistakeholder investigation of whether nanospecific amendments are needed to the current technical guidelines on chemical safety assessment. This has to develop, among others, specific advice on

1. How REACH information requirements on intrinsic properties of nanomaterials can be fulfilled
2. The appropriateness of the relevant test methods for nanomaterials
3. The possible specific testing strategies, if relevant
4. Information needed for safety evaluation and risk management of nanomaterials (especially, information beyond the current information requirements under REACH)
5. How to do exposure assessment for nanomaterials, hazard, and risk characterization for nanomaterials[3,4]

The latter will involve threshold/non-threshold considerations, analysis of existing evidence related to setting limit values for nanomaterials, identification of critical items for dose description (mass, number concentration, surface area, particle size(s) etc.), whether and how no-effect-levels for health and the environment could be established, and finally development of recommendations on the feasibility of whether categorization of nanomaterials (e.g., different types of CNTs) in the hazard assessment is compatible with the exposure assessment parameters/metrics in order to prepare a meaningful risk characterization.[3,4]

In regard to novel food, EFSA published a scientific opinion in 2008 on the potential risks arising from the use of nanotechnology in food, concluding that nanotechnology aspects shall be considered when risk assessment guidance documents in the food and feed area are reviewed, and, among others, recommend that risk assessment of nanomaterials in the food and feed areas should consider the specific properties of nanomaterials in addition to those common to the equivalent non-nanoforms.[70] Recently, EFSA closed for public comments on a draft guidance on risk assessment concerning potential risks arising from applications of nanoscience and nanotechnologies to food and feed. This guidance holds practical advice on how to complete risk assessments of nanomaterials used in food and food products.[71]

In the light of the limitations of chemical risk assessment, a number of alternative or complementary tools and methods, such the precautionary matrix[72] and multicriteria decision analysis,[73] have been proposed recently. Many of them hold great promises, but they need further evaluation and validation.[74]

Conclusion

In this entry, we presented a number of major pieces of legislation such as REACH, the WFD, and the Novel Foods Regulation, and discussed their relevance and limitations in regard to nanomaterials. Only a limited number of EU regulations, directives, etc., actually mention nanotechnology and/nanomaterials. In general, there seem to be three overall challenges when it comes to current regulation of nanomaterials: 1) limitations in regard to terminology and definitions of key terms such as a "substance," "novel food," etc.; 2) safety assessment requirements triggered by thresholds values not tailored to the nanoscale, but based on bulk material and; 3) limitations related to lack of metrological tools, (eco) toxicological data, and environmental exposure limits as required by, e.g., REACH, WFD, the pharmaceuticals regulation, and the recast of the Novel Foods Regulation. Chemical risk assessment provides a fundamental element in support of existing legislation. Risk assessment is normally said to consist of four elements, i.e., hazard identification, dose–response assessment, exposure assessment, and risk characterization. Each of these four elements holds a number of limitations that are not easily overcome, although a lot of effort is being put into investigating the applicability of each of them. However, political decisions to revise substance definition and the current thresholds that trigger safety evaluation are still needed as these are not tailored to the nanoscale, but based on bulk material.

References

1. European Parliament and Council of the European Union (EP & CEU). Regulation (EC) No 1223/2009 of the European Parliament and of the Council of 30 November 2009 on cosmetic products (1). Off. J. Eur. Union L342, vol. 52, 22 December 2009. ISSN 1725–2555. L 342/59 L 342/209.
2. Bowman, D.M.; Calster, G.v.; Friedrichs, S. Nanomaterials and regulation of cosmetics. Nat. Nanotechnol. **2009**, 5, 92.
3. Safenano. REACH-NanoInfo: Rip-oN 2, 2010a. Available at http://www.safenano.org/REACH nanoInfo.aspx (accessed October 2010).
4. Safenano. REACH-NanoHazEx: Rip-oN 3, 2010b. Available at http://www.safenano.org/ REACHnanoHazEx.aspx (accessed October 2010).
5. European Parliament and the Council of the European Union (EP & CEU). Regulation (EC) No 1907/2006 of the European Parliament and of the Council of 18 December 2006 concerning the Registration, Evaluation, Authorisation and Restriction of Chemicals (REACH), establishing a European Chemicals Agency, amending Directive 1999/45/ EC and repealing Council Regulation (EEC) No 793/93 and Commission Regulation (EC) No 1488/94 as well as Council Directive 76/769/ EEC and Commission Directives 91/155/EEC, 93/67/EEC, 93/105/EC and 2000/21/EC. 30.12.2006. Off. J. Eur. Union L 2006, 396/1-L 396/849.
6. Führ, M.; Hermann, A.; Merenyi, S.; Moch, K.; Möller, M. Legal Appraisal of Nanotechnologies, 2007. Available at http://www.umweltdaten.de/publikationen/fpdf-l/3198.pdf (accessed December 2010). Umwelt Bundes Amt UBA-FB 000996. ISSN 1862–4804.
7. C&EN. Carbon losses its exemption statues under REACH. Chem. Eng. News 2008, June 23, p.9.
8. Chaundry, Q.; Blackburn, J.; Floyd, P.; George, C.; Nwaogu, T.; Boxall, A.; Aitken, R. *A Scoping Study to Identify Gaps in Environmental Regulation for the Products and Applications of Nanotechnologies;* Department for Environment, Food and Rural Affairs: London, 2006.
9. Hansen SF. *Regulation and Risk Assessment of Nanomaterials—Too Little, Too Late?,* PhD Thesis; Technical University of Denmark: Kgs. Lyngby, Denmark, 2009. Available at http://www2.er.dtu. dk/publications/fulltext/2009/ENV2009-069.pdf (accessed April 2009).

10. Commission of the European Communities (CEC). *Communication from the Commission to the European Parliament, the Council and the European Economic and Social Committee Regulatory Aspects of Nanomaterials, [Sec(2008) 2036] Com(2008), 366 final;* Commission of the European Communities: Brussels, Belgium, 2008a.

11. Scientific Committee for Emerging and Newly Identified Health Risks (SCENIHR). *The Appropriateness of the Risk Assessment Methodology in Accordance with the Technical Guidance Documents for New and Existing Substances for Assessing the Risks of Nanomaterials;* European Commission of Health and Consumer Protection Directorate-General: Brussels, Belgium, 2007.

12. European Parliament and the Council of the European Union (EP & CEU). Directive 2000/60/EC of the European Parliament and of the Council of 23 October 2000 establishing a framework for Community action in the field of water policy. Off. J. Eur. Commun. **2000**, *L 327,* 1–72, Brussels, Belgium.

13. Baun, A.; Hartmann, N.B.; Greiger, K.D.; Hansen, S.F. Setting the limits for engineered nanoparticles in European surface waters. J. Environ. Monit. **2009**, *11,* 1774–1781.

14. Li, Q.; Mahendra, S.; Lyon, D.Y.; Liga, M.V.; Li, D.; Alvarez, P. Antimicrobial nanomaterials for water disinfection and microbial control: Potential applications and implications. Water Res. **2008**, *42,* 4591–4602.

15. Nano Iron. *Technical Data Sheet.* NANOFER 25S. Ra- jhrad, Czech Republic, 2009. http://www.nanoiron.cz/en/?f1/4technical_data (accessed February 2011).

16. Li, X.; Elliott, D.W.; Zhang, W.X. Zero-valent iron nanoparticles for abatement of environmental pollutants: Materials and engineering aspects. Crit. Rev. Solid State Mater. Sci. **2006**, *31,* 111–122.

17. Hartmann, N.B.; Von der Kammer, F.; Hofmann, T.; Baalousha, M.; Ottofuelling, S.; Baun, A. Algal testing of titanium dioxide nanoparticles—Testing considerations, inhibitory effects and modification of cadmium bioavailability. Toxicology **2010**, *269* (2010), 190–197.

18. Stone, V.; Hankin, S.; Aitken, R.; Aschberger, K.; Baun, A.; Christensen, F.; Fernandes, T.; Hansen, S.F.; Hartmann, N.B.; Hutchinson, G.; Johnston, H.; Micheletti, G.; Peters, S.; Ross, B.; Sokull-Kluettgen, B.; Stark, D.; Tran, L. Engineered Nanoparticles: Review of Health and Environmental Safety (ENRHES), 2010. Available at http://nmi.jrc.ec.europa.eu/project/ENRHES.htm (accessed February 2010).

19. Scientific Committee for Emerging and Newly Identified Health Risks (SCENIHR). *Risk Assessment of Products of Nanotechnologies;* Scientific Committee on Emerging and Newly Identified Health Risks, European Commission of Health and Consumer Protection Directorate-General: Brussels, Belgium, 2009.

20. Tiede, K. *Detection and Fate of Engineered Nanoparticles in Aquatic Systems, PhD Thesis, University of York,* Environment Department and Central Science Laboratory: York, U.K., 2008.

21. Hassellöv, M.; Readman, J.W.; Ranville, J.F.; Tiede, K. Nanoparticle analysis and characterization methodologies in environmental risk assessment of engineered nanoparticles. Ecotoxicology **2008**, *17,* 344–361.

22. Lead, J.R.; Wilkinson, K.J. Natural aquatic colloids: Current knowledge and future trends. Environ. Chem. **2006**, *3,* 159–171.

23. European Parliament and the Council of the European Union (EP & CEU).Regulation (EC) No 726/2004 of the European Parliament and of the Council of 31 March 2004 Laying Down Community Procedures for the Authorisation and Supervision of Medicinal Products for Human and Veterinary Use and Establishing a European Medicines Agency, 2004.

24. Council of the European Communities (CEC). Council Directive 90/385/EEC of 20 June 1990 on the approximation of the laws of the Member States relating to active implantable medical devices. Off. J. Eur. Commun. **1990**, *L 189,* 17–36.

25. Council of the European Communities (CEC). Council Directive 93/42/EEC of 14 June 1993 concerning medical devices. Off. J. Eur. Commun. **1993**, *L 169,* 1–43.

26. Council of the European Communities (CEC). Directive 2001/83/EC of the European Parliament and of the Council of 6 November 2001 on the Community code relating to medicinal products for human use. Off. J. Eur. Commun. **2001**, *L 311,* 67–128

27. Editorial. Regulating nanomedicine. Nat. Mater. **2007**, *6*, 249.
28. D'Silva, J.; Van Calster, G. Regulating Nanomedicine: A European Perspective, 2008. Available at http://ssrn.com/abstract=1286215 (accessed November 2008).
29. European Group on Ethics in Science and New Technologies (EGE). Opinion 21—On the Ethical Aspects of Nanomedicine, The European Group on Ethics in Science and New Technologies (EGE), 2007. Retrieved March 15, 2008, available at http://ec.europa.eu/european_group_ethics/publications/docs/final_publication_%20op21_en.pdf.
30. N&ET Working Group. Report on Nanotechnology to the Medical Devices Expert Group: Findings and Recommendations, 2007. Available at http://ec.europa.eu/enterprise/medical_devices/net/entr-2007-net-wg-report-nanofinal.pdf (accessed December 2008).
31. European Medicines Agency (EMEA). *European Medicines Agency: Reflection Paper on Nanotechnology-Based Medicinal Products for Human Use, EMEA/CHMP/70769/2006;* European Medicines Agency: London, 2006a.
32. European Medicines Agency (EMEA). *Guideline on the Environmental Risk Assessment of Medicinal Products for Human Use,* EMEA/CHMP/SWP/4447/00; European Medicines Agency: London, 2006b.
33. Zhang, X.; Sun, H.; Zhang, Z.; Niu, Q.; Chen, Y.; Crittenden, J.C. Enhanced bioaccumulation of cadmium in carp in the presence of titanium dioxide nanoparticles. Chemo- sphere **2007**, *67,* 160–166.
34. Baun, A.; Hartmann, N.B.; Grieger, K.; Kusk, K.O. Ecotoxicity of engineered nanoparticles to aquatic invertebrates—A brief review and recommendations for future toxicity testing. Ecotoxicology **2008a**, *17* (5), 387–395.
35. Baun, A.; Hansen, S.F. Environmental challenges for nanomedicine. Nanomedicine **2008**, *2* (5), 605–608.
36. European Parliament and the Council of the European Union (EP & CEU). Regulation (EC) No 178/2002 of the European Parliament and of the Council of 28 January 2002 laying down the general principles and requirements of food law, establishing the European Food Safety Authority and laying down procedures in matters of food safety. Off. J. Eur. Commun. **2002**, L 31/1-L 31/24.
37. Friends of the Earth (FOE). *Out of the Laboratory and on to Our Plates Nanotechnology in Food and Agriculture*; Friends of the Earth: Australia, Europe, and U.S.A, 2008.
38. Halliday J. EU Parliament Votes for Tougher Additives Regulation, Foodnavigator.com, July 12, 2008. Available at http://www.foodnavigator.com/Legislation/EU-Parlia-mentvotes-for-tougher-additives-regulation (accessed November 2008).
39. European Parliament and the Council of the European Union (EP & CEU). Directive 2008/98/EC of the European Parliament and of the Council of 19 November 2008 on waste and repealing certain directives. Off. J. Eur. **2008**, Union L 312/3-L 312/30.
40. Commission of the European Communities (CEC). *Proposal for a Regulation of the European Parliament and of the Council on Novel Foods and Amending Regulation (EC) No. XXX/XXXX#* [common procedure] (presented by the Commission) [SEC(2008) 12] [SEC(2008) 13]. COM(2007), 872 final 2008/0002 (COD). 14-1 2008; Commission of the European Communities: Brussels, Belgium, 2008b.
41. European Parliament. Draft Recommendation for Second Reading on the Council Position at First Reading for Adopting a Regulation of the European Parliament and of the Council on Novel Foods, Amending Regulation (EC) No 1331/2008 and Repealing Regulation (EC) No 258/97 and Commission Regulation (EC) No 1852/2001 (11261/2/2009–C7-0000/2010–2008/0002(COD)), 2010.
42. C&EN. Wrangling Over Substance ID Hits REACH Nano Project. Lack of Consensus will Lead to Decision Driven by Policy Not Science, 2011. Available at http://chemicalwatch.com/6324/wran-gling-over-substance-id-hits-reach-nano-project?q=Wrangling%20over%20substance%20ID%20hits%20REACH%20nano%20project (accessed January 2011).

43. ECHA. Guidance Documents. Guidance on Information Requirements and Chemical Safety Assessment, 2010. Available at http://guidance.echa.europa.eu/docs/guidance_document/information_ requirements_en.htm?time=1288375681 (accessed October 2010).

44. Poland, C.A.; Duffin, R.; Kinloch, I.; Maynard, A.; Wallace, W.A.H.; Seaton, A.; Stone, V.; Brown, S.; Macnee, W.; Donaldson, K. Carbon nanotubes introduced into the abdominal cavity of mice show asbestos-like pathogenicity in a pilot study. Nat. Nanotechnol. **2008**, *3*, 423–428.

45. Shvedova, A.A.; Kisin, E.R.; Mercer, R.; Murray, A.R.; Johnson, V.J.; Potapovich, A.I.; Tyurina, Y.Y.; Gorelik, O.; Arepalli, S.; Schwegler-Berry, D.; Hubbs, A.F.; Antonini, J.; Evans, D.E.; Ku, B.K.; Ramsey, D.; Maynard, A.; Kagan, V.E.; Castranova, V.; Baron, P. Unusual inflammatory and fibrogenic pulmonary responses to single-walled carbon nanotubes in mice. Am. J. Physiol. Lung Cell. Mol. Physiol. **2005**, *289*, 698–708.

46. Kagan, V.E.; Tyurina, Y.Y.; Tyurin, V.A.; Konduru, N.V.; Potapovich, A.I.; Osipov, A.N.; Kisin, E.R.; SchweglerBerry, D.; Mercer, R.; Castranova, V.; Shvedova, A.A. Direct and indirect effects of single walled carbon nanotubes on RAW 264.7 macrophages: Role of iron. Toxicol. Lett. **2006**, *165* (1): 88–100.

47. Ma-Hock, L.; Treumann, S.; Strauss, V.; Brill, S.; Luizi, F.; Mertler, M.; Wiench, K.; Gamer, A.O.; van Ravenzwaay, B.; Landsiedel, R. Inhalation toxicity of multi-walled carbon nanotubes in rats exposed for 3 months. Toxicol. Sci. **2009**, *112*, 273–275.

48. Pauluhn, J. Subchronic 13-week inhalation exposure of rats to multiwalled carbon nanotubes: Toxic effects are determined by density of agglomerate structures, not fibrillar structures. Toxicol. Sci. **2010**, *113* (1), 226–242.

49. Oberdorster E. Manufactured nanomaterials (fullerenes, C60) induce oxidative stress in juvenile largemouth bass. Environ. Health Perspect. **2004**, *112*, 1058–1062.

50. Zhu, S.Q.; Oberdorster, E.; Haasch, M.L. Toxicity of an engineered nanoparticle (fullerene, C-60) in two aquatic species, *Daphnia* and fathead minnow. Mar. Environ. Res. **2006**, *62*, S5–S9.

51. Henry, T.B.; Menn, F.M.; Fleming, J.T.; Wilgus, J.; Compton, R.N.; Sayler, G.S. Attributing effects of aqueous C_{60} nano-aggregates to tetrahydrofuran decomposition products in larval zebrafish by assessment of gene expression. Environ. Health Perspect. **2007**, *115* (7), 1059–1065.

52. Templeton, R.C.; Ferguson, P.L.; Washburn, K.M.; Scrivens, W.A.; Chandler, G.T. Life-cycle effects of single-walled carbon nanotubes (SWNTS) on an estuarine meiobenthic copepod. Environ. Sci. Technol. **2006**, *40* (23), 7387–7393.

53. Baun, A.; Sorensen, S.N.; Rasmussen, R.F.; Hartmann, N.B.; Koch, C.B. Toxicity and bioaccumulation of xenobiotic organic compounds in the presence of aqueous suspensions of aggregates of nano-C60. Aquat. Toxicol. **2008b**, *86*, 379–387.

54. Oberdorster, G.; Maynard, A.; Donaldson, K.; Castranova, V.; Fitzpatrick, J.; Ausman, K.; Carter, J.; Karn, B.; Kreyling, W.; Lai, D.; Olin, S.; Monteiro-Riviere, N.; Warheit, D.; Yang, H. Principles for characterizing the potential human health effects from exposure to nanomaterials: Elements of a screening strategy. Part. Fibre Toxicol. **2005**, *2*, 8.

55. Oberdorster, G. Significance of particle parameters in the evaluation of exposure dose–response relationships of inhaled particles. Part. Sci. Technol. **1996**, *14*, 135–151.

56. Oberdorster, G.; Stone, V.; Donaldson, K. Toxicology of nanoparticles: A historical perspective. Nanotoxicology **2007**, *1* (1), 2–25.

57. Stoeger, T.; Reinhard, C.; Takenaka, S.; Schroeppel, A.; Karg, E.; Ritter, B.; Heyder, J.; Schultz, H. Instillation of six different ultrafine carbon particles indicates surface area threshold dose for acute lung inflammation in mice. Environ. Health Perspect. **2006**, *114*(3), 328–333.

58. Stoeger, T.; Schmid, O.; Takenaka, S.; Schulz, H. Inflammatory response to TiO_2 and carbonaceous particles scales best with BET surface area. Environ. Health Perspect. **2007**, *115* (6), A290–A291.

59. Wittmaack, K. In search of the most relevant parameter for quantifying lung inflammatory response to nanoparticle exposure: Particle number, surface area, or what? Environ. Health Perspect. **2007a**, *115*, 187–194.

60. Wittmaack, K. Dose and response metrics in nanotoxicology: Wittmaack responds to Oberdoerster et al. and Stoeger et al. Environ. Health Perspect. **2007b**, *115* (6), A290–291.

61. Warheit, D.B.; Webb, T.R.; Colvin, V.L.; Reed, K.L.; Sayes, C.R. Pulmonary bioassay studies with nanoscale and fine-quartz particles in rats: Toxicity is not dependent upon particle size but on surface characteristics. Toxicol. Sci. **2007a**, *95* (1), 270–280.

62. Warheit, D.B.; Webb, T.R.; Reed, K.L.; Frerichs, S.; Sayes, C.M. Pulmonary toxicity study in rats with three forms of ultrafine-TiO_2 particles: Differential responses related to surface properties. Toxicology **2007b**, *230,* 90–104.

63. CCA. *Small Is Different: A Science Perspective on the Regulatory Challenges of the Nanoscale;* The Council of Canadian Academies: Ottawa, Canada, 2008.

64. Boxall, A.B.A.; Chaudhry, Q.; Sinclair, C.; Jones, A.; Aitken, R.; Jefferson, B.; Watts, C. *Current and Future Predicted Environmental Exposure to Engineered Nanoparticles;* Central Science Laboratory: York, U.K., 2008.

65. Hansen, S.F.; Michelson, E.; Kamper, A.; Borling, P.; Stuer-Lauridsen, F.; Baun, A. Categorization framework to aid exposure assessment of nanomaterials in consumer products. Ecotoxicology **2008,** *17* (5), 438–447.

66. Luoma, S.N. *Silver Nanotechnologies and the Environment Old Problems or New Challenges?* PEN 15. Project on Emerging Nanotechnologies; Woodrow Wilson International Center for Scholars: Washington, D.C., 2008.

67. Park, B.; Donaldson, K.; Duffin, R.; Tran, L.; Kelly, F.; Mudway, I.; Morin, J-P.; Guest, R.; Jenkinson, P.; Samaras, Z.; Giannouli, M.; Kouridis, H.; Martin, P. Hazard and risk assessment of a nanoparticulate cerium oxide-based diesel fuel additive—A case study. Inhal. Toxicol. **2008,** *20* (6), 547–566.

68. Shinohara, N. *Risk Assessment of Manufactured Nanomaterials—Fullerene (C60),* Interim Report issued on October 16, 2009, Executive Summary. The Research Institute of Science for Safety and Sustainability, AIST available at http://www.aistriss.jp/main/modules/product/nano_rad.html.

69. Mueller, N.; Nowack, B. Exposure modeling of engineered nanoparticles in the environment. Environ. Sci. Technol. **2008,** *42,* 4447–4453.

70. European Food Safety Authority (EFSA). Draft Opinion of the Scientific Committee on the Potential Risks Arising from Nanoscience and Nanotechnologies on Food and Feed Safety (Question No EFSAQ-2007-124). European Food Safety Authority: Parma, Italy, 2008.

71. European Food Safety Authority (EFSA). Endorsed for Public Consultation Draft Scientific Opinion Guidance on Risk Assessment Concerning Potential Risks Arising From Applications of Nanoscience and Nanotechnologies to Food and Feed. EFSA Scientific Committee, European Food Safety Authority (EFSA): Parma, Italy, 2011.

72. Höck, J.; Epprecht, T.; Hofmann, H.; Höhner, K.; Krug, H.; Lorenz, C.; Limbach, L.; Gehr, P.; Nowack, B.; Rie- diker, M.; Schirmer, K.; Schmid, B.; Som, C.; Stark, W.; Studer, C.; Ulrich, A.; von Götz, N.; Wengert, S.; Wick, P.; Guidelines on the Precautionary Matrix for Synthetic Nanomaterials. Federal Office of Public Health and Federal Office for the Environment, Berne 2010, Version 2. Available at http://www.bag.admin.ch/themen/chemika-lien/00228/00510/05626/index.html?lang=en (accessed 17 May 2010).

73. Linkov, I.; Satterstrom, F.K.; Steevens, J.; Ferguson, E.; Pleus, R.C. Multi-criteria decision analysis and environmental risk assessment for nanomaterials. J. Nanopart. Res. **2007,** *9* (4), 543–554.

74. Grieger, K.D.; Linkov, I.; Hansen, S.F.; Baun, A. A review of alternative frameworks and approaches for assessing environmental risks of nanomaterials. Nanotoxicology, **2011,** *in press.*

75. Harrington, R. Nano Risk Assessment: A Work in Progress, 2010. Available at http://www. foodproductiondaily.com/Quality-Safety/Nano-risk-assessment-a-work-in-progress/?c=lIxHi8 W7vkWyj1EO6E28rw%3D%3D&utm_source=newsletter_daily&utm_medium=email&utm_campaign=Newsletter%2BDaily (accessed June 2010).

76. Höck, J.; Epprecht, T.; Hofmann, H.; Höhner, K.; Krug, H.; Lorenz, C.; Limbach, L.; Gehr, P.; Nowack, B.; Riediker M. et al. *Guidelines on the Precautionary Matrix for Synthetic Nanomaterials,* Version 2; Swiss Federal Office for Public Health and Federal Office for the Environment: Berne, Switzerland, 2010.

77. Lam, C.W.; James, J.T.; McCluskey, R.; Hunter R.L. Pulmonary toxicity of single-wall carbon nanotubes in mice 7 and 90 days after intratracheal instillation. Toxicol. Sci. **2004**, *77*, 126–134.

78. Paik, S.Y.; Zalk, D.M.; Swuste, P. Application of a pilot control banding tool for risk level assessment and control of nanoparticle exposures. Ann. Occup. Hyg. **2008**, *52* (6), 419–428.

79. Royal Society and the Royal Academy of Engineering. *Nanoscience and Nanotechnologies: Opportunities and Uncertainties;* Royal Society: London, 2004.

VI

ENT: Environmental Management Using Environmental Technologies

IV

40

Industrial Waste: Soil Pollution and Remediation

W. Friesl-Hanl,
M.H. Gerzabek,
W.W. Wenzel, and
W.E.H. Blum

Introduction

Anthropogenic activities such as mining and smelting, combustion of waste and fossil fuels, as well as the use of organic and inorganic chemicals and radionuclides in agriculture and industry entail risks for the overall environment and human beings. The awareness of these risks due to industrial wastes emerged through numerous cases of severe environmental impacts many years after their disposal. Notorious examples include hazardous waste disposal in the Love Canal, New York, United States[1]; Lekkerkerk near Rotterdam, Netherlands[2]; the collapsed tailing dams containing acidic pyrite sludge in Donana, Spain[3]; cyanide- and metal-rich liquid waste disposals near Baia Mare, Romania[4]; and many other accidents.[5] A report describing the world's worst-polluted industrial sites was updated in 2016 with the focus on the top polluting industries identified as (1) used lead–acid battery recycling, (2) mining and ore processing, (3) lead smelting, (4) tanneries, (5) artisanal small-scale gold mining, (6) industrial dumpsites, (7) industrial estates, (8) chemical manufacturing, (9) product manufacturing, and (10) dye industry. These industries put over 32 million people at risk and account for 7 million to 17 million Disability-Adjusted Life Years (DALYs) in low- and middle-income countries.[6] Heavily polluted sites are distributed across all continents, and a former report highlighted only the tip of an iceberg, listing sites such as Dzerzinsk (chemical weapons, Russia), Linfen (coal industry, China), Ranipet (tanning industry, India), Kabwe (mining and smelting of Pb, Zambia), Haina (recycling of Pb, Dominican Republic), and La Oroya (polymetallic smelter, Peru).[7] While full remediation of these sites is economically and technically impossible, low-cost in situ technologies for managing and monitoring aiming at risk reduction are becoming increasingly available. More importantly, we need to learn our lessons and avoid similar pollution in the future.

Besides these negative effects of industrial waste, their reuse potential for soil amelioration, remediation, and recultivation also should be considered. Future perspectives and visions for handling industrial waste must be discussed in terms of the potential environmental risks and how to minimize them. The proclaimed Sustainable Development Goals (SDGs) by the United Nations (UN) should be a guidance for these next steps.[8]

Industrial Waste Overview

Generally, industrial activities generate three categories of waste: (1) gaseous emissions, (2) wastewater, and (3) solid waste. Gaseous pollutants or particulates are generated by combusting fossil fuels and wastes; they are found both close to the emitter and far away, especially in forest ecosystems, due to their filter capacity.[9] Dispersion and long-term transport of heavy metals, radionuclides, and persistent organic pollutants occur.[10] Acidification and eutrophication are due to the transboundary distribution of nutrients often involved in combustion processes (e.g., S, N compounds). Wastewater derives from domestic and industrial establishments; combined, they are known as municipal wastewater. Surface water and groundwater are the most threatened sinks of pollutants, besides irrigated areas. Upon removal of water, wastewater turns to sludge and later to solid waste. Generally, solid waste generation starts with mining and petroleum production and with the subsequent processing of these raw materials to goods for consumption. Emerging waste includes mineral ores, tailings, slags, and oil-contaminated soils and residues that stem from the processing of these materials.

Industrial activities generate mostly mixed wastes, yielding transient waste types. Pollutants can be released in gaseous, particulate, aqueous, or solid form, depending on the industrial processes involved. They emanate from point[11] or diffuse[12] sources. Generally, pollution tends to be highest close to emitters and declines with increasing distance. However, elevated concentrations of persistent organic chemicals have been found in environments far from pollution sources, such as in the Arctic, in distant mountain regions, and in remote forest soils.[13]

Soil quality is prone to degradation through physical, chemical, and biological processes, resulting in soil erosion, reduced productivity, and soil and groundwater contamination. Detrimental effects of industrial waste on soil comprise the following: (1) siltation and compaction through application of fine-grained and salt-rich materials; (2) any type of pollution, acidification, or landfilling; and (3) loss of macro- and microfaunal diversity by intoxication. Heavy metals (Pb, Hg, Cd, Cr), metalloids (As, Sb), and organic compounds (e.g., dioxins and other persistent organic pollutants) are the most relevant soil pollutants. Since 1950, more than 140,000 new chemicals and pesticides have been synthesized whereof 5,000 are produced in large quantities, dispersed in the environment, and responsible for human exposure.[14] Waste generation has dramatically increased from chemical manufacturing, wastewater treatment, agriculture, mining and smelting, food processing, energy production, pulp and paper production, and other industries. The subsequent waste–environmental interactions have potential consequences on soil and water qualities. Estimates show that up to 75% of the materials used in industrial processes do not end up in final products.[15]

Prevention of industrial waste production, enhanced by optimization of processes reducing the amount of generated waste and the concentration of pollutants in waste streams, has a higher priority than end-of-pipe technologies such as soil remediation. However, emerging technologies to filter and retain pollutants in solid industrial waste materials require close monitoring as this may limit the re-use of such materials, for example, in soil remediation, due to increased pollutant concentrations. For instance, novel clean coal technologies lead to changed coal combustion residues (CCR) composition as volatile elements (e.g., Hg and Se) are more efficiently retained while due to concurrent changes in pH the solubility (Hg, Cd, Cu) and volatilization (Hg, NH_3) of certain pollutants are likely enhanced.[16]

Potential Use of Industrial Waste

Soil Amelioration

In the case of metals, humankind has imbalanced biogeochemical cycles through the use of metals. Considerable amounts are lost in technical processes into the above-mentioned waste categories. Electronic waste is one of the most rapidly growing problems.[17] E-waste volume estimation for 2013 is given in Table 1. Instead of exporting this waste[18] from "developed" into "developing" countries and thereby contaminating soil and water due to the lack of legal regulations there, the strategy should be to promote recycling and reuse of metals. Besides heavy metal-contaminated arable land, nutrient-deficient areas are widespread over the world. Here, the reuse of Cu, Mn, Mo, Fe, and Zn from electronic waste as nutrients could contribute toward more sustainable systems.

Gravel sludge, a fine-grained by-product of the gravel industry—applied on sandy soils poor in nutrients—can improve soil physical properties and crop production.

On the other hand, the use of industrial waste materials as soil amendments has been estimated to contaminate thousands of hectares of productive agricultural land in countries throughout the world. This is a particular problem in the Asia Pacific region, where land-based disposal of untreated wastes is still being practiced due to the lack of regulatory guidelines.[19] If crops are cultivated on contaminated soils, they may become less competitive on the world market, provided efficient quality control is implemented. Figure 1 gives a schematic overview of waste reuse possibilities.

Soil Remediation

In situ fixation of labile metals in soils by addition of amendments is one option to remediate heavy metal-contaminated soils. The aim is to reduce the risk of contamination by balancing ecological and economical needs. Several industrial wastes or by-products have been screened for their ability to immobilize heavy metals. Red mud is produced by the alumina industry around the world in countries like Jamaica, Australia, the United States, and Germany[20] (Table 1). Results of greenhouse, outdoor, and field experiments indicate immobilization of many heavy metals in soil after red

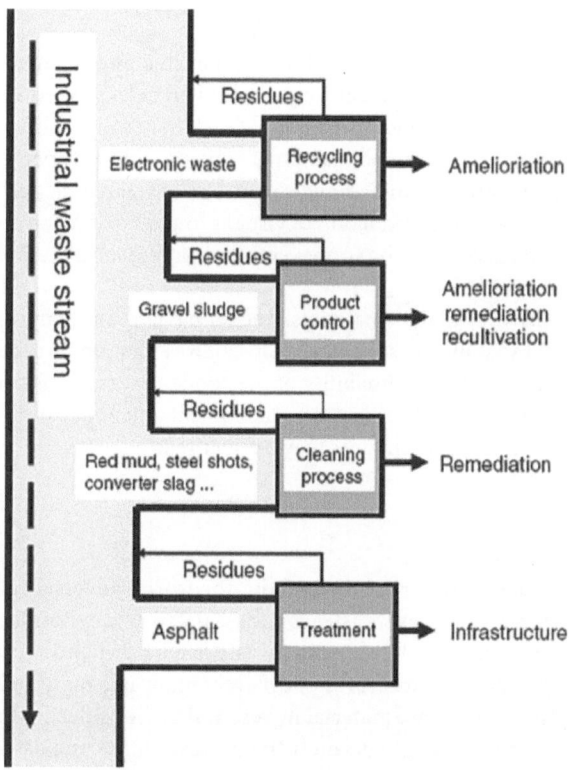

FIGURE 1   A schematic overview of waste reuse possibilities.

TABLE 1   Industrial Waste Quantities

Industry	Industrial Waste	Produced Amount (Mg/a)
Electronic industry	Electronic waste	20–50 Mio, worldwide[16]
Alumina industry	Red mud	120 Mio, worldwide[20]
Pulp and paper industry	Paper mill sludge	16 Mio, US[24], CA[25]
Iron	Slags (25%–30% of crude iron)	300–360 Mio,[a] worldwide[29]
Steel industry	Slags (10%–15% of crude steel)	190–290 Mio,[b] worldwide[29]
Energy production	CCP	40 Mio, EU15, (2016)[30]
	CCP	110 Mio, US (2017)[31]

[a] Calculated from produced amount of iron.
[b] Calculated from produced amount of steel. CCP, coal combustion products (including fly ash, flue-gas desulfurization gypsum, bottom ash); US, United States; CA, Canada.

mud addition.[21–23] Similar results were obtained for steel shots, a waste product of the metal shaping industry; converter and basic slag from the steel and iron industry; and "Iron Rich," generated by TiO_2 production.[26] Fly ash, a coal combustion residue, and beringite, a modified aluminosilicate that originates from the fluidized bed burning of coal refuse from a former coal mine in Beringen (Belgium), also have potential for metal immobilization.[27] A further example could be the reuse of the tailings of the world's largest siderite mining site (Erzberg, Austria).[28] The tailings consist of high amounts of fine-textured iron and carbonates, and it has been demonstrated in pot and field experiments that they hold promise for in situ inactivation of heavy metals in polluted soils and associated improvement of crop yield and quality if applied to arable land.[32] The composition of these amendments varies considerably, and the immobilizing effect is mostly due to Al-, Fe-, Mn-oxides, phosphates, silicates, and alkaline materials.

An emerging amendment for soil remediation/amelioration could be biochar. These materials derive from different sources. If organic residues/waste by-products will be pyrolyzed and applied on soils, different functions could be improved. Besides the storage of carbon in soil,[33] also soil physical (e.g., water holding capacity[34]) or chemical (e.g., immobilization of contaminants) properties can be improved.[35]

Application rates of these mentioned amendments up to 5% (w/w) are practicable and result in establishing a vegetation cover on bare land, reduction of visual symptoms of metal phytotoxicity, and a significant reduction of the metal accumulation in aboveground biomass. An example of a field application is given in Friesl-Hanl et al. (2017).[23]

Advantages of these easily available materials, including contaminant inactivation and often low costs,[36] are opposed by some disadvantages or failures that may occur due to indigenous contamination.[37] Monitoring of the sustainability of the applied measures is necessary and should be mandatory.[24] The recommendation is to adjust the amendment and application rate on a case-by-case basis. In certain cases, removal of contaminants may be required.

Conclusions

Pollution by industrial waste is widespread throughout the world, adversely affecting soil and environment qualities. Management of these wastes is necessary to achieve a more sustainable system. After exhausting the two approaches of preventing waste generation and recycling suitable materials, the use of wastes, by-products, or mixtures of different materials for amelioration, remediation, or recultivation is one option to regulate material fluxes. Besides considering the advantages of using waste materials, we also have to take into account the disadvantages and assess them on a case-by-case basis.

References

1. Manhan, S.E. *Environmental Chemistry*, 6th Ed.; Lewis Publisher: New York, 1994.

2. Alloway, B.J.; Ayres, D.C. *Chemical Principles of Environmental Pollution*; Chapman and Hall: London, 1993.

3. Meharg, A.A.; Osborn, D.; Pain, D.J.; Sanchez, A.; Naveso, M.A. Contamination of Doñana food chain after the Aznalcollar mine disaster. *Environ. Pollut.* 1999, *105* (3), 387–390.

4. Premysl, S.; Pavonic, M.; Boucek, J.; Kokes, J. Baia Mare accident – Brief ecotoxicological report of Czech experts. *Ecotoxicol. Environ. Safety* 2001, *49*, 255–261.

5. Johnson, J. Industrial safety – The uneven world of chemical accident investigation. *Chem. Eng. News* 2016, 18–20.

6. Pure Earth Blacksmith Institute and Green Cross Switzerland, World's Worst Pollution Problems. The Toxics beneath Our Feet, 2016, 54, available at http://www.pureearth.org (accessed March 2019).

7. The Blacksmith Institute. The World's Worst Polluted Places—The Top Ten. P 57, available at http://www.blacksmithinstitute.org (accessed March 2019).

8. UN SDGs (Sustainable development Goals); https://www.un.org/sustainabledevelopment/news/communications-material/ (accessed on 27 March 2019). Transforming our world. The 2030 Agenda for sustainable development. A/RES/70/1; p 37.

9. Schulte, A.; Blum, W.E.H. Schwermetalle im Waldökosystem. In Geochemie und Umwelt; Matschullat, J., Tobschall, H.J.; Voigt, H.J., Eds.; Springer: Berlin, Heidelberg, 1997; 53–74.

10. Blum, W.E.H. Soil degradation caused by industrialization and urbanization. In *Towards Sustainable Land Use*; Blume, H.P.; Eger, H.; Fleischhauer, E.; Hebel, A.; Reij, C.; Steiner, K.G., Eds.; Catena: Reiskirchen, 1998; Vol. I, Advances in Geoecology 31, 755–766.

11. Naidu, R.; Megharaj, M.; Dillon, P.; Kookana, R.; Correll, R.; Wenzel, W.W. Soils and point source pollution. In *Encyclopedia of Soil Science*; Lal, R., Ed.; Marcel Dekker, Inc.: New York, 2002; 1012–1017.

12. Naidu, R.; Megharaj, M.; Dillon, P.; Kookana, R.; Correll, R.; Wenzel, W.W. Soils and non-point source pollution. In *Encyclopedia of Soil Science*; Lal, R., Ed.; Marcel Dekker, Inc.: New York, 2002; 872–874.

13. Wania, F. On the origin of elevated levels of persistent chemicals in the environment. *Environ. Sci. Pollut. Res.* 1999, *6* (1), 11–19.

14. Landrigan, P.J.; Fuller, R.; Acosta, N.J.R.; Adeyi, O.; Arnold, R.; Baldé, A.B.; Bertollini, R.; Bose-O'Reilly, S.; Boufford, J.I.; Breysse, P.N.; Chiles, T. The Lancet Commission on pollution and health. *The Lancet* 2018, *391*, 462–512.

15. Onida, M. Innovation in prevention of quantitative and qualitative waste production. In *Innovation in Waste Management*, Proceedings of the IV European Waste Forum, Milano, Italy, Nov 30–Dec 1, 2000; Centro de Ingegneria per la Protezione dell' Ambiente (C.I.P.A.), Eds.; European Waste Forum: Milano, 2000; 91–105.

16. Dellantonio, A.; Reepmann, F.; Fitz, W.J.; Wenzel, W.W. Disposal of coal combustion residues in terrestrial systems: Contamination and risk management. *J. Environ. Qual.* 2010, *39*, 761–775. doi:10.2134/jeq2009.0068

17. Wang, F.; Huisman, J.; Stevels, A.; Baldé, C.P. Enhancing e-waste estimates: Improving data quality by multivariate input-output analysis. *Waste Manag.* 2013, *33*, 2397–2407.

18. Puckett, J.; Byster, L.; Weservelt, S.; Gutierrez, R.; Davis, S.; Hussain, A.; Dutta, M. *Exporting Harm: The High-Tech Trashing of Asia*; The Basel Action Network: Washington, DC, 2002.

19. Ullah, S.M.; Nuruzzaman, M.; Gerzabek, M.H. Heavy metal and microbiological pollution of water and sediments by industrial wastes, effluents and slums around Dhaka City. In *Tropical Limnology III*; Timotius, K.H.; Goltenboth, F., Eds.; Satya Waeann University Press: Salatiga, Indonesia, 1995; 179–186.

20. Hua, Y.; Heal, K.V.; Friesl-Hanl, W. The use of red mud as an immobiliser for metal/metalloid-contaminated soil: A review. *J. Hazard. Mater.* 2017, *325*, 17–30; doi:10.1016/j.jhazmat.2016.11.073

21. Friesl, W.; Lombi, E.; Horak, O.; Wenzel, W.W. Immobilization of heavy metals in soils using inorganic amendments in a greenhouse study. *J. Plant Nutr. Soil Sci.* 2003, *166*, 191–196.

22. Friesl, W.; Horak, O.; Wenzel, W. Immobilization of heavy metals in soils by the application of bauxite residues: Pot experiments under field conditions. *J. Plant Nutr. Soil Sci.* 2004, *167* (1), 54–59.

23. Friesl-Hanl, W.; Platzer, K.; Riesing, J.; Horak, O.; Waldner, G.; Watzinger, A.; Gerzabek, M.H. Non-destructive soil amendment application techniques on heavy metal-contaminated grassland: Success and long-term immobilizing efficiency. *J. Environ. Manag.* 2017, *186*, 167–174; doi:10.1016/j.jenvman.2016.08.068.

24. Geng, X.; Zhang, S.Y.; Deng, J. Characteristics of paper mill sludge and its utilization for the manufacture of medium density fiberboard. *Wood Fiber Sci. 39* (2), 2007, 345–351.

25. Bonomo, L.; Higginson, A.E. *International Overview on Solid Waste Management*; Academic Press: London, 1988.

26. Knox, A.S.; Seaman, J.C.; Mench, M.J.; Vangronsveld, J. Remediation of metal- and radionuclides-contaminated soils by in situ stabilization techniques. In *Environmental Restoration of Metal-Contaminated Soils*; Iskandar, I.K., Ed.; Lewis Publishers: Boca Raton, 2000; 21–60.

27. Vangronsveld, J.; Van Assche, F.; Clijsters, H. Reclamation of a bare industrial area contaminated by non-ferrous metals: In situ metal immobilization and revegetation. *Environ. Pollut.* 1995, *87*, 51–59.

28. Siderit, available at http://de.wikipedia.org/wiki/Siderit (accessed March 2019).

29. U.S. Geological Survey, Mineral commodity summaries 2019: U.S. Geological Survey, 2019, 200, https://doi.org/10.3133/70202434.

30. ECOBA European Coal Combustion Products Association, available at http://www.coal-ash.co.il/docs/ECOBA_Stat_2016_EU15.pdf (accessed March 2019).

31. ACAA American Coal Ash Association, available at https://www.acaa-usa.org/publications/productionusereports.aspx (accessed March 2019).

32. Touceda-Gonzalez, M.; Brader, G.; Antonielli, L.; Ravindran, V.B.; Waldner, G.; Friesl-Hanl, W.; Corretto, E.; Campisano, A.; Pancher, M.; Sessitsch, A. Combined amendment of immobilizers and the plant growth-promoting strain Burkholderia phytofirmans PsJN favours plant growth and reduces heavy metal uptake. *Soil Biol. Biochem.* 2015, *91*, 140–150.

33. Schmidt, H.-P.; Anca-Couce, A.; Hagemann, N.; Werner, C.; Gerten, D.; Lucht, W.; Kammann, C. Pyrogenic carbon capture and storage; *GCB Bioenergy.* 2018, 1–19. doi:10.1111/gcbb.12553

34. Ojeda, G.; Mattana, S.; Àvila, A.; Alcañiz, J.M.; Volkmann, M.; Bachmann, J. Are soil–water functions affected by biochar application? *Geoderma* 2015, *249–250*, 1–11, doi:10.1016/j.geoderma.2015.02.014

35. Uchimiya, M.; Wartelle, L.H.; Klasson, K.T.; Fortier, C.A.; Lima, I.M. Influence of pyrolysis temperature on biochar property and function as a heavy metal sorbent in soil. *J. Agric. Food Chem.* 2011, *59*, 2501–2510. doi:10.1021/jf104206c

36. Puschenreiter, M.; Horak, O.; Friesl, W.; Hartl, W. Low-cost agricultural measures to reduce the heavy metal transfer into human food chain – A review. *Plant Soil Environ.* 2005, *51* (1), 1–11.

37. Friesl-Hanl, W.; Platzer, K.; Horak, O.; Gerzabek, M.H. Immobilising of Cd, Pb, and Zn contaminated arable soils close to a former Pb/Zn smelter: A field study in Austria over 5 years. *Environ. Geochem. Health* 2009, *31*, 581–594.

41

Pest Management: Crop Diversity

Maria R. Finckh and
Jan Henrik Schmidt

Introduction

Improvement in crop management using modern machinery equipment, fertilizers, and pesticides and progress in plant breeding within a few selected crops has led in recent decades to highly specialized agricultural practices on farms. Out of more than 300,000 plant species existing, about 7,000 crops are known to be cultivated for human or animal diet; however, 60% of the word's energy source is based on only three crops, namely wheat, maize, and rice [1]. As a consequence, a general impoverishment of plant diversity and a high degree of genetic erosion in several important crops for human nutrition is documented since the end of World War II [2].

Simplification of the ecosystem by using one-sided crop rotations, monoculture, and crop plants of uniform genotype and the elimination of weeds with herbicides results in a strong selection for adapted pests, pathogens, and weeds leading to frequent resistance breakdown and severe weed infestations [3–5] and, importantly, a general loss of biodiversity [6].

Increasing the diversity of selection pressures acting on pests and weeds can reduce their ability to develop resistances to pesticides. Diverse crop ecosystems with adapted crops in sequential cropping including cover crops and living mulches, crop populations, and species mixtures are generally less damaged by pests, pathogens, and weeds compared to monocultures [2]. Interestingly, weeds are less abundant but weed species diversity is often higher within such diverse crop rotations. Hence, outbreaks of weed, pest, and disease epidemics and the probability of losses are reduced [4,7,8].

In this chapter, diversity by planting sequence (crop rotation), crop-border diversity, and crop-weed diversity are discussed.

Influence of Crop Rotation on Diseases, Insect Pests, and Weeds

The positive effects of a diverse crop rotation on yield and agricultural stability have been well known since ancient times. Awareness has built that much of the rotation effects lie in the plants themselves as they release considerable amounts of organic compounds into the soil in order to influence the microbiome, the solubility of nutrients and to interact with other plants. Most pathogens and pests are weak competitors outside their hosts and barely survive. If soil life is abundant, then it will outcompete and often directly live on the resting stages of pests and pathogens and weeds [9].

Crop rotation is most effective in the case of specialized pathogens and pests that are dependent on a host crop or have a narrow host range. For example, many soil-borne pathogens or insects survive on root or plant residues and require the cultivation of a susceptible crop for continuous survival. The length of host crop interval for insect or disease control will depend on how quickly the insect pest or pathogen can be destroyed by starvation and/or by other antagonistic effects [10,11].

With respect to weeds, important criteria for their control are variation in planting time and timing as well as techniques of seed bed preparation [12,13]. To prevent weeds from germinating simultaneously with the cash crop or at least from seed shed, the whole cropping system needs to be considered [14]. Main effects are caused by the sowing time of crops (winter or spring crops), sequence and placement of crops in the crop rotation, competitive ability of different crops against weeds, and direct weed control (herbicides and/or mechanical control) [15]. Allelopathic effects, that is, the exudation of specific compounds from the roots of certain crops may reduce germination of different weeds [15,16]. For perennial weeds, methods that deplete the weed storage organs through competition, for example, due to living mulches and perennial forage crops, and/or frequent mowing can be useful to reduce number and biomass of weeds [17-19]. These methods combined with fragmentation of weed rhizomes that can reduce the viability of storage organs have proven a high efficacy in Quackgrass (*Elymus repens*) control in Sweden [20]. In monoculture and pure grain rotations, weed infestation can reach a high level and weeds are more difficult to control compared to a rotation in alternation with dicotyledonous plants, for example, field beans, potatoes, or canola [21,22]. Especially the degree of infestation of problem weeds, for example, *Apera spica venti*, *Viola tricolor arvensis*, and *Matricaria* spp., increases considerably [23,24].

When developing a crop rotation, certain criteria should thus be considered: adequate change in tillage practices and timing, maintenance of high levels of organic matter in the soil, inclusion of crops which do not stimulate subsequent growth of the pathogen or crops that have direct negative effects on pests and pathogens (e.g., producing toxins) [11]. For example, oats and brassicas, such as various mustards, are known to exude substances that are directly suppressive to many pathogens, pests, and weeds and therefore justifying their recent popularity in rotations [25–27]. However, they are also hosts to some broad-range pathogens such as *Fusarium avenaceum*. Such broad host range pathogens need to be managed through the enhancement of soil suppressiveness [28,29]. To achieve this, soil protection from damaging compaction through heavy machinery, radiation, drought, and heavy rainfall by high aboveground living and dead biomass, such as occurring in minimum tillage systems, is mandatory [28,30]. In addition, this will foster the microbial activity in the topsoil layers and result, in the long term, in sustainable production systems.

Influence of Decoy and Trap Crops on Pests

Decoy crops are non-host crops that are sown to stimulate the activation of dormant propagules of the pathogen in the absence of the host. In this way, the soil-borne pathogens waste their inoculum potential. For example, *Lolium* spec., *Papaver rhoeas*, and *Reseda odorata* can act as decoy crops for the pathogen *Plasmodiophora brassicae* in Brassica [31]. In the case of trap crops, the crop is host to the pathogen often nematodes. The trap crop attracts nematodes to infect, but the crop is harvested or destroyed before the nematode can complete its life cycle. A famous, still effective method had been developed by Julius Kühn in the 1840s in Germany when certain crucifers are sown and plowed before the beet cyst nematode (*Heterodera schachtii*) can fully develop its life cycle [26]. Recent breeding efforts resulted in *H. schachtii*-resistant oilseed radish varieties that are used as winter cover crops prior to growing sugar beets. Root exudates of resistant varieties induce hatching of young nematode larvae from eggs and cysts with subsequent penetration of the radish roots. However, due to the resistance mechanism, *H. schachtii* juveniles are not capable of building large feeding sites within the radish roots, which prevents the forming of new cysts [32].

Influence of Weed and Border Diversity on Pests and Diseases

Weeds not only compete with crops; they may also be intermediate hosts for diseases and parasites while at the same time offering food and refuge to beneficial insects within the agricultural ecosystem. Within monocropped fields, weeds are important sources of biodiversity and may be useful to improve the stability of the agricultural ecosystem. They play an important ecological role by supporting a complex of beneficial arthropods that aid in suppressing pest populations and thus the pest damage [33–36]. Strip management with weeds or flowering crops is by now standard in many sustainable perennial cropping systems such as orchards (e.g., [37–39]). Weed strips or living mulches, such as grasses or clover species, sown between winter cereals increase ground beetle densities and the number of species considerably by providing these beneficial arthropods with better food supplies and more suitable overwintering sites, from which they can colonize cereals in spring [40].

The abundance and diversity of entomophagous insects within a field are closely related to the character of the surrounding vegetation but also to the field size [6,41]. There are many examples that indicate that crops cultivated near hedgerows, grassy margins, or uncultivated fields with flowering weeds sustain less damage by insect pests than crops cultivated in the absence of such flowering vegetation [33–36]. Nevertheless, it is important to consider the complete life cycle and feeding habit of pathogens and insect pests. Certain weeds and structure elements will serve as alternate hosts to crop pathogens. Carrot flies (*Psila rosae*) need hedges nearby to protect themselves from heat and many aphids overwinter in woody plants. There, adequate distances to fields or the inclusion of specific border strips as trap crops need to be considered [7]. In most cases, field margins, consisting of hedges, sown grasses, and flower strips, for example, provide more beneficial effects through natural enemies regulating pest populations than detrimental effects through pest habitat provision [42].

Future Concerns

Diversity provides an essential key to reduce the risk of losing crop yield to pest damage. Of immediate concern are the effects of climate vagaries and change that will lead to the permanent change and invasions of new pests to new areas [43]. Many pesticides may disappear from the market in the near future due to their detrimental effects. Thus, massive insecticide use in combination with increasing field sizes and simplified cropping sequences were made responsible for the drastic insect decline in the past decades [44], while glyphosate and other herbicides are antibiotics and suspected of being responsible for the development of multi-resistance with direct implications on human health [45–48] including in bee decline [46]. The realization of greater crop diversity by crop rotation and trap crops and surrounding vegetation to stabilize overall crop yields will be of even greater importance in the face of climate change and the need to reduce pesticide use. The most effective way to increase general biodiversity is to simply reduce overall field sizes, though [6]. The implementation of concepts based on crop diversity will preserve long-term stability and productivity of agricultural land and minimize environmental problems caused by intensive agriculture, for example, biodiversity loss, soil erosion, groundwater and air pollution with nutrients and pesticides.

See also *Intercropping for Pest Management*.

References

1. Li, X. and K.H.M. Siddique, *Future Smart Food: Rediscovering Hidden Treasures of Neglected and Underutilized Species for Zero Hunger in Asia*. 2018, Bangkok: FAO. 242.
2. Storkey, J., et al., The future of sustainable crop protection relies on increased diversity of cropping systems and landscapes, in *Agroecosystem Diversity*, G. Lemaire, et al., Editors. 2019, London: Academic Press. pp. 199–209.

3. Heap, I. *The International Survey of Herbicide resistant Weeds.* https://www.weedscience.com, 2019.

4. Lamichhane, J.R., et al., Advocating a need for suitable breeding approaches to boost integrated pest management: a European perspective. *Pest Management Science,* 2018. **74**(6): pp. 1219–1227.

5. Ostergaard, H., et al., Time for a shift in crop production: embracing complexity through diversity at all levels. *Journal of the Science of Food and Agriculture,* 2009. **89**: pp. 1439–1445.

6. Sirami, C., et al., Increasing crop heterogeneity enhances multitrophic diversity across agricultural regions. *Proceedings of the National Academy of Sciences,* 2019. **116**(33): pp. 16442–16447.

7. Finckh, M.R. and M.S. Wolfe, Biodiversity enhancement, in *Plant Diseases and Their Management in Organic Agriculture,* M.R. Finckh, A.H.C.V. Bruggen, and L. Tamm, Editors. 2015, St. Paul, MN: APS. pp. 153–174.

8. Bedoussac, L., et al., Grain legume–cereal intercropping systems, in *Achieving Sustainable Cultivation of Grain Legumes,* S. Sivasankar, et al., Editors. 2018, Cambridge: Burleigh Dodds Science Publishing. pp. 243–255.

9. Cook, R.J. and K.F. Baker, *The Nature and Practice of Biological Control of Plant Pathogens.* 1983, St. Paul, MN: APS Press. pp. 1–539.

10. Finckh, M.R., et al., Organic temperate legume disease management, in *Plant Diseases and Their Management in Organic Agriculture,* M.R. Finckh, A.H.C. van Bruggen, and L. Tamm, Editors. 2015, St. Paul, MN: APS Press. pp. 293–310.

11. Leoni, C., W. Rossing, and A.H.C. van Bruggen, Crop rotation, in *Plant Diseases and Their Management in Organic Agriculture,* M.R. Finckh, A.H.C. van Bruggen, and L. Tamm, Editors. 2015, St. Paul, MN: APS Press. pp. 127–140.

12. Bond, W. and A.C. Grundy, Non-chemical weed management in organic farming systems. *Weed Research,* 2001. **41**(5): pp. 383–405.

13. Brandsæter, L.O., et al., Control of perennial weeds in spring cereals through stubble cultivation and mouldboard ploughing during autumn or spring. *Crop Protection,* 2017. **98**: pp. 16–23.

14. Liebman, M. and A.S. Davis, Integration of soil, crop and weed management in low-external-input farming systems. *Weed Research.* 2000. **40**: pp. 27–47.

15. Liebman, M.L. and E. Dyck, Crop rotation and intercropping strategies for weed management. *Ecological Applications,* 1993. **3**: pp. 92–122.

16. Belz, R. and K. Hurle, Weed suppression by crops – Which role play allelochemicals? *Mitteilungen aus der Biologischen Bundesanstalt für Land- und Forstwirtschaft.* 2000. p. 487.

17. Kolberg, D., et al., Effect of rhizome fragmentation, clover competition, shoot-cutting frequency, and cutting height on quackgrass (Elymus repens). *Weed Science,* 2017: pp. 1–11.

18. Schipanski, M.E., et al., Balancing multiple objectives in organic feed and forage cropping systems. *Agriculture, Ecosystems & Environment,* 2017. **239**: pp. 219–227.

19. Tautges, N.E., et al., Competitive ability of rotational crops with weeds in dryland organic wheat production systems. *Renewable Agriculture and Food Systems,* 2017. **32**(1): pp. 57–68.

20. Bergkvist, G., et al., Control of *Elymus repens* by rhizome fragmentation and repeated mowing in a newly established white clover sward. *Weed Research,* 2017. **57**(3): pp. 172–181.

21. Blackshaw, R.E., et al., Tillage intensity and crop rotation affect weed community dynamics in a winter wheat cropping system. *Canadian Journal of Plant Science,* 2001. **81**(4): pp. 805–813.

22. Ruisi, P., et al., Weed seedbank size and composition in a long-term tillage and crop sequence experiment. *Weed Research,* 2015: pp. 1–9.

23. Schmidt, J.H., S. Junge, and M.R. Finckh, Cover crops and compost prevent weed seed bank buildup in herbicide-free wheat–potato rotations under conservation tillage. *Ecology and Evolution,* 2019. **9**(5): pp. 2715–2724.

24. Davis, A.S., K.A. Renner, and K.L. Gross, Weed seedbank and community shifts in a long-term cropping systems experiment. *Weed Science,* 2005. **53**(3): pp. 296–306.

25. Chellemi, D.O., A.H.C. van Bruggen, and M.R. Finckh, Direct control of soilborne diseases, in *Plant Diseases and their Management in Organic Agriculture*, M.R. Finckh, A.H.C. van Bruggen, and L. Tamm, Editors. 2015, St. Paul, MN: APS Press. pp. 217–226.

26. Hallmann, J. and S. Kiewnick, Diseases caused by nematodes in organic agriculture, in *Plant Diseases and their Management in Organic Agriculture*, M.R. Finckh, A.H.C. van Bruggen, and L. Tamm, Editors. 2015, St. Paul, MN: APS Press. pp. 91–105.

27. Kato-Noguchi, H., J. Mizutani, and K. Hasegawa, Allelopathy of oats. II. Allelochemical effect of L-tryptophan and its concentration in oat root exudates. *Journal of Chemical Ecology*, 1994. **20**(2): pp. 315–319.

28. Schlatter, D., et al., Disease suppressive soils: new insights from the soil microbiome. *Phytopathology*, 2017. **107**(11): pp. 1284–1297.

29. Weller, D.M., et al., Microbial populations responsible for specific soil suppressiveness to plant pathogens. *Annual Review of Phytopathology*, 2002. **40**:pp. 309–348.

30. Kinkel, L.L., M.G. Bakker, and D.L. Schlatter, A coevolutionary framework for managing disease-suppressive soils. *Annual Review of Phytopathology*, 2011. **49**: pp. 47–67.

31. Chaube, H.S. and U.S. Singh, Adjustment of crop culture to minimize disease, in *Plant Disease Management: Principles and Practices*, H.S. Chaube and U.S. Singh, Editors. 1991, CRC Press: Boca Raton, FL. pp. 199–214.

32. Müller, J., Der Einfluß von Larvenschlupf und Weibchenentwicklung bei Heterodera schachtii auf die Resistenzbewertung kreuzblütiger Zwischenfrüchte. *Nachrichtenblatt des Deutschen Pflanzenschutzdienstes*, 1989. **41**(4): pp. 56–56.

33. Altieri, M.A., *Biodiversity and Pest Management in Agroecosystems*. 1994, Haworth Press: New York.

34. Andow, D.A., Vegetational diversity and arthropod population responses. *Annual Review Of Entomology*, 1991. **36**: pp. 561–586.

35. Tooker, J.F. and S.D. Frank, Genotypically diverse cultivar mixtures for insect pest management and increased crop yields. *Journal of Applied Ecology*, 2012. **49**(5): pp. 974–985.

36. Letourneau, D.K., et al., Does plant diversity benefit agroecosystems? A synthetic review. *Ecological Applications*, 2011. **21**: pp. 9–21.

37. Tamm, L., I. Pertot, and D. Gubler, Organic grape disease management, in *Plant Disease Management in Organic Agriculture*, M.R. Finckh, A.H.C. van Bruggen, and L. Tamm, Editors. 2015, St. Paul, MN: APS Press. pp. 335–350.

38. Holb, I. and B. Heijne, Organic apple disease management, in *Plant Disease Management in Organic Agriculture*, M.R. Finckh, A.H.C. van Bruggen, and L. Tamm, Editors. 2015, St. Paul, MN: APS Press. pp. 316–334.

39. Tamm, L., H. Willer, and A.H.C. van Bruggen, Organic perennial crop production, in *Plant Diseases and their Management in Organic Agriculture*, M.R. Finckh, A.H.C. van Bruggen, and L. Tamm, Editors. 2015, St. Paul, MN: APS Press. pp. 33–41.

40. Hartwig, N.L. and H.U. Ammon, Cover crops and living mulches. *Weed Science*, 2002. **50**(6): pp. 688–699.

41. Martínez, E., et al., Habitat heterogeneity affects plant and arthropod species diversity and turnover in traditional cornfields. *PLoS One*, 2015. **10**(7): p. e0128950.

42. Marshall, E.J.P. and A.C. Moonen, Field margins in northern Europe: their functions and interactions with agriculture. *Agriculture, Ecosystems & Environment*, 2002. **89**(1): pp. 5–21.

43. Juroszek, P. and A.V. Tiedemann, Plant pathogens, insect pests and weeds in a changing global climate: a review of approaches, challenges, research gaps, key studies, and concepts. *Journal of Agricultural Science*, 2013. **151**: pp. 163–188.

44. Hallmann, C.A., et al., More than 75 percent decline over 27 years in total flying insect biomass in protected areas. *PLoS One*, 2017. **12**(10): p. e0185809.

45. Van Bruggen, A.H.C., et al., Environmental and health effects of the herbicide glyphosate. *Science of the Total Environment*, 2018. **616–617**: pp. 255–268.

46. Motta, E.V.S., K. Raymann, and N.A. Moran, Glyphosate perturbs the gut microbiota of honey bees. *Proceedings of the National Academy of Sciences*, 2018. **115**: pp. 10305–10310.

47. Kurenbach, B., et al., Agrichemicals and antibiotics in combination increase antibiotic resistance evolution. *Peer-reviewed Journal*, 2018. **6**: p. e5801.

48. Kurenbach, B., et al., Sublethal exposure to commercial formulations of the herbicides dicamba, 2,4-dichlorophenoxyacetic acid, and glyphosate cause changes in antibiotic susceptibility in *Escherichia coli* and *Salmonella enterica serovar typhimurium*. mBio, 2015. **6**(2): pp. e00009–e00015.

42

Pest Management: Intercropping

Maria R. Finckh

Introduction

Until the past few hundred years, agricultural systems were based on large numbers of different crops, crop varieties, and landraces that were heterogeneous in their genetic make-up [1]. In addition, farming systems included both animals and plants further increasing diversity. As a result of increasing specialization, mechanization, and modern plant breeding, diversity on the farming system level and crop level has been drastically reduced worldwide at an ever-accelerating speed especially over the past 100 years. Fewer and fewer varieties that are genetically homogeneous are being grown in ever-larger fields [2]. The large-scale monoculture agricultural practices relying on pesticide and mineral fertilizer inputs have also led to a general decline in soil organic carbon and with this in soil microbial diversity and activity [3,4] with consequent effects on soil fertility and health [5].

Monoculture is usually understood to be the continuous use of a single crop species over a large area. However, with respect to plant pathogens and pests, it is important to differentiate between monoculture at the level of *species*, *variety*, or *resistance genes* (see Table 1) [1,6]. For example, within a species, there may be many different genotypes with different resistances to a specific pest or pathogen and great variation with respect to competitiveness with weeds and other crops. Depending on the breeding system, within a variety, diversity for resistance or morphological traits can be non-existent or high. Especially clonally reproduced or strictly inbreeding species may contain no diversity within a given

TABLE 1 Possibilities for Intercropping at Three Levels of Uniformity on Which Monocultures Are Commonly Practiced [1]

Level of Uniformity	Intercropping Possibilities
Species: Different individuals may differ in genetic make-up (resistance, morphology, etc.)	Arrangements among and within species, varieties, and resistances using intercropping
Variety: Usually genetically uniform, the same gene(s) in the same genetic background	Arrangements among and within varieties and resistances—includes variety mixtures, multilines, and populations
Resistance gene: The same gene may exist in different genetic backgrounds	Arrangements among resistances—multilines and populations

variety. Resistance gene monocultures can arise if different varieties all possess the same resistance (or susceptibility) gene(s). For example, in the late 1960s, virtually all hybrid maize cultivars in the south-eastern United States possessed the cytoplasmically inherited Texas male sterility (*Tms*). Unfortunately, *Tms* is closely linked to susceptibility to certain strains of the pathogen *Cochliobolus carbonum* (syn. *Helminthosporium maydis*). The monoculture for susceptibility (while different varieties had been planted) led to selection for these strains and in 1970 the pathogen caused more than \$1 billion (= 10^9) losses [7]. Currently, Europe is experiencing the decline of the European ash (*Fraxinus excelsior*) due to the invasion of a novel pathogen (*Hymenoscyphus fraxineus*) for which there is almost no resistance present, that is, a monoculture of susceptibility [8].

Intercropping [9] can be practiced at the species, variety, and gene level (Table 1) with effects on pathogens [1,6,10], insect pests [11,12], and weeds [13–15] (Table 2). One of the most important considerations for the successful design of intercropping systems for pest control is the achievement of *functional diversity*, that is, diversity that limits pathogen and pest expansion and that is designed to make use of knowledge about host–pest/pathogen interactions to direct host and pathogen evolution [1,10]. Moreover, functional diversity is also a matter of complementary use of resource niches, for example, deep versus shallow rooting, legume versus non-legume crops. A famous example, not only in terms of functional diversity but also of human diet, is the successful intercropping of the "Three sisters", namely maize, bean, and winter squash, in Central and North America since 3500 B.C. [16].

Protection Mechanisms Acting in Intercropped Systems

Pathogens, insect pests, and weeds differ fundamentally in their biology and their effects on crops, and different protection mechanisms act with respect to these organisms (Table 2).

Pathogens are mostly dispersed through wind, water splash, soil, and animals (vectors). In intercropped systems, the most important mechanisms for disease control are mechanical distance and barrier effects and changes in microclimatic conditions due to differences in plant architecture. In addition, resistance reactions induced by avirulent pathogen strains may prevent or delay infection by virulent strains. A large percentage of the reduction of airborne diseases such as the powdery mildews and rusts in cereal cultivar mixtures has been shown to be due to induced resistance [17,18]. The protection mechanisms are universal with respect to airborne, splashborne, and some soilborne, diseases and they may be enhanced by pathogen diversity that, in turn, is enhanced by plant diversity [19]. Mixtures of plants varying in reaction to a range of diseases will lead to a multitude of additional interactions and the overall response in such populations will tend to correlate with the disease levels of the components that are most resistant to these diseases. In addition, less affected plants may compensate for yield losses due to reduced competition from diseased neighbors [1].

In contrast to pathogens for which passive or vectored dispersal is the norm, insects often search actively for their hosts. Thus, behavioral, visual, and olfactory cues play an important role. While environmental factors and landing on a non-host is likely the most important mortality factor for pathogens, natural enemies are at least as important for insect population dynamics [11,12]. Host dilution may affect an insect's ability to see and/or smell its hosts. Predators and parasitoids are dependent on the constant presence of prey and alternative food sources, such as pollen and nectar, in the absence of the hosts and for effective control of insect pests, the presence of sufficient numbers of natural enemies is critical. The importance of natural enemies was often only recognized after insecticide applications induced pest resurgence due to the destruction of natural enemy populations. Intercrops and weeds therefore can play an important role in regulating insect pests. Plant–insect communication also plays a role. For example, plants may signal their neighbors about insect attacks leading to the production of antinutritive compounds or attractants for natural enemies [20,21].

Weeds usually are early successional plants adapted to colonize open, nutrient-rich spaces. Intercrops, especially cover and mulch crops, directly compete with weeds for these spaces and also for light. As many weeds are adapted to certain crops and cropping patterns, changing these patterns (e.g., rotations)

TABLE 2 Mechanisms Affecting Pathogens, Insect Pests, and Weeds in Intercropped Systems and Selected Additional Interactions of Importance [1,13,19,29]

Mechanisms Reducing Disease
Increased distance between susceptible plants
Barrier effects of intercrop
Variation in plant architecture may lead to less humid microclimate
Induced resistance
Selection for the most resistant and/or competitive genotypes
Interactions among pathogen strains on host plants
Increased microbial diversity and activity above and below ground
Mechanisms Reducing Insect Pests
Enhancement of natural enemies
Reduction of host density (reduced resource concentration)
Reduction of plant apparency (visual or olfactory cues reduced)
Alteration of host quality (with respect to the insect pest) through plant–plant and plant–microbe interactions
Increased microbial diversity and activity above and below ground
Mechanisms Reducing Weeds
Reduction of bare soil and layering of crops (increased competition for light, water, and nutrients)
Variation in tillage needs and operations of intercrops may disturb weeds
Other Beneficial Interactions
Yield enhancement through niche differentiation of hosts
Compensation for yield losses by less affected hosts
Better soil cover with intercrop (soil and water conservation, microclimatic effects)
Possible Unwanted Interactions
Weeds may serve as alternate hosts for pathogens and insects
Interactions among virus vectors and weeds
Greater difficulty to specifically reduce weeds with herbicides or mechanically
Overall denser intercropped stands may produce a more humid microclimate that may enhance certain problems

and management operations (e.g., sowing time) connected with different kinds of crops within the same field impede the adaptation and dominance of (problem) weeds. Also, filling the spaces that usually would be taken up by weeds with useful or more neutral plants will reduce weed habitat and help outcompeting them [13,14,22]. An important consideration is that plants may be weeds only during certain phases of crop development. At other stages, the presence of the same "weeds" may be beneficial because they may provide food and habitat for beneficial insects and erosion control.

Besides the many positive effects of intercrops, it is important to keep in mind that weeds may serve as alternative hosts for insect pests and pathogens and that insects often are disease vectors, especially for viruses that may reside symptomless in certain weeds [1].

In order to understand the many interactions in intercropped systems, it is indispensable not only to consider plants, insects, and pathogens in the system but also the whole microbiome that interacts with them [23]. Plants not only take up nutrients from the soil, but they also actively release organic carbon-based chemicals such as organic acids and sugars into the soil. Roughly speaking about 30%–50% of the carbon that plants assimilate through photosynthesis is released into the soil [24]. These compounds provide the energy (carbon) and organic acids necessary for the microorganisms in order to function as well as for supporting the weathering of soil minerals making them available as plant nutrients. In addition, most plants live in symbiosis with root infecting mycorrhizal fungi that usually are very specific and genetically diverse [25]. The diversity of root exudates and in turn the diversity of the soil microbiome is greatly enhanced by plant diversity including intraspecific diversity (e.g., [26]), and by this, many of the interactive processes between crops, weeds, insects, and pathogens are influenced. For example, high microbial diversity and activity in the soil usually supports resistance induction and direct disease suppression [19,27]. Also, suppression of the production of certain secondary metabolites by plants can be triggered by certain rhizosphere bacteria. In turn, this may reduce their attractiveness to certain insects. In the case of cucumbers, this is helpful in suppressing the spotted and striped cucumber beetles (*Diabrotica undecimpunctata* and *Acalymma vittata*, respectively) that act as vectors (transmitting agents) of bacterial wilt of cucumber due to *Erwinia tracheiphila* [28].

Intercropping in Practice

Variety mixtures and multilines are used mainly to control diseases. For example, they are used in cereals on a commercial scale in the United States, Denmark, Finland, Poland, and Switzerland to control rusts, mildews, and certain soilborne diseases (e.g., *Cephalosporium* stripe). When barley cultivar mixtures were used on more than 360,000 ha in the former German Democratic Republic, powdery mildew of barley and consequently fungicide input was reduced by 80% within 5 years [1]. Wheat cultivar mixtures and multilines are grown on several hundred thousand hectares in the United States in the Pacific Northwest and in Kansas to protect against diseases and abiotic stresses. In Colombia, coffee multilines are grown on more than 700,000 ha to control coffee rust [1].

Attention has also been called to possible beneficial effects of greater intravarietal diversity in the oat-frit fly (*Oscinella frit* L.) system. The flies can attack the host plants only at a particular growth stage and a higher degree of variability within an oat crop could allow for escape from attack and subsequent compensation.

Cereal species mixtures for feed production have been reported as important practice in Poland and have been shown consistently to restrict diseases. In Switzerland, the "maize-ley" system (i.e., maize planted without tillage into established leys), which is being promoted to reduce soil losses and nutrient leaching, has been shown to reduce smut disease and attacks by European stem borer and aphids [30].

The deliberate planting or maintenance of flowering weeds and grass in established vineyards and orchards in Switzerland and Germany greatly increases natural enemies while reducing soil erosion. This practice is becoming increasingly popular world-wide if water is not the limiting factor in the system.

The required reduction in insect populations for effective reduction of insect-transmitted diseases may be beyond that which can be achieved by diversity alone. However, simultaneous reduction of insect vectors, for example, by enhancing natural enemies and plant resistance and diversity and disease inoculum, for example, by dilution of the susceptible hosts through intercropping with resistant varieties or non-host species can be effective [31,32]; in addition, interactions with the soil microbiome making hosts less attractive to their insect pests as described above [28] may play a role.

Future Concerns

There are many reasons why intercropping is not practiced more widely. First, modern crops are bred to be grown in monoculture and may not necessarily be well adapted to intercropping. Efforts of breeders to produce breeding lines adapted to intercropping need to be strengthened. Breeding for intra-and inter-specific diversity has recently gained momentum [10,33]. EU legislation has been put into place to support these activities at least for the organic sector [34], and there is a need to expand this into the agricultural sector as a whole. Second, while intercropping clearly provides a means for reducing pesticide needs, there is a lack of adapted machinery allowing for efficient management of intercropped cultures. Third, successful intercropping strategies have to be carefully designed as preventive measures that include the whole growing system and rotational cycle [23], while monoculture cropping is very simple due to the availability of (very) cheap fertilizer and pesticides that can be applied once a problem occurs. However, pesticides are more and more criticized for their unintended side-effects causing insect decline [35] and affecting human and environmental health [36,37] and alternative approaches based on applied agroecology are needed [38]. Fourth, a concern often raised is the quality of products raised as intercrops such as varietal mixtures of cereals. In some countries, there is resistance in the food processing industry to such products. However, such problems could be overcome if breeders, producers, and processors work together. For example, in the 1980s, in the German Democratic Republic, first-quality malting barley was produced in large-scale mixtures in collaboration with breeders, growers, and processors [39]. Also, some of the highest-quality coffee in the world from Colombia is produced from multilines demonstrating that quality produce is not dependent on genetic uniformity. For the use of products grown from species mixtures, technological advances (e.g., color- or shape-based optical and/or mechanical separation devices during harvest or processing) need to be implemented.

References

1. Finckh, M.R. and M.S. Wolfe, Biodiversity enhancement, in *Plant Diseases and Their Management in Organic Agriculture*, M.R. Finckh, A.H.C.V. Bruggen, and L. Tamm, Editors. 2015, St. Paul, MN: APS. pp. 153–174.
2. Agardy, T., et al., *Ecosystems and Human Well-Being: Biodiversity Synthesis. Millennium Ecosystem Assessment*. 2005, Washington, DC: World Resources Institute.
3. Bellamy, P.H., et al., Carbon losses from all soils across England and Wales 1978–2003. *Nature*, 2008. **437**: pp. 245–248.
4. Capriel, P., Trends in organic carbon and nitrogen contents in agricultural soils in Bavaria (south Germany) between 1986 and 2007. *European Journal of Soil Science*, 2013. **64**: pp. 445–454.
5. Cook, R.J., Mangement of resident plant growth-promoting rhizobacteria with the cropping system: a review of experience in the US Pacific Northwest. *European Journal of Plant Pathology*, 2007. **119**: pp. 255–264.
6. Finckh, M.R. and M.S. Wolfe, Diversification strategies, in *The Epidemiology of Plant Diseases*, B.M. Cooke, D. Gareth Jones, and B. Kaye, Editors. 2006, Heidelberg: Springer-Verlag. pp. 269–307.
7. Ullstrup, A.J., The impacts of the southern corn leaf blight epidemics of 1970–1971. *Annual Review Phytopathology*, 1972. **10**: pp. 37–50.
8. Landolt, J., et al., Ash dieback due to *Hymenoscyphus fraxineus*: what can be learnt from evolutionary ecology? *Plant Pathology*, 2016. **65**(7): pp. 1056–1070.
9. Vandermeer, J.H., *The Ecology of Intercropping*. 1989, Cambridge, New York, Melbourne: Cambridge University Press. pp. 1–235.
10. Finckh, M.R., Integration of breeding and technology into diversification strategies for disease control in modern agriculture. *European Journal of Plant Pathology*, 2008. **120**: pp. 399–409.
11. Tooker, J.F. and S.D. Frank, Genotypically diverse cultivar mixtures for insect pest management and increased crop yields. *Journal of Applied Ecology*, 2012. **49**(5): pp. 974–985.
12. Letourneau, D.K., et al., Does plant diversity benefit agroecosystems? A synthetic review. *Ecological Applications*, 2011. **21**: pp. 9–21.
13. Liebman, M.L. and E. Dyck, Crop rotation and intercropping strategies for weed management. *Ecological Applications*, 1993. **3**: pp. 92–122.
14. Liebman, M.L. and E.R. Gallandt, *Many little hammers: ecological management of crop-weed interactions*, in *Ecology in Agriculture*, L.E. Jackson, Editor. 1997, San Diego, CA, New York, London: Academic Press. pp. 291–343.
15. Poggio, S.L., Structure of weed communities occurring in monoculture and intercropping of field pea and barley. *Agriculture Ecosystems & Environment*, 2005. **109**: pp. 48–58.
16. Hart, J.P., Evolving the three sisters: the changing histories of maize, bean, and squash in New York and the Greater Northeast, in *Current Northeast Paleoethnobotany II*, J.P. Hart, Editor. 2008, Albany, NY: The University of the State of New York. pp. 87–99.

17. Calonnec, A., H. Goyeau, and C. Devallavieillepope, Effects of induced resistance on infection efficiency and sporulation of *Puccinia striiformis* on seedlings in varietal mixtures and on field epidemics in pure stands. *European Journal of Plant Pathology*, 1996. **102**: pp. 733–741.

18. Walters, D.R., Are plants in the field already induced? Implications for practical disease control. *Crop Protection*, 2009. **28**: pp. 459–465.

19. Finckh, M.R. and L. Tamm, Organic management and airborne diseases, in *Plant Diseases and their Management in Organic Agriculture*, M.R. Finckh, A.H.C.V. Bruggen, and L. Tamm, Editors. 2015, St. Paul, MN: APS Press. pp. 53–66.

20. Ninkovic, V., R. Glinwood, and J. Pettersson, Communication between undamaged plants by volatiles: the role of allelobiosis, in *Communication in Plants*, F. Baluka, S. Mancuso, and D. Volkmann, Editors. 2006, Berlin: Springer. pp. 421–434.

21. Ninkovic, V., U. Olsson, and J. Pettersson, Mixing barley cultivars affects aphid host plant acceptance in field experiments. *Entomologia Experimentalis et Applicata*, 2002. **102**(2): pp. 177–182.

22. Finckh, M.R. and A.H.C.V. Bruggen, Organic production of annual crops, in *Plant Diseases and their Management in Organic Agriculture*, M.R. Finckh, A.H.C.V. Bruggen, and L. Tamm, Editors. 2015, St. Paul, MN: APS Press. pp. 25–32.

23. Bruggen, A.H.C.V., et al., One health—Cycling of diverse microbial communities as a connecting force for soil, plant, animal, human and ecosystem health. *Science of the Total Environment*, 2019. **664**: pp. 927–937.

24. Marschner, H., *Marschner's Mineral Nutrition of Higher Plants (Third Edition)*, P. Marschner, Editor. 2012, San Diego, CA: Academic Press.

25. Smith, S.E. and D.J. Read, *Mycorrhizal Symbiosis*. 2010, New York: Elsevier Science.

26. Mazzola, M. and Y.-H. Gu, Wheat genotype-specific induction of soil microbial communities suppressive to disease incited by *Rhizoctonia solani* anastomosis group (AG)-5 and AG-8. *Phytopathology*, 2002. **92**: pp. 1300–1307.

27. Schlatter, D., et al., Disease suppressive soils: new insights from the soil microbiome. *Phytopathology*, 2017. **107**(11): pp. 1284–1297.

28. Zehnder, G.W., et al., Application of rhizobacteria for induced resistance. *European Journal of Plant Pathology*, 2001. **107**: p. 39–50.

29. Andow, D.A., Vegetational diversity and arthropod population responses. *Annual Review of Entomology*, 1991. **36**: pp. 561–586.

30. Finckh, M.R., et al., Cereal variety and species mixtures in practice, with emphasis on disease resistance. *Agronomie*, 2000. **20**: pp. 813–837.

31. Power, A.G., *Competition between viruses in a complex plant-pathogen system. Ecology*, 1996. **77**: pp. 1004–1010.

32. Finckh, M.R. and M.S. Wolfe, *The use of biodiversity to restrict plant diseases and some consequences for farmers and society*, in *Ecology in Agriculture*, L.E. Jackson, Editor. 1997, San Diego, CA: Academic Press. pp. 199–233.

33. Baćanović-Šišić, J., D. Dennemoser, and M.R. Finckh. *Symposium on breeding for diversification. A joint meeting of the EUCARPIA section, organic and low-input agriculture*, ECO-PB, LIVESEED, INSUSFAR, DIVERSify, HealthyMinorCereals, ReMIX, and Wheatamix University of Kassel, 19th–21st February 2018, Witzenhausen, Germany. 2018. Kassel University Press.

34. EC, Commission implementing decision of 9 October 2018 amending implementing decision 2014/150/EU on the organisation of a temporary experiment providing for certain derogations for the marketing of populations of the plant species wheat, barley, oats and maize pursuant to Council Directive 66/402/EEC. *Official Journal of European Union*, 2018. **L**: p. L256/65–L256/66.

35. Hallmann, C.A., et al., More than 75 percent decline over 27 years in total flying insect biomass in protected areas. *PLoS One*, 2017. **12**(10): p. e0185809.

36. Bruggen, A.H.C.V., et al., Environmental and health effects of the herbicide glyphosate. *Science of the Total Environment*, 2018. **616**: pp. 255–268.

37. Kurenbach, B., et al., *Agrichemicals and antibiotics in combination increase antibiotic resistance evolution. Peer-Reviewed Journal*, 2018. **6**: p. e5801.
38. Finckh, M.R., et al., Disease and pest management in organic farming: a case for applied agroecology, in *Improving Organic Crop Cultivation*, Vol. 1, U. Köpke, Editor. 2018, Cambridge: Burleigh Dodds. pp. 271–301.
39. Wolfe, M.S., Barley diseases: maintaining the value of our varieties, in *Barley Genetics VI*, L. Munk, Editor. 1992, Copenhagen: Munksgaard International Publishers, Ltd. pp. 1055–1067.

43

Precision Agriculture: Water and Nutrient Management

Robert J. Lascano,
Timothy S. Goebel,
and J.D. Booker

Introduction

Management of agronomic inputs, such as water and fertilizers, to cropping systems requires information on when and how much of each input to apply. The correct management decision is essential to maintain the productivity of the cropping system, and a management strategy is to use crop production functions that describe (mathematically) maximum economic yield (profit) as a function of agronomic inputs. Essentially, the strategy is to apply the least amount of input to produce the maximum economic crop yield in the farming operation. In conventional cropping systems, inputs are generally applied uniformly across a field regardless of their need, and the amount applied is normally based on average responses of these inputs to crop yield across the field. Since the 1970s, increased costs of crop production and emphasis on production efficiency revolutionized production agriculture. Environmental concerns, including quality and safety of harvested products and impact of the cropping procedures on the ecosystem, are also considered in the decision-making process. Developments in spatial statistics and computer hardware (i.e., increased microprocessor speed and decreased cost), and increased availability of soil, elevation, and weather data have contributed to improved concepts and procedures to address spatial and temporal variability in cropping systems.

Precision agriculture (PA) and precision farming are generic terms that describe the way production management inputs (e.g., water, nutrients, harvest aids, and herbicides) are managed in response to cropping system variability. In contrast to a uniform blanket application of an input across the field, each input is applied according to specific needs across the field. PA is an integrated crop management system that attempts to match the kind and amount of inputs with the actual crop needs for small areas within a farm field. Perhaps a better descriptor for this type of farming is site-specific management, which manages an agricultural crop at a spatial scale smaller than the entire field by dividing it into management zones, defined by topography, soil type, and level of a particular nutrient, such as nitrogen and phosphorus.

Driving forces that have contributed to producers implementing and adopting PA procedures are advances in computer hardware and software, electronics, and equipment technology (sensors); decreased profit margins due to increased costs of production; and environmental awareness. For example, advances in crop growth simulation models, variable-rate application equipment, adoption of soil sampling for nutrients across the field and as a function of soil depth, and integration of crop yield monitors with global positioning satellite systems have contributed to the use of PA concepts to manage crops. The cost of agronomic inputs continues to increase, for example, petroleum-based inputs such as fertilizer, diesel, insecticides, and herbicides. Awareness of environmental concerns related to nutrient contamination of groundwater and surface water and the quality and safety of food and fiber are factors that impact how crops are produced and delivered to the consumer. Operating a farm requires management strategies that consider both economic and environmental consequences, and PA provides the concepts that can be used to achieve this goal.

In recent years, the development of sensors to measure soil properties[1,2] and of using remote sensing techniques[3,4] has increased the availability of georeferenced digital data used in PA. These large databases are generically referred to as "Big Data", and examples of their application to PA are given by Sonka[5] and by Fulton and Port.[6] Further, the availability of real-time weather information via networks known as Mesonets[7] and the use of software applications that are accessed via smart phones[2,8,9] has both modernized and revolutionized crop production and gives producers information to manage their farming operation more efficiently. Examples of these applications for nutrient and water management are given by Hedley,[10] Abit et al.[11] and by Neupane and Gao.[12] In PA, the rate of adoption and use of this technology will continue to increase. The factors that contribute to this implementation are the decreased cost of computer hardware, increased availability of input data, and development of computer software providing tools to manage the farming enterprise more efficiently. Also, tailoring production inputs to site-specific locations contributes to improve the sustainability of the food supply.[13]

This chapter gives a general overview of PA with emphasis on crop water and nutrient use, and crop yield using the state-space analysis to describe, for example, how cotton lint yield varies at a landscape scale. This entry is divided into five parts. First, we give a general overview of PA. Second, we describe the relation between crop yield and water use. Third, as examples, we show measurements of cotton water use along two 700 m transects on a 60-ha field and the use of a large-scale model to calculate the temporal and spatial variability of soil water content in a 70-ha field. Fourth, we use geostatistical tools to quantify cotton lint yield as a function of nitrogen, topography, and soil water. Fifth, we describe how the future of PA will leverage knowledge from PA research, such as that presented here, to further support the development and use of large-scale cropping system models.

Precision Agriculture

PA, also known as site-specific management,[13,14] refers to the practice of applying agronomic inputs across a farm, mainly fertilizers and other chemicals, at variable rates based on soil nutrients or chemical tests, soil textural changes, weed pressures, and/or yield maps for each field in the farm.[13–15] In large fields (e.g., >40 ha), crop yield and thus crop water and nutrient use are notoriously variable.[12,14,16] The sources of this variation are related to soil physical and chemical properties, pests, microclimate, genetic and phenological responses of the crop and their interactions.[16] The technology for crop yield mapping is more advanced than current methodologies for determining and understanding the causes of yield variability. Prevailing and traditional management practices treat fields uniformly as one unit. However, reports[15–18] show that to understand underlying soil processes that explain crop yield variability, research must be done at the landscape level and using appropriate statistical tools for large-scale studies.[18]

Crop Yield and Water Use

There is a linear relation between crop yield and water use when the only limiting factor is water[19]; however, root water uptake is synergistically related to nutrient uptake, and the two processes cannot be separated. Precision farming has the potential to improve water and nutrient use efficiency on large fields provided there is quantitative understanding of what factors affect crop water and nutrient use and how they vary across the field.[16,20,21] It is known that crop water and nutrient use are a function of many biotic and abiotic variables, including managed inputs, and harvestable yield is a manifestation of how these variables and inputs interact and are integrated during the growing season. However, it is difficult to determine a hierarchy on the contribution of each input and variable to the measured crop yield using classical statistics.[22-25] Often, variables that affect water and nutrient supply to the plant contribute to crop yield at a high level assuming an adequate plant stand and weed control. The cause-and-effect relation between a single state variable and crop yield is site specific and is difficult to establish without considerable sampling of the soil and/or crop. The establishment of response functions, that is, crop water and nutrient use as a function of variable x_i, gives only a partial answer to explain crop water use, nutrient use, and yield based on inputs. The general idea of PA is to optimize input application to the measured crop yield at each sampling location using the law of the optimum formulated by Liebscher, which states that a production factor that is in minimum supply contributes more to production, the closer other production factors are to their optimum.[26] This is a simple premise; however, the decisions for variable-rate application of any agronomic input must consider temporal and spatial variability of the soil's properties affecting crop growth, water and nutrient use, and yield. Soil factors that affect water storage, such as depth of the root-restricting layer and soil textural differences, must be considered in any precision farming operations that attempt to improve crop water use and yield related to agronomic inputs. Similarly, to improve the use of any micro- and macronutrient by the crop, the overall cycle of the nutrient must be considered, including its availability in the soil and demand by the crop. Examples of managing nitrogen fertilization and irrigation at a site-specific and farm scale for cotton production is given by Bronson et al.,[17] Li et al.,[18] and Booker et al.[27]

Precision farming must incorporate the inherent spatial and temporal variability of soil physical, chemical, and biological factors within a field for input management. Accurate representation of spatial and temporal variability in a field requires taking and analyzing many samples. Sampling is normally done on a grid with a scale that can vary from one to several hundred meters.[22,23] Once properties are measured, geostatistical tools (e.g., semivariogram, kriging, cokriging), and other spatial statistical tools (e.g., autocorrelation, cross-correlation, state-space analysis), can be used to establish statistical relations in space and to minimize the number of soil samples to characterize and map fields.[22,23,25] The number of samples required *a priori* to determine spatial and temporal variability is perhaps the single largest deterrent in the application of precision farming practices to manage and improve crop water and nutrient use. Collection of field data to characterize the spatial variability must remain a priority for any study that attempts to understand how to maximize crop yield across the landscape. The trend of using data generated from pedotransfer functions[28] rather than field measured data is of concern and results thus obtained are preliminary at best and should be verified using measured field data. Further, input data generated from pedotransfer functions and used to assess the spatial variability of soil properties can be used as a guide to establish a sampling scheme to measure the properties of interest. Input data generated from pedotransfer functions are not a substitute to field measured data.

There is not much information published on combined crop water and nutrient use across large fields at the landscape level and in the context of precision farming.[17,29-31] An exception is the study by Li et al.[18] where cotton water and total nitrogen uses were measured along a 700-m transect to illustrate the landscape pattern of cotton water and total nitrogen uses and to determine the underlying soil processes governing cotton lint yield variability. In this study, state-space analysis[18,22,25] was used to formulate management decisions that may improve crop water and nitrogen use and lint yield using precision farming practices. An additional study regarding variable-rate nitrogen at different locations within a

48-ha field is given by Bronson et al.,[17] and in a 14-ha field, by Booker et al.[27] A global-scale assessment showing that global yield crop variability is a function of fertilizer use, irrigation, and climate is given by Mueller et al.[32] and a review of integrated nutrient management to sustain crop productivity while minimizing the environmental impact is given by Wu and Ma.[33]

Landscape Crop Water Use

The concept of crop water and nitrogen use in a 60-ha field study is given by Li et al.[18] and Li and Lascano.[34] In 1999, a field experiment was conducted near Lamesa, Texas, on a research farm of Texas A&M University on the southern edge of the High Plains of Texas. The soil was classified as an Amarillo sandy loam. The field was 60 ha with slopes ranging between 0.3% and 6.3%.[18] To evaluate the effect of soil water, nitrate-nitrogen (NO_3-N), and topography on cotton lint yield across the landscape, two irrigation levels were used. The irrigation treatments consisted of water applications at a 50% and 75% grass reference evapotranspiration (ET_o) with a center-pivot low-energy precision application irrigation system.[35] At each irrigation level, one transect was established following the circular pattern of the center pivot. The two transects were instrumented with 50 neutron access tubes, each 15 m apart, and soil volumetric water content (θ_v) was measured periodically throughout the growing season. At each point, θ_v was measured in 0.3 m depth increments to 2.0 m depth using a neutron probe calibrated for this soil. In addition, at each transect point soil texture, soil and plant NO_3-N, leaf area index, cotton lint yield, slope, plant density, and other parameters were measured.[18,34]

A comprehensive study to evaluate site-specific management of cotton production systems at a landscape scale was done by Booker et al.[36,37] using the Precision Agricultural Landscape Modeling System (PALMS),[38] which was integrated with the cotton simulation model GOSSYM.[39] The combined model, PALMSCot, was applied to simulate a cotton crop irrigated with a 400-m center pivot system and covering a ¼-section of land (~65 ha). The PALMSCot model is grid-based (10-m resolution) and defines up to 26 soil depth layers and for each "cell" the water, energy, nutrient, and carbon balance is calculated on a ¼-h time interval for the length of the growing season. This model provides a detailed calculation of the water and nitrogen use for each cell across the field. For example, for an irrigated field with a 400-m-long center pivot sprinkler system, PALMSCot calculates the cotton lint yield, evapotranspiration, and nitrogen use for each of the 5,026 cells defined by the 10×10 m grid system (length and width) and the assigned soil depth.

Statistical Calculations

It has been shown that the use of classical statistics, such as regression analysis and analysis of variance, is designed to describe the strength of the covariance structure between variables and fails to completely explain the cause and effect between, for example, crop yield and measured soil variables in precision farming experiments.[18,22,23,25,31] These techniques, in general, account for spatial and temporal variability through blocking and do not describe the spatial and temporal structure. Instead, there are other more appropriate statistical tools for relating the variability of soil and plant parameters measured in space and time. For example, the structure of the spatial (or less often, temporal) variance between measurements may be derived from the sample semivariogram, which is the average variance between neighboring measurements separated by the same distance. Spatial or temporal structure between variables is often determined using autocorrelation and cross-correlation functions. Although these techniques can be used to evaluate the temporal variability structure, they are most often used in PA to analyze spatial variables. Autocorrelation measures the linear correlation of a variable in space along a transect. Cross-correlation is the comparison of two variables measured along a transect and is used to describe the spatial correlation between two landscape variables, that is, where one variable, the tail variable, lags behind the head variable by some distance. The spatial association between several variables can be described using state-space analysis, which is a multivariate autoregressive technique.[18,22]

Spatial Analysis of Crop Water Use

To illustrate the variability of crop water use or crop evapotranspiration (ET_c), values measured along the 50% irrigation transect were selected.[18] In Figure 1, the relation between the scaled ET_c and elevation, both as a function of distance along the transect, is shown. The ET_c data are scaled to the maximum value of 426 mm of water, which was measured 210 m from the south end of the transect. These results show, as expected and, that in general higher ET_c was measured at lower elevations and that ET_c tended to decrease at the higher elevations. Higher elevations are eroded and the depth of the root zone is shallower holding less water. In contrast, lower elevations tend to have more clay and thus hold more water.

Calculated values of soil volumetric water content for the surface 0.05 m for a center pivot irrigated cotton field in Floydada, Texas, are shown in Figure 2.[36,37] The four circles shown in Figure 2 illustrate the corresponding wetted soil area (blue) by the center pivot, with values of 45% soil water content over a 6-day period, starting on the 10 July and ending on the 16 July. This is an example of the spatial resolution, 20×20 m, and temporal variability that can be used to analyze crop water use across the landscape that can be obtained with a model such as PALMSCot.

Spatial cross-correlation between cotton lint yield and soil water, cotton lint yield and site elevation, and soil water and site elevation are shown in Figure 3. For a 95% confidence interval, the cotton lint yield was positively cross-correlated with soil θ_v across a lag distance of ±30 m. Cotton lint yield and θ_v were negatively cross-correlated with elevation at a lag distance of ±30 m. These results show the effect of topography on the θ_v and of crop water use measured along the transect. Similar results are given in other reports.[17,30,31,40] In this example, the cross-correlation between θ_v and elevation shows the spatial structure of measured variables and, further, that more water was stored in lower elevations, resulting in higher ET_c.

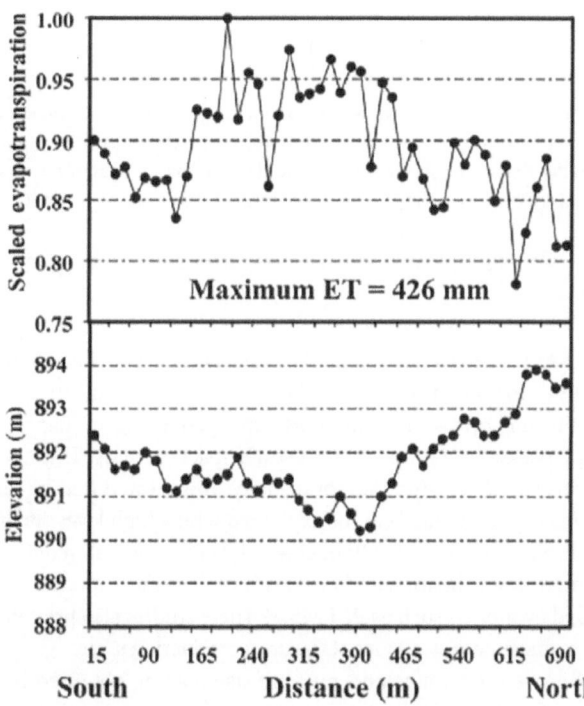

FIGURE 1 Scaled crop evapotranspiration and elevation as a function of distance along a 700-m transect.

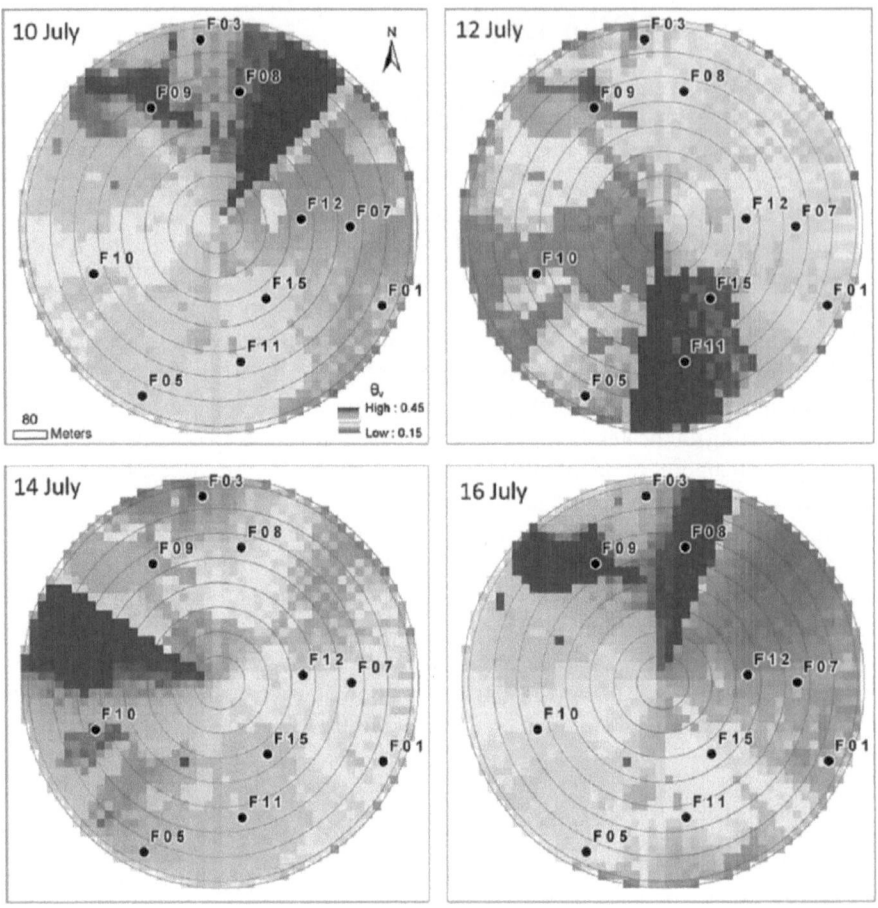

FIGURE 2   Calculated values of soil volumetric water content for the surface 0.05 m obtained with the PALMSCot model using 20×20 m grid cells for the period from 10 to 16 July 2011. This is a center pivot irrigated cotton production field located in Floydada, Texas. The soils at the site are classified as a Pullman clay loam with 0%–1% slope and an Olton clay loam with a 1%–3% slope. The blue zones correspond to irrigation water applied by the center pivot sprinkler system.
Source: Booker et al.[37]

Linear regression analysis between θ_v and cotton lint yield and relative site elevation is shown in Figure 4, and the state-space analysis for the relation between cotton lint yield and three measured parameters is shown in Figure 5. Results in Figure 4 show the shortcomings of using an inappropriate statistical tool to understand underlying processes explained with the state-space analysis. This analysis (Figure 5) quantified how cotton lint yields varied as a function of distance and showed that by using θ_v, soil NO_3-N and elevation the variation in cotton lint yield can be explained with a high level of confidence. While studies like those of Li et al.,[18] Bronson et al.,[17] and Booker et al.[27] are empirical in design, the relationships that are evaluated provide important validation and field testing of the more mechanistic mass and energy balance accounting provided by models such as PALMS.[38] These studies also provide foundational information for developing PA management strategies at the crop production scale.

Benefits of PA to improve crop water and nutrient use may be obtained by an economic analysis of maximizing crop yield as a function of application of nitrogen fertilizer and irrigation water as given by the state-space equation.[18,34] In the example given, the decision can be made to apply more

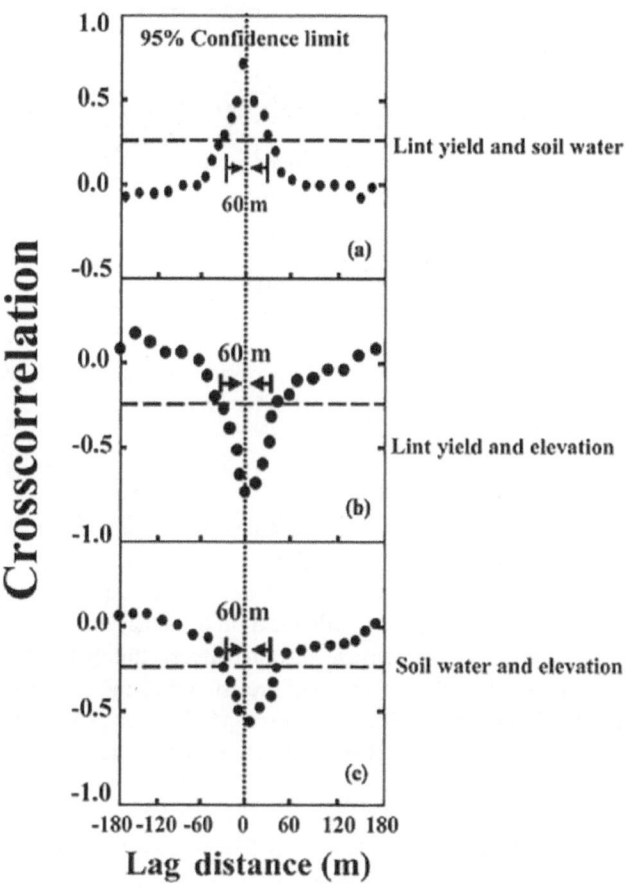

FIGURE 3 Cross-correlation as a function of lag distance. (a) Lint yield and soil water, (b) lint yield and elevation, and (c) soil water and elevation. Shown is the 95% confidence for the cross-correlation distance.
Source: Li et al.[18]

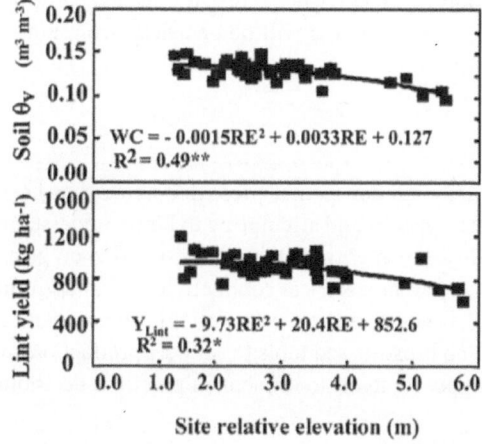

FIGURE 4 Soil water content (θ_v) and cotton lint yield as a function of site relative elevation.

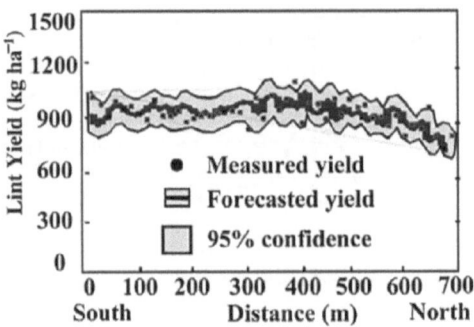

FIGURE 5   State-space equation relating cotton lint yield (Y) to water content (W), nitrogen level (N), and elevation (E) as a function of distance and location (i) along a 700-m transect. **Source:** Li et al.[18]

nitrogen fertilizer to the lower areas of the field that also hold more water and increase crop water use and nitrogen and lint yield. With the introduction of variable rate planters, it is possible to discriminate site locations and plant more "drought"-tolerant varieties or change the seeding rate in areas that are prone to have less soil water. This implies the delineation of management zones[14,15] within a field that are defined based on potential crop water use and their interaction with other input variables to maximize economic yield across the field. This type of precision farming is slowly being adopted, and wide use remains within the realm of possibilities that this type of farming has to offer. The introduction and use of computer models of cropping systems will likely expedite and facilitate the adoption of PA management techniques.[38]

$$Y_{(50\%ET)i} = -0.201\ Y_{i-1} + 1.107\ W_{i-1} + 0.332\ N_{i-1}D49.54\ E_{i-1} + \varepsilon_i$$

A final consideration is the cost/benefit of PA practices and its impact on agriculture. Currently, hardware for variable-rate application of agronomic inputs is relatively expensive and in many cases unavailable; however, with increased adoption and use of these practices, the cost will be reduced. For example, tractor guidance systems[41,42] were quickly adopted by producers, and high demand reduced its cost. Further, environmental and material cost concerns for a given area will probably place limits on the amount of certain nutrients, for example, nitrogen fertilizer, used for crop production. This will force producers to apply nitrogen and other nutrients across the field according to site-specific needs and position along the landscape. These practices will be beneficial from both an environmental and an economical point of view.[10,11,20,33]

Future of PA

Considerable PA-related research, similar to that presented above, has been conducted over the past decade, studying empirical relationships and attempting to better understand the underlying processes controlling crop yield variability. Much of this research has focused on grid soil and crop sampling,[43] surface characterization (e.g., apparent electrical conductivity),[44] and ground- or aerial-based remote sensing.[3,4,45,46] Such research has described numerous useful process relationships but has been somewhat less successful in providing broad-based tools to support production-scale management. The lack of development of decision-support systems to implement precision decisions is the major impediment to the adoption of PA.[47]

Recent advances in the availability of soil data provided by the U.S. Department of Agriculture, elevation data provided by the U.S. Geological Survey, and weather data from weather Mesonets (networks) provide the necessary input to model the water, energy, nutrient, and carbon balance of large-scale agricultural fields. An example of such a model is the Precision Agricultural-Landscape Modeling System (PALMS) given by Molling et al.[38] The integration of PALMS with crop growth models[36,37] provides a framework whereby site-specific management of crops is an achievable goal.

The concept of using simulation models to manage crops was introduced in the 1970s. Many of the theoretical algorithms related to model crop photosynthesis and transpiration were formulated by C.T. de Wit and coworkers at Wageningen University, Netherlands.[48] An example of such model is the simulation of field water use and crop yield given by Feddes et al.[49] An example of a crop-specific model, that is, cotton, known as GOSSYM/COMAX was developed by McKinion et al.[39] This model was used by crop consultants in the Texas High Plains to provide services on irrigation scheduling and application of nitrogen fertilizer, growth regulators, and chemicals to terminate the crop.[50] The biggest drawback ofthe application of these models was that the required inputs, soil and weather, were both difficult and expensive to obtain. Furthermore, these models provided only an average estimate of crop yield for the entire field regardless of size. The models could be run separately for different parts of a field, but this increased the demand on limited computer and input resources and even then did not represent the interaction between various parts of the field. In retrospect, we now recognize that these models were ahead of their time. Given the current availability of soil and weather data that is required by these models, along with the increased computer speed and reduced cost, a resurgence in the application of simulation models to manage crops and cropping systems is anticipated.

In coming years, a likely scenario to emerge to manage cropping systems will be based on the combination of three factors. First, is the realization from site-specific management that shows that crop yield varies temporally and spatially and that increases in crop yield are possible by targeting different amounts of an input, for example, irrigation water and fertilizer, to specific parts of the field. Second, crop management, from planting to harvesting, is complex, and simulation models can be used as a decision aid. Use of crop simulation models, developed in the 1970s–1980s, is facilitated due to the increased availability of required soil, elevation, and weather input data to execute the models and reduced cost of computer hardware. The third factor is an increased awareness of producers on production efficiency and environmental concerns. For example, in many agricultural areas, the amount of nitrogen fertilizer that can be applied is restricted and linked to the residual nitrogen in the soil and its potential effect on contamination of surface water and groundwater.[11,13,14,33,51] Advances in management information systems, development of computer software, and communication via the Internet provide us with the tools to manage a crop in real time.[1,6,36,52] We are currently working toward the development of a PA model that includes all of the above factors.[52,53]

The integration of a landscape-scale model such as PALMS[38] with a cotton growth model[39] using site-specific management of water and nitrogen[17,18] can give us the tools to manage, for example, a 50-ha irrigated cotton field. The model provides three key features important to real-time production-scale management: (1) it represents the variability in space and time within the entire field and accounts for hydrologic interactions between areas within the field; (2) it provides water, energy, nutrient, and carbon balance information without reliance on field-installed hardware that must be avoided during management operations; and (3) it can provide predictive information that can support various what-if scenario evaluations (something that physical field measurements do not provide). For example, this field can be divided into 5,000 (100 m^2) or 20,000 (25 m^2) cells, and the model will calculate a cotton lint yield value for each cell, using weather data collected at or near the field and previously collected soil and elevation data, both of which are stable and once developed can be used for many growing seasons. Further, the estimate of cotton lint yield is based on interactions of soil–plant–weather parameters, and the model itself can be used to explore what combination of inputs (e.g., water and nitrogen) would give the highest economic yield while minimizing leaching of nitrogen below the root zone. This is a current topic of research of our cropping system research unit.[52,53]

The input and output terms of the water balance of a cropping system are usually well quantified except for information on the input variable rain, that is, frequency and rate of rainfall events across the landscape.[53] Rainfall events are normally measured at a single point in space and seldom is the rate of individual rain events measured and/or recorded. Determining, the amount of water from a rain event that is stored in the soil, that is, effective rain, and runs off is key to correctly model the water balance across the landscape. For example, in the semiarid Texas High Plains, about 86% of annual rain events are <13 mm[54] and a large portion of this rain is lost as evaporation from the soil before it can be used by the crop as transpiration.[55] To correctly calculate the water balance across the landscape we need to determine the amount for any rain event that is stored in the soil and is used by the crop as transpiration.[55,56] Rainfall is variable in space and an appropriate cluster of rain gauges is needed across the landscape to measure effective rain and correctly determine the water balance across the landscape.

Conclusion

PA is a generic term that describes the way that agronomic inputs to a farming operation are managed and each input is applied according to specific needs across the field. An outcome of PA is the recognition that crop yields vary temporally and spatially. Crop management, from planting to harvest, is complex and requires management strategies that consider economic and environmental consequences. Recently, nutrient contamination and quality and safety of food delivered to consumers are also factors that have received consideration. A summary of general concepts learned from PA experiments follows.

First, agronomic experimental work needs to be done at the scale of application. For example, results from small research plots are normally not transferable to large fields.[22,25,40] The underlying principle is to take advantage of the inherent spatial variability of soil properties across the landscape. In PA experiments, the spatial variability of a given property is used as the source of variation instead of imposing treatments (e.g., levels of nitrogen fertilizer) to obtain variation. Second, the use of classical statistics fails to explain the cause and effect between variables. Spatial structure between variables can be quantified using specialized statistical tools such as autocorrelation, cross-correlation, and state-space analysis.[22,25] Third, given the complexity of current cropping systems, crop models[39,48] introduced in the 1980s are being combined with geographic information data of soils and elevation, real-time weather, and management information systems to provide a framework to manage crops. These models can provide three key features important to real-time production-scale management: (1) representation of variability in space and time within the entire field and hydrologic interactions between areas within the field; (2) water, energy, nutrient, and carbon balance information without reliance on field-installed hardware that must be avoided during management operations; and (3) predictive information to support various what-if scenario evaluations. As a result, we now have the capability of tracking different layers of data (soil, plant, and weather) and, for example, a 50-ha field may be divided into 5,000 cells, each cell 10 m wide, 10 m long, and 2.0 m deep.[36,37,52,53] The water, energy, carbon, and nitrogen balance (input equal to output) of each cell is calculated, and the change of any one variable on crop yield can be evaluated. This is a powerful management tool that will assist producers on how to maximize economic crop yield while minimizing nutrient losses to surface or groundwater.[52,53] Adoption of PA practices will continue to rise given the increased constraints of production, advances in technology, and demand for safe food and fiber of high quality.[13,33,46,47]

References

1. Kumar, S.A.; Ilango, P. The impact of wireless sensor network in the field of precision agriculture: A review. *Wireless Pers. Commun.* **2018**, *98*, 685–698.
2. Rogovska, N.; Laird, D.A.; Chiou, Chien-Ping; Bond, L.J. Development of field mobile soil nitrate sensor technology to facilitate precision fertilizer management. *Precis. Agric.* **2019**, *20*, 40–55.
3. Mulla, D.J. Twenty five years of remote sensing in precision agriculture: Key advances and remaining knowledge gaps. *Biosyst. Eng.* **2013**, *114*, 358–371.

4. Ferguson, R.; Rundquist, D. Remote sensing for site-specific crop management. In *Precision Agriculture Basics*; Shannon, D.K.; Clay, D.E.; Kitchen, N.R., Eds.; ASA-CSSA-SSSA: Madison, WI, **2018**, 8, 103–117.

5. Sonka, S.T. Big data: Fueling the next evolution of agricultural innovation. *J. Innov. Manag.* **2016**, 4(1), 114–136.

6. Fulton, J.P.; Port, K. 830. In *Precision Agriculture Basics*; Shannon, D.K.; Clay, D.E.; Kitchen, N.R., Eds. ASA-CSSA-SSSA: Madison, WI, **2018**, 12, 169–188.

7. Mahan, J.R.; Lascano, R.J. Irrigation analysis based on long-term weather data. *Agriculture* **2016**, 6, 42, doi:10.3390/agriculture6030042

8. Pongnumkul, S.; Chaovalit, P.; Surasvadi, N. Applications of smartphone-based sensors in agriculture: A systematic review of research. *J. Sensors* **2015**, Article ID 195308, 18, doi:10.1155/2015/195308

9. Mesas-Carrascosa, F.J.; Castillejo-González, I.L.; Sánchez de la Orden, M.; García-Ferrer, A. Real-time mobile phone application to support land policy. *Comput. Electron. Agric.* **2012**, 85, 109–111.

10. Hedley, C. The role of precision agriculture for improved nutrient management on farms. *J. Sci. Food Agric.* **2015**, 95, 12–19.

11. Abit, M.J.M.; Arnall, D.B.; Phillips, S.B. Environmental implications of precision agriculture. In *Precision Agriculture Basics*; Shannon, D.K.; Clay, D.E.; Kitchen, N.R., Eds.; ASA-CSSA-SSSA: Madison, WI, **2018**; 14, 210–220.

12. Neupane, J.; Gao, W. Agronomic basis and strategies for precision water management: A review. *Agronomy* **2019**, 9, 87, doi:10.3390/agronomy9020087

13. Gebbers, R.; Adamchuk, V.I. Precision agriculture and food security. *Science* **2010**, 327 (5967), 828–831.

14. Bongiovanni, R.; Lowenberg-DeBoer, J. Precision agriculture and sustainability. *Precis. Agric.* **2004**, 5, 359–387.

15. Plant, R.E. Site-specific management: The application of information technology to crop production. *Comput. Electron. Agric.* **2001**, 30, 9–29.

16. Monzon, J.P.; Calviño, P.A.; Sadras, V.O.; Zubiaurre, J.B., Andrade, F.H. Precision agriculture based on crop physiological principles improves whole-farm yield and profit: A case study. *Eur. J. Agron.* **2018**, 99, 62–71.

17. Bronson, K.F.; Booker, J.D.; Bordovsky, J.P.; Keeling, J.W.; Wheeler, T.A.; Boman, R.K.; Parajulee, M.N.; Segarra, E.; Nichols, R.L. Site-specific irrigation and nitrogen management for cotton production in the Southern High Plains. *Agron. J.* **2006**, 98, 212–219.

18. Li, H.; Lascano, R.J.; Booker, J.; Wilson, L.T.; Bronson, K.F. Cotton lint yield variability in a heterogeneous soil at a landscape scale. *Soil Tillage Res.* **2001**, 58, 245–258.

19. Sinclair, T.R.; Tanner, C.B.; Bennett, J.M. Water-use efficiency in crop production. *BioScience* **1984**, 34(1), 36–40.

20. Fixen, P.; Brentrup, F.; Bruulsema, T.W.; Garcia, F.; Norton, R.; Zingore, S. Nutrient/fertilizer use efficiency: Measurement, current situation and trends. In *Managing Water and Fertilizer for Sustainable Agricultural Intensification*; Drechsel, P.; Heffer, P.; Magen, H.; Mikkelsen, R.; Wichel, D. Eds.; International Fertilizer Industry Association, International Water Management Institute, International Plant Nutrition Institute, and International Potash Institute: Paris, France, First edition, **2015**, 2, 8–38.

21. Sharma, B.; Molden, D.; Cook, S. Water use efficiency in agriculture: Measurement, current situation and trends. In *Managing Water and Fertilizer for Sustainable Agricultural Intensification*; Drechsel, P.; Heffer, P.; Magen, H.; Mikkelsen, R.; Wichel, D. Eds.; International Fertilizer Industry Association, International Water Management Institute, International Plant Nutrition Institute, and International Potash Institute: Paris, France, First edition, **2015**, 3, 39–64.

22. Wendroth, O.; Al-Oman, A.M.; Kirda, C.; Reichardt, K.; Nielsen, D.R. State-space approach to spatial variability of crop yield. *Soil Sci. Soc. Am. J.* **1992**, 56, 801–807.

23. Nielsen, D.R.; Wendroth, O.; Pierce, F.J. Emerging concepts for solving the enigma of precision farming research. In *Precision Agriculture, Proceedings of the Fourth International Conference*, Minneapolis, MN, July 19–22, 1998; Robert, P.C., Rust, R.H., Larson, W.E., Eds.; **1999**; 303–318.

24. Shumway, R.H.; Stoffer, D.S. *Time Series Analysis and Its Application*; Springer Verlag: New York, 2000; 549 pp.

25. Nielsen, D.R.; Wendroth, O. *Spatial and Temporal Statistics*. Catena Verlag: Reiskirchen, Germany, 2003; 398 pp.

26. Kho, R.M. On crop production and the balance of available resources. *Agric. Ecosyst. Environ.* **2000**, *80*, 71–85.

27. Booker, J.D.; Bronson, K.F.; Keeling, J.W.; Bordovsky, J.P.; Segarra, E.; Velandia-Parra, M. Farm-scale testing of site-specific irrigation and nitrogen fertilization for cotton production in the Southern High Plains of Texas, USA. In *Precision Agriculture '05*; Stafford, J.V., Ed.; Wageningen Academic Publishers: Wageningen, Netherlands, 2005; 951–957.

28. Van Looy, K.; Bouma, J.; Herbst, M.; Koestel, J.; Minasny, B.; Mishra, U.; Montzka, C.; Nemes, A.; Pachepsky, Y.A.; Padarian, J.; Schaap, M.G.; Tóth, B.; Verhoef, A.; Vanderborght, J.; Van der Ploeg, M.J.; Weihermüller, L.; Zacharias, S.; Zhang, Y.; Vereecken, H. Pedotransfer functions in earth system science: Challenges and Perspectives. *Rev. Geophys.* **2017**, *55*, 1199–1256.

29. Halvorson, G.A.; Doll, E.C. Topographic effects on spring wheat yield and water use. *Soil Sci. Soc. Am. J.* **1991**, *55*, 1680–1685.

30. Hanna, A.Y.; Harlan, P.W.; Lewis, D.T. Soil available water as influenced by landscape position and aspect. *Agron. J.* **1982**, *74*, 999–1004.

31. Timlin, D.J.; Pachepsky, Y.A.; Snyder, V.A.; Bryant, R.B. Spatial and temporal variability of corn grain yield on a hillslope. *Soil Sci. Soc. Am. J.* **1998**, *62*, 764–773.

32. Mueller, N.D.; Gerber, J.S.; Johnston, M.; Ray, D.K.; Ramankutty, N.; Foley, J.A. Closing yield gaps through nutrient and water management. *Nature* **2012**, *490*, 253–257.

33. Wu, W.; Ma, B. Integrated nutrient management (INM) for sustaining crop productivity and reducing environmental impact: A review. *Sci. Total Environ.* **2015**, *512–513*, 415–427.

34. Li, H.; Lascano, R.J. Deficit irrigation for enhancing sustainable water use: Comparison of cotton nitrogen uptake and prediction of lint yield in a multivariate autoregressive state-space model. *Environ. Exp. Bot.* **2011**, *71*, 224–231.

35. Lyle, W.M.; Bordovsky, J.P. Low energy precision application (LEPA) irrigation system. *Trans. ASAE.* **1981**, *24*, 1241–1245.

36. Booker, J.D.; Lascano, R.J.; Evett, S.R., Zartman, R.E. Evaluation of a landscape-scale approach to cotton modeling. *Agron. J.* 2014, *106*, 2263–22794.

37. Booker, J.D.; Lascano, R.J.; Molling, C.C.; Zartman, R.E.; Acosta-Martinez, V. Temporal and spatial simulation of production-scale irrigated cotton systems. *Precis. Agric.* **2015**, *16*, 630–653.

38. Molling, C.C.; Strikwerda, J.C.; Norman, J.M.; Rodgers C.A.; Wayne R.; Morgan, C.L.S.; Diak, G.R.; Mecikalski, J.R. Distributed runoff formulation designed for a precision agricultural landscape modeling system. *J. Am. Water Res. Assoc.* **2005**, *41*, 1289–1313.

39. McKinion, J.M.; Baker, D.N.; Whisler, F.D.; Lambert, J.R. Application of the GOSSYM/COMAX system to cotton crop management. *Agric. Syst.* **1989**, *31*, 55–65.

40. Cassel, D.K.; Wendroth, O.; Nielsen, D.R. Assessing spatial variability in an agricultural experiment station field: Opportunities arising from spatial dependence. *Agron. J.* **2000**, *92*(4), 706–714.

41. Wilson, J.N. Guidance of agricultural vehicles—A historical perspective. *Comput. Electron. Agric.* **2000**, *25*, 3–9.

42. Bell, T. Automatic tractor guidance using carrier-phase differential GPS. *Comput. Electron. Agric.* **2000**, *25*, 5366.

43. Bronson, K.F.; Keeling, J.W.; Booker, J.D.; Chua, T.T.; Wheeler, T.A.; Boman, R.K.; Lascano, R.J. Influence of landscape position, soil series, and phosphorus fertilizer on cotton lint yield. *Agron. J.* **2003**, *95*, 949–957.

44. Bronson, K.F.; Booker, J.D.; Officer, S.J.; Lascano, R.J.; Maas, S.J.; Searcy, S.W.; Booker, J. Apparent electrical conductivity, soil properties and spatial covariance in the U.S. Southern High Plains. *Precis. Agric.* **2005**, *6*, 297–311.

45. Stafford, J.V. Implementing precision agriculture in the 21st century. *J. Agric. Eng. Res.* **2000**, *76*, 267–275.

46. Zhang, N.; Wang, M.; Wang, N. Precision agriculture—A worldwide overview. *Comput. Electron. Agric.* **2002**, *36*, 113–132.

47. McBratney, A.; Whelan, B.; Ancev, T.; Bouma, J. Future directions of precision agriculture. *Precis. Agric.* **2005**, *6*, 7–23.

48. Bouman, B.A.M.; Van Keulen, H.; Van Laar, H.H., Rabbinge, R. The "School of de Wit' crop growth simulation models: A pedigree and historical overview. *Agric. Syst.* **1996**, *52*(2/3), 171–1793.

49. Feddes, R.A.; Kowalik, P.J.; Zaradny, H. *Simulation of Field Water Use and Crop Yield*. Simulation Monographs; Centre for Agricultural Publishing and Documentation: Wageningen, Netherlands, 1978; 188 pp.

50. Staggenborg, S.A.; Lascano, R.J.; Krieg, D.R. Determining cotton water use in a semiarid climate with the GOSSYM cotton simulation model. *Agron. J.* **1996**, *88*, 740–745.

51. Bronson, K.F.; Malapati, A.; Booker, J.D.; Scanlon, B.R.; Hudnall, W.M.; Schubert, A.M. Residual soil nitrate in irrigated southern High Plains cotton fields and Ogallala groundwater nitrate. *J. Soil Water Conserv.* **2009**, *64*, 98–104.

52. Nelson, J.R.; Lascano, R.J.; Booker, J.D.; Zartman, R.E.; Goebel, T.S. Evaluation of the precision agricultural landscape modeling system (PALMS) in the semiarid Texas Southern High Plains. *Open J. Soil Sci.* **2013**, *3*, 169–181.

53. Lascano, R.J.; Nelson, J.R. Circular planting to enhance rainfall capture in dryland cropping systems at a landscape scale: Measurement and simulation. *Adv. Agric. Syst. Models* **2014**, *5*, 85–111.

54. Jones, O.R.; Eck, H.V.; Smith, S.J.; Coleman, G.A.; Hauser, V.L. Runoff, soil, and nutrient losses from rangeland and dry-farmed cropland in the Southern High Plains. *J. Soil Water Conserv.* **1985**, *40*(1), 161–164.

55. Kool, D.; Agam, N.; Lazarovitch, N.; Heitman, J.L.; Sauer, T.J.; Ben-Gal, A. A review of approaches for evapotranspiration partitioning. *Agric. Forest Meteorol.* **2014**, *184*, 56–70.

56. Goebel, T.S.; Lascano, R.J. Rainwater use by cotton under subsurface drip and center pivot irrigation. *Agric. Water Manage.* **2019** *215*, 1–7.

VII

PRO: Basic Environmental Processes

44

Green Processes and Projects: Systems Analysis

Abhishek Tiwary

Introduction

Given the level of inadvertent impacts generated from human breakthroughs in science and technology post industrial revolution, be it from the combustion engines based on coal/fossil fuel or the use of plastic in modern day commodities, it is imperative to develop a systems perspective on the sustainability of a process or project. This chapter introduces a holistic approach to environmental appraisal for minimizing the impacts from transboundary interactions between the material flows of a system, thereby ensuring efficient management of any potential environmental impacts. This approach combines application of "dynamic science-based" geo-spatial analysis with "static inventory-based" life cycle assessment (LCA) to understand the full-scale of impacts generated from a proposed process or project at the systems level, thereby ascertaining their true "greening" potential. Hereafter a "process" implies a set of activities targeting a specific output (e.g., 1 MW electricity, 1 ton of a material), whereas a "project" implies a combination of cross-cutting processes to achieve a specific objective (e.g., constructing a new building, greening of transport network). Two case studies are presented, one each for a green process and a green project, to enable the readers to get a grasp of the usefulness of this tool in global environmental management, more so in the context of supply chain spread diffusively over numerous geographical regions.

Systems Approach in Environmental Management

In the context of fool-proofing human existence from untoward climate change feedbacks, management of both built-up and natural environments has gained center stage in recent years. It is widely agreed now that effective environmental management involves striking a fine balance between the technological innovations underpinning its viability (essential for economic growth) and the constraints to long-term sustainability (Figure 1). This chapter deals with how a systems approach can be applied to achieve this goal.

Traditionally, systems approach has its roots in industrial (or laboratory) processes that simply comprise a series of related activities aimed at optimizing production (and hence the profit). However, in the face of climate change, much of the environmental research conducted over the last decade has changed our notion of the scale of management problem we face. Take, for instance, the case of cheap biofuel provisioning in the first decade of the 21st century—the demand for palm oil in the Western world has caused massive deforestation in Indonesia and Amazonia. Having a wider scope to environmental management, beyond what can be considered as local activities within a country's borders, has given way to the systems approach in environmental management. Although simple in principle and rigorously tested in processing industries, its adaptation to the wider environmental problems comes with extreme operational challenges. The latter is mainly highlighted in terms of data availability, essential to the success of this vision. Despite this, it is considered as an efficient framework to facilitate a more holistic management of material, energy, and pollution across a range of related activities, usually spread over large geographical areas.

The main advantage of using a systems approach is to figure out the hotspots of environmental concerns (greenhouse gas emissions, ecotoxicity, air pollution, etc.) within a system. For this purpose, the scope of environmental management has now crossed local boundaries to a systems scale, encompassing the whole range of supply-chain spreading over several countries to ensure global sustainability. These associated activities form what is known as a "life cycle" in terms of environmental appraisal. It facilitates mapping of the "stocks" and "flows" of emissions (also called environmental burdens) through different stages. The concept has been used to develop the LCA and Material Flow Analysis (MFA) methods. Both these techniques have proven to be useful diagnostic tools. Detailed texts describing the LCA/MFA methods exist in the literature. Interested readers can find some useful sources listed in different parts of this chapter; in particular, *The Hitch Hiker's Guide to LCA*[1] serves as a good starting point to develop fundamental understanding of these concepts.

The main objective of applying LCA/MFA is to assess the critical environmental burdens and impacts contributing to the adverse effects on human and ecological health. These critical pollutants can then

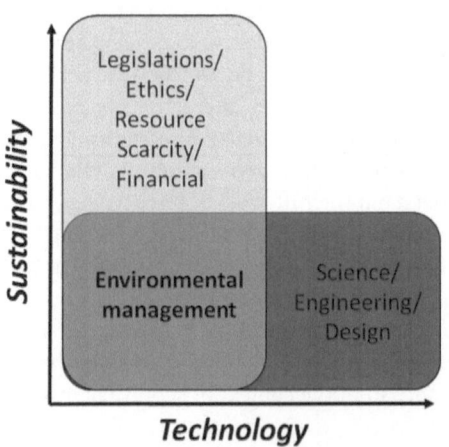

FIGURE 1 Constraints to environmental management at a systems scale.

be dealt with, and their impacts mitigated, through implementation of effective control measures. This ensures a solution to an environmental problem at the local scale while considering its footprints at the global scale, literally over the whole life cycle. As a consequence, this approach offers a robust framework for implementing holistic sustainability.

Best Practice Guidelines in Systems Analysis

Standard Protocols

In essence, the International Organization for Standardization ISO 14044[2] specifies four mandatory steps for quality assurance in LCA, namely—goal and scope definition, inventory analysis, impact assessment, and interpretation. To ensure that the environment as a whole is protected, the process chain forming the system can be scoped in two ways. The first comprises pre-chains involved in excavation of the resources and their transportation to the industrial site, the end-of-pipe emissions from the plants, and the disposal of the wastes at the end of the cycle. This approach in life cycle thinking is usually referred to as "cradle-to-gate" since it follows an activity from the extraction of raw materials (i.e., cradle) to the delivery of the product or service (i.e., the exit gate). The other, in addition to accounting for all the above, also includes one-off construction and demolition of the infrastructures or end-of-life processing and disposal of the equipment/commodities used. Appropriately so, it is then called a "cradle-to-grave" system.

As a standard practice, the resources utilized and the emissions added to the environment in all these activities are usually modeled on a unit basis, typically annual turnover for the industrial products and annual usage for the services. This is termed as the "functional unit" of analysis in LCA. Typical functional unit for an assessment can be "activities over 1 year," representative of the quantitative metric of the output of products or services that the system is expected to deliver. Likewise, the typical timescale for an assessment could range from hours to 100 years. The latter is specifically relevant for assessing long-term environmental impacts such as global warming potential and acidification.

The spatial context of setting up a life cycle model is also essential in order to establish its system boundaries. Typical spatial scales range from local (e.g., urban) to wider environment (e.g., remote excavation site, mines) and possibly representing a global scale. This aspect of the analysis is useful for consistency checks, comparing different LCA results, and ensuring they have been conducted on comparable system boundaries. Given that life cycle approach facilitates assessment of a full spectrum of process chains involved in a process for all the activities, it is recommended to group the burdens and impacts obtained from an analysis into two distinct categories: (1) arising in the direct environment from the main activity/ies under investigation (commonly known as "foreground") and (2) arising from a whole series of linked pre- and post-chain activities in the wider global environment (known as "background"). This is clearly shown in Figure 2, which shows a series of activities associated with the production of electricity from biofuel through a schematic representing a power generation system. In this figure, the main activity (i.e., the cogeneration plant) is shown as the foreground (shaded region), whereas the peripheral activities, associated primarily with the sourcing of the raw materials, shipment, and final disposal are shown in the background. On the one hand, this approach enables clear accounting of the environmental impacts and, on the other, it offers insight into the hotspots at each stage of process chain, facilitating effective management of the problem through visualizing the entire system at the same time. This is considered superior to resource- and cost-intensive piecemeal solutions encountered in traditional management approaches, specifically so in the context of achieving global sustainability.

The environmental impacts from LCA are usually calculated from mass balances of the input/output flows on the basis of the problem-oriented (midpoint) approach. It is a baseline method that provides a list of impact assessment categories grouped into obligatory impacts, used in most LCAs. A baseline indicator is considered suitable for simplified studies since it utilizes the principle of best available practice when several methods for obligatory impact categories are available.

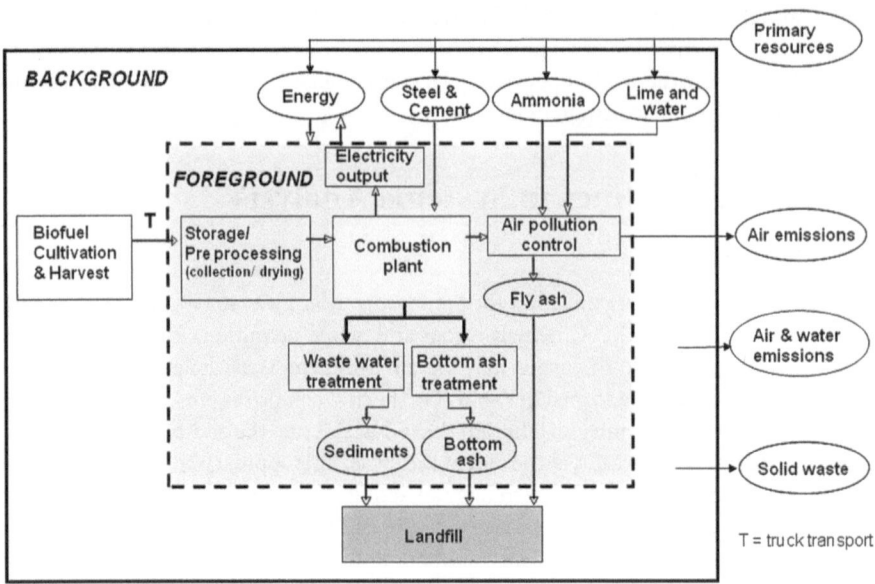

FIGURE 2   Split between foreground and background activities from a systems perspective.
Source: Tiwary and Colls.[3]

Uncertainty Assessment

To decide on the future courses of action based purely on results of LCA analysis, one needs to account for the uncertainty in these results—this enables reaching better decisions. The information on uncertainty in LCA is usually contained in the assessment of probability of realistic representativeness of the results. This is expressed in terms of the confidence bounds on LCA results, which illustrate the region within which the true values have an estimated likelihood of falling.

This section covers only the fundamentals of uncertainty inherent in LCA. Readers will be guided to use dedicated literature to pursue a more in-depth uncertainty assessment exercise.

A number of approaches have been suggested in the literature for the integrated consideration of both technical and valuation uncertainties involved in LCA. The former is associated with the uncertainties in data collection, while the latter is inherent to the impact assessment method used. The reason we need to care about uncertainty in LCA is because the statements or assertions we tend to make about the world on the basis of direct LCA outputs may be wrong. These errors have been mainly associated with either uncertainty or variability in the outputs. Whereas variability is considered to be inherent in the real world, uncertainty is mainly associated with inaccurate measurements, lack of data, and/or model assumptions.[4] The following can be considered as a rigorous (although not an exhaustive) list of categories of uncertainties identified in the literature:

- *Parameter uncertainty*: This is due to the uncertainty within the large number of parameters used in LCA models. It leads to uncertainty in the final output from the LCA exercise. Empirical inaccuracy, non-representativeness (incomplete or outdated measurements), and lack of data are common sources of parameter uncertainty.
- *Epistemological uncertainty*: The use of the information in databases for life cycle inventory (LCI) introduces epistemological uncertainty since the system where the data is to be used to model a process may differ from the system where the data was generated.[5]
- *Model uncertainty*: In situations where an LCA model suffers from uncertainty in the underlying model assumptions and the basic model calculations (departure from the default linear

programming), the results of a parameter uncertainty analysis can be misleading and hence provide no useful information.

- *Spatial variability*: Application of spatially averaged data to model specific processes in certain parts of the world leads to these discrepancies.
- *Temporal variability*: These are mainly attributed to time-dependent variations in emissions and other technical process characteristics.

Handling Uncertainty in Systems Analysis

It has been suggested that not all the methods of statistical analyses of uncertainty can be applied to LCA, primarily because the underlying LCA data are not based on random samples, that is, we are not strictly dealing with random variables that follow some known frequency distribution. However, one can apply subjective probability distributions to quantify the uncertainties involved. The established procedure to achieve this consists of the following five steps: selection of essential parameters, determination of probability distributions, Monte Carlo simulation, significance analysis, and interpretation of the results.[4,6]

Recent approaches have tried to integrate consideration of both technical and valuation uncertainties during decision making on the basis of the results provided by LCA. These are to be used in conjunction with established decision support tools based on multiple criteria decision analysis since it has been demonstrated that the structure of LCA has parallels with a decision analysis approach to decision making.[7] Key elements of this approach include "distinguish ability analysis" to determine whether the uncertainty in the performance information is likely to make it impossible to distinguish between the activities under consideration, and the use of a multivariate statistical analysis approach, such as principal component analysis. The latter enables rapid analysis of large numbers of parallel sets of results, thereby allowing for the identification of choices (options) that lead to similar and/or opposite evaluation of activities.

Application of Subjective Probability Distributions Approach

This method of uncertainty analysis in LCA has inherent attributes since they are an alternative to simple point estimates, which allow for the use of a range within which we expect the true value to lie rather than using a single number to estimate the results of some real-world quantity (e.g., tons of sheet of steel required to make 1 ton of an output product). In this manner, by developing subjective probability distribution, the methods of uncertainty analysis can be used in LCA to account for the uncertainty in the results. This further enables accounting for the reliability of the decisions reached on the basis of the LCA outcomes, and if the reliability of the conclusions is not sufficient for our decision-making needs, then uncertainty analysis helps identify which data uncertainties are most significant, that is, influential in the process chain. Furthermore, applying the inverse method, this step can help the decision maker determine the levels of reduction in data uncertainty required to reach a specific level of confidence in the results. For example, to attain "90% confidence" in the results, one would need to assure that using the subjective probability distribution approach if the LCA analysis of a process is repeated several times, each time using new and equally probable randomly selected values for the uncertain input quantities, the conclusions would be correct 90% of the time.

Uncertainty Analysis versus Data Quality Characterization

The LCA is very dependent on data of good quality. In most cases, LCA practitioners rely on generic databases provided by different sources. The commonly available databases to date are the European Union's (EU) European Reference Life Cycle Data System (ELCD), Swiss National LCI Database (ecoinvent), the U.S. LCI Database created by the National Renewable Energy Laboratory and its partners, the Canadian Raw Materials Database (CRMD), the Swedish National LCA Database (SPINE@ CPM), the Danish Food Database, and the Korean National LCI Database (KNCPC). Apart from these, currently Australia and Japan also have ongoing initiatives to generate national LCA databases.

Owing to a limited number of sources providing information while generating the LCI data, quite often the uncertainty related to the amount of a specific input or output cannot be derived from the available information. For such circumstances Frischknecht et al.[8] have developed a simplified standard procedure to quantify the uncertainty. This simplified approach includes a qualitative assessment of data quality indicators on the basis of a pedigree matrix from published literature. Basic uncertainty factors are used for the kind of input and output considered. For example, it is assumed that CO_2 emissions generally show a much lower uncertainty as compared with CO emissions since the former is usually calculated from fuel input, whereas the latter depends on numerous parameters such as boiler characteristics, engine maintenance, and load factors. Table 1 provides a list of proposed uncertainty factors in a pedigree matrix, which have been based on expert judgments.

TABLE 1 Examples of Basic Uncertainty Factors (Dimensionless) Applied for Technosphere Inputs and Outputs and for Elementary Flows

Input/Output Group	c	p	a
Demand of			
Thermal energy, electricity, semifinished products,	1.05	1.05	1.05
working material, waste treatment services	2.00	2.00	2.00
Transport services (t km)	3.00	3.00	3.00
Infrastructure			
Resources			
Primary energy carriers, metals, salts	1.05	1.05	1.05
Land use, occupation	1.50	1.50	1.50
Land use, transformation	2.00	2.00	2.00
Pollutants Emitted to Air			
CO_2	1.05	1.05	
SO_2	1.05		
NMVOC total	1.50		
NO_x, N_2O	1.50		1.40
CH_4, NH_3	1.50		1.20
Individual hydrocarbons	1.50	2.00	
PM >10 micron	1.50	1.50	
PM_{10}	2.00	2.00	
$PM_{2.5}$	3.00	3.00	
Polycyclic aromatic hydrocarbons (PAH)	3.00		
CO, heavy metals	5.00		
Inorganic emissions, others		1.50	
Radionuclides (e.g., radon-222)		3.00	
Pollutants emitted to water			
BOD, COD, DOC, TOC, inorganic compounds		1.50	
(NH_4, PO_4, NO_3, Cl, Na)		3.00	
Individual hydrocarbons, PAH		5.00	1.80
Heavy metals			1.50
Pesticides			1.50
NO_3, PO_4			
Pollutants emitted to soil			
Oil, hydrocarbon total		1.50	
Heavy metals		1.50	1.50
Pesticides			1.20

Source: Data from Frischknecht et al.[8].

c, combustion emissions; p, process emissions; a, agricultural emissions.

Case Study I—System Analysis of Biofuel-Based Electricity Generation Process

This example demonstrates the application of a systems approach in assessing the merits of biofuels as a green process. The LCA generates a profile of all the steps involved in the process chain—both onsite and upstream—to provide quantitative information on potential impacts of an industrial activity (i.e., a system). Usually, this is done in terms of the released emissions (i.e., burdens), using a linear model. It takes into account the emission factors of all the known pollutants, inventoried during controlled analysis following standard monitoring protocols. These are then scaled by the volume of the industrial activity to provide their actual burdens. Also, in the case that LCA data are available for multiple similar installations, their respective operational performances can be benchmarked and links between operational efficiency and environmental impacts can be established.[9] A number of LCA studies available in the literature compare the environmental impacts of energy production (heat and/or electricity) from co-firing different biofuels in an existing coal-fired power plant, showing their overall greenhouse gas benefits. These cover the use of agricultural residues such as straw and residual wood, short rotation coppiced (SRC) wood,[10] and perennial rhizomatous grasses.[11,12] A more recent study quantifies the airborne emissions from different biomass-based electricity production systems using different technologies, feedstocks, and scales in order to establish the extent to which offsite emissions may contribute to overall environmental impact.[13]

In LCA, the emission factors are adjusted according to the pollution abatement technologies used in the industry to compensate for the release of acidic gases, although they do not reflect the fate of the emissions once they are out of the stack. Thus, this approach remains capable of providing realistic emission scenarios as long as the fate of the released emissions is not altered significantly by the dispersion and chemical transformation in the surrounding media (air, water, or soil). Whereas it allows successful prediction of greenhouse gas emissions in a fairly straightforward manner (assuming minimal phase alterations), the modeling of the gas–particle interactions leading to quantification of total particulate matter (PM) loading is far from complete. For example, the PM emissions calculated in the LCA from a biofuel combustion plant represent mainly the dust emissions from fly ash.[14] Any additional aerosols generated from gas-phase interactions of the resulting emissions, either during biofuel cultivation or combustion, would not be adequately quantified within this approach.

This case study utilizes the power of LCA as a diagnostic tool to track the pollutants and carbon emissions over the whole life cycle and mainly focuses on feasible management options for mitigating secondary aerosol generation potential from photochemical neutralization of the acidic emissions with ammonia using the following precursor chemistry[15]:

$$NH_{3(g)} + HCl_{(g)} \leftrightarrow NH_{4(aq)}^+ + Cl_{(aq)}^-$$

A cradle-to-gate system is applied to all the energy systems modeled in this study, accounting for all the relevant flows involved in the extraction of resources (renewable/non-renewable) and cultivation of biofuels leading up to production of the required amount of energy outputs. As shown in Figure 2, the system comprises a series of foreground and background activities. The foreground activities are considered as the focal point of the system. Shown in the central part as shaded region in the figure, it involves storage and preprocessing of the fuel and its combustion to produce electricity. The background activities primarily include the cultivation and harvesting of the biofuel, production of required chemicals, and their transport to the power plant. The other end of the process chain accounts for the disposal of wastes generated. The atmospheric emissions from all these stages have been accounted for in the models.

Electricity production from different renewable biofuel sources has been modeled using scenario analysis. Representative biofuel options currently feasible have been considered. The analysis in this study is based on the functional unit defined as "1 terajoule (TJ) electricity produced from biomass and delivered to the grid." It is achieved by using available LCI data for a 50-MW electricity steam turbine/back pressure cogeneration plant firing solid biomass.[16] It is important to note that, in this study, the modeled system has not been credited for the cogenerated heat in the process. The following five different electricity production scenarios, representative of the technology for 2010 (base scenarios), have been considered, each using different types of biofuels:

- *Scenario A*: Perennial rhizomatous grass (*Miscanthus giganteas*).
- *Scenario B*: SRC chips.
- *Scenario C*: Residual/waste wood.
- *Scenario D*: SRC chips–grass blend (by energy); SRC chips (80%) and perennial rhizomatous grass (20%).
- *Scenario E*: Waste wood–grass blend (by energy); waste wood (80%) and perennial rhizomatous grass (20%).

In scenarios D and E, a fuel mix of perennial grass and wood in 1:4 ratio (by energy) has been considered. This is meant to improve the combustion quality and minimize excessive atmospheric emissions. In all the base scenarios, the biofuels have been assumed to be sourced locally (50 km distance to the combustion plant) and transported from their origin to the energy production site using 40-ton payload trucks. This is mainly aimed to investigate potential local air quality degradation from interactions of emissions from the power plant and the cultivation sites.

The atmospheric burdens for almost all the base scenarios (A–E) are dominated by pollutant emissions from the power plant (Figure 3). Overall, there is no clear winner among the scenarios. For example, scenarios C and E (waste wood) are best for CO_2, NH_3, non-methylated volatile organic carbons (NMVOCs), and CH_4, and scenario B (SRC chips) has the lowest HCl emissions. Scenario A (miscanthus) is best for N_2O and PM; however, it is worst for HCl and SO_2. For both miscanthus and SRC wood (scenarios A, B, and D), only biomass cultivation is considered to be the main source of NH_3 release to the local environment. However, it can be noted that a considerable amount of CO_2 is emitted during biomass production, transport, storage, and drying processes for these scenarios, which are mainly associated with the use of fossil-based energy sources in these stages. It has been reported that the forest logging operations involved in harvesting SRC wood contribute to significant atmospheric emissions

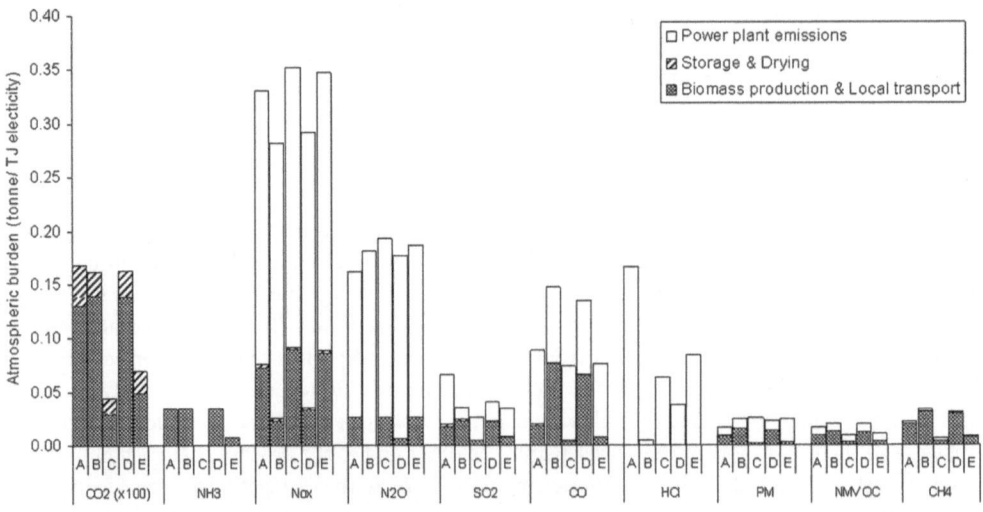

FIGURE 3 Environmental burdens of atmospheric emissions for base scenarios A–E.
Source: Tiwary and Colls.[3]

of CO_2, CO, NMVOCs, and PM due to fuel, chainsaw, and hydraulic oil consumption by heavy-duty diesel engine vehicles (Athanassiadis, 2000). This is reflected for the base scenarios B and D whose emissions of these pollutants from the "biomass production and local transport" stage are much higher compared with the rest of the scenarios. In case of waste wood, no mechanical chipping was assumed to be involved and hence the corresponding emissions from scenarios C and E have been relatively much lower. However, the power plant loadings of NO_x, SO_2, HCl, and PM in these two scenarios are found to be much higher than in scenarios B and D. This could be due to their incomplete combustion, which has been reported in earlier studies to result in highly variable emissions.[17,18]

Emissions of CO and NO_x per terajoule electricity output from power plant alone show comparable values for all the base scenarios but on a life cycle basis, that is, including the biofuel sourcing and storage stages, waste wood (scenarios C and E) seems to have up to 25% higher NO_x burdens, whereas SRC wood (scenarios B and D) seems to have up to 100% higher CO burdens. The corresponding emissions of SO_2 and HCl from miscanthus plant (scenario A) are higher by as much as 120% and 350%, respectively, compared with the rest of the scenarios. Therefore, scenario A poses the maximum likelihood of secondary aerosol generation potential through interactions of the acidic gas emissions from the power plant with the ammonia released from nearby harvest fields. On the other hand, SRC wood combustion plant (scenario B) has much lower HCl emissions (<5 kg T/J electricity) and, therefore, despite having considerable contributions to NH_3 release from the harvesting stage, its secondary aerosol generation potential is over 35-fold lower. Results indicate that co-firing waste wood sourced locally (scenarios C and E) would have the least secondary aerosol generation potential from photochemical neutralization, despite having higher HCl loadings than SRC wood, as it shows negligible NH_3 emissions over its fuel life cycle. Nevertheless, this is balanced by higher direct PM releases and life cycle NO_x releases from waste wood combustion. Although in this study heavy metals have not been included in the analysis owing to limited information, separate studies have reported wood combustion to be enhancing emissions of heavy metals.[19,20] The latter highlights additional air quality problems associated with large-scale implementation of wood-burning installations in the future.

The LCA model offers the flexibility to assess reduction in aerosol formation from different mitigation options, both for material (and cost) and energy, which is an invaluable screening exercise in order to come up with a viable and cost-effective management solution. For example, in this case study, the following three mitigation measures have been assessed.

Biomass Gasification and Direct Firing

Gasification technology is considered to provide efficient and clean power generation from biomass.[21] In this study, gasification of three different types of biomass (miscanthus, SRC, and waste wood) through air injection into the fuel in a circulating fluidized bed has been considered. The synthesis gas generated is compressed and injected into the combustion chamber of a cogeneration plant to produce steam that subsequently drives a turbine. As in the base scenarios, the system has not been credited for the cogenerated heat in the process for consistency.

Relative changes in the environmental burdens from application of biomass integrated gasification/combined cycle technology have been estimated with reference to the power plant emissions from the base scenarios (Figure 4). Results suggest significant reductions (up to 100% in some cases) in the emissions of acidic gases in the direct vicinity of the power plant. This is mainly due to much lower release of HCl and SO_2 from the plant for all scenarios considered. Likewise, PM emission in the foreground is also reduced as a consequence of gasification. In addition, N_2O emission is also lowered in all cases, whereas little change is observed in NO_x emission, except for a prominent decrease in the case of residual waste wood combustion (scenario C). However, while achieving the significant reduction in acidic gas burdens, this approach apparently increases end-of-pipe emissions of CH_4 and CO. Both these pollutants are produced during gasification of the biofuel, and their atmospheric emissions are associated with unintended (fugitive) release from the gasifier plant.[22] The main consideration for no obvious relative change (%) in

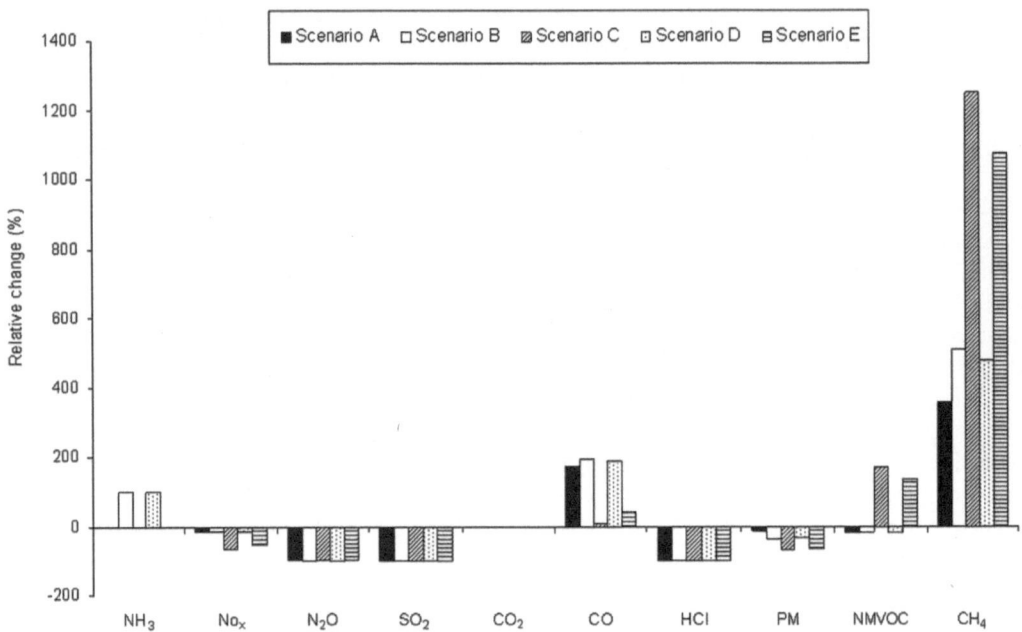

FIGURE 4   Relative change in environmental burdens by application of gasification option to base scenarios A–E.
Source: Tiwary and Colls.[3]

CO_2 emission in gasification options relative to base case is due to offsetting of the emissions enhancement from processing of additional feedstock with the increased efficiency in combustion of the gasified fuel. The increase in CH_4 emissions for scenarios A, B, and D ranges between 200% and 500%; however, for scenarios C and E, it appears to be exceptionally high (up to 1,250%). The latter two scenarios also show increased emissions of NMVOCs. As both CH_4 and NMVOCs contribute to global warming, the two scenarios (C and E) do not seem to present sustainable alternatives. Excessive release of methane could be potentially minimized by optimizing the gasification temperature and circulating the fugitive methane into the combustion process.[22] It is also noticed that power production from SRC wood biomass from this approach leads to production of NH_3, which makes scenarios B and D unviable too. Therefore, although gasification for scenario A enhances its CO emissions (up to 200% increase) relative to the corresponding base scenario, it presents the best option for reducing secondary aerosol generation potential.

Delaying the Harvest of Perennial Energy Crop

Delaying the harvest of miscanthus from late autumn (current common practice) to early spring has been reported to foster prolonged soil–vegetation interactions by allowing the stems to dry during the winter months. In effect, it allows for leaching of a substantial portion of the ash, chlorine, and potassium contents from the biomass through roots,[12] thereby reducing the adverse environmental impacts later during combustion. Delaying the harvest also allows for extended recycling of nutrients by their rhizome systems, which enables them to have low demand for nutrient inputs, especially for nitrogen. This lowers their fertilization needs. However, it has been reported that this delaying approach inadvertently leads to substantial yield losses (of up to 35%) and reduction in calorific value of the biofuel. The latter entails additional emissions from extra biomass combustion for the same energy throughput.

This study used the data provided by recent studies on reduction in dry matter yield, water content, and concentrations of specific toxic components, including ash, nitrogen, chlorine, and sulfur, as a consequence of delaying the harvest.[12] It mainly focused on assessing the influence of delaying the harvest of the biomass and its linked effects on storage and drying, whereas additional effects of topography and soil properties for differences in site locations could not be quantified at this stage. The ideal late harvest is recommended at a water content of 20% or less, as this minimizes the cost of harvesting and drying while keeping a high biomass quality of the crop. For the purpose of estimating the environmental burdens from drying of harvested biomass during storage, we have used literature data (Table 2).

Although the emission per unit biomass consumption is reduced in this manner, the resulting biofuel has deteriorated in calorific value. The requirement of additional biomass supply to compensate the lost calorific value of the fuel leads to considerable increase in the life cycle environmental burdens for NH_3, N_2O, and CO and slight increments for SO_2, PM, NMVOCs, and CH_4 (Figure 5). Delaying the harvest leads to reduction in the water content of the biomass. Interestingly, this allows for mitigating CO_2 emissions at two stages of the fuel life cycle: (1) reduced demands for fossil fuel in the storage and drying process and (2) increased energy conversion efficiency from enhanced combustion in the power plant. However, despite offering some reduction in emissions of CO_2, NO_x, and HCl, this technique does not seem to provide a definitive and reliable option to mitigate the secondary aerosol generation potential from chemical interactions of the precursor gases.

TABLE 2 Model Parameters for Technical Drying of Energy Crop in Storage

Harvest Type	Energy Use in Drying	Water Content
Autumn harvest (>40 wt.%)[a]	68 kWh heating oil + 9.7 kWh electricity per ton of water removal	15 wt.%[b]
Early winter harvest (20–40 wt.%)[a]	144 kWh electricity per ton of water removal	15 wt.%[b]
Late winter/early spring harvest (<20 wt.%)[a]	No drying is carried out	<20 wt.%[b]

Source: Data from Tiwary and Colls.[3]
[a] Water content of freshly harvested biomass.
[b] Water content of biomass before combustion.

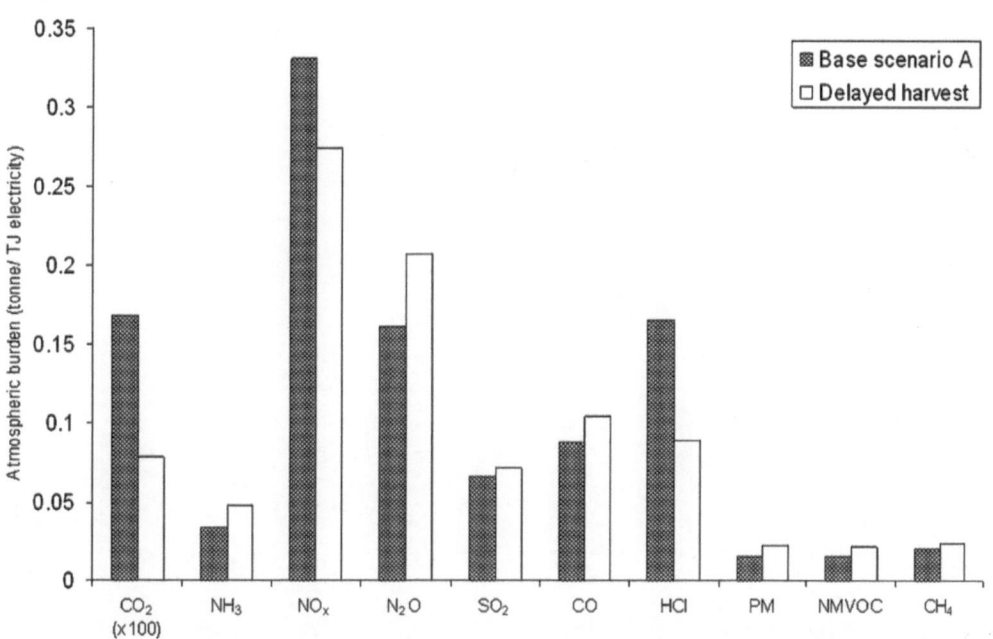

FIGURE 5 Comparison of the life cycle environmental burdens for base scenario A and delayed harvest of miscanthus.
Source: Tiwary and Colls.[3]

Increasing the Separation of Biomass Plant from Harvest Site (Importing the Biofuel)

As preliminary results from base scenarios (as shown in Figure 3) suggest, ammonia emissions from harvesting miscanthus and SRC wood are high. Assuming that the harvest sites in the new locations would be remotely rural with low acidic releases to the ambient air, a third possible mitigation option of reducing the impact of ammonia by displacing the cultivation sites has been considered. This option may also become necessary in the case of an eventual shift to larger production and use of biomass, as the EU is going to need large areas of agricultural land for its production.[23] In the United Kingdom alone, an estimated 125,000 ha of energy cropping would be required by 2010 to meet the government's target for electricity generation from biomass (DEFRA, 2001). In such situations, importing biomass from other European countries that offer favorable climatic conditions, mainly higher solar radiation and soil water contents, has been recommended as a more sustainable option.[24]

As considering all probable transportation means to move biofuels from one country to another gives rise to numerous possible scenarios,[25] some practical assumptions, based on the geographical location of the electricity generation site (here assumed to be in South England), have been made. Each option includes transport from production sites to a central gathering point, followed by a combination of international transport and local transport to a storage location near the generation plant. It is noteworthy that this assessment does not account for the additional emissions from the storage requirements of the biofuel crops, either in the country of their origin or during their long-haul transport.

As shown in Figure 6, the corresponding transport-related emissions for all the pollutants considered from this transport chain are much higher, which makes it less favorable from environmental perspectives. To a larger extent, this is solely dependent on the assumption used in this study to source miscanthus from land-locked Czech Republic and transport it overland across the stretch of European mainland on a freight train to the nearest ferry port. Had the biofuel been sourced from a coastal country and shipped directly to a port, the emission regimes would be very different. Therefore, one message that comes out of this study is that additional emissions from transporting the biofuel are very much site dependent, and this can be optimized.

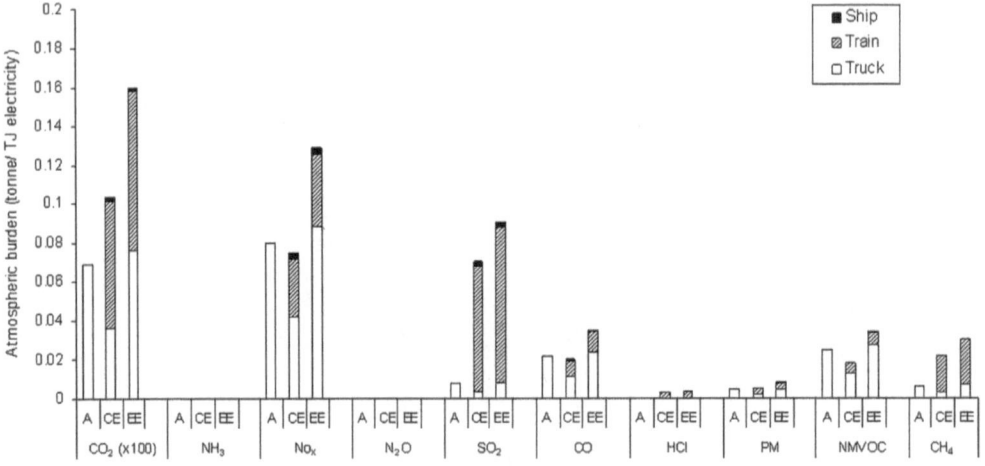

FIGURE 6 Transport-related burdens for different options associated with the transport of miscanthus (A—miscanthus in base scenario, CE—miscanthus imported from Central Europe, EE—miscanthus imported from Eastern Europe).
Source: Tiwary and Colls.[3]

Case Study II—System Analysis of a Project Integrating Green Initiatives

This section demonstrates the application of a systems approach in assessing the environmental impacts of low-carbon urban planning as a green project. Interactions between a suite of cross-sectoral green initiatives have been considered for two scenarios—status quo and aggressive, including decarbonizing of road transport, decentralizing energy production through biomass plants, and increasing the urban green vegetation cover. The study utilized real and projected information for the North East region of the United Kingdom to assess air quality implications on a systems scale. Using a geo-spatial analysis framework, it captures both the direct and the second-order environmental impacts arising from the interplay between different anthropogenic and biogenic components of green initiatives to ensure sustainable development through amelioration of local (and regional) air quality while minimizing climate change impacts. The assessment is mainly confined to pollutants that are currently of particular concern, including both primary and secondary PM (considered here as the combined pool of PM_{10}, $PM_{2.5}$, i.e., particles with aerodynamic diameters <10 and <2.5 μm, respectively), ozone (O_3) and nitrogen dioxide (NO_2).

Cross-Sectoral Assessment Framework for a Green Project

A systems scale assessment framework, encompassing plausible combinations of emerging green initiatives that will be implemented over the next 10–20 years (around 2020/2030 horizons) is presented in Figure 7. It draws together the evidence-base from available literature on cross-disciplinary climate change and urban sustainability research, applying a cross-sectoral approach to three broad categories of green initiatives, including (a) use of vegetation, (b) low-emission personal transport, and (c) renewable energy from biomass. Each initiative (shown in boldface text) is characterized by a set of positive and negative environmental impacts (shown in italicized text) with their resulting air quality implications. These depend on the activities involved and their influence on either formation or removal of air pollutants. This was considered as an essential first step toward scoping the systems framework of landscape interactions between biogenic emissions (primarily biogenic volatile organics, bVOCs) and the anthropogenic emissions from fossil fuel combustion (mainly from transport and energy sources) in

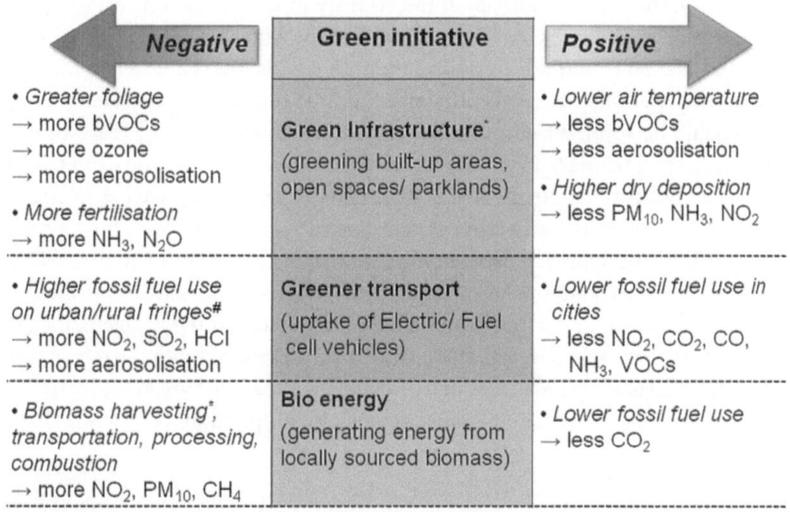

Negative	Green initiative	Positive
• *Greater foliage* → more bVOCs → more ozone → more aerosolisation • *More fertilisation* → more NH_3, N_2O	**Green Infrastructure**[*] (greening built-up areas, open spaces/ parklands)	• *Lower air temperature* → less bVOCs → less aerosolisation • *Higher dry deposition* → less PM_{10}, NH_3, NO_2
• *Higher fossil fuel use on urban/rural fringes*[#] → more NO_2, SO_2, HCl → more aerosolisation	**Greener transport** (uptake of Electric/ Fuel cell vehicles)	• *Lower fossil fuel use in cities* → less NO_2, CO_2, CO, NH_3, VOCs
• *Biomass harvesting*[*], *transportation, processing, combustion* → more NO_2, PM_{10}, CH_4	**Bio energy** (generating energy from locally sourced biomass)	• *Lower fossil fuel use* → less CO_2

FIGURE 7 Theoretical framework illustrating the positive and negative implications for air quality from emerging green initiatives (2020/2030 horizons).
Source: Tiwary et al.[26]

future Green Cities. This step incorporates both the positive and the negative effects arising from such interactions, in order to assess the overall sustainability implications of the green initiatives.

Uptake of greener transportation technologies, through a combination of low-emitting internal combustion and electric/fuel-cell traction, is projected to reduce primary emissions of CO, NO_x and PM from vehicle use. However, such initiatives can be considered green only to the extent that the source of energy supply is renewable. Fuel cell-powered vehicles may still be associated with pollutant emissions in peri-urban regions if the hydrogen is generated by fossil fuel sources, which would contribute to additional aerosols from atmospheric reactions of SO_2, NO_x, NH_3 and VOCs, originating from the refineries. Incorporating biomass into the future energy mix is meant to keep the decarbonized energy generation affordable but at the cost of air quality. Systems-level assessments of different fuel mixes have found increased N_2O, NH_3 and primary PM from the harvest phase and enhanced NO_x, CH_4 and secondary aerosol (SA) formation potentials from the combustion phase.

The effect of land cover modification on ambient temperature is another key driver to energy demand (for cooling/heating) and corresponding pollutant emissions (both in terms of primary components from associated activities as well as the photochemical precursors, i.e., chemicals that lead to tropospheric ozone production). While climate change effects are projected to contribute to aggravation of the urban heat island effects (and an overall increase in ambient temperature), lower air temperatures (resulting from vegetation cover modifications and retrofitting initiatives to enhance evapotranspiration and albedo-effects) have shown reduction in cooling electricity demand. The implications appear to be varied at global levels depending on the green initiatives pursued and more deterministic scenario modeling would be required to inform the outcomes.

Conclusion

This chapter introduced some basic concepts of system analysis of "greener" processes and projects, specifically in the context of holistic environmental management. It explained the implementation of transboundary material flows and life cycle thinking in terms of ensuring global sustainability, by avoiding unacceptable implications for interventions in wider process chains—both in the supply streams and in the waste streams.

It is noted that uncertainty within systems analysis can be neither neglected nor simply addressed. This is mainly because the information about uncertainty in science-driven data cannot be fully captured within the inventoried databases. This chapter also provides an overview of different sources of uncertainly and the available methods to resolve them. A significant share of this uncertainty arises in practice, based on the relationship between the data and the intended reality being modeled. Software and algorithms should be applied rigorously by researchers on a breadth of real-world case studies in order to closely identify which parameters are the major sources of uncertainty in LCA results.

Two example case studies illustrate the role of LCA and geo-spatial approaches in assessing the greenhouse gases and air quality emissions across system boundaries, respectively, for a process and a project. It shows the merit of this approach in revealing the hotspots contributing to potential environmental impacts while devising green solutions viable at systems scale. Some discussion is presented to develop possible mitigation for transboundary environmental management of the problem, including all associated activities in the process chain. Overall, this chapter provides insight into the role of systems approach in efficient environmental management, through assessment of the "true green credentials" of a process or a project in terms of its greenhouse gas emissions, ecotoxicity, air pollution, etc. within the system.

References

1. Baumann, H.; Tillman, A. *The Hitch Hiker's Guide to LCA*; Studentlitteratur AB: Lund, Sweden, 2004.
2. ISO 14044. *Environmental Management—Life Cycle Assessment—Requirements and Guidelines*; International Organisation for Standardisation: Geneva, Switzerland, 2006.

3. Tiwary, A.; Colls J. Mitigating secondary aerosol generation potentials from biofuel use in the energy sector. *Sci. Total Environ.* **2010**, *408*, 607–616.

4. Sonnemann, G.W.; Schuhmacher, M.; Castells, F. Uncertainty assessment by a Monte Carlo simulation in a life cycle inventory of electricity produced by a waste incinerator. *J. Clean. Prod.* **2003**, *11*, 279–292.

5. von Bahr, B.; Steen, B. Reducing epistemological uncertainty in life cycle inventory. *J. Clean. Prod.* **2004**, *12*, 369–388.

6. Maurice, B.; Frischknecht, R.; Coelho-Schwirtz, V.; Hungerbuhler, K. Uncertainty analysis in life cycle inventory. Application to the production of electricity with French coal power plants. *J. Clean. Prod.* **2000**, *8*, 95–108.

7. Hertwich, E.G.; Hammitt, J.K. A decision-analytic framework for impact assessment. Part 1: LCA and decision analysis. *Int. J. Life Cycle Assess.* **2001**, *6* (1), 5–12.

8. Frischknecht, R., et al. The ecoinvent database: Overview and methodological framework. *Int. J. Life Cycle Assess.* **2005**, *10* (1), 3–9.

9. Lozano, S.; Iribarren, D.; Moreira, M.T.; Feijoo, G. The link between operational efficiency and environmental impacts: A joint application of life cycle assessment and data envelopment analysis. *Sci. Total Environ.* **2009**, *407* (5), 1744–1754.

10. Heller, M.C.; Keoleian, G.A.; Mann, M.K.; Volk, T.A. Life cycle energy and environmental benefits of generating electricity from willow biomass. *Renew. Energy* **2004**, *29* (7), 1023–1042.

11. Kaltschmitt, M.; Reinhardt, G.A.; Stelzer, T. Life cycle analysis of biofuels under different environmental aspects. *Biomass Bioenergy* **1997**, *12*, 21–34.

12. Lewandowski, I.; Heinz, A. Delayed harvest of miscanthus—Influences on biomass quantity and quality and environmental impacts of energy production. *Eur. J. Agron.* **2003**, *19* (1), 45–63.

13. Thornley, P. Airborne emissions from biomass based power generation systems. *Environ. Res. Lett.* **2008**, *3*, 6–8.

14. Faaij, A., et al. Externalities of biomass based electricity production compared with power generation from coal in the Netherlands. *Biomass Bioenergy* **1998**, *14*, 125–147.

15. Ryu, C., et al. Effect of fuel properties on biomass combustion: Part I. Experiments—Fuel type, equivalence ratio and particle size. *Fuel* **2006**, *85* (7–8), 1039–1046.

16. Öko-Institut. *Global Emission Model of Integrated System (GEMIS) Database*; Öko-Institut: Darmstadt, Germany, 2006; available at http://www.oeko.de.

17. McDonald, J.D., et al. Fine particle and gaseous emission rates from residential wood combustion. *Environ. Sci. Technol.* **2000**, *34* (11), 2080–2091.

18. Rivela, B.; Moreira, M.T.; Muñoz, I.; Rieradevall, J.; Fei-joo, G. Life cycle assessment of wood wastes: A case study of ephemeral architecture. *Sci. Total Environ.* **2006**, *357* (1–3), 1–11.

19. Damen, K.; Faaij, A. *A Life Cycle Inventory of Existing Biomass Import Chains for "Green" Electricity Production, NW&S-E-2003-01*; Utrecht University: Utrecht, The Netherlands, 2003.

20. Wierzbicka, A., et al. Particle emissions from district heating units operating on three commonly used biofuels. *Atmos. Environ.* **2005**, *39* (1), 139–150.

21. Rodrigues, M.; Faaij, A.; Walter, A. Techno-economic analysis of co-fired biomass integrated gasification/combined cycle systems with inclusion of economies of scale. *Energy* **2003**, *28* (12), 1229–1258.

22. Lucas, C.; Szewczyk, D.; Blasiak, W.; Mochida, S. High-temperature air and steam gasification of densified biofuels. *Biomass Bioenergy* **2004**, *27* (6), 563–575.

23. Faaij, A. Bio-energy in Europe: Changing technology choices. *Energy Policy* **2006**, *34* (3), 322–342.

24. Lewandowski, I., et al. The potential biomass for energy production in the Czech Republic. *Biomass Bioenergy* **2006**, *30* (5), 405–421.

25. Hamelinck, C.N.; Suurs, R.A.; Faaij, A. *International Bioenergy Transport Costs and Energy Balance*; Utrecht University, Copernicus Institute Science Technology Society: Utrecht, The Netherlands, 2003.

26. Tiwary, A., et al. Systems scale assessment of the sustainability implications of emerging green initiatives. *Environ. Pollut.* **2013**, *183*, 213–223.

45

Green Products: Production

Puangrat
Kajitvichyanukul,
Jirapat
Ananpattarachai,
and Apichon
Watcharenwong

Introduction

Since 1990, there has been a wide concern for environmental product development and how to achieve green products. Products affect the environment in several ways. Once a product has been designed and generated from the manufacturing process, its environmental attributes are largely fixed along its life cycle. Environmental requirements have been made by several attempts to ultimately obtain environmentally friendly products. For example, design processes are aimed to minimize raw material and energy consumption, waste generation, health and safety risk, and ecological degradation. Several terminologies of product design integrating environmental concerns were introduced in the late 1990s. These terms include design for the environment,[1–3] life cycle analysis,[2] green chemistry,[4–6] pollution prevention,[7–10] environmentally conscious manufacturing,[11,12] and sustainable development.[13] The aggregation of these terms is called "industrial ecology."[14] In 1996, the Office of Pollution Prevention and Toxics of the U.S. Environmental Protection Agency (EPA) set up two programs, Design for the Environment (DfE) and green chemistry, under the industrial ecology scope. The DfE has been defined as follows: "Design for environment is the systematic process by which firms design products and processes in an environmentally conscious way."[15] The broad definition of green chemistry has been also provided as follows: "Green chemistry is the design of chemical products and processes that reduce or eliminate the use and generation of hazardous substances and seek to reduce and prevent pollution at source."[16] Since 1995, green chemistry has played an important role in green product development, especially chemical products. Nowadays, green chemistry has become an important tool in achieving sustainable development. The implementation of green chemistry—the design of chemical products and processes that reduce or eliminate the use and generation of hazardous substances—has become essential for increasing the standard of human living without having a negative impact on the environment. In addition, green chemistry provides solutions to critical global challenges such as climate change, sustainable agriculture, energy, toxics in the environment, and depletion of natural resources.

In this entry, the concepts in green chemistry for producing green products, especially involving the chemical industrial process, are reviewed. It is noted that chemical products are largely created using energy-intensive processes and non-renewable, petroleum-based resources as feedstocks.[17] Many of these industrial processes use hazardous materials or generate wastes that are harmful to human health and the environment. Thus, introducing green chemistry concepts through these chemical products is a particularly good example of developments in this field that may inspire future sustainable development.

Principles of Green Chemistry

The principles of green chemistry have been refined from a diverse set of practices and emerging research. The EPA defines green chemistry as the use of chemistry for pollution prevention at the molecular level.[18] The mission of green chemistry is to promote innovative chemical technologies that reduce or eliminate the use or generation of hazardous substances in the design, manufacture, and use of chemical products.[19] They address alternative starting and target materials; alternative reagents, solvents, and catalysts; and improved processes and process control in producing chemical products. Green chemistry consists of chemicals and chemical processes designed to reduce or eliminate negative environmental impacts.

The principles of green chemistry were originally published by Pual Anastas and John Warner in *Green Chemistry: Theory and Practice* (Oxford University Press: New York, 1998). The 12 principles of green chemistry provide a road map to implement green chemistry. Following Anatas and Warner,[20] the principles are as follows:

1. *Prevention.* It is better to prevent waste than to treat or clean up waste after it has been created.
2. *Atom economy.* Synthetic methods should be designed to maximize the incorporation of all materials used in the process into the final product.
3. *Less hazardous chemical syntheses.* Wherever practicable, synthetic methods should be designed to use and generate substances that possess little or no toxicity to human health and the environment.
4. *Designing safer chemicals.* Chemical products should be designed to effect their desired function while minimizing their toxicity.
5. *Safer solvents and auxiliaries.* The use of auxiliary substances (e.g., solvents, separation agents) should be made unnecessary wherever possible and innocuous when used.
6. *Design for energy efficiency.* Energy requirements of chemical processes should be recognized for their environmental and economic impacts and should be minimized. If possible, synthetic methods should be conducted at ambient temperature and pressure.
7. *Use of renewable feedstocks.* A raw material or feedstock should be renewable rather than depleting whenever technically and economically practicable.
8. *Reduce derivatives.* Unnecessary derivatization (use of blocking groups, protection/deprotection, temporary modification of physical/chemical processes) should be minimized or avoided if possible because such steps require additional reagents and can generate waste.
9. *Catalysis.* Catalytic reagents (as selective as possible) are superior to stoichiometric reagents.
10. *Design for degradation.* Chemical products should be designed so that at the end of their function, they break down into innocuous degradation products and do not persist in the environment.
11. *Real-time analysis for pollution prevention.* Analytical methodologies need to be further developed to allow for real-time, in-process monitoring and control before the formation of hazardous substances.
12. *Inherently safer chemistry for accident prevention.* Substances and the form of a substance used in a chemical process should be chosen to minimize the potential for chemical accidents, including releases, explosions, and fires.

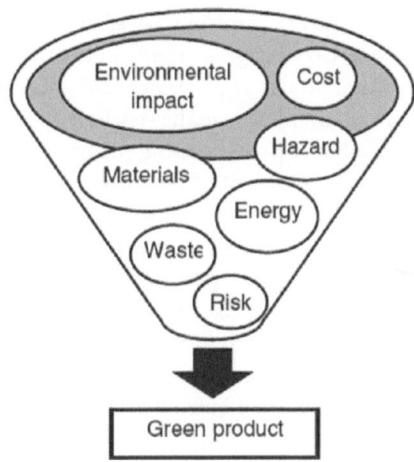

FIGURE 1 Objectives of green chemistry.

Green chemistry addresses hazards, whether physical (flammability, explosivity), toxicological (carcinogenicity, endocrine disruption), or global (ozone depletion, climate change) as an inherent property of a molecule.[21] Green chemistry also addresses all aspects of chemical designing manufacturing.[22,23] The four general objectives derived from the green chemical philosophy are as follows:[24]

- Reduction of use and generation of polluting chemicals in the chemical process
- Reduction of use of dangerous chemicals in the chemical process
- Reduction of the harmful effects of final products.
- Reduction of the use of exhaustible feedstock materials and of scarce resources

The objectives of green chemistry are shown in Figure 1. These concepts lead the movement on the front lines of research, education, and incentive and research programs of the EPA, National Science Foundation, U.S. Department of Energy, U.S. Department of Agriculture, and international organizations worldwide.

Approaching Sustainability through Principles

During the latter half of the 20th century, an exponentially increasing number of environmental regulations were enacted in developed countries. In the United States, the Clean Air Act of 1970, the Clean Water Act of 1972, the Resource Conservation and Recovery Act of 1976, and their amendments, have set the fundamental regulations covering air, water, and hazardous waste pollutants, respectively. In 2001, the EPA promulgated pharmaceutical maximum achievable control technology standards that regulate 187 hazardous air pollutants, including solvents commonly used in the pharmaceutical industry. In Europe, the European Union has a set of common rules for permitting and controlling industrial installations in the Integrated Pollution Prevention and Control (IPPC) Directive of 1996.[25,26] The IPPC Directive is about minimizing pollution from industrial sources throughout the European Union. In 2007, REACH, the new Regulation on Registration, Evaluation, Authorization, and Restriction of Chemicals, was launched and considered a tool for green chemistry. This regulation improves the former legislative framework on chemicals of the European Union.[27] One of the main aims of REACH is to improve the protection of human health and the environment from the risks that can be posed by chemicals. The major concern of REACH is chemical safety and in that it is one of the green chemistry metrics.[28]

As environmental regulation has become more stringent owing to climate concern, industries have to look for innovative ways to obtain purer final products using a reduced amount of raw materials and energy and generating a smaller amount of waste. Concepts of green chemistry have become a solution for the industrial requirement. Many "green" terminologies have emerged and been defined for use in environmental product development. Green engineering has been defined as follows: "Green engineering is the design, commercialization and use of processes that are feasible and economic, reduce the generation of pollution at the source, and minimize the risk to human health and the environment."[16] Green technology is also widely used as it involves using science to create technologies that conserve natural resources and lessen the human impact on the environment.

Green chemistry, green engineering, and green technology have been focused on and proposed by several scientists and researchers worldwide to respond to the need of industry. Collaboration among industries, academic institutes, and government agencies were initiated to push green chemistry to the manufacturing process. In 2003, the collaboration of industry, the EPA, and the U.S. Department of Commerce's Manufacturing Extension Partnership, called the Green Suppliers Network, was officially launched in the automotive industry, and is currently in the aerospace, pharmaceutical and health-care, and office furniture industries. The established outcomes of this program are the effective processes and products with higher profits while minimizing environmental impact. In 2005, the Committee on Grand Challenges for Sustainability in the Chemical Industry arranged a workshop on identifying research needs to enable the industry to develop products and processes for achieving the goal of sustainability. In addition, the pharmaceutical industry and the American Chemical Society Green Chemistry Institute collaborated to encourage innovation while catalyzing the integration of green chemistry and green engineering in the pharmaceutical industry. To support the collaboration in green chemistry, the Green Chemistry Research and Development Act of 2005[29] was proposed and passed by the House Science Committee in April 2005. This bill promotes green chemistry by authorizing a coordinated green chemistry research and development program.

The emergence of these collaborations is the driving force behind the development of green business. It is reported that green chemistry is drawing many investors from different companies to invest in producing green product lines. For example, Wilmar International, a Singaporean company, plans to launch production of derivatives of glycerol through a green chemistry process. Mitsubishi Chemical and Thai Company PTT plc are jointly conducting a study focused on the production of biosuccinic acid and biopolybutylene succinate. Moreover, the Japanese trading house Mitsui & Co. and Toyota Tsusho are also involved in green chemistry investments in the Asian regions.[30]

Nowadays, concepts of green chemistry have widely spread to the manufacturing process with the understaning that it is a superior, innovative chemistry that is cost-effective and has minimal impact on the environment. The applications of the concepts through the manufacturing process are shown in Figure 2. The types of products and processes adopting green chemistry principles include medicine, food, and energy production; packaging materials; household and commercial cleaning products; electronics and automotive chemicals; and a wide range of consumer goods.[31]

Advances in Green Chemistry to Produce Green Products

Catalysts

Catalysts play a significant role in green chemistry by decreasing energy requirements, increasing selectivity, and permitting the use of less hazardous reaction conditions. The first example in applying green chemistry to the catalytic process is the synthesis of maleic anhydride.[32-34] Originally, the benzene process was used; however, it was substituted by the butane process for the following reasons:

- The loss of two carbon atoms (starting from benzene) is avoided.
- Using butane, the by-products are carbon oxides and a small amount of acetic acid. With benzene, several byproducts are formed.
- The toxicity aspects related to the use of benzene are avoided, reducing costs related to safety systems and benzene handling.

FIGURE 2 Applications of concepts through the manufacturing process.

The benefits of using the butane process following the green chemistry concept are as follows: 1) use of non-toxic reactants; 2) improved atom economy; 3) a complex multistep transformation (the reaction is a 14 e⁻ oxidation, with abstraction of 8 H atoms from the butane molecule and insertion of 3 O atoms) is realized in one single step, without using solvents; and 4) waste formation is minimized.

Synthesizing adipic acid directly from cyclohexene using an aqueous 30% hydrogen peroxide solution, a catalytic amount of Na_2WO_4, and a phase-transfer catalyst is another green chemistry application.[35] Previously, adipic acid was derived from the reduction of benzene to cyclohexene under high temperature and pressure resulting in nitrous oxide, a greenhouse gas and ozone-depleting substance, as an undesirable product. Using the new process, adipic acid is produced with a high yield without the use of organic solvents and without the generation of nitrous oxide.[36] Using the concept of on-site production is another alternative way in green chemistry that is already used in some new production processes to avoid the storage of toxic or dangerous chemicals. The on-site production of HCN in the manufacturing process of toluene diisocyanate has already been implemented. This was made possible by the combined development of new catalysts and new reactor engineering solutions that allow an economic production even in a small size plant.[34]

Nowadays, biocatalysts, both isolated-enzyme and whole-cell systems, are increasingly being used to assist in synthetic routes to complex molecules. The biggest role for biocatalysis is found in the pharmaceutical sector.[37–39] Biocatalysis is one of the greenest technologies for the synthesis of chiral molecules due to exquisite regioselectivity and stereoselectivity in water under mild conditions. Recently, Tao and Xu[40] reviewed the development of novel biocatalytic pharmaceutical processes to replace chemical routes with poorer process efficiency and higher manufacturing costs. It is emphasized that to empower the green chemistry feature of biocatalysis, it is essential to integrate enzymatic transformations strategically into chemical transformations at the retrosynthetic level. Enzymatic catalysis can provide a new dimension for route redesign to meet both the process and green chemistry metrics. It is underscored that recent advances in large-scale DNA sequencing and enzyme-directed evolution rendered biocatalysis a much more practical technology to provide green chemistry solutions for the industrial production of chemicals.

Nanomaterials

In the search for environmentally friendly materials, recently green chemistry has shown its role in nanomaterials, especially in the preparation and synthesizing steps. A solvent is considered an important chemical that is widely used in synthesizing nanomaterials. It is an important parameter that determines the green nature of the nanoparticle catalysis, regarding the fact that the solvent accounts for 50% of the greenhouse gas emissions during post-treatment and 60% of the energy used in pharmaceutical processes.[41] Recently, there have been several attempts to use other chemicals to substitute these harmful reducing agents and solvents to obtain engineered nanoparticles. For solvent substitution, ionic liquids are very promising replacements for the traditional volatile organic solvents owing to their high mobility, low melting points, negligible vapor pressure, thermal stability, low toxicity, large electrochemical window, non-flammability, and ability to dissolve a variety of chemicals.[42–45] Ionic liquids are versatile solvents that retain their liquid states over a wide range of conditions, but have a minimal vapor pressure. They also possess interesting properties that are a combination of the characteristics of water and organic solvents. The use of ionic liquids as stabilizers for nanoparticles was first demonstrated by Dupont et al. in 2002.[46] Ionic liquids can be functionalized and attached with weak coordination groups to further enhance the stability as well as catalytic activity in the synthesis reaction. Zhao et al.[47] used nitrile- functionalized ionic liquids to protect Pd nanoparticles with an average diameter of 5 nm. The resulting Pd nanoparticles were excellent recyclable catalysts, and an optimum balance between stability and reactivity can be achieved. Ionic liquids widely used in nanomaterial synthesis are BMI(BF$_4$), BMI(NTf$_2$), BMI(PF$_6$), and C$_2$OHMIM(BF$_4$). Nowadays, various nanostructured materials, such as iridium,[46] palladium,[47,48] gold,[49] tellurium,[50] TiO$_2$,[51,52] ZnO,[53–56] and CoPt[57] have been synthesized in ionic liquids.

One of the concerns in the process of preparation of nanomaterials is the choice of the reducing agent. zIt is reported in the literature that in the preparation of super-paramagnetic nanoparticles, the reductants used to date include hydrazine,[58] sodium borohydride,[59] carbon monoxide,[60] and dimethylformamide.[61] All of these are highly reactive chemicals and pose potential environmental and biological risks.[62] As appeared in recent literature reviews, many interesting methods are currently being applied to the green preparation of nanoparticles, called biosynthesis of nanoparticles by glucose,[62] starch,[63–65] plant leaf broth,[66] edible mushroom extract,[67] apiin,[68] and by latex of *Jatropha curcas*.[69] These biological extracts are used as reducing agents in the green synthesis of nanoparticles, especially in metal nanoparticle synthesis. Raveendran et al.[70] reported a completely "green" synthesis of Ag nanoparticles with sizes of 1–8 nm using starch as a capping agent and α-D-glucose as a reducing agent. Mallikarjuna and Varma[71] described a simple method for the shape-controlled synthesis of nanostructures (with size 2–15 nm) of noble metals such as Au, Ag, Pd, and Pt by microwave-assisted spontaneous reduction of noble metal salts using α-D-glucose, sucrose, and maltose in aqueous solution. Fructose, glucose, and sucrose were used as reducing and capping agents for the preparation of Au, Ag, Pd, Pt, and Au–Ag nanoparticles by heating to dryness for 2 h the solutions in a hot water bath.[72] The effect of two different sugars on the sizes and surface morphology for Ag nanoparticle synthesis was also reported by Filippo et al.[73] Furthermore, Nersisyan et al.[74] developed an effective way of preparing a nanosized (10–50 nm) colloidal dispersion of Ag using glucose as a reducing agent. Lu et al.[62] also reported a facile and green method to synthesize superparamagnetic Fe$_3$O$_4$ nanoparticles with α-D-glucose as the reductant, which is a mild, renewable, inexpensive, and non-toxic reducing agent, and without any additional stabilizer and dispersant.

Solvents

Solvents play an important role in chemical synthesis. Considering green chemistry requirements, water is an extremely attractive solvent choice. It is inexpensive and readily available, non-toxic, non-carcinogenic, and nonflammable. In addition, water is probably the least expensive among all available solvents.[75]

Besides water, supercritical fluids (SCFs) draw attention from chemists for their several commercial applications. These SCFs are gases that are compressed until their density approaches that of liquids. Compression can occur only above the "critical temperature" of the fluid, since at lower temperatures the fluid liquefies upon compression. Therefore, an SCF is a gas that displays some of the properties of a liquid. The advantages of using SCFs are related to the absence of toxic residues, the relatively low temperature needed for the extraction process, and their higher solvent power.[34] The SCFs are characterized by high diffusivity, low viscosity, and intermediate density. Among the SCFs, supercritical CO_2 (SC-CO_2) is the most widely used solvent in many applications. It is non-toxic, noncarcinogenic, and non-flammable. Oakes et al.[76] investigated the catalysis of the Diels–Alder reaction between n-butyl acrylate and cyclopentadiene in SC-CO_2. Scandium tris (trifluoromethanesulfonate) was selected as the Lewis acid catalyst. By varying the pressure of the solvent, a significant improvement over selectivity is achieved in conventional solvent. It was reported that rate enhancement was observed under SC-CO_2 conditions.

High boiling alcohols, including glycerol and polyethylene glycol, are also considered as "green solvents." They have recently attracted much attention as reaction media.[77,78] These high boiling alcohols are of low toxicity and volatility, and are biodegradable, non-expensive, and easily functionalized. These properties represent important environmentally benign characteristics. Glycerol is a natural product widely used as a cosmetic and food additive. Polyethylene glycol is also non-toxic and is widely used in the pharmaceutical,[79] semiconductor,[80] and food industries.[81]

Application of Green Chemistry in the Manufacturing Process

There are several ways to apply green chemistry to the manufacturing process for industries. The concepts can be applied through waste minimization, waste or material recycling, solvent selection, atom utilization, intensive processing, and clean synthesis. According to the Presidential Green Chemistry Challenge from the EPA, three basic methods are currently used in industrial processes:

1. Applications of greener synthetic pathways—Many green pathways for a new chemical product are widely used. It involves using a novel, green pathway to redesign the synthesis of an existing chemical product. The synthetic pathways include
   - Use greener feedstocks that are innocuous or renewable (e.g., biomass, natural oils)
   - Use novel reagents or catalysts, including biocatalysts and microorganisms
   - Apply natural processes, such as fermentation or biomimetic synthesis
   - Use atom-economical systems
   - Use convergent syntheses

An example of applying the synthetic pathway in the manufacturing process is from Dow and BASF, winners of the 2010 Greener Synthetic Pathways Award (EPA). Both companies have jointly developed a new route to make propylene oxide with hydrogen peroxide that eliminates most of the waste and greatly reduces water and energy use.

2. Use of greener reaction conditions—This application is mainly about improving conditions other than the overall design or redesign of a synthesis, including greener analytical methods. The greener reaction conditions include the following:
   - Substitute hazardous solvents with solvents that have a reduced impact on human health and the environment
   - Use solventless reaction conditions and solid-state reactions
   - Use novel processing methods
   - Eliminate energy- or material-intensive separation and purification steps
   - Improve energy efficiency, including reactions running closer to ambient conditions

A successful example of applying green reaction conditions is from Merck and Codexis. Both companies have developed a second-generation green synthesis of sitagliptin, the active ingredient in Januvia™, a treatment for type 2 diabetes. This collaboration has led to an enzymatic process that reduces waste, improves yield and safety, and eliminates the need for a metal catalyst. This work also received the 2010 Greener Reaction Conditions Award from the EPA.

3. Design of greener chemicals—Applications include the design of chemical products that are less hazardous than the products or technologies they replace. These chemical products are:
   - Less toxic than current products
   - Inherently safer with regard to accident potential
   - Recyclable or biodegradable after use
   - Safe for the atmosphere (e.g., do not deplete ozone or form smog)

An example of the design of green chemicals is spinosad. It is an environmentally safe pesticide but is not stable in water and therefore cannot be used to control mosquito larvae. Clarke, the winner of the 2010 Designing Greener Chemicals Award from the EPA, has developed a way to encapsulate spinosad in a plaster matrix, allowing it to be released slowly in water and provide effective control of mosquito larvae. This pesticide, Natular™, replaces organophosphates and other traditional, toxic pesticides and is approved for use in certified organic farming.

Conclusion

The concepts of green chemistry for producing green products and examples of green chemistry implementation in industrial processes have been discussed in this entry. It is known that many industrial processes use hazardous materials or generate wastes that are harmful to human health and the environment. Implementation of green chemistry to the design of chemical products and processes that reduce or eliminate the use and generation of hazardous substances has become essential to increase the standard of human living without the adverse effect on the environment. Using the green production system in the industry can gain the benefit of enabling chemical production that is more profitable, less wasteful, less damaging to the environment, and more socially acceptable. In this entry, the applications of green chemistry through green product production have been reviewed. The process of green products in industrial manufacturing can be applied through several green concepts, including waste minimization, waste or material recycling, solvent selection, atom utilization, intensive processing, and clean synthesis. The challenge for the industrial research and development sector is to develop products, processes, and services to improve the quality of life, the environment, and the industry competitiveness with the sustainable aspect toward the future. Recently, green chemistry has been implemented in several industrial products, including medicines, food, energy, packaging materials, household and commercial cleaning products, electronic and automotive chemicals, and a wide range of consumer goods. Biocatalysts, nanoparticles, and green solvents are the next challenge in green synthesis to produce high-quality merchandise to serve the need of consumers. Beyond green chemistry, green engineering and green technology have been focused on and proposed by several scientists and researchers worldwide. They are in the process of reinforcing steps to create cleaner production in industrial processes in the near future.

References

1. Allenby, B.R.; Richards, D. *The Greening of Industrial Ecosystems*; National Academy Press: Washington, D.C., 1994.
2. Graedel, T.E.; Allenby, B.R. *Industrial Ecology*; Prentice Hall: New York, 1995.
3. Graedel, T E.; Allenby, B.R. *Design for Environment*; Prentice Hall: New York, 1996.

4. Anastas, P.T. Benign by design chemistry. In *Benign by Design: Alternative Synthetic Design for Pollution Prevention*; Anastas, P.T., Farris, C.A., Eds.; ACS Symposium Series 577; American Chemical Society: Washington, D.C., 1994; 2–22.

5. Anastas, P.T.; Williamson, T.C. Green chemistry: An overview. In *Green Chemistry: Designing Chemistry for the Environment*; Anastas, P.T., Williamson, T.C., Eds.; ACS Symposium Series 626; American Chemical Society: Washington, D.C., 1996; 1–17.

6. Garrett, R.L. Pollution prevention, green chemistry, and the design of safer chemicals. In *Designing Safer Chemicals: Green Chemistry for Pollution Prevention*; ACS Symposium Series 640; DeVito, S.C., Garrett, R.L., Eds.; American Chemical Society: Washington, D.C., 1996; 2–15.

7. Socolow, R.; Andrews, C.; Berkhout, F.; Thomas, V. *Industrial Ecology and Global Change*; Cambridge University Press: Cambridge, U.K., 1994.

8. Frosch, R.; Gallopoulos, N. Strategies for manufacturing. Sci. Am. **1989**, *261* (3), 144–152.

9. Becker, M.; Ashford, N.A. Exploiting opportunities for pollution prevention in EPA enforcement agreements. Environ. Sci. Technol. **1995**, *29* (5), 220A–226A.

10. Breen, J.J.; Dellarco, M.J. Pollution prevention: The new environmental ethic. In *Pollution Prevention in Industrial Processes: The Role of Process Analytical Chemistry*; Breen, J.J., Dellarco, M.J., Eds.; ACS Symposium Series 508; American Chemical Society: Washington, D.C., 1992; 2–12.

11. Baccini, P.; Brunner, P.H. *Metabolism of the Anthropo sphere*; Springer-Verlag: New York, 1991.

12. Fiksel, J. Introduction. In *Design for Environment: Creating Eco-Efficient Products and Processes*; Fiksel, J., Ed.; McGraw-Hill: New York, 1996; 2–8.

13. National Science and Technology Council. *Technology for a Sustainable Future: A Framework for Action*; Office of Science and Technology: Washington, D.C., 1994.

14. Anastas, P.T.; Breen, J.J. Design for the environment and Green Chemistry: The heart and soul of industrial ecology. J. Cleaner Prod. **1997**, *5* (1–2), 97–102.

15. Lenox, M.; Jordan, B.; Ehrenfeld J. In *The Diffusion of Design for Environment: A Survey of Current Practice*, Proceedings of the IEEE International Symposium on Electronics and the Environment, Dallas, TX, May 6–8, 1996; 25–30.

16. Mason, T.J. Sonochemistry and the environment—Providing a "green" link between chemistry, physics and engineering. Ultrason. Sonochem. **2007**, *14* (4), 476–483.

17. Nameroff, T.J.; Garant, R.J.; Albert, M.B. Adoption of green chemistry: An analysis based on US patents. Res. Policy **2004**, *33* (6–7), 959–974.

18. U.S. Environmental Protection Agency, Green Chemistry, available at http://www.epa.gov/greenchemistry (accessed May 14, 2012).

19. Anastas, P.T., Heine, L.G., Williamson, T.C. Green chemical syntheses and processes: Introduction. In *Green Chemical Syntheses and Processes*; Anastas, P.T., Heine, L.G., Williamson, T.C., Eds.; ACS Symposium Series 767; American Chemical Society: Washington, D.C., 2000; 1–17.

20. Anastas, P.T.; Warner, J.C. *Green Chemistry: Theory and Practice*; Oxford University Press: New York, 1998.

21. Anastas, P.T. Meeting the challenges to sustainability through green chemistry. Green Chem. **2003**, G29–G34.

22. Clark J.H. Part 1 Green chemistry for sustainable development. In *Green Separation Processes*; Afonso, C.A.M., Crespo, J.G., Eds.; Wiley-VCH Verlag GmbH & Co. KGaA: Weinheim, Germany, 2005; 1–31.

23. Poliakoff, M., Licence, P. Sustainable technology: Green chemistry. Nature **2007**, *450* (7171), 810–812.

24. Mestres R. A brief structured view of green chemistry issues. Green Chem. **2004**, (1), G10–G12.

25. European Commission. *Directive 91/61/EC Concerning Integrated Pollution Prevention and Control, No. L2J7, HMSO*; Official Journal of the European Communities, European Union: London, 1996.

26. European Commission, Joint Research Centre, Institute for Prospective Technological Studies. Reference document. Available at http://eippcb.jrc.es/reference/ (accessed May 14, 2012).
27. European Commission. Alphabetical index. Available at http://ec.europa.eu/atoz_en.htm (accessed March 14, 2012).
28. Demirci, U.B. How green are the chemicals used as liquid fuels in direct liquid-feed fuel cells? Environ. Int. **2009**, *35* (3), 626–631.
29. *Boehlert, S.L. H.R. 1215 (109th): Green Chemistry Research and Development Act of 2005;* Report 109–82; Committee on Science and Technology, U.S. House of Representatives: Washington, D.C., 2005; 1–65.
30. Markets and Business. Southeast Asia attracting green-chemistry investments. Focus Catalysts **2010**, *2010* (1), 2.
31. Manley, J.B.; Anastas, P.T.; Cue, B.W., Jr. Frontiers in green chemistry: Meeting the grand challenges for sustainability in R&D and manufacturing. J. Clean. Prod. **2008**, *16* (6), 743–750.
32. Centi, G.; Cavani, F.; Trifirò, F. *Selective Oxidation by Heterogeneous Catalysis;* Kluwer Academic Publishers/Plenum Press: New York, 2000.
33. Centi, G. Vanadyl pyrophosphate—A critical overview. Catal. Today **1993**, *16* (1), 5–26.
34. Centi, G.; Perathoner, S. Catalysis and sustainable (green) chemistry. Catal. Today **2003**, *77* (4), 287–297.
35. Sato, K.; Aoki, M.; Noyori, R. A "green" route to adipic acid: Direct oxidation of cyclohexenes with 30 percent hydrogen peroxide. Science **1998**, *281* (5383), 1646–1647.
36. Kirchhoff, M.M. Promoting sustainability through green chemistry. Resour. Conserv. Recycling **2005**, *44* (3), 237–243.
37. Patel, R.N. Microbial/enzymatic synthesis of chiral pharmaceutical intermediates. *Curr. Opin. Drug Discov. Devel.* **2003**, *6* (6), 902–920.
38. Buckland, B.C.; Robinson, D.K.; Chartrain, M. Biocatalysis for pharmaceuticals—Status and prospects for a key technology. *Metab. Eng.* **2000**, *2* (1), 42–48.
39. Pesti, J.A.; DiCosimo, R. Recent progress in enzymatic resolutions and desymmetrization of pharmaceuticals and their intermediates. *Curr. Opin. Drug Discov. Devel.* **2003**, *6* (6), 884–901.
40. Tao, J.; Xu, J.-H. Biocatalysis in development of green pharmaceutical processes. Curr. Opin. Chem. Biol. **2009**, *13* (1), 43–50.
41. Liu, S.; Xiao, J. Toward green catalytic synthesis—Transition metal-catalyzed reactions in nonconventional media. J. Mol. Catal. A Chem. **2007**, *270* (1–2), 1–43.
42. Welton, T. Room-temperature ionic liquids. Solvents for synthesis and catalysis. Chem. Rev. **1999**, *99* (8), 2071–2084.
43. Wasserscheid, P.; Keim, W. Ionic liquids—New "solutions" for transition metal catalysis. Angew. Chem. Int. Ed. Engl. **2000**, *39* (21), 3772–3789.
44. Dupont, J.; de Souza, R.F.; Suarez, P.A.Z. Ionic liquid (molten salt) phase organometallic catalysis. Chem. Rev.**2002** *102* (10), 3667–3692.
45. Xu, W.; Cooper, E.I.; Angell, C.A. Ionic liquids: Ion mobilities, glass temperatures and fragilities. J. Phys. Chem. *B* **2003**, *107* (25), 6170–6178.
46. Dupont, J.; Fonseca, G.S.; Umpierre, A.P.; Fichtner, P.F.P.; Teixeira, S.R. Transition-metal nanoparticles in imidazo- lium ionic liquids: Recyclable catalysts for biphasic hydrogenation reactions. J. Am. Chem. Soc. **2002**, *124* (16), 4228–4229.
47. Zhao, C.; Wang, H.-Z.; Yan, N.; Xiao, C.-X.; Mu, X.-D.; Dyson, P.J.; Kou, Y. Ionic-liquid-like copolymer stabilized nanocatalysts in ionic liquids: II. Rhodium-catalyzed hydrogenation of arenes. J. Catal. **2007**, *250* (1), 33–40.
48. Huang, J.; Jiang, T.; Han, B.X.; Gao, H.X.; Chang, Y.H.; Zhao, G.Y.; Wu, W.Z. Hydrogenation of olefins using ligand-stabilized palladium nanoparticles in an ionic liquid. Chem. Commun. **2003**, *14*, 1654–1655.

49. Kim, K.S.; Demberelnyamba, D.; Lee, H. Size-selective synthesis of gold and platinum nanoparticles using novel thiolfunctionalized ionic liquids. Langmuir **2004**, *20* (3), 556–560.

50. Zhu, Y.J.; Wang, W.W.; Qi, R.J.; Hu, X.L. Microwave-assisted synthesis of single-crystalline tellurium nanorods and nanowires in ionic liquids. Angew. Chem. Int. Ed. Engl. **2004**, *43* (11), 1410–1414.

51. Zhou, Y.; Antonietti, M. Synthesis of very small TiO_2 nanocrystals in a room-temperature ionic liquid and their selfassembly toward mesoporous spherical aggregates. J. Am. Chem. Soc. **2003**, *125* (49), 14960–14961.

52. Nakashima, T.; Kimizuka, N. Interfacial synthesis of hollow TiO_2 microspheres in ionic liquids. J. Am. Chem. Soc. **2003**, *125* (21), 6386–6387.

53. Cao, J.M.; Wang, J.; Fang, B.Q.; Chang, X.; Zheng, M.B.; Wang, H.Y. Microwave-assisted synthesis of flower-like ZnO nanosheet aggregates in a room-temperature ionic liquid. Chem. Lett. **2004**, 33 (10), 1332–1333.

54. Wang, J.; Cao, J.M.; Fang, B.Q.; Lu, P.; Deng, S.G.; Wang, H.Y. Synthesis and characterization of multipod, flowerlike, and shuttle-like ZnO frameworks in ionic liquids. Mater. *Lett.* **2005**, *59* (11), 1405–1408.

55. Goharshadi, E.K.; Ding, Y.; Nancarrow, P. Green synthesis of ZnO nanoparticles in a room-temperature ionic liquid 1-ethyl-3-methylimidazolium bis(trifluoromethylsulfonyl)imi de. J. Phys. Chem. Solids **2008**, *69* (8), 2057–2060.

56. Goharshadi, E.K.; Ding, Y.; Jorabchi, M.N.; Nancarrow, P. Ultrasound-assisted green synthesis of nanocrystalline ZnO in the ionic liquid (hmim)(NTf₂). Ultrason. Sonochem. **2009**, *16* (1), 120–123.

57. Wang, Y.; Yang, H. Synthesis of CoPt nanorods in ionic liquids. J. Am. Chem. Soc. **2005**, *127* (15), 5316–5317.

58. Hou, Y.; Kondoh, H.; Shimojo, M.; Sako, E.O.; Ozaki, N.; Kogure, T.; Ohta, T. Inorganic nanocrystal self-assembly via the inclusion interaction of β-cyclodextrins: Toward 3D spherical magnetite. J. Phys. Chem. B **2005**, *109* (11), 4845–4852.

59. Cain, J.L.; Harrison, S.R.; Nikles, J.A.; Nikles, D.E. Preparation of α-Fe particles by reduction of ferrous ion in lecithin/cyclohexane/water association colloids. J. Magn. Mater. **1996**, *155* (1–3), 67–69.

60. Mondal, K.; Lorethova, H.; Hippo, E.; Wiltowski, T.; Lalvani, S.B. Reduction of iron oxide in carbon monoxide atmosphere—Reaction controlled kinetics. Fuel Process. Technol. **2004**, *86* (1), 33–47.

61. Jian, P.; Yahui, H.; Yang, W.; Linlin, L. Preparation of polysulfone–Fe_3O_4 composite ultrafiltration membrane and its behavior in magnetic field. J. Membr. Sci. **2006**, *284* (1–2), 9–16.

62. Lu, W.; Shen, Y.; Xie, A.; Zhang, W. Green synthesis and characterization of superparamagnetic Fe_3O_4 nanoparticles. J. Magn. Magn. Mater. **2010**, *322* (13), 1828–1833.

63. Chairam, S.; Poolperm, C.; Somsook, E. Starch vermicelli template-assisted synthesis of size/shape-controlled nanoparticles. Carbohydr. Polym. **2009**, *75* (4), 694–704.

64. Kassaee, M.Z.; Akhavan, A.; Sheikh, N.; Beteshobabrud, R. γ-Ray synthesis of starch-stabilized silver nanoparticles with antibacterial activities. Radiat. Phys. Chem. **2008**, *77* (9), 1074–1078.

65. Vigneshwaran, N.; Nachane, R.P.; Balasubramanya, R.H.; Varadarajan, P.V. A novel one-pot "green" synthesis of stable silver nanoparticles using soluble starch. Carbohydr. Res. **2006**, *341* (12), 2012–2018.

66. Shankar, S.S.; Rai, A.; Ahmad, A.; Sastry, M. Rapid synthesis of Au, Ag, and bimetallic Au core–Ag shell nanoparticles using neem *(Azadirachta indica)* leaf broth. J. Colloid Interface Sci. **2004**, *275* (2), 496–502.

67. Philip, D. Biosynthesis of Au, Ag and Au–Ag nanoparticles using edible mushroom extract. Spectrochim. Acta A Mol. Biomol. Spectrosc. **2009**, *73* (2), 374–381.

68. Kasthuri, J.; Veerapandian, S.; Rajendiran, N. Biological synthesis of silver and gold nanoparticles using apiin as reducing agent. Colloids Surf. B Biointerfaces **2009**, *68* (1), 55–60.

69. Bar, H.; Bhui, D.K.; Sahoo, G.P.; Sarkar, P.; De, S.P.; Misra, A. Green synthesis of silver nanoparticles using latex of *Jatropha curcas.* Colloids Surf. A Physicochem. Eng. Asp. **2009**, *339* (1–3), 134–139.

70. Raveendran, P.; Fu, J.; Wallen, S.L. Completely "green" synthesis and stabilization of metal nanoparticles. J. Am. Chem. Soc. **2003**, *125* (46), 13940–13941.

71. Mallikarjuna, N.N.; Varma, R.S. Microwave-assisted shape- controlled bulk synthesis of noble nanocrystals and their catalytic properties. Cryst. Growth Des. **2007**, *7* (4) 686–690.

72. Panigrahi, S.; Kundu, S.; Ghosh, S.K.; Nath, S.; Pal, T. Sugar assisted evolution of mono- and bimetallic nanoparticles. Colloids Surf. A Physicochem. Eng. Asp. **2005**, *264* (1–3), 133–138.

73. Filippo, E.; Serra, A.; Buccolieri, A.; Manno, D. Green synthesis of silver nanoparticles with sucrose and maltose: Morphological and structural characterization. J. Non. Cryst. Solids **2010**, *356* (6–8), 344–350.

74. Nersisyan, H.H.; Lee, J.H.; Son, H.T.; Won, C.W.; Maeng, D.Y. A new and effective chemical reduction method for preparation of nanosized silver powder and colloid dispersion. Mater. Res. Bull. **2003**, *38* (6), 949–956.

75. Li, C.-J. Water as a benign solvent for chemical syntheses. In *Green Chemistry: Frontiers in Benign Chemical Syntheses and Processes*; Anastas, P.T., Williamson, T.C., Eds.; Oxford University Press: New York, 1998; 234.

76. Oakes, R.S.; Heppenstall, T.J.; Shezad, N.; Clifford, A.A.; Rayner, C.M. Use of scandium tris(trifluoromethanesulfona te) as a Lewis acid catalyst in supercritical carbon dioxide: Efficient Diels–Alder reactions and pressure dependent enhancement of endo:exo stereoselectivity. Chem. Commun. **1999**, *16*, 1459–1460.

77. Heldebrant, D.J.; Jessop, P.G. Liquid poly(ethylene glycol) and supercritical carbon dioxide: A benign biphasic solvent system for use and recycling of homogeneous catalysts. J. Am. Chem. Soc. **2003**, *125* (19), 5600–5601.

78. Hou, Z.; Theyssen, N.; Brinkmann, A.; Leitner, W. Biphasic aerobic oxidation of alcohols catalyzed by poly(ethylene glycol)-stabilized palladium nanoparticles in supercritical carbon dioxide. Angew. Chem. Int. Ed. Engl. **2005**, *44* (9) 1346–1349.

79. Zalipsky, S.; Harris, J.M. Introduction to chemistry and biological applications of poly(ethylene glycol). In *Poly(Ethylene Glycol): Chemistry and Biological Applications*; Harris, J.M., Zalipsky, S., Eds.; ACS Symposium Series 680; American Chemical Society: Washington, D.C., 1997; 1–13.

80. Kajitvichyanukul, P.; Amornchat, P. Effects of diethylene glycol on TiO_2 thin film properties prepared by sol–gel process. Sci. Technol. Adv. Mater. **2005**, *6* (3–4), 344–347.

81. Code of Federal Regulations, *Food and Drug, 2001,* Title 21, vol. 3, CITE: 21CFR172.820; Food and Drug Administration (FDA): Washington, D.C., 2001.

Index